Studies in Classification, Data
and Knowledge Organiza~

More information about this series at http://www.springer.com/series/1564

Tadashi Imaizumi • Akinori Okada •
Sadaaki Miyamoto • Fumitake Sakaori •
Yoshiro Yamamoto • Maurizio Vichi
Editors

Advanced Studies in Classification and Data Science

 Springer

Editors

Tadashi Imaizumi
School of Management
and Information Sciences
Tama University
Tokyo, Japan

Akinori Okada
Rikkyo University
Tokyo, Japan

Sadaaki Miyamoto
University of Tsukuba
Tsukuba, Japan

Fumitake Sakaori
Department of Mathematics
Chuo University
Tokyo, Japan

Yoshiro Yamamoto
Department of Mathematics
Tokai University
Hiratsuka-shi, Japan

Maurizio Vichi
Department of Statistical Sciences
Sapienza University of Rome
Roma, Italy

ISSN 1431-8814 ISSN 2198-3321 (electronic)
Studies in Classification, Data Analysis, and Knowledge Organization
ISBN 978-981-15-3310-5 ISBN 978-981-15-3311-2 (eBook)
https://doi.org/10.1007/978-981-15-3311-2

This Springer imprint is published by the registered company Springer Nature Singapore Pte Ltd.
The registered company address is: 152 Beach Road, #21-01/04 Gateway East, Singapore 189721, Singapore

Preface

This volume contains revised versions of selected papers presented at the biennial conference of the International Federation of Classification Societies, which was held in Tokyo from August 8–10, 2017. Japanese Classification Society and Department of Mathematics of Tokai University organized this IFCS conference. Tadashi Imaizumi (Tama University) chaired the Scientific Program Committee and Sadaaki Miyamoto (University of Tsukuba) served as the vice chair of the Scientific Program Committee. Yoshiro Yamamoto (Tokai University) chaired the Local Organizing Committee.

We are grateful to the members of the Scientific Program Committee: Brian C. Franczak (CSS), Salvatore Ingrassia (CLADAG), Hans Kestler (GfKl), Éva Laczka (HSA-CMSG), Sugnet Lubbe (SASA-MDAG), Berthold Lausen (IFCS President-Elect), Fionn Murtagh (BCS), Christian Hennig (IFCS Secretary), Mohamed Nadif (SFC), Paul McNicholas (IFCS Publication Officer), Heon Jin Park (KCS), Nema Dean (IFCS Treasurer), Józef Pociecha (SKAD), Vladimir Batagelj (SSS), Abderrahmane Sbihi (MCS), Theodoros Chadjipantelis (GDSA), José Fernando Vera (SEIO-AMyC), Sonya Coleman (IPRCS), Jeroen K. Vermunt (VOC), José Conclaves Dias (CLAD), Patrick Groenen (IASC Delegate), and Carlos Cuevas Covarrubias (SoCCCAD).

Over 300 scholars from 26 countries attended the conference. More than 250 contributions were organized into special sessions, contributed paper sessions, and one poster session. Moreover, five keynote lectures were given on different topics of data analysis and classification.

The volume is a collection of full papers submitted after the conference. Papers were selected after a peer-review process, according to the high-quality standards of the series. The volume is complied with 40 contributions organized in five parts including contributions on:

- Classification and Cluster Analysis
- Multidimensional Analysis and Visualization
- Statistical Methods
- Statistical Data Analysis
- Statistical Data Analysis for Social Science

Acknowledgements

We are indebted to many people who allowed the success of the IFCS 2017 conference with their commitment. This volume is the final scientific outcome of all the works done for the organization of the conference during the days of the conference and after the end of it. First, we are grateful to the Department of Mathematics of Tokai University who hosted the conference. In particular, our thanks are addressed to the members of the local organizing committee: Takafumi Kubota (Tama University), Miki Nakai (Ritsumeikan University), Atsuho Nakayama (Tokyo Metropolitan Univesity), Makiko Oda (National Defence Academy), Takuya Ohmori (Tama University), Kosuke Okusa (Kyshu University), Kumiko Shiina (National Center for University Entrance Examinations), Akinobu Takeuchi (Jissen Women's University), Makoto Tomita (Tokyo Medical and Dental University), Hiroshi Yadohisa (Doshisha University), and Satoru Yokoyama (Teikyo University)

And we cordially appreciate Mr. Yutaka Hirachi and Ms. Vaishnavi Venkatesh at Springer Nature for assisting us in publishing the current volume.

Tokyo, Japan   Tadashi Imaizumi
Tokyo, Japan   Akinori Okada
Tsukuba, Japan   Sadaaki Miyamoto
Tokyo, Japan   Fukitake Sakaori
Hiratsuka-shi, Japan   Yoshiro Yamamoto
Rome, Itally   Maurizio Vichi
January 2020

Contents

About the Editors

Tadashi Imaizumi School of Management and Information Sciences, Tama University, Tokyo, Japan

Akinori Okada Rikkyo University, Tokyo, Japan

Sadaaki Miyamoto University of Tsukuba, Ibaraki, Japan

Fumitake Sakaori Department of Mathematics, Chuo University, Tokyo, Japan

Yoshiro Yamamoto Department of Mathematics, Tokai University, Kanagawa, Japan

Maurizio Vichi Department of Statistical Sciences, Sapienza University of Rome, Rome, Italy

Part I
Classification and Cluster Analysis

Multilevel Model-Based Clustering: A New Proposal of Maximum-A-Posteriori Assignment

Silvia Bacci, Francesco Bartolucci, and Fulvia Pennoni

Abstract We deal with the problem of latent variable prediction in the context of multilevel latent class models for categorical responses provided by individuals nested in groups. In particular, we propose a posterior assignment rule that jointly predicts the individual- and group-level latent variables. This proposal is alternative to the common maximum-a-posteriori rule, which is based on first predicting the latent variables at cluster level and, then, those at individual level. To illustrate the proposal, we show the results of two simulation studies and two applications on data related to the national and the international assessment of student skills.

1 Introduction

Multilevel models (Goldstein 2011) are employed when sample units are clustered in groups, as is typical in educational studies, to account for the unobserved heterogeneity between and within groups. These models take into account that group-level unobservable factors may affect in a similar way the behavior of all individuals in the same group. This behavior is observed through the response variables considered in the study. Consequently, the dependence between the responses provided by individuals in the same group is suitably considered and it is possible to compare groups and produce league tables (Goldstein and Spiegelhalter

S. Bacci
Department of Statistics, Computer Science, Applications "Giuseppe Parenti",
University of Florence, Florence, Italy
e-mail: silvia.bacci@unifi.it

F. Bartolucci
Department of Economics, University of Perugia, Perugia, Italy
e-mail: francesco.bartolucci@unipg.it

F. Pennoni (✉)
Department of Statistics and Quantitative Methods, University of Milano-Bicocca, Milano, Italy
e-mail: fulvia.pennoni@unimib.it

© Springer Nature Singapore Pte Ltd. 2020   3
T. Imaizumi et al. (eds.), *Advanced Studies in Classification and Data Science*,
Studies in Classification, Data Analysis, and Knowledge Organization,
https://doi.org/10.1007/978-981-15-3311-2_1

1996). For instance, schools or classes are compared at national and international level on the basis of the students' acquired knowledge assessed by test scores.

In particular, we focus on models based on discrete latent variables (Skrondal and Rabe-Hesketh 2013) to account for the multilevel structure of the data and on the case of categorical responses. The latent variables are associated with each group and to each individual, with that at group level characterized by a finite number of support points (and related weights) identifying *latent clusters*. Groups belonging to the same latent cluster share a common effect on the units within the group. Similarly, the components of the discrete latent variable at individual level define homogenous classes of individuals, named *latent classes*, so that individuals in the same latent class have the same behavior. The approach at issue relies on the Multilevel Latent Class (MLC) model (Vermunt 2003), which is an extension of the classical latent class model (Lazarsfeld and Henry 1968; Goodman 1974; Pennoni 2014). It has been applied by many authors in the educational context (Gnaldi et al. 2016; Vermunt 2008).

While maximum likelihood estimation of the MLC model parameters through the Expectation-Maximization (EM) algorithm (Baum et al. 1970; Dempster et al. 1977) is already well established, an issue that still deserves attention concerns the prediction of the latent variables at individual and group level on the basis of the estimated parameters. In the literature about latent class models and in that about multilevel models based on discrete latent variables, the *Maximum-A-Posteriori* (MAP) approach is considered as the standard rule to assign individuals to the latent classes (Goodman 1974, 2007). This rule is also very popular for finite-mixture models in general (McLachlan and Peel 2000). It consists in selecting, for every individual, the latent class having the highest posterior probability, which corresponds to the conditional distribution of the latent variable given the observed responses. The MAP allocation has been proved superior with respect to the method of the expected proportions, as stated by Goodman (2007), and to the method of bagging proposed by Dias and Vermunt (2008). More recently, extending the method proposed in Bandeen-Roche et al. (1997), Bray et al. (2015) discussed an alternative approach for multiple categorical responses, which is based on multiple pseudo-class draws. This method is similar to the MAP, but it randomly assigns individuals to latent classes for a repeated number of times according to the posterior probabilities. The authors proved that the MAP assignment is still superior in terms of bias.

For MLC models, the MAP rule must be applied to each hierarchical level, allocating first each group to the latent cluster having the highest posterior probability and then allocating individuals to the latent classes. The latter may be applied in two different ways named in the following as *marginal* MAP and *conditional* MAP. In the marginal MAP each latent variable at individual level is considered separately with respect to the latent variable at group level and, then, each individual is assigned to the latent class with the highest marginal posterior probability. In the conditional MAP the latent class is predicted for each individual, conditionally on the value predicted for the corresponding group-level latent variable.

From a certain point of view, both marginal and conditional MAP rules lead to suboptimal solutions because the allocation of individuals and groups may not correspond to the joint MAP probability of the latent variables. We propose a variant of MAP that is inspired by a similar problem arising with longitudinal data in the context of hidden Markov (HM) models (Bartolucci et al. 2013; Zucchini et al. 2016). In detail, we propose to adapt the Viterbi algorithm (Juang and Rabiner 1991; Viterbi 1967) to the MLC model as it is simple to implement by forward and backward recursions and has a linear complexity.

The remainder of this article is organized as follows. In Sect. 2 we describe the MLC model, focusing on its formulation based on discrete latent variables at group and individual level. In Sect. 3 we provide technical details on the proposed Viterbi-based allocation algorithm. In Sect. 4 we show the results of two simulation studies aimed at comparing the performance of the clustering algorithms. In Sect. 5 we provide the results of the proposal applied to two different datasets on education and, finally, in Sect. 6 we summarize the main conclusions.

2 Multilevel Latent Class Model

Let Y_{hij} denote the categorical response variable for individual i within group h and item j, with $h = 1, \ldots, H, i = 1, \ldots, n_h, j = 1, \ldots, r$, let $\mathbf{Y}_{hi} = (Y_{hi1}, \ldots, Y_{hir})'$ be the vector of responses referred to this individual, and let $\mathbf{Y}_h = (\mathbf{Y}'_{h1}, \ldots, \mathbf{Y}'_{hn_h})'$ be the collection of all responses referred to group h. The corresponding observed values are denoted in lower case, that is, by y_{hij}, $\mathbf{y}_{hi}$, and $\mathbf{y}_h$, respectively. In the binary case, each response variable may assume values 0 or 1, otherwise it may assume l_j values from 0 to $l_j - 1$. In the latter case, the responses typically come from ordinal polytomous items. When available, we consider covariates at group and individual level, which are collected in the vectors $\mathbf{w}_h$ and $\mathbf{x}_{hi}$, respectively.

We assume two discrete latent variables, U_h and V_{hi}, identifying *latent clusters* and *latent classes*, respectively (in the following we use a different text color to distinguish between them). The distribution of each latent variable has a finite number of support points. For each U_h, the support points are k_U, with corresponding mass probabilities $\lambda_{h,u} = p(U_h = u|\mathbf{w}_h)$, $u = 1, \ldots, k_U$, which possibly depend on the fixed covariates $\mathbf{w}_h$. The support points of each V_{hi} are k_V and the mass probabilities are denoted by $\pi_{hi,v|u} = p(V_{hi} = v|U_h = u, \mathbf{x}_{hi})$, $v = 1, \ldots, k_V$, at individual level depending on the group-specific latent variable U_h and, possibly, on the fixed covariates $\mathbf{x}_{hi}$. Note that, if no covariate is available, these probabilities are equal across groups and individuals. On the other hand, when these covariates are available, we adopt a multinomial logit parameterization:

$$\log \frac{\lambda_{h,u}}{\lambda_{h,1}} = \mathbf{w}'_h \boldsymbol{\phi}_u, \quad u = 2, \ldots, k_U,$$

$$\log \frac{\pi_{hi,v|u}}{\pi_{hi,1|u}} = \mathbf{x}'_{hi} \boldsymbol{\psi}_{uv}, \quad u = 1, \ldots, k_U, \ v = 2, \ldots, k_V,$$

for suitable parameter vectors $\boldsymbol{\phi}_u$ and $\boldsymbol{\psi}_{uv}$; see also Dayton and Macready (1988), Formann (2007).

A crucial assumption of the MLC model is that of *local independence*, according to which the responses in $\mathbf{Y}_{hi}$ are conditionally independent given the latent variable V_{hi} for every individual i in group h. This enforces the interpretation of the latent variable as the only factor affecting the response variables. Moreover, in agreement with the typical structure of multilevel models, for each group h the latent variables V_{hi}, $i = 1, \ldots, n_h$, are conditionally independent given U_h. This implies the following *manifest distribution* for the responses referred to the same group h:

$$p(\mathbf{Y}_h = \mathbf{y}_h) = \sum_{u=1}^{k_U} \lambda_{h,u} \, p(\mathbf{Y}_h = \mathbf{y}_h | U_h = u),$$

$$p(\mathbf{Y}_h = \mathbf{y}_h | U_h = u) = \prod_{i=1}^{n_h} p(\mathbf{Y}_{hi} = \mathbf{y}_{hi} | U_h = u),$$

$$p(\mathbf{Y}_{hi} = \mathbf{y}_{hi} | U_h = u) = \sum_{v=1}^{k_V} \pi_{hi,v|u} \, p(\mathbf{Y}_{hi} = \mathbf{y}_{hi} | V_{hi} = v),$$

$$p(\mathbf{Y}_{hi} = \mathbf{y}_{hi} | V_{hi} = v) = \prod_{j=1}^{r} p(Y_{hij} = y_{hij} | V_{hi} = v),$$

where the dependence on the covariates is not explicitly indicated as these covariates are assumed to be fixed and given. We also assume independence between the group-level latent variables, so that the overall manifest distribution is

$$p(\mathbf{Y}_1 = \mathbf{y}_1, \ldots, \mathbf{Y}_H = \mathbf{y}_H) = \prod_{h=1}^{H} p(\mathbf{Y}_h = \mathbf{y}_h).$$

Given observed data referred to a sample of n subjects collected in H groups, the log-likelihood of the MLC model is

$$\ell(\boldsymbol{\theta}) = \sum_{h=1}^{H} \log p(\mathbf{Y}_h = \mathbf{y}_h),$$

where $\boldsymbol{\theta}$ is the vector of free model parameters. These parameters are estimated by maximizing $\ell(\boldsymbol{\theta})$ through the EM algorithm (Baum et al. 1970; Dempster et al. 1977). As usual, this algorithm relies on the complete data log-likelihood, denoted by $\ell^*(\boldsymbol{\theta})$, which corresponds to the log-likelihood that we would compute if we knew the values of the latent variables describing clusters and classes. The algorithm alternates two steps (E and M) until convergence. The E-step computes the conditional expected value of $\ell^*(\boldsymbol{\theta})$ given the observed data and the current value of the parameters. This expected value is maximized at the M-step with respect to $\boldsymbol{\theta}$ and the estimate of this parameter vector is then updated until convergence.

After estimation, an important task is the prediction of the latent variables at group and individual level. The conditional MAP approach provides the following two predicted components:

$$\hat{u}_h = \underset{u=1,\ldots,k_U}{\operatorname{argmax}} \hat{a}_{h,u}, \quad h = 1, \ldots, H, \tag{1}$$

$$\hat{v}_{hi} = \underset{v=1,\ldots,k_V}{\operatorname{argmax}} \hat{b}_{hi,v|\hat{u}_h}, \quad i = 1, \ldots, n_h, \quad h = 1, \ldots, H, \tag{2}$$

where $\hat{a}_{h,u}$ and $\hat{b}_{hi,v|u}$ are the estimated posterior probabilities at both levels. These quantities are obtained by substituting the parameter estimates in the following posterior probabilities:

$$a_{h,u} = p(U_h = u | \mathbf{Y}_h = \mathbf{y}_h) = \frac{\lambda_{h,u} p(\mathbf{Y}_h = \mathbf{y}_h | U_h = u)}{p(\mathbf{Y}_h = \mathbf{y}_h)},$$

$$b_{hi,v|u} = p(V_{hi} = v | U_h = u, \mathbf{Y}_{hi} = \mathbf{y}_{hi}) = \frac{\pi_{hi,v|u} p(\mathbf{Y}_{hi} = \mathbf{y}_{hi} | V_{hi} = v)}{p(\mathbf{Y}_{hi} = \mathbf{y}_{hi} | U_h = u)}.$$

It is worth noting that this rule does not provide an optimal solution in the sense that, for a given cluster h, it does not maximize the following posterior probability

$$p(U_h = u, V_{h1} = v_1, \ldots, V_{hn_h} = v_{n_h} | \mathbf{Y}_h = \mathbf{y}_h)$$

$$= \frac{\lambda_{h,u} \prod_{i=1}^{n_h} \pi_{hi,v_i|u} p(\mathbf{Y}_{hi} = \mathbf{y}_{hi} | V_{hi} = v_i)}{p(\mathbf{Y}_h = \mathbf{y}_h)} \tag{3}$$

jointly with respect to $(u, v_1, \ldots, v_{n_h})'$. In fact, using the terminology of HM models (Bartolucci et al. 2013; Zucchini et al. 2016), the method based on solutions (1) and (2) is a sort of *local decoding*.

In the following section, we introduce an algorithm to maximize (3) with respect to $(u, v_1, \ldots, v_{n_h})'$ that is based on a simplified version of the Viterbi algorithm (Juang and Rabiner 1991; Viterbi 1967). This algorithm is used for *global decoding* in dealing with HM models.

3 Proposed Latent Variable Prediction Algorithm

We introduce a unidimensional discrete latent variable Z_{hi} derived from the discrete latent variables (U_h, V_{hi}), having a number of support points (or *latent states* according to the terminology adopted in the HM literature) equal to the product of k_U and k_V; these support points are denoted in the following by z, with $z =$

Table 1 Example: support points of Z_{hi} when $k_U = 2$ and $k_V = 3$

u	1	1	1	2	2	2
v	1	2	3	1	2	3
z	1	2	3	4	5	6

$1, \ldots, k_Z$. Table 1 shows an example for the case of $k_U = 2$ and $k_V = 3$ to illustrate how the values of Z_{hi} are coded depending on those of U_h and V_{hi}.

Obviously, the latent variables in the sequence $Z_{h1}, \ldots, Z_{hn_h}$ are not independent, differently from the original latent variables U_h and V_{hi}. In particular, it is important noting that each sequence $Z_{h1}, \ldots, Z_{hn_h}$ follows a Markov chain of first-order with initial probabilities

$$\rho_{h,z} = p(Z_{h1} = z) = \lambda_{h,u(z)} \pi_{h1,v(z)|u(z)},$$

where $u(z)$ and $v(z)$ are the categories of the underlying latent variables U_h and V_{hi} corresponding to Z_{hi}, and transition probabilities

$$\rho_{hi,z|\bar{z}} = p(Z_{hi} = z | Z_{h,i-1} = \bar{z}) = \pi_{hi,v(z)|u(z)}, \quad \text{iff} \quad u(z) = u(\bar{z}), \tag{4}$$

and $\rho_{hi}(z|\bar{z}) = 0$ when $u(z) \neq u(\bar{z})$. The latter constraint guarantees that the latent cluster of every group is constant for all individuals within the same group. Besides, given the arbitrary order of individuals within groups, we assume that the transition probability from $\bar{z}$ to z does not depend on $\bar{z}$. For the example illustrated in Table 1, this leads to the following matrix of the transition probabilities, defined as in (4), which has a block diagonal structure:

$$\begin{pmatrix} \rho_{hi,1|1} & \rho_{hi,2|1} & \rho_{hi,3|1} & 0 & 0 & 0 \\ \rho_{hi,1|2} & \rho_{hi,2|2} & \rho_{hi,3|2} & 0 & 0 & 0 \\ \rho_{hi,1|3} & \rho_{hi,2|3} & \rho_{hi,3|3} & 0 & 0 & 0 \\ 0 & 0 & 0 & \rho_{hi,4|4} & \rho_{hi,5|4} & \rho_{hi,6|4} \\ 0 & 0 & 0 & \rho_{hi,4|5} & \rho_{hi,5|5} & \rho_{hi,6|5} \\ 0 & 0 & 0 & \rho_{hi,4|6} & \rho_{hi,5|6} & \rho_{hi,6|6} \end{pmatrix}.$$

Given that the overall multilevel model resembles the HM model, we propose to use a Viterbi algorithm (Juang and Rabiner 1991; Viterbi 1967) to jointly allocate individuals and groups by maximizing the posterior probability in (3), so as to obtain a predicted sequence $\tilde{\mathbf{z}}_h = (\tilde{z}_{h1}, \ldots, \tilde{z}_{hn_h})$ given the observed responses in $\mathbf{y}_h$. In the end, we get the predicted latent cluster $\tilde{u}_h = u(\tilde{z}_{hi})$ and the latent class $\tilde{v}_{hi} = v(\tilde{z}_{hi})$ for $i = 1, \ldots, n_h$.

For a given group h and for $z = 1, \ldots, k_Z$, let

$$p_{h1}(z, \mathbf{y}_h) = p(Z_{h1} = z, \mathbf{Y}_{h1} = \mathbf{y}_{h1}),$$

and

$$p_{hi}(z, \mathbf{y}_h) = \max_{z_1,\dots,z_{i-1}} p(Z_{h1} = z_1, \dots, Z_{h,i-1} = z_{i-1}, Z_{hi} = z,$$

$$\mathbf{Y}_{h1} = \mathbf{y}_{h1}, \dots, \mathbf{Y}_{hi} = \mathbf{y}_{hi}),$$

with $i = 2, \dots, n_h$. The Viterbi algorithm is based on the following forward-backward recursions:

1. for $i = 1$ and $z = 1, \dots, k_Z$ compute

$$p_{h1}(z, \mathbf{y}_h) = \rho_{h1}(z) p(\mathbf{Y}_{h1} = \mathbf{y}_{h1} | Z_{h1} = z);$$

2. for $i = 2, \dots, n_h$ and $z = 1, \dots, k_Z$ compute

$$\hat{p}_i(z, \mathbf{y}_h) = p(\mathbf{Y}_{hi} = \mathbf{y}_{hi} | Z_{hi} = z) \max_{\bar{z}=1,\dots,k_Z} (p_{h,i-1}(\bar{z}, \mathbf{y}_h) \rho_{hi,z|\bar{z}});$$

3. for $i = n_h$ find the optimal state

$$\tilde{z}_{h,n_h} = \operatorname*{argmax}_{z=1,\dots,k_Z} p_{hn_h}(z, \mathbf{y}_h);$$

4. for $i = n_h - 1, \dots, 1$ predict the optimal state

$$\tilde{z}_{hi}(\mathbf{y}_h) = \operatorname*{argmax}_{z=1,\dots,k_Z} [p_{hi}(z, \mathbf{y}_h) \rho_{h,i+1,\tilde{z}_{h,i+1}|z}].$$

It is worth noting that the resulting algorithm has a numerical complexity that linearly increases with the overall sample size as it must be repeatedly applied for all clusters.

4 Simulation Study

In order to compare the proposal illustrated above, we performed two separate simulation studies of the allocation accuracy for individuals and groups: in the first study (Study I) the model is correctly specified whereas, in the second (Study II), it is misspecified. Each study is based on different scenarios; 50 random samples are generated under each scenario. In all cases, the comparison between the two approaches is based on the rates of different allocations ($DISagreement$, indicated as DIS in the following sections) of groups in latent clusters (denoted by DIS_U) and of individuals in latent classes (denoted by DIS_{V1} for the marginal MAP and DIS_{V2} for the conditional MAP).

A summary of the results of the two studies is provided in Table 2, whereas details on each design and the discussion of the results are provided in the following two subsections. The complete results are available from authors upon request.

4.1 Study I

The first simulation study is based on 20 different scenarios resulting from the following design:

- $r = 8$ binary response variables (items);
- $H = 50, 100$ (groups) with $n_h = 10, 50$ (individuals).

We generated binary responses by assuming a discrete distribution for U and V, with a number of latent classes and clusters equal to 3 (i.e., $k_U = 3$ and $k_V = 3$). Their components are defined as the quadrature nodes of a Gaussian distribution with null mean and standard deviations σ_U and σ_V, respectively. To consider distinctive latent clusters and classes we fixed their values as illustrated in Table 3. Moreover, the weights of U correspond to the weights for the Gaussian quadrature, whereas the weights of V are affected by the components of U through an inverse logit transformation.

As shown in Table 2 (Study I), under the correct model specification both MAP and Viterbi approaches tend to be very coherent in the allocation of groups and individuals. More precisely, all the statistics computed for DIS_U in Study I are smaller (except for the standard deviation) than the corresponding measures in Study II: disagreement rates of latent clusters range from 0.00% (vs 6.0%) to 62.7% (vs 73.6%), with mean equal to 15.8% (vs 28.1%) and median equal to 8.9% (vs 24.5%). The level of concordance between the two approaches is also higher in the allocation of individuals. These results confirm the lower discordance between conditional MAP and Viterbi rather than between marginal MAP and Viterbi allocations.

Table 2 Main results of the simulation studies (Study I and Study II): disagreement rates between Viterbi and MAP approaches in the allocation of groups (DIS_U) and individuals (DIS_{V1} for marginal MAP and DIS_{V2} for conditional MAP)

	Study I			Study II		
Statistic	DIS_U	DIS_{V1}	DIS_{V2}	DIS_U	DIS_{V1}	DIS_{V2}
Min.	0.000	0.010	0.000	0.060	0.020	0.009
q_1	0.006	0.018	0.001	0.141	0.071	0.030
Median	0.089	0.076	0.010	0.245	0.130	0.055
q_3	0.134	0.123	0.045	0.380	0.197	0.107
Max.	0.627	0.470	0.453	0.736	0.473	0.271
Mean	0.158	0.126	0.096	0.281	0.146	0.077
St. dev.	0.199	0.145	0.167	0.166	0.100	0.061

Table 3 Simulation design for the variability of the components of U and V

Scenario	Type	Variability	
1	Low	$\sigma_U = 0.2$	$\sigma_V = 0.2$
2	Intermediate	$\sigma_U = 1.0$	$\sigma_V = 1.0$
3	High	$\sigma_U = 2.0$	$\sigma_V = 2.0$
4	Mixed type 1	$\sigma_U = 0.2$	$\sigma_V = 1.0$
5	Mixed type 2	$\sigma_U = 1.0$	$\sigma_V = 0.2$

Regarding their performance, the allocation improves when latent classes and clusters are highly separated. For instance, under scenario 1 in Table 3, rates of misclassified groups range from 0.9 to 1.5% for the proposed Viterbi algorithm and behave similarly for MAP approach, whereas rates of misclassified individuals range around 4.5% with a slightly worse behavior of marginal MAP. Coherently, the rates of correct global allocations (i.e., allocation of groups and individuals at the same time) range from 14.8 to 14.9% in the presence of many individuals within groups and from 62.1 to 62.2% in the presence of a few individuals within each group.

4.2 Study II

The second simulation study is based on 72 different scenarios resulting from the combination of the following criteria:

- $r = 8, 10$ binary response variables (items);
- $H = 50, 100$ (groups) with $n_h = 10, 25, 50$ (individuals);
- $U_h \sim N(0, 1)$ (latent variables at cluster level) and $V_{hi} \sim N(\mu_u, 1)$ (latent variables at individual level).

As shown in Table 2 (Study II), the disagreement rates on the latent clusters (DIS_U) range from 6.0 to 73.6% with mean equal to 28.1% and median to 24.5%. Concerning the allocation of individuals to the latent classes, the two MAP approaches have quite different behaviors. Disagreement rates between marginal MAP and Viterbi (DIS_{V_1}) range from 2.0 to 47.3%, whereas the range for conditional MAP (see values of DIS_{V_2}) is definitely shifted toward lower values, from 0.9 to 27.1%. Coherently, the 14.6% of individuals are in average differently allocated through the marginal MAP compared to the Viterbi approach, against the 7.7% of individuals allocated through the conditional MAP (median values are 13.0% and 5.5%, respectively).

Overall, we observe that the disagreement between the MAP approach and the proposed Viterbi algorithm increases when the number of groups (H), the number of individuals within groups (n_h), and the number of latent clusters and classes (k_U and k_V) increase, whereas it decreases when the number of items (r) increases. We also monitored the trend of three different normalized entropy criteria for the degree of separation within the latent classes and clusters and between them (see Celeux and

Soromenho (1996) and Lukocienė and Vermunt (2010) for details). We found that disagreement rates tend to follow the trend of the corresponding entropy measures.

5   Examples on Educational Data

The first example we use to illustrate our proposal is based on data derived from the Italian and Mathematics tests elaborated by the Italian National Institute for the Evaluation of the Educational System (INVALSI) and the second is based on the Italian data of the International Association for the Evaluation of Educational Achievement provided by the Trends in International Mathematics and Science Study (TIMSS) and Progress on International Reading Literacy Study (PIRLS).

In both cases, we account for the presence of covariates and we compare the results of the MAP and Viterbi approaches in terms of rates of schools and students that are differently assigned to latent clusters and classes. Similarly to the simulation studies illustrated in Sect. 4, the discordance in the allocations are denoted respectively by DIS_U, DIS_{V1}, and DIS_{V2}. The complete results are shown in Tables 4 and 5 in the Appendix.

5.1   Example 1: Application to INVALSI Data

The INVALSI tests were administered in June 2009, at the end of the pupils' compulsory educational period in Italy, which is stated to be of 10 years according to current Italian legislation. The sample includes 16,877 students nested in 1302 schools. The Mathematics test consists of 27 items covering four main content domains (Numbers, Shapes and Figures, Algebra, and Data and Previsions), whereas the Italian test includes two sections: Reading Comprehension and Grammar. The first section is made of 30 items, which require students to demonstrate a range of abilities and skills in constructing meaning from the two written texts. The Grammar section is made of 10 items, which measure the ability of understanding the morphological and syntactic structure of sentences within a text. All items are of multiple choice type, with one correct answer and three distractors, and are dichotomously scored (assigning 1 point to correct answers and 0 otherwise). For a more detailed description and analysis on these data see Gnaldi et al. (2016).

We estimate three types of MLC models for responses on: (*i*) Mathematics without considering covariates; (*ii*) Mathematics accounting for the effect of gender on the weights at student level and for the effect of geographical area at school level; and (*iii*) Mathematics, Reading, and Grammar tests including the effects of gender and geographical area. Each model is estimated for different combinations of the number of support points of the latent components: $k_U = 4, 5$ and $k_V = 3, 4, 5, 6$ (see Table 4 in the Appendix).

Overall, the entropy measures suggest that latent classes and, mostly, latent clusters are weakly separated (all values of the entropy measure range from 0.30 to 0.50). Table 4 shows high rates of discordant classifications between the MAP approach and the proposed Viterbi algorithm. Indeed, as concerns the allocation of schools, DIS_U ranges in the interval 4.1 to 8.1%, corresponding to a number of schools that are differently allocated ranging from 53 to 105 with a mean equal to 82 (6.3%). Regarding the allocation of students, the marginal MAP performs much worse than conditional MAP, in terms of comparison with Viterbi approach. In fact, DIS_{V1} ranges from a minimum of 3.3% (560 students) to a maximum of 12.9% (2184 students) with a mean of 8.4% (1418 students), whereas the same values under the conditional MAP are substantially reduced, ranging from 0.3% (54 students) to 1.8% (300 students) with a mean equal to 1.2% (198 students).

5.2   Example 2: Application to TIMSS&PIRLS Italian Data

The second example concerns the combined Italian data collected within the 2011 large-scale studies TIMSS&PIRLS. For more details on these two international assessments on Mathematics, Science, and Literacy provided by the International Association for the Evaluation of Educational Achievement (IEA) see Foy (2013).

We consider the achievement scores at the fourth grade when the Italian pupils are 9 to 10 years old. The sample includes 3741 students nested in 200 schools. The analysis considers 5 ordinal responses for Mathematics, 5 for Reading, and 5 for Science. Each variable has the following ordered categories: 0, *below* the low International Benchmark (IB); 1, *at or above* the *low* IB, but *below the intermediate* IB; 2, *at or above* the *intermediate* IB, but *below the high* IB; 3, *at or above* the *intermediate* IB, but *below the high* IB; 4, *at or above* the *advanced* IB.

The covariates are collected from the background parents' questionnaires, the principals' questionnaire, and from external data archives. We consider four covariates referred to the pupils and four to the schools: *gender*; home *resources* for learning; early *literacy/numeracy* tasks of the pupil; dummy variable for the Italian *language* spoken at home (1 if yes); school *adequate* environment and resources; school *safety* and orderly; socio-economic *condition* of the area where the school is located (gross value added at province level, from an external data archive) and dummy variables for five Italian *geographical* areas (North-West, North-East, Center, South, South-Islands). A more detailed description and analysis of these data can be found in Grilli et al. (2016).

We compare the allocation algorithms by considering the following MLC models for responses on: (*i*) Mathematics; (*ii*) Mathematics with covariates at student and school level; (*iii*) Mathematics, Reading, and Science; (*iv*) Mathematics, Reading, and Science with covariates at student and school level. Each model is estimated for increasing values of the latent components (see the first two columns of Table 5 in the Appendix). Then, the two allocation rules are evaluated similarly to the previous example.

The estimated entropy values suggest that latent classes are well-separated (the entropy values for the latent classes range from 0.03 to 0.12), whereas the latent clusters are less separated (the entropy for the latent clusters ranges from 0.20 to 0.50). As shown in Table 5 for models (*i*) and (*ii*), the disagreement rates referred to the schools' allocations lie in the interval 0.005 to 0.05. This corresponds to a number of schools that are differently classified from 2 to 14 with an average value equal to 5 (2.5%). These rates are lower for models (*iii*) and (*iv*): the number of schools that are differently classified ranges from 2 to 8. We also observe that the rates of disagreement are not particularly influenced by the presence/absence of covariates at both levels.

Regarding the students' allocation, the marginal MAP performs much worse than the conditional MAP, when compared with the proposed Viterbi algorithm. The disagreement rates under the marginal MAP range from a minimum of 35 students to a maximum of 149 students when models (*i*) and (*ii*) are considered, whereas they are lower under models (*iii*) and (*iv*), for which the number of students that are differently allocated ranges from 10 to 77. Under the conditional MAP the differences substantially reduce, ranging from 0 to 11 students for models (*i*) and (*ii*) and from 0 to 3 students for models (*iii*) and (*iv*).

6   Conclusions

In the Latent Class (LC) literature the most accredited method to allocate individuals to the classes is the Maximum-A-Posteriori (MAP) approach. In this contribution, we study the behavior of the MAP method in the context of the Multilevel Latent Class (MLC) model, which represents a generalization of the standard LC model to account for hierarchical data structures. More in detail, we compare the MAP rule with an alternative approach based on a suitable adaptation of the Viterbi algorithm employed in the hidden Markov literature for global decoding and, as such, representing a gold standard.

Our analyses based on simulation studies and real data show that a different classification of individuals and groups may be found between the standard MAP approach and the proposed one, mainly when the latent components (latent classes and latent clusters) are poorly separated. This is an important aspect as the allocation of individuals and groups can have relevant consequences from a practical point of view, mainly when the identification of classes of individuals or clusters of groups performing poorly or highly on a certain latent variable of interest is translated into specific decisions (e.g., incentives for the best performers and penalties for the wrong performers).

As a possible development of the proposed approach we suggest the extension to more complex models in terms of hierarchical structure and parameterization adopted for the conditional distribution of the responses given the latent variables. In fact, depending on the specific context of application, this conditional distribution may be parameterized as in certain item response theory models (Bartolucci et al.

2015) to take into account, for instance, the ordinal structure of the responses on the basis of individual and item parameters.

Acknowledgments We acknowledge the financial support from the grant "Finite mixture and latent variable models for causal inference and analysis of socio-economic data" (FIRB—Futuro in ricerca) funded by the Italian Government (RBFR12SHVV).

Appendix

Table 4 Results of models (*i*), (*ii*), and (*iii*) estimated for the INVALSI data with an increasing number of support points of U and V

k_U	k_V	# U	#V1	#V2	DIS_U	DIS_{V1}	DIS_{V2}
Mathematics (*i*)							
4	3	95	1159	206	0.073	0.069	0.012
4	4	72	1420	196	0.055	0.084	0.012
4	5	83	2144	280	0.064	0.127	0.017
5	4	105	1320	212	0.081	0.078	0.013
5	5	92	2157	271	0.071	0.128	0.016
5	6	102	2184	256	0.078	0.129	0.015
Mathematics with covariates (*ii*)							
4	3	93	1221	190	0.071	0.072	0.011
4	4	62	1305	180	0.048	0.077	0.011
4	5	80	1871	241	0.061	0.111	0.014
5	4	81	1445	181	0.062	0.086	0.011
5	5	89	1935	265	0.068	0.115	0.016
5	6	88	1933	262	0.068	0.115	0.016
Mathematics, reading, grammar with covariates (*iii*)							
4	3	53	560	54	0.041	0.033	0.003
4	4	64	831	93	0.049	0.049	0.006
4	5	66	835	95	0.051	0.049	0.006
5	4	100	1037	300	0.077	0.061	0.018
5	5	72	855	103	0.055	0.051	0.006
5	6	87	1307	182	0.067	0.077	0.011

Table 5 Results of models (i), (ii), (iii), and (iv) estimated for the TIMSS&PIRLS combined data with an increasing number of support points of U and V

k_U	k_V	#U	#$V1$	#$V2$	DIS_U	DIS_{V1}	DIS_{V2}
Mathematics (i)							
4	3	3	90	5	0.015	0.024	0.001
4	4	6	51	3	0.030	0.014	0.001
4	5	2	105	0	0.010	0.028	0.000
5	4	11	61	1	0.055	0.016	0.000
5	5	12	111	5	0.060	0.030	0.001
5	6	14	107	9	0.070	0.029	0.002
Mathematics with covariates (ii)							
4	3	4	80	6	0.020	0.021	0.002
4	4	8	35	2	0.040	0.009	0.001
4	5	4	102	2	0.020	0.027	0.001
5	4	3	53	4	0.015	0.014	0.001
5	5	10	97	4	0.050	0.026	0.001
5	6	7	149	11	0.035	0.040	0.003
Mathematics, reading, science (iii)							
3	2	3	23	1	0.015	0.006	0.000
3	3	2	34	2	0.010	0.009	0.001
3	4	3	16	0	0.015	0.004	0.000
4	3	2	32	1	0.010	0.009	0.000
4	4	2	30	1	0.010	0.008	0.000
4	5	3	27	0	0.015	0.007	0.000
5	4	4	10	2	0.020	0.003	0.001
5	5	8	27	3	0.040	0.007	0.001
5	6	6	50	0	0.030	0.013	0.000
6	5	8	50	2	0.040	0.013	0.001
Mathematics, reading, science with covariates (iv)							
3	2	1	22	0	0.005	0.006	0.000
3	3	5	30	2	0.025	0.008	0.001
3	4	2	22	0	0.010	0.006	0.000
4	3	2	26	1	0.010	0.007	0.000
4	4	0	49	0	0.000	0.013	0.000
4	5	2	30	1	0.010	0.008	0.000
5	4	2	41	2	0.010	0.011	0.001
5	5	4	21	3	0.020	0.006	0.001
5	6	6	77	1	0.030	0.021	0.000
6	5	5	32	5	0.025	0.009	0.001

References

Bandeen-Roche, K., Miglioretti, D.L., Zeger, S.L., Rathouz, P.J.: Latent variable regression for multiple discrete outcomes. J. Am. Stat. Assoc. **92**, 1375–1386 (1997)

Bartolucci, F., Farcomeni, A., Pennoni, F.: Latent Markov Models for Longitudinal Data. Chapman and Hall/CRC Press, Boca Raton (2013)

Bartolucci, F., Bacci, S., Gnaldi, M.: Statistical Analysis of Questionnaires: A Unified Approach Based on R and Stata. Chapman and Hall/CRC Press, Boca Raton (2015)

Baum, L.E., Petrie, T., Soules, G., Weiss, N.: A maximization technique occurring in the statistical analysis of probabilistic functions of Markov chains. Ann. Math. Stat. **41**, 164–171 (1970)

Bray, B.C., Lanza, S.T., Tan, X.: Eliminating bias in classify-analyze approaches for latent class analysis. Struct. Equ. Model. Multidiscip. J. **22**, 1–11 (2015)

Celeux, G., Soromenho, G.: An entropy criterion for assessing the number of clusters in a mixture model. J. Classif. **13**, 195–212 (1996)

Dayton, C.M., Macready, G.B.: Concomitant-variable latent-class models. J. Am. Stat. Assoc. **83**, 173–178 (1988)

Dias, J.G., Vermunt, J.K.: A bootstrap-based aggregate classifier for model-based clustering. Comput. Stat. **23**, 643–659 (2008)

Dempster, A.P., Laird, N.M., Rubin, D.B.: Maximum likelihood from incomplete data via the EM algorithm (with discussion). J. R. Stat. Soc. Ser. B **39**, 1–38 (1977)

Formann, A.K.: Mixture analysis of multivariate categorical data with covariates and missing entries. Comput. Stat. Data Anal. **51**, 5236–5246 (2007)

Foy, P.: TIMSS and PIRLS 2011 User Guide for the Fourth Grade Combined International Database. TIMSS & PIRLS International Study Center, Boston College, Chestnut Hill (2013)

Gnaldi, M., Bacci, S., Bartolucci, F.: A multilevel finite mixture item response model to cluster examinees and schools. ADAC **10**, 53–70 (2016)

Goldstein, H.: Multilevel Statistical Models. John Wiley & Sons, Chichester (2011)

Goldstein, H., Spiegelhalter, D.J.: League tables and their limitations: statistical issues in comparisons of institutional performance. J. R. Stat. Soc. Ser. A **3**, 385–443 (1996)

Goodman, L.A.: Exploratory latent structure analysis using both identifiable and unidentifiable models. Biometrika **61**, 215–231 (1974)

Goodman, L.A.: The analysis of systems of qualitative variables when some of the variables are unobservable. Part I-A modified latent structure approach. Am. J. Sociol. **79**, 1179–1259 (1974)

Goodman, L.A.: On the assignment of individuals to latent classes. Sociol. Methodol. **37**, 1–22 (2007)

Grilli, L., Pennoni, F., Rampichini, C., Romeo, I.: Exploiting TIMSS and PIRLS combined data: multivariate multilevel modelling of student achievement. Ann. Appl. Stat. **10**, 2405–2426 (2016)

Juang, B.H., Rabiner, L.R.: Hidden Markov models for speech recognition. Technometrics **33**, 251–272 (1991)

Lazarsfeld, P.F., Henry, N.W.: Latent Structure Analysis. Houghton Mifflin, Boston (1968)

Lukocienė, O., Vermunt, J.K.: Determining the number of components in mixture models for hierarchical data. In: Fink, A., Lausen, B., Seidel, W. and Ultsch, A. (eds.) Advances in Data Analysis, Data Handling and Business Intelligence, pp. 241–249. Springer, Berlin (2010)

McLachlan, G.J., Peel, D.: Finite Mixture Models. Wiley, New York (2000)

Pennoni, F.: Issues on the Estimation of Latent Variable and Latent Class Models. Scholars' Press, Saarbücken (2014)

Skrondal, A., Rabe-Hesketh, S.: Generalized Latent Variable Modelling: Multilevel, Longitudinal and Structural Equation Models. Chapman and Hall/CRC Press, Boca Raton (2013)

Vermunt, J.K.: Multilevel latent class models. Sociol. Methodol. **33**, 213–239 (2003)

Vermunt, J.K.: Multilevel latent variable modeling: an application in education testing. Austrian Journal of Statistics **37**, 285–299 (2008)

Viterbi, A.J.: Error bounds for convolutional codes and an asymptotically optimum decoding algorithm. IEEE Trans. Inf. Theory **13**, 260–269 (1967)

Zucchini, W., MacDonald, I.L., Langrock, R.: Hidden Markov Models for Time Series: An Introduction Using R. Chapman and Hall/CRC Press, Boca Raton (2016)

Multi-Criteria Classifications in Regional Development Modelling

Beata Bal-Domańska

Abstract The article presents the discussion regarding the influence of taking selected approaches to the classification of regions on estimation results of Solow–Swan growth models (case of convergence processes). As the existing modelling effects, obtained in the area of regional development, show the estimation results of the production function parameters' assessment, based on Solow–Swan growth model or beta-convergence models, are divided depending on the period and units covered by the study (countries, regions). The article presents modelling results of the development of economies in the European Union regions at NUTS 2 level, depending on the development factors in line with the extended MRW (Mankiw–Romer–Weil) growth model in the selected period for various groups of regions. Each time the econometric analysis was conducted in the groups of NUTS 2 regions, separated in terms of the level of smart specialization. However, every time, along with the change in the grouping method, the dividing boundary of the inclusion of regions into a given group was changing. It allowed the assessment of estimations stability depending on the inclusion of the selected regions into particular groups of smart specialization.

To sum up, the conducted analysis allowed for presenting conclusions regarding the stability of the obtained estimations in the groups of the European Union regions in the period of 2003–2015 characterized by a different level of smart specialization in the defined groups. The more favourable results were obtained based on models developed for two groups and using the k-means method.

B. Bal-Domańska (✉)
Wrocław University of Economics, Department of Regional Economics, Jelenia Góra, Poland
e-mail: beata.bal-domanska@ue.wroc.pl

© Springer Nature Singapore Pte Ltd. 2020

T. Imaizumi et al. (eds.), *Advanced Studies in Classification and Data Science*,
Studies in Classification, Data Analysis, and Knowledge Organization,
https://doi.org/10.1007/978-981-15-3311-2_2

1 Introduction

The research covering regional development and convergence has been among the mainstream interests of scientists and politicians for years (Mankiw et al. 1992; Sala-i-Martin 1996; Kliber and Malaga 2002; Próchniak and Witkowski 2006; Batóg and Batóg 2006; Pukeliene and Butkus 2012; Smętkowski and Wójcik 2012; James and Campbell 2013). They are often carried out using econometric models. In the course of regional studies the frequently encountered problem is related to the selection of variables, methods and assessment criteria, which allow analysing the research problem in the most optimal way.

The article is focused on the problem of absolute (unconditional) and conditional convergence modelling in the groups of regions presenting a similar profile of smart specialization. The basic problem is related to the assessment of cluster analysis methods selection and the number of groups as an initial stage of analysing relationships and processes occurring in the defined groups of units.

2 The Assumptions and Research Procedure

While developing the procedure for assessing regional development one often comes across the following problems:

- how to define regional groups in order to obtain unambiguous results allowing the diagnosis of similarities in the occurring development processes in the particular groups of regions—this is the question about the selection of cluster analysis methods and the number of groups of regions,
- is the cluster analysis method selection important for the obtained results? In terms of this problem not just the change of the value of model structural parameters is important (it is obvious with almost every modification of the structure or scope of the model objects), but the stability of the drawn conclusions as to the direction of dependence, the strength of the relationship (rate of changes), the significance of the obtained parameters in the regions characterized by the same profile.

The discussion presented below was focused on answering these two questions. The analysis was carried out based on convergence models developed for the NUTS-2 European Union groups of regions, which were previously grouped according to the level of smart specialization, i.e. in the knowledge-intensive and high-technologies based sectors. The assessment was performed for the panel of 267 regions in the period of 2003–2015.

Research procedure:

1. Defining smart specialization characteristics of NUTS-2 EU regions
   Smart specialization emphasizes the actual size and role of knowledge-based sectors (manufacturing and service) in the employment structure of particular

countries. Smart specialization was defined by means of two basic diagnostic indicators presenting, in terms of smart specialization, the most important parts of economy:

- *HT*—employment in high and medium high-technology manufacturing as the share of total employment (%) (NACE Rev 2 Code: 20, 21, 26–30),
- *KIS*—employment in knowledge-intensive services as the share of total employment (%) (NACE Rev 2 Code: 50–51, 58–66, 69–75, 78, 80, 84–93), and additional variables, presenting employment in the other sectors:
- *LHT*—employment in high and medium high-technology manufacturing as the share of total employment (%) (NACE Rev 2 Code: 10–19, 22–25, 31–33),
- *LKIS*—employment in knowledge-intensive services as the share of total employment (%) (NACE Rev 2 Code: 45–47, 49, 52, 53, 55, 56, 68, 77, 79, 81, 82, 94–99),
- *Other*—not listed above.

2. Defining time range and spatial extent of the research

According to the data availability the analysis was conducted based on the panel covering 267 (among 276 EU regions) EU NUTS-2 regions in the period 2003–2015.

The cluster analysis of regions in terms of smart specialization level was carried out based on the data collected in 2015.

3. Defining groups of regions presenting a similar level of smart specialization

One of the research stages was to determine the number of groups, and also to identify the group profile (i.e. distinguished at the background of other in the smart specialization sector, with particular focus on *HT* and *KIS* sectors).

It was assumed that:

- **the number of groups** should not be too large, so as to ensure the appropriate number of objects in groups and thus the degree of freedom relevant for model quality. Ultimately, the regions were divided into 2 or 4 groups of smart specialization. The classification into two groups allowed assessing the stability of estimation results in the groups of similar size. It was followed by dividing these groups into smaller ones and, simultaneously, featuring the higher level of specialization (less internally differentiated ones),
- **grouping variables**: the division was based on one (*HT* or *KIS*), two (*HT, KIS*), four (*HT, KIS, LKIS, Other*) or five (*HT, LHT, KIS, LKIS, Other*) indicators of smart specialization defined in the first stage of the research. Each of the indicator sets included at least one of the key variables to determine the level of regional smart specialization (key variables: *KIS* or *HT*).

When defining the character (profile) of each group, special attention was paid to the level of key variables, *KIS* and *HT*. It was shown in the group name by indicating which of the variables (and specifically their mean value) reached the highest level in the given group of regions. For example, *hHT* (high level of the mean for *HT* variable) means that in this group the average employment level in the sector of medium and high-technology in manufacturing was the highest among all analysed groups. In the situation when mean values of one of the key variables adopted a very similar level in all groups (the difference was less than 1% point), the name showed the level of the second key variable (significantly differentiating the groups), e.g. for two groups, in both groups the level of employment in *HT* was similar, but the level of employment in *KIS* sector was significantly different in each of them, therefore group (class) 1. received the name—*hKIS* (high level of the mean for *KIS* variable), and group 2. *lKIS* (low level of the mean for *KIS* variable).

- **grouping methods**: three methods were used in grouping the analysed regions: (A) method based on descriptive statistics (arithmetic mean and median), (B) hierarchical clustering Ward method, (C) *k*-means which aims at grouping *n* observations into *k* clusters in which each observation belongs to the cluster with the nearest mean. Each of these three methods generated two classifications with two or four groups. Based on this, it was possible, within each method, to compare the stability of modelling results with slight changes in the assumptions, such as e.g. the change of distance measurement method within the same cluster analysis method. Finally, the following variants of groups were obtained:

  (A) Descriptive statistics—the division performed based on the value of arithmetic mean or median; this approach was used to define groups of regions based on one or two key variables (*HT, KIS*).

   The regions classified using this method have a clearly defined position in the group, which results from obtaining the highest or the lowest values of a given variable. The outlier groups (presenting the highest or the lowest values) clearly aggregate the regions with precisely defined values of variables. In case of the grouping variant with a median based on one variable the advantage is the division into two *n*-equal groups.

  (B) Ward method—the division made using Euclidean distance (in table marked as "Euclid.") or square of Euclidean distance ("sq. Euclid."); this approach was used to define the classes of regions based on two (*HT, KIS*), four (*HT, KIS, LKIS, Other*) or five variables (*HT, LHT, KIS, LKIS, Other*).

  (C) *k*-means method—the division performed following different assumptions as to the initial clustering vector, in case of "sorted" variant—the option was to "sort the distances and take observations at a constant interval," and for "max. distance" variant—"select observations as to maximize cluster distances"; this approach was used to define groups of regions based on 2, 4 or 5 variables of smart specialization.

   The cluster analyses for (B) and (C) variants were calculated in STATISTICA program.

4. Estimation of regional development and convergence models in the groups of regions

The analysis covered convergence models presenting the following structure:

- absolute (unconditional):

$$\ln GDP_{it} = \left(1 + \frac{1 - e^{-\beta T}}{T}\right) \ln GDP_{i(t-1)} + \alpha_i + \varepsilon_{it} \tag{1}$$

- conditional:

$$\ln GDP_{it} = \left(1 + \frac{1 - e^{-\beta T}}{T}\right) \ln GDP_{i(t-1)} + \lambda_1 \ln \Delta EMPL_{it}$$
$$+ \lambda_2 \ln TETR_{it} + \alpha_i + \varepsilon_{it} \tag{2}$$

where: β—the parameter defining the convergence rate against the long-term equilibrium (the distance covered in a year); $\ln GDP$—logarithm of gross domestic product in t-th year ($t = 1, 2, \ldots, T$) and i-th object (region) ($i = 1, 2, \ldots, N$); $\ln TETR$—logarithm of employment among workers (15 years or over) with tertiary education (level 5–8) as the percentage of total employment (Eurostat: *lfst*); $\ln \Delta EMPL$—logarithm of employment growth rate (15 years or over) and the deprecation rate plus δ and technological progress rate g ($\delta + g$)—it was assumed to be equal to 0.5; $\ln S$—logarithm of investment rate (due to lack of data for a group of regions, this variable was eliminated from further analysis), λ—parameter defining the impact of j-th ($j = \ln TETR, \ln \Delta EMPL$) variable in l-th group of regions ($l = 1, 2$ or $l = 1, 2, 3, 4$); α_i—specific for each region, fixed in time, individual effects; ε_{it}—random term.

In order to obtain estimates of the convergence model (1) and (2) parameters the system General Method of Moments estimator (Arellano and Bover 1995; Blundell and Bond 1998; Bond et al. 2012) was used.

The calculations of convergence models were performed in STATA 11 program.

The conclusions drawn from the analysis result directly from the realization of the studied population variables and can be sensitive to outliers, which in case of regional studies, where all analysis objects (regions) are included, means that including or excluding even a single region from a group can affect the obtained results.

The following assumptions were made for the analysis:

- convergent conclusions are expected for the groups covering similar regions in terms of the level of characteristics describing the structure of employment in smart specialization sectors, including, in particular, the key innovation sectors, i.e. *HT* and *KIS*,
- in case of grouping based on descriptive statistics (mean and median) the differences in cluster analysis results for two groups are only expected for the population of regions presenting asymmetrical distribution (for symmetrical

distribution the results should be similar both for the groups distinguished based on the mean and median),
- in case of grouping based on measures clearly identifying the phenomenon level (mean, median) similar conclusions are expected for at least the outlier classes (the highest or the lowest variable values),
- more consistent results are expected for models obtained for a smaller number of groups and/or variables, distinguished by critical (strict) criteria (e.g. lowest/highest level groups of development among four groups).

3   Analysis of the Results

The analysis results were different depending on the number of identified groups. In case of two groups of regions, presenting different smart specialization levels, high consistency of results was achieved for conditional and unconditional (absolute) convergence models compared within the framework of groups (A, B or C) defined in terms of the same convergence method (see Table 1). The consistency of obtained results referred to both structural parameters signs, their values and their significance. It referred to all groups regardless of the number of grouping variables used. Moreover, for groups distinguished based on mean measures and the majority of group variants obtained based on cluster analysis methods (Ward and k-means) in most cases the differences in parameter values of $\ln\text{GDP}_{it-1}$, for both conditional and unconditional convergence models, did not exceed 2% (see Table 1).

In case of k-means method in two cases two identical grouping were obtained. In this case the measure of comparability equals 0.0 suggesting "no differences in parameters value".

In case of Ward method significant differences in estimates were visible. The achieved result can be related to the differences in number of regions in groups. For the variant "Ward; four variables", the large differences in parameters for the appropriate profiles (classes) were noted, ranged 1.1–7.8%. According to variant based on 5 variables and "Ward method and Euclidean distance" the cluster analysis results were characterized by a similar level of employment in HT sector and simultaneously by significant differences in employment in KIS sector. As a result the groups of high and low KIS profile were developed including 136 and 131 regions, respectively. In case of grouping using variant 2 "Ward method with squared Euclidean distance" different group profiles were developed—a relatively small group of high HT profile (68 regions) and a large group of high KIS profile (199 regions). Such extensive differences in profiles and numbers resulted, naturally, in different results (processes).

Summing up the collected results, the speed of convergence in the same profile groups was evaluated. In case of convergence speed (estimated based on $\ln GDP_{it-1}$

Table 1 Parameters of β absolute (1) and conditional (2) convergence models and their compatibility—case of two groups

Grouping variables	Cluster analysis method	Groups	Estimation res. (p-value) Model (1) a	Model (2) a	b	c	No. observ.	Comparability % Model (1) a	Model (2) a	b	c
All	x	All	0.863	0.805	0.627	0.088	267	x			
HT	Mean	hHT	0.915	0.873	0.666	0.076	104	−0.1	−1.7	−6.6	14.4
		lHT	0.823	0.769	0.588	0.082	163	−1.9	−0.4	5.5	−17.4
	Median	hHT	0.913	0.859	0.625	0.089	133	x			
		lHT	0.807	0.765	0.622	0.070	134	x			
KIS	Mean	hKIS	0.868	0.796	0.595	0.067	134	−0.9	−1.0	0.5	3.5
		lKIS	0.903	0.912	0.717	0.036	133	0.0	−0.2	−0.8	2.9
	Median	hKIS	0.860	0.788	0.598	0.069	130	x			
		lKIS	0.902	0.911	0.712	0.037	137	x			
HT KIS	Ward	hHT	0.914	0.921	0.716	0.033*	124	0.1	−0.6	0.1	16.1
	Euclid.	hKIS	0.874	0.796	0.648	0.074	143	−0.3	−0.3	−1.9	−2.6
	Ward	hHT	0.914	0.916	0.717	0.039	136	x			
	sq. Euclid.	hKIS	0.871	0.793	0.636	0.072	131	x			
	k-mean	hHT	0.908	0.913	0.71	0.038	130	0.0	0.0	0.0	0.0
	max dist.	hKIS	0.870	0.789	0.626	0.076	137	0.0	0.0	0.0	0.0
	k-mean	hHT	0.908	0.913	0.71	0.038	130	x			
	sorted	hKIS	0.870	0.789	0.626	0.076	137	x			
HT KIS LKIS Other	Ward	hHT	0.9	0.908	0.718	0.035	147	−1.1	−2.6	−5.7	33.0
	Euclid.	hKIS	0.858	0.776	0.601	0.075	120	3.7	7.8	−1.8	−109
	Ward	hHT	0.890	0.886	0.679	0.052	169	x			
	sq. Euclid.	hKIS	0.890	0.842	0.590	0.036	98	x			
	k-mean	hKIS	0.869	0.801	0.622	0.071	149	0.0	0.0	0.0	0.0
	max dist.	lKIS	0.891	0.907	0.702	0.030	118	0.0	0.0	0.0	0.0
	k-means	hKIS	0.869	0.801	0.622	0.071	149	x			
	sorted	lKIS	0.891	0.907	0.702	0.030	118	x			
HT LHT KIS LKIS Other	Ward	hKIS	0.875	0.807	0.617	0.060	136	−1.8	−3.8	21.1	32.4
	Euclid.	lKIS	0.875	0.889	0.687	0.038	131	No corresponding group			
	Ward	hKIS	0.858	0.777	0.783	0.089	199	x			
	sq. Euclid.	hHT	0.924*	0.94	0.565	0.006	68	x			
	k-mean	hHT	0.913	0.921	0.688	0.029*	111	−1.1	−0.9	1.8	15.4
	max dist.	hKIS	0.874	0.789	0.626	0.076	156	−0.3	1.0	1.3	−6.3
	k-mean	hHT	0.903	0.913	0.701	0.035	119	x			
	sorted	hKIS	0.871	0.797	0.634	0.071	148	x			

Source: Own calculation in STATISTICA and STATA 11

a $\ln GDP_{it-1}$, b $\ln\Delta EMPL_{it}$, c $\ln TETR_{it}$

( ) if not marked, parameter statistical significant at the level of 0.003

*Parameter statistical significant at the level of 0.08

Table 2 Comparison of rate of absolute convergence based of model (1)—case of two groups

	Minimum value within groups			Maximum value within groups		
	Param. of $\ln GDP_{it-1}$	Speed of Converg. β	Half time to Convergence	Param. of $\ln GDP_{it-1}$	Speed of Converg. β	Half time to Convergence
Groups obtained using (A) (descriptive statistics) method						
hHT	**0.913**	9.1	7.6	**0.915**	8.9	7.8
lHT	**0.807**	21.4	3.2	**0.823**	19.5	3.5
Groups obtained using (A) descriptive statistics and (C) k-means methods; four variables variant						
hKIS	**0.860**	15.1	4.6	**0.875**	13.4	5.2
lKIS	**0.875**	13.3	5.2	**0.903**	10.2	6.8
Groups obtained using (B) Ward and (C) k-means methods; 2 and 5 variables variant						
hHT	**0.890**	11.6	6.0	**0.924***	7.9	8.8
hKIS	**0.858**	15.4	4.5	**0.890**	11.6	6.0

Source: author's compilation
( ) if not marked, parameter statistical significant at the level of 0.003
*Parameter statistical significant at the level of 0.08

parameters) in absolute convergence models (1) the following was observed (see Table 2):

- when comparing groups of high and low *HT* profile the $\ln GDP_{it-1}$ parameter was higher, and thus lower convergence speed in *hHT* groups (these groups were present in the cluster analysis according to (A) variant descriptive statistics),
- in case of high and low *KIS* groups higher $\ln GDP_{it-1}$ parameter (and lower convergence speed) was obtained for *lKIS* group (these groups were present in the cluster analysis according to (A) variant descriptive statistics and (C) "k-means, for four variables"),
- in case of *hHT* and *hKIS* groups higher $\ln GDP_{it-1}$ parameter (and thus lower convergence speed) was obtained for *hHT* group (these groups were present in the cluster analysis performed according to Ward method (variant B) and (C) "k-means, for 2 and 5 variables").

The results presented in Table 2. indicate the presence of absolute beta-convergence processes within each of defined groups. The occurrence of beta convergence means that poorer regions catch up with the richer ones. The speed of convergence (until the situation of long-term balance is reached) in high and low *HT* regions reached the level of about 8.9–9.1% in hHT groups and about 19.5–21.4% in *lHT* ones. It means that the low *HT* regions growth at faster rates. In the high *HT* groups of regions about 7.6–7.8 years to reduce the development gap by half, while in low *HT* groups from 3.2 to 3.5 years are needed.

More demanding models were developed for four groups of smart specialization regions (Table 3). This was due to the fact that the groups obtained the same division method, but on the basis of various smart specialization variables, they differed in numbers and profile.

According to the comparison results of models developed for four groups, it should be observed that in most cases the statistically significant parameters were

Table 3 Selected estimation results of β absolute and conditional convergence models—case of four groups

Grouping variables	Cluster analysis method	Groups	Estimation results (p-value)				No. of observ.
			Model (1)	Model (2)			
			a	a	b	c	
HT KIS LKIS Other	Ward Euclid.	hHT	0.951*	0.966*	0.51*	0.001	87
		hKIS	0.858*	0.776*	0.608*	0.075*	120
		med. high	0.831	0.854	0.1.01	0.027	48
		med. low	0.888	0.834	0.443	0.050	12
	Ward sq. Euclid.	hHT	0.912*	0.899*	0.541*	0.05*	120
		hKIS	0.890*	0.842*	0.590*	0.036*	98
		med. high	0.789	0.868	0.980	0.024	34
		med. low	0.88	0.834	0.489	0.038	15
	k-mean max dist.	hHT	0.917*	0.911*	0.648*	0.045*	84
		hKIS	0.877*	0.819*	0.567*	0.052*	118
		lHT	0.759	0.834	0.950	0.028	42
		lKIS	0.896	0.867	0.391	0.052	23
	k-mean sorted	hHT	0.918*	0.909*	0.665*	0.048*	83
		hKIS	0.876*	0.815*	0.568*	0.054*	117
		lHT	0.759	0.835	0.950	0.028	42
		lKIS	0.89	0.856	0.383	0.059	25

Source: Own calculation in STATISTICA and STATA 11
a $\ln\text{GDP}_{it-1}$, b $\ln\Delta\text{EMPL}_{it}$, c $\ln\text{TETR}_{it}$
*Parameters significant at 0.02 level

not obtained. While the values of structural parameter, e.g. in absolute convergence models (1) for groups of the same profile, can be considered as close in the vast majority of cases, then the conclusions drawn for them differed considerably. Often, for one of them, the parameter close to the value was statistically significant, while for the counterpart it was statistically insignificant. Contrary to expectations, the smallest consistency was presented by the models based on descriptive statistics. High consistency of models for four groups was obtained for models of regional groups defined based on four variables (*HT, KIS, LKIS, Other*) (see Table 4). An interesting case was recorded for the groups identified using k-means method. Out of four groups one (including 42 low *HT* regions) was the same as its equivalent obtained in both the "k-means sorted" variant and "max.distance" variant.

It should be noted that parameters of $\ln GDP_{it-1}$ were characterized by higher stability (comparability). It was observed in case of models for both 2 and 4 groups of smart specialization regions. The differences in their values most often, for both conditional and unconditional convergence models, did not exceed 2% (see Tables 1 and 4 of "Comparability (%)" column). Similar results were achieved for $\ln\Delta EMPL_{it}$ variable. Much more significant differences were observed, however, for $\ln TETR_{it}$ variable, recorded even for models regarded as comparable in terms of $\ln GDP_{it-1}$ parameter values.

Table 4 Compatibility of β absolute and conditional convergence parameters—case of four groups

Grouping variables	Cluster analysis method	Groups	Comparability (%)			
			Model (1)	Model (2)		
			a	a	b	c
HT	Ward Euclid.	hHT	4.3	7.6	5.8	98.0
KIS	vs.	hKIS	3.7	7.8	3.0	109.8
LKIS	Ward	Med. high	5.3	1.6	3.4	12.7
Other	sq. Euclid.	Med. low	1.0	0.0	9.4	31.3
	k-means	hHT	0.1	0.2	2.6	7.1
	Max. dist.	hKIS	0.2	0.5	0.2	3.7
	vs.	lHT	0.0	0.0	0.0	0.0
	k-means sort.	lKIS	0.6	1.3	2.1	12.8

Source: Own calculation in STATISTICA and STATA 11
$a \ln\text{GDP}_{it-1}$, $b \ln\Delta\text{EMPL}_{it}$, $c \ln\text{TETR}_{it}$

4 Conclusions

The presented results offer a preliminary analysis aimed at the application possibilities of various methods and tools in the regional analyses. While answering the question asked at the beginning of the presented discussion, it should be pointed out that better results were obtained based on models developed for two groups. The number of variables, on the basis of which the groups were defined, does not seem to affect the formulated conclusions. Among the cluster analysis methods used, good results (high consistency of results) were achieved in the groups (in particular for case of four groups) obtained using k-means method.

Acknowledgement The study was funded within the framework of the National Science Centre project 2015/17/B/HS4/01021.

References

Arellano, M., Bover, O.: Another look at the instrumental variables estimation of error-components models. J. Econ. **68**, 29–51 (1995)

Batóg, J., Batóg, B.: Analysis of income convergence in the Baltic Sea region. In: Baltic Business Development. Regional Development. SME Management and Entrepreneurship, pp. 33–40. University of Szczecin (2006)

Blundell, R., Bond, S.: Initial conditions and moment restriction in dynamic panel data models. J. Econ. **87**, 115–143 (1998)

Bond, S., Hoeffler, A., Temple, J.: GMM estimation of empirical growth models. Economics Group, Nuffield College, University of Oxford, Economics Papers Nr 2001-W21 (2012)

James, R.D., Campbell, H.S.: The effects of space and scale on unconditional beta convergence: test results from the United States, 1970–2004. GeoJournal **78**, 803 (2013). https://doi.org/10. 1007/s10708-012-9467-5

Kliber, P., Malaga, K.: On the convergence of growth path towards steady-states in OECD countries in Solow–Swan type model. In Charemza, W., Strzaå, K. (eds.) East European Transition and EU Enlargement, a Quantitative Approach. Physica Verlag, Heidelberg, New York (2002)

Mankiw, N., Romer, D., Weil, D.: A contribution to the empirics of economic growth. Q. J. Econ. **107**(2), 407–437 (1992)

Próchniak, M., Witkowski, B.: Modelowania realnej konwergencji w skali międzynarodowej. Gospodarka Narodowa **10**(182), 1–32 (2006)

Pukeliene, V., Butkus, M.: Evaluation of regional beta convergence in UE countries NUTS-3 level. Ekonomika **91**(2), 22–37 (2012)

Sala-i-Martin, X.X.: The classical approach to convergence analysis. Econ. J. **106**(437), 1019–1036 (1996)

Smętkowski, M., Wójcik, P.: Regional convergence in Central and Eastern European countries—a multidimensional approach. Eur. Plan. Stud. **20**(6), 923–939 (2012)

Non-parametric Latent Modeling and Network Clustering

François Bavaud

Abstract The paper exposes a non-parametric approach to latent and co-latent modeling of bivariate data, based upon alternating minimization of the Kullback–Leibler divergence (EM algorithm) for complete models. For categorical data, the iterative algorithm generates a soft clustering of both rows and columns of the contingency table. Well-known results are revisited, and new procedures are presented. In particular, the consideration of square contingency tables induces a clustering algorithm for weighted networks, differing from spectral clustering or modularity maximization techniques. Also, we present a novel co-clustering algorithm, distinct from the Baum–Welch algorithm, applicable to HMM models for unrestricted bigrams counts. Three case studies illustrate the theory.

1 Introduction: Parametric and Non-parametric Mixtures

Two variables can be dependent, yet conditionally independent given a third one, that is $X \perp Y|G$ but $X \not\perp Y$: in bivariate *latent models* of dependence M, joint bivariate probabilities $P(x, y)$ express as

$$P(x, y) = \sum_{g=1}^{m} p(x, y, g) = \sum_{g=1}^{m} p(g)p(x|g)p(y|g) \qquad (1)$$

where x, y, g denote the values of X, Y, G, and $p(x, y, g)$ their joint probability.

Bivariate data, such as summarized by normalized contingency tables $F(x, y) = \frac{n(x,y)}{n(\bullet,\bullet)}$, where $n(x, y)$ counts the number of individuals in $x \in X$ and $y \in Y$, can be approached by latent modeling, consisting in inferring a suitable model $P(x, y) \in M$ of the form (1), typically closest to the observed frequencies $F(x, y)$ in the

F. Bavaud (✉)
University of Lausanne, Lausanne, Switzerland
e-mail: fbavaud@unil.ch

© Springer Nature Singapore Pte Ltd. 2020
T. Imaizumi et al. (eds.), *Advanced Studies in Classification and Data Science*,
Studies in Classification, Data Analysis, and Knowledge Organization,
https://doi.org/10.1007/978-981-15-3311-2_3

maximum-likelihood sense, or in the least squares sense. Mixture (1) also defines *memberships* $p(g|x) = p(x|g)p(g)/p(x)$ and $p(g|y)$; hence latent modeling also performs *model-based clustering*, assigning observations x and y among groups $g = 1, \ldots, m$.

Latent modeling and clustering count among the most active data-analytic research trends of the last decades. The literature is simply too enormous to cite even a few valuable contributions, often (re-)discovered independently among workers in various application fields. Most approaches are parametric, typically defining $p(x|g)$ and $p(y|g)$ as exponential distributions of some kind, such as the multivariate normal (continuous case) or the multinomial (discrete case) (see e.g. Govaert and Nadif 2013 and references therein). Parametric modeling allows further hyperparametric Bayesian processing, as in latent Dirichlet allocation (Blei et al. 2003).

By contrast, we focus on non-parametric models specified by the whole family of log-linear *complete models* $\mathcal{M}$ corresponding to $X \perp Y|G$, namely (see e.g. Christensen 2006)

$$\mathcal{M} = \{p \mid \ln p(x, y, g) = a(x, g) + b(y, g) + c\}$$

Equivalently,

$$\mathcal{M} = \{p \mid p(x, y, g) = \frac{p(x, \bullet, g)\, p(\bullet, y, g)}{p(\bullet, \bullet, g)}\}$$

where "$\bullet$" denotes the summation over the replaced argument. The corresponding class of bivariate models M of the form (1) simply reads $M = \{P \mid P(x, y) = \sum_g p(x, y, g) \equiv p(x, y, \bullet)$, for some $p \in \mathcal{M}\}$.

Observations consist of the joint empirical distribution $F(x, y)$, normalized to $F(\bullet, \bullet) = 1$. In latent modeling, one can think of the observer as a color-blind agent perceiving only the margin $f(x, y, \bullet)$ of the complete distribution $f(x, y, g)$, but not the color (or group) g itself (see Fig. 1). Initially, any member f of the set

$$\mathcal{D} = \{f \mid f(x, y, \bullet) = F(x, y)\}$$

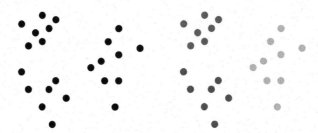

Fig. 1 Left: observed data, where (x, y) are the object coordinates. Right: complete data (x, y, g), where the group g is labeled by a color. In psychological terms, (x, y) is the *stimulus*, and (x, y, g) the *percept*, emphasizing the EM algorithm as a possible model for cognition

seems equally compatible with the observations F, and the role of a clustering algorithm precisely consists in selecting a few good candidates $f \in \mathcal{D}$, or even a unique one, bringing color to the observer.

This paper exposes a non-parametric approach to latent and co-latent modeling of bivariate data, based upon alternating minimization of the Kullback–Leibler divergence (EM algorithm) for complete log-linear models (Sect. 2). For categorical data, the iterative algorithm generates a soft clustering of both rows and columns of the contingency table. Well-known facts are re-exposed, and new results are presented. In particular, the consideration of square contingency tables induces a clustering algorithm for weighted networks, differing from spectral clustering or modularity maximization techniques (Sect. 3). Also, we present a novel co-clustering algorithm, distinct from the Baum–Welch algorithm, applicable to HMM models for unrestricted bigrams counts.

Three case studies illustrate the theory: latent (co-)betrayed clustering of a term-document matrix (Sect. 2.3), latent clustering of spatial flows (Sect. 3.2), and latent co-clustering of bigrams in French (Sect. 3.4).

2 EM Latent Clustering: A Concise Derivation from First Principles

The *alternating minimization* procedure (Csiszár and Tusnády 1984) provides an arguably elegant derivation of the EM algorithm; see also e.g. Cover and Thomas (1991) or Bavaud (2009). The maximum-likelihood model $\hat{P} \in M$ of the form (1) minimizes the Kullback–Leibler divergence $K()$

$$\hat{P} = \arg\min_{P \in M} K(F \| P) \qquad K(F \| P) = \sum_{x,y} F(x, y) \ln \frac{F(x, y)}{P(x, y)}$$

where $F(x, y)$ denotes the empirical bivariate distribution. On the other hand, the complete Kullback–Leibler divergence $K(f \| p) = \sum_{x,y,g} f(x, y, g) \ln \frac{f(x,y,g)}{p(x,y,g)}$, where $f(x, y, g)$ is the empirical "complete" distribution (see Fig. 1), enjoys the following properties (see e.g. Bavaud 2009 for the proofs, standard in Information Theory):

$$\hat{p}(x, y, g) := \arg\min_{p \in \mathcal{M}} K(f \| p) = \frac{f(x, \bullet, g) \, f(\bullet, y, g)}{f(\bullet, \bullet, g)} \qquad \textbf{M-step} \quad (2)$$

$$\tilde{f}(x, y, g) := \arg\min_{f \in \mathcal{D}} K(f \| p) = \frac{p(x, y, g)}{p(x, y, \bullet)} F(x, y) \qquad \textbf{E-step} \quad (3)$$

Furthermore, $\min_{f \in \mathscr{D}} K(f\|p) = K(F\|P)$, and thus

$$\min_{P \in M} K(F\|P) = \min_{p \in \mathscr{M}} \min_{f \in \mathscr{D}} K(f\|p)$$

Hence, starting from some complete model $p^{(0)} \in \mathscr{M}$, the EM-sequence $f^{(t+1)} := \tilde{f}[p^{(t)}]$ defined in (3) and $p^{(t+1)} := \hat{p}[f^{(t+1)}]$ defined in (2) converges towards a *local minimum* of $K(f\|p)$. Observe the margins to coincide after a single EM cycle in the sense $p^{(t)}(x, \bullet, \bullet) = F(x, \bullet)$ and $p^{(t)}(\bullet, y, \bullet) = F(\bullet, y)$ for all $t \geq 1$.

For completeness sake, note that $\mathscr{D}$ and $\mathscr{M}$ are *closed* in the following sense, as they are in other instances of the EM algorithm in general. Critically and crucially:

1. $\mathscr{D}$ is convex, that is closed under additive mixtures $\lambda f_1 + (1 - \lambda) f_2$; this turns out to be the case for maximum entropy problems in general.
2. $\mathscr{M}$ is log-convex, that is closed under multiplicative mixtures $p_1^\lambda p_2^{(1-\lambda)}/Z(\lambda)$ where $Z(\lambda)$ is a normalization constant; this is the case for exponential models, as well as for non-parametric log-linear models in general.

2.1 Latent Co-clustering

Co-clustering describes the situation where each of the observed variables is attached to a *distinct* latent variable, the latter being mutually associated. That is, $X \perp Y|(U, V)$, $X \perp V|U$, and $Y \perp U|V$ while $X \not\perp Y$, and $U \not\perp V$ in general. Equivalently, $X \to U \to V \to Y$ forms a "Markov chain," in the sense of Cover and Thomas (1991). Bivariate joint probabilities express as

$$P(x, y) = \sum_{u=1}^{m_1} \sum_{v=1}^{m_2} p(x, y, u, v) = \sum_{u,v} p(u, v) p(x|u) p(y|v) \qquad (4)$$

Complete models $\mathscr{M}$, restricted models M, and complete empirical distributions $\mathscr{D}$ are

$$\mathscr{M} = \{p \mid p(x, y, u, v) = \frac{p(x \bullet u\bullet)\, p(\bullet y \bullet v)\, p(\bullet \bullet uv)}{p(\bullet \bullet u\bullet)\, p(\bullet \bullet \bullet v)}\} \qquad (5)$$

$$M = \{P \mid P(x, y) = p(x, y, \bullet, \bullet) \text{ with } p \in \mathscr{M}\} \qquad (6)$$

$$\mathscr{D} = \{f \mid f(x, y, \bullet, \bullet) = F(x, y)\} \qquad (7)$$

where $F(x, y)$ denotes the observed empirical distribution. The steps of the former section apply again, yielding the EM algorithm

$$\hat{p}(x, y, u, v) := \arg\min_{p \in \mathscr{M}} K(f\|p) = \frac{f(x \bullet u\bullet)\, f(\bullet y \bullet v)\, f(\bullet \bullet uv)}{f(\bullet \bullet u\bullet)\, f(\bullet \bullet \bullet v)} \qquad \textbf{M-step}$$

$$(8)$$

$$\tilde{f}(x, y, u, v) := \arg\min_{f \in \mathscr{D}} K(f\|p) = \frac{p(x, y, u, v)}{p(x, y, \bullet, \bullet)} F(x, y) \qquad \textbf{E-step}$$

$$(9)$$

where $K(f\|p) = \sum_{x,y,u,v} f(x, y, u, v) \ln \frac{f(x,y,u,v)}{p(x,y,u,v)}$ measures the divergence of the complete observations from the complete model.

2.2   Matrix and Tensor Algebra for Contingency Tables

The material of Sects. 2 and 2.1 holds *irrespectively of the continuous or discrete nature of X and Y*: in the continuous case, integrals simply replace sums. In the discrete setting, addressed here, categories are numbered as $i = 1, \ldots, n$ for X, as $k = 1, \ldots, p$ for Y and as $g = 1, \ldots, m$ for G. Data consist of the relative $n \times p$ contingency table F_{ik} normalized to $F_{\bullet\bullet} = 1$.

2.2.1   Latent Co-clustering

Co-clustering models and complete models express as

$$P_{ik} = \sum_{u=1}^{m_1} \sum_{v=1}^{m_2} c_{uv} \, a_i^u \, b_k^v \qquad\qquad p_{ikuv} = c_{uv} \, a_i^u \, b_k^v \qquad (10)$$

- where $c_{uv} = P(U = u, V = v) = p(\bullet \bullet uv)$, obeying $c_{\bullet\bullet} = 1$, is the *joint latent distribution* of row, respectively column groups u and v
- $a_i^u = p(i \bullet u\bullet)/p(\bullet \bullet u\bullet)$ (with $a_\bullet^u = 1$) is the *row distribution conditionally to the row group* $U = u$, also referred to as *emission probability* (Sect. 3)
- $b_k^v = p(\bullet k \bullet v)/p(\bullet \bullet \bullet v)$ (with $b_\bullet^v = 1$) is the column distribution or emission probability conditionally to the column group $V = v$.

Hence, a complete model p is entirely determined by the triple (C, A, B), where $C = (c_{uv})$ is $m_1 \times m_2$ and normalized to unity, $A = (a_i^u)$ is $n \times m_1$ and $B = (b_k^v)$ is $p \times m_2$, both row-standardized.

It is straightforward to show that the successive application of the E-step (9) and the M-step (8) to $p \equiv (C, A, B)$ yields the new complete model $\ddot{p} \equiv (\ddot{C}, \ddot{A}, \ddot{B})$ with

$$\ddot{c}_{uv} = c_{uv} \sum_{jl} \frac{F_{jl}}{P_{jl}} a_j^u \, b_l^v \qquad (11)$$

$$\ddot{a}_i^u = a_i^u \frac{\sum_{lv'} c_{uv'} \frac{F_{il}}{P_{il}} b_l^{v'}}{\sum_{jlv'} c_{uv'} \frac{F_{jl}}{P_{jl}} a_j^u b_l^{v'}} \tag{12}$$

$$\ddot{b}_k^v = b_k^v \frac{\sum_{ju'} c_{u'v} \frac{F_{jk}}{P_{jk}} a_j^{u'}}{\sum_{jlu'} c_{u'v} \frac{F_{jl}}{P_{jl}} a_j^{u'} b_l^v} \tag{13}$$

Also, after a single EM cycle, margins are respected, that is $\ddot{P}_{i\bullet} = F_{i\bullet}$ and $\ddot{P}_{\bullet k} = F_{\bullet k}$.

In hard clustering, rows i are attached to a single group denoted $u[i]$, that is $a_i^u = 0$ unless $u = u[i]$; similarly, $b_k^v = 0$ unless $v = v[k]$. Restricting P in (10) to hard clustering yields *block clustering*, for which $K(F\|P) = I(X : Y) - I(U : V)$, where $I()$ is the mutual information (e.g. Kullback 1959; Bavaud 2000; Dhillon et al. 2003).

The set M of models P of the form (10) is convex, with extreme points consisting of hard clusterings. $K(F\|P)$ being convex in P, its minimum is attained for convex mixtures of hard clusterings, that is for *soft clusterings*.

2.2.2   Latent Clustering

Setting $m_1 = m_2 = m$ and C diagonal with $c_{gh} = \rho_g \, \delta_{gh}$ yields the latent model

$$P_{ik} = \sum_{g=1}^{m} \rho_g \, a_i^g \, b_k^g \qquad p_{ikg} = \rho_g \, a_i^g \, b_k^g \tag{14}$$

together with the corresponding EM-iteration $p \equiv (\rho, A, B) \rightarrow \ddot{p} \equiv (\ddot{\rho}, \ddot{A}, \ddot{B})$, namely

$$\ddot{\rho}_g = \rho_g \, \kappa_g \qquad \ddot{a}_i^g = a_i^g \frac{\sum_l b_l^g \frac{F_{il}}{P_{il}}}{\kappa_g} \qquad \ddot{b}_k^g = b_k^g \frac{\sum_j a_j^g \frac{F_{jk}}{P_{jk}}}{\kappa_g} \tag{15}$$

where $\kappa_g = \sum_{jl} a_j^g b_l^g \frac{F_{jl}}{P_{jl}}$. Similar (if not equivalent) updating rules have been proposed in information retrieval and natural language processing (Saul and Pereira 1997; Hofmann 1999), as well as in the *non-negative matrix factorization* framework (Lee and Seung 2001; Finesso and Spreij 2006).

By construction, families of latent models (14) M_m with m groups are nested in the sense $M_m \subseteq M_{m+1}$.

The case $m = 1$ amounts to *independence models* $P_{ik} = a_i b_k$, for which the fixed point $\ddot{a}_i = F_{i\bullet}$ and $\ddot{b}_k = F_{\bullet k}$ is, as expected, reached after a single iteration, irrespectively of the initial values of a and b.

By contrast, $m \geq \mathrm{rank}(F)$ generates *saturated models*, exactly reproducing the observed contingency table. For instance, assume that $m = p = \mathrm{rank}(F) \leq n$; then taking $a_i^g = F_{ig}/F_{\bullet g}$, $b_k^g = \delta_{kg}$ and $\rho_g = F_{\bullet g}$ (which already constitutes a fixed point of (15)) evidently satisfies $P_{ik} = F_{ik}$.

2.3   Case Study I: Reuters 21578 Term-Document Matrix

The $n \times p = 20 \times 1266$ document-term normalized matrix F, constituting the *Reuters 21578* dataset, is accessible through the R package **tm** (Feinerer et al. 2008). The co-clustering algorithm (11), (12), and (13) is started by randomly assigning uniformly each document to a single row group $u = 1, \ldots, m_1$, and by uniformly assigning each term to a single column group $v = 1, \ldots, m_2$. The procedure turns out to converge after about 1000 iterations (Fig. 2), yielding a locally minimal value $K_{m_1 m_2}$ of the Kullback–Leibler divergence. By construction, $K_{m_1 m_2}$ decreases with m_1 and m_2. Latent clustering (15) with m groups is performed analogously, yielding a locally minimal value K_m.

Experiments with three or four groups yield the typical results $K_3 = 1.071180 > K_{33} = 1.058654 > K_{43} = 1.038837 > K_{34} = 1.036647 > K_4 = 0.877754 > K_{44} = 0.873071$. The above ordering is expected, although inversions are frequently observed, under differing random initial configurations. Model selection procedures, not addressed here, should naturally consider in addition the degrees of freedom, larger for co-clustering models. The latter do not appear as particularly rewarding here (at least for the experiments performed, and in contrast to the results associated with case study III of Sect. 3.4): indeed, joint latent distributions C turn out to be "maximally sparse," meaning that row groups u and column groups v are essentially the same. Finally, each of the 20 documents of the Reuters 2157 dataset happens to belong to a single row group (hard clusters), while only a minority of the 1266 terms (say about 20%) belong to two or more column groups (soft clusters).

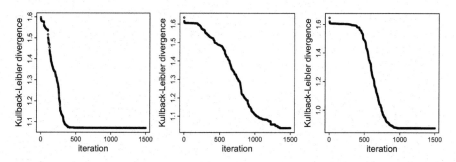

Fig. 2 Case study I: convergence of the latent and co-latent iterating procedure. Left: latent model with $m = 3$. Middle: co-latent model with $(m_1, m_2) = (3, 4)$. Right: co-latent model with $(m_1, m_2) = (4, 4)$

3 Network Clustering

When the observed categories x and y belong to the same set indexed by $i, j = 1, \ldots, n$, the relative square contingency table F_{ij} defines a *directed weighted network* on n vertices: F_{ij} is the weight of edge (ij), $F_{i\bullet}$ is the outweight of vertex i (relative outdegree), and $F_{\bullet i}$ its inweight i (relative indegree), all normalized to unity. Frequently, F_{ij} counts the relative number of units initially at vertex i, and at vertex j after some fixed time. Examples abound in spatial migration, spatial commuting, social mobility, opinion shifts, confusion matrices, textual dynamics, etc.

A further restriction, natural in many applications of latent network modeling, consists in identifying the row and column emission probabilities, that is in requiring $b_i^g = a_i^g$. This condition generates four families of nested latent network models of increasing flexibility, namely

$$P_{ij} = \sum_{g=1}^{m} \rho_g \, a_i^g \, a_j^g \qquad\qquad \text{latent (symmetric) network model} \quad (16)$$

$$P_{ij} = \sum_{u,v=1}^{m} c_{uv} \, a_i^u \, a_j^v \ \text{ with } c_{uv} = c_{vu} \qquad \text{co-latent symmetric network model} \quad (17)$$

$$P_{ij} = \sum_{u,v=1}^{m} c_{uv} \, a_i^u \, a_j^v \ \text{ with } c_{u\bullet} = c_{\bullet u} \qquad \text{co-latent MH network model} \qquad (18)$$

$$P_{ij} = \sum_{u,v=1}^{m} c_{uv} \, a_i^u \, a_j^v \qquad\qquad \text{co-latent general network model} \quad (19)$$

By construction, $P = P'$ in models (16) and (17), making latent and co-latent *symmetric* clustering suitable for *unoriented weighted networks* with $F_{ij} = F_{ji}$. By contrast, unrestricted co-latent models (19) describes general *oriented weighted networks*. Symmetric matrices $F = (F_{ij})$ appear naturally in reversible random walks on networks, or in spatial modeling where they measure the spatial interaction between regions (spatial weights), and constitute a weighted version of the adjacency matrix, referred to as an *exchange matrix* by the author (Bavaud 2014 and references therein; see also Berger and Snell 1957).

Latent models (16) are positive semi-definite or *diffusive*, that is endowed with *non-negative eigenvalues*, characteristic of a continuous propagation process from one place to its neighbors. In particular, the diagonal part of P in (16) cannot be too small. In contrast, co-latent symmetrical network models (17) are flexible enough to describe phenomena such as bipartition or periodic alternation, implying negative eigenvalues.

The condition (18) of *marginal homogeneity* (MH) on the joint latent distribution C is inherited by the restricted models, in the sense $P_{i\bullet} = P_{\bullet i}$. They constitute appropriate models for the bigram distributions of single categorical sequences (of length N, constituted of n types), for which $F_{i\bullet} = F_{\bullet i} + O(N^{-1})$; see the case study III of Sect. 3.4. Formulation (18) describes m hidden states related by a Markov transition matrix $p(v|u) = c_{uv}/c_{u\bullet}$, as well as n observed states related to the hidden ones by the *emission probabilities* $a_i^u = p(i|u)$. Noticeably enough, (18) precisely encompasses the ingredients of the *hidden Markov models* (HMM) (see e.g. Rabiner 1989).

3.1   Network Latent Clustering

Approximating F by P in (16) amounts in performing a *soft network clustering*: the *membership* of vertex i in group g (of weight ρ_g) is

$$z_{ig} = p(i|g) = \frac{p(i)p(g|i)}{p(g)} = \frac{f_i\, a_i^g}{\rho_g} \quad \text{with} \quad f_i = F_{i\bullet} = F_{\bullet i} \quad \text{and} \quad \rho_g = \sum_{i=1}^{n} f_i\, z_{ig}$$

EM-updating rules for memberships (instead of emission probabilities, for a change)

$$P_{ij} = f_i f_j \sum_{g=1}^{m} \frac{z_{ig} z_{jg}}{\rho_g} \qquad \ddot{z}_{ig} = z_{ig} \sum_{j} \frac{F_{ij}}{P_{ij}} \frac{f_j z_{jg}}{\rho_g} \qquad \ddot{\rho}_g = \sum_{i} f_i\, \ddot{z}_{ig} \tag{20}$$

define a soft clustering iterative algorithm for unoriented weighted networks, presumably original.

3.2   Case Study II: Inter-Cantonal Swiss Migrations

Consider the $n \times n$ matrix $N = (N_{ij})$ of inter-cantonal migratory flows in Switzerland, counting the number of people inhabiting canton i in 1980 and canton j in 1985, $i, j = 1, \ldots, n = 26$, for a total of sum$(N) = 6'039'313$ inhabitants, 93% of which lie on the diagonal (stayers). The symmetric, normalized matrix $F = \frac{1}{2}(N + N')/N_{\bullet\bullet}$ is diffusive, largely dominated by its diagonal. As a consequence, direct application of algorithm (20) from an initial random cantons-to-groups assignation produces somewhat erratic results: a matrix $F = (F_{ij})$ too close to the identity matrix $I = (\delta_{ij})$ cannot by reasonably approximated by the latent model (16), unless $m = n$, where each canton belongs to its own group.

Here, the difficulty lies in the shortness of the observation period (5 years, smaller than the average moving time), making the off-diagonal contribution $1 - \text{trace}(F)$ too small. Multiplying the observation period by a factor $\lambda > 1$ generates, up to $O(\lambda^2)$, a modified relative flow $\tilde{F}_{ij} = \lambda F_{ij} + (1 - \lambda)\delta_{ij} f_i$, where $f_i = F_{i\bullet} = F_{\bullet i}$ is the weight of canton i. The modified $\tilde{F}$ is normalized, symmetric, possesses unchanged vertex weights $\tilde{F}_{i\bullet} = f_i$, and its off-diagonal contribution is multiplied by λ. Of course, λ cannot be too large, in order to insure the non-negativity of $\tilde{F}$ ($\lambda \leq 6.9$ here) as well as its semi-positive definiteness ($\lambda \leq 6.4$ here).

Typical realizations of (20), with $\lambda = 5$, are depicted in Figs. 3 and 4: as expected, spatially close regions tend to be regrouped.

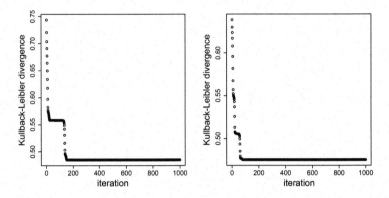

Fig. 3 Decrease of the Kullback–Leibler divergence for the two realizations of Fig. 4, respectively. Horizontal plateaux correspond to metastable minima in the learning of the latent structure, followed by the rapid discovery of a better fit

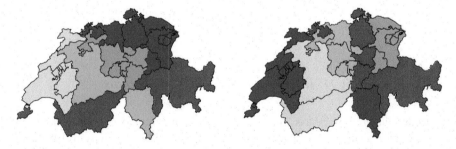

Fig. 4 Case study II: two realizations of the network latent clustering algorithm (20), applied to the modified flow matrix $\tilde{F}$, with random initial assignment to $m = 6$ groups, and final hard assignment of canton i to group $\arg\max_g z_{ig}$

3.3 Network General Co-clustering

Latent co-clustering (19) applies to contingency tables F of general kind, possibly asymmetric or marginally inhomogeneous, and possibly exhibiting diffusivity, alternation, or a mixture of them. Implementing the common emission probabilities constraint in the M-step (8) yields together with (19) the updating rule

$$\ddot{c}_{uv} = c_{uv} \sum_{ij} \frac{F_{ij}}{P_{ij}} a_i^u a_j^v \qquad \ddot{a}_i^u = a_i^u \frac{\sum_{j'v'} (c_{uv'} \frac{F_{ij'}}{P_{ij'}} + c_{v'u} \frac{F_{j'i}}{P_{j'i}}) a_{j'}^{v'}}{\sum_{i'j'v'} (c_{uv'} \frac{F_{i'j'}}{P_{i'j'}} + c_{v'u} \frac{F_{j'i'}}{P_{j'i'}}) a_{i'}^u a_{j'}^{v'}}$$

(21)

Let us recall that the classical Baum–Welch algorithm handles the HMM modeling of a single sequence of tokens, whose bigram counts are, up to the first and last token, marginally homogeneous by construction. By contrast, the presumably original iterative algorithm (21) can handle marginally inhomogeneous data as well, such as those resulting from concatenation of multiple sequences. Further experimentations are needed at this stage to gauge the generality of model (19), and the efficiency of the hidden transition parameters estimates by means of (21).

For symmetric data $F = F'$, the symmetric model (17) can be tackled by (21) above, with the simplifying circumstance that the additive symmetrization occurring in the numerator and denominator of $\ddot{a}_i^u$ is not necessary anymore, provided that the initial joint probability c_{uv} is symmetrical, a circumstance which automatically insures the symmetry of further iterates $\ddot{c}_{uv}$.

3.4 Case Study III: Modeling Bigrams

We consider the first chapters of the French novel "La Bête humaine" by Zola (1890). After suppressing all punctuation, accents, and separators with exception of the blank space, and converting upper-case letters to lower-case, we are left with a sequence of $N = 725'000$ tokens on $n = 27$ types (the alphabet + the space), containing 724'999 pairs of successive tokens or *bigrams*. The resulting $n \times n$ normalized contingency table $F = (F_{ij})$ is far from symmetric (for instance, the bigram **qu** occurs 6'707 times, while **uq** occurs only 23 times), but almost *marginally homogenous*, that is $F_{i\bullet} \cong F_{\bullet i} + 0(N^{-1})$ (and exactly marginally homogenous if one starts and finishes the textual sequence with the same type, such as a blank space).

Symmetrizing F as $F^s = (F + F')/2$ does not make it diffusive, and hence unsuitable by latent modeling (16), because of the importance of large negative eigenvalues in F^s, betraying *alternation*, typical in linguistic data—think in particular of the vowels–consonants alternation (e.g. Goldsmith and Xanthos 2009). This being said, symmetric co-clustering of F^s (17) remains a possible option.

Table 1 results from the general co-clustering algorithm (21) applied on the original, asymmetric bigram counts F itself. Group 4 mainly emits the vowels, group 3 the blank, group 2 the s and t, and group 1 other consonants. Alternation is betrayed by the null diagonal of the Markov transition matrix W—with the exception of group 2.

The property of marginal homogeneity $F_{i\bullet} = F_{\bullet i}$ permits in addition to obtain the *memberships Z from the emissions A*, by first determining the solution ρ of $\sum_g \rho_g\, a_i^g = f_i$, where $f_i = F_{i\bullet} = F_{\bullet i}$ is the relative frequency of letter i, and then by defining $z_{ig} = \rho_g\, a_i^g / f_i$.

Table 1 Case study III: emission probabilities A (left), memberships Z (middle), joint latent distribution C (right, top), latent probability transition matrix W (right, middle) and its corresponding stationary distribution π (right, bottom). All values are multiplied by 100 and rounded to the nearest integer

Emission probabilities A:

Group	1	2	3	4
–			61	
a				25
b	3			
c	10	1	1	
d	12	2		
e	1			45
f	4			
g	3	1		
h	2			1
i	1	3	13	5
j	2			
k				
l	17			4
m	8		2	
n	17	9		
o				13
p	8		1	
q		5		
r	10	6	4	3
s	6	28		
t	7	26		
u		8	10	3
v	7			
w				
x		2		
y		1		
z		1		

Memberships Z:

Group	1	2	3	4
–			100	
a				100
b	89		11	
c	87	5	8	
d	90	10		
e	2		1	97
f	87			13
g	71	26		3
h	65			35
i	4	9	61	26
j	100			
k		100		
l	77	2		21
m	76		24	
n		53	47	
o				100
p	89		9	2
q		88		12
r	40	21	22	17
s	20	80		
t	26	74		
u		27	57	16
v	100			
w	100			
x		100		
y	60	40		
z		100		

Joint latent distribution C:

	1	2	3	4
1	0	0	1	21
2	0	3	12	2
3	17	7	0	7
4	5	8	17	0

Latent probability transition matrix W:

	1	2	3	4
1	0	0	7	93
2	0	15	71	14
3	54	24	0	22
4	17	27	57	0

Stationary distribution π:

1	22
2	18
3	31
4	30

4 Conclusion

The non-parametric approach to clustering and co-clustering presented in this study circumvents the necessity of specifying a particular family of generating models together with associated parameters—a considerable simplification. Its final outcomes and numerical efficiency yet remain to be carefully compared to parametric approaches currently in use. The parsimony of the formal assumptions, firmly rooted in the alternating minimization paradigm of Information Theory (aka EM algorithm for complete models), must be underlined. Many important issues and well-identified paradigms (log-linear models, latent semantic indexing, non-negative matrix factorization, block clustering, HMM modeling) emerge naturally in this framework, a promising circumstance regarding the formal coherence of clustering methods.

The procedure for network clustering, outlined in Sect. 3.1, should be further investigated and presumably improved, either by prior preprocessing of the network data of by refinement of the formalism, and compared to alternative existing approaches, such as spectral clustering or modularity maximization. Also, the Bayesian extension of the non-parametric formalism, taking into account prior information, should be addressed as well, and compared to the topic modeling parametric approach.

References

Bavaud, F.: An information theoretical approach to factor analysis. In: Proceedings of the 5th International Conference on the Statistical Analysis of Textual Data (JADT 2000), pp. 263–270 (2000)

Bavaud, F.: Information theory, relative entropy and statistics. In: Sommaruga, G. (ed.) Formal Theories of Information. Lecture Notes in Computer Science, vol. 5363, pp. 54–78. Springer, Berlin (2009)

Bavaud, F.: Spatial weights: Constructing weight-compatible exchange matrices from proximity matrices. In Duckham, M., et al. (eds.) GIScience 2014. Lecture Notes in Computer Science, vol. 8728, pp. 81–96. Springer, Berlin (2014)

Berger, J., Snell, J.L.: On the concept of equal exchange. Syst. Res. Behav. Sci. **2**, 111–118 (1957)

Blei, D.M., Ng, A.Y., Jordan, M.I.: Latent Dirichlet allocation. J. Mach. Learn. Res. **3**, 993–1022 (2003)

Christensen, R.: Log-Linear Models and Logistic Regression. Springer, Berlin (2006)

Cover, T.M., Thomas, J.A.: Elements of Information Theory. Wiley, London (1991)

Csiszár, I., Tusnády, G.: Information geometry and alternating minimization procedures. In: Dedewicz, E.F. (ed.) Statistics and Decisions. Supplement Issue, vol. 1, pp. 205–237 (1984)

Dhillon, I.S., Mallela, S., Modha, D.S.: Information-theoretic co-clustering. In: Proceedings of the Ninth ACM SIGKDD International Conference on Knowledge Discovery and Data Mining, pp. 89–98 (2003)

Feinerer, I., Hornik, K., Meyer, D.: Text Mining Infrastructure in R. J. Stat. Softw. **25**(5), 1–54 (2008)

Finesso, L., Spreij, P.: Nonnegative matrix factorization and I-divergence alternating minimization. Linear Algebra Appl. **416**, 270–287 (2006)

Goldsmith, J., Xanthos, A.: Learning phonological categories. Language **85**, 4–38 (2009)

Govaert, G., Nadif, M.: Co-Clustering. Wiley, London (2013)

Hofmann, T.: Probabilistic latent semantic indexing. In: Proceedings of the 22nd Annual International ACM SIGIR Conference on Research and Development in Information Retrieval, pp. 50–57 (1999)

Kullback, S.: Information Theory and Statistics. Wiley, London (1959)

Lee, D.D., Seung, H.S.: Algorithms for non-negative matrix factorization. In: Advances in Neural Information Processing Systems, pp. 556–562 (2001)

Rabiner, L.R.: A tutorial on hidden Markov models and selected applications in speech recognition. Proc. IEEE **77**(2), 257–286 (1989)

Saul, L., Pereira, F.: Aggregate and mixed-order Markov models for statistical language processing. In: Proceedings of the 2nd International Conference on Empirical Methods in Natural Language Processing (1997)

Zola, E.: La bête humaine. Retrieved from https://www.gutenberg.org/ebooks/5154 (1890)

Efficient, Geometrically Adaptive Techniques for Multiscale Gaussian-Kernel SVM Classification

Guangliang Chen

Abstract Single-scale Gaussian-kernel support vector machines (SVM) have achieved competitive accuracy in many practical tasks; however, a fundamental limitation of the underlying model is its use of a single bandwidth parameter which essentially assumes that the training and test data has a uniform scale everywhere. In cases of data with multiple scales, only one parameter may be unable to fully capture the heterogeneous scales present in the data. In this paper, we present two efficient approaches to constructing multiscale Gaussian kernels for SVM classification by following the multiple-kernel learning research by Gonen and Alpaydin (J Mach Learn Res 12: 2211–2268, 2011) and a self-tuning spectral clustering procedure introduced by Zelnik-Manor and Perona (Advances in Neural Information Processing Systems 17:1601–1608, 2004) in the unsupervised setting, respectively. The resulting kernels adapt to the different scales of the data and are directly computable from the training data, thus avoiding expensive hyperparameter tuning tasks. Numerical experiments demonstrate that our multiscale kernels lead to superior accuracy and fast speed.

1 Introduction

Given training data $\mathbf{x}_1, \ldots, \mathbf{x}_n \in \mathbb{R}^d$ with binary labels $y_i = \pm 1$, the Gaussian-kernel Support Vector Machine (GSVM) Boser et al. (1992), Cortes and Vapnik (1995) trains a nonlinear classifier

$$f(\mathbf{x}) = \text{sgn}\left(\sum \lambda_i y_i \kappa_\sigma(\mathbf{x}_i, \mathbf{x}) + b\right), \quad \forall \mathbf{x} \in \mathbb{R}^d \tag{1}$$

G. Chen (✉)
Department of Mathematics and Statistics, San José State University, San José, CA, USA
e-mail: guangliang.chen@sjsu.edu

© Springer Nature Singapore Pte Ltd. 2020   45
T. Imaizumi et al. (eds.), *Advanced Studies in Classification and Data Science*,
Studies in Classification, Data Analysis, and Knowledge Organization,
https://doi.org/10.1007/978-981-15-3311-2_4

by solving the following quadratic optimization problem:

$$\max_{\lambda_1,\ldots,\lambda_n} \sum_i \lambda_i - \frac{1}{2} \sum_i \sum_j \lambda_i \lambda_j y_i y_j \kappa_\sigma(\mathbf{x}_i, \mathbf{x}_j)$$

$$\text{subject to} \quad 0 \leq \lambda_i \leq C \text{ and } \sum \lambda_i y_i = 0. \tag{2}$$

Here, κ_σ represents the Gaussian radial basis function (RBF) kernel

$$\kappa_\sigma(\mathbf{z}_1, \mathbf{z}_2) := e^{-\|\mathbf{z}_1 - \mathbf{z}_2\|^2/(2\sigma^2)} = e^{-\gamma\|\mathbf{z}_1 - \mathbf{z}_2\|^2}, \quad \forall \, \mathbf{z}_1, \mathbf{z}_2 \in \mathbb{R}^d \tag{3}$$

with the Euclidean distance and associated bandwidth parameter σ, or $\gamma = 1/(2\sigma^2)$ in an alternative formulation, that affects the smoothness of the decision boundary. The variable C in (2), on the other hand, is a tradeoff parameter (which can be seen in the primal formulation of the problem) that controls the size of the margin between the two classes in a feature space. The two hyperparameters γ, C together often sensitively affect the predictive accuracy and are typically tuned by numerically solving an expensive optimization problem that maximizes the cross validation (CV) accuracy over the first quadrant of $\mathbb{R}^2$ (Hsu and Lin 2002; Bergstra and Bengio 2012; Lin et al. 2008; Snoek et al. 2012):

$$\max_{\gamma>0, \, C>0} \text{CVaccuracy}(\gamma, C). \tag{4}$$

For example, the popular Grid-Search algorithm (Hsu and Lin 2002) selects the best combination (γ, C) from a two-dimensional grid obtained through a finite discretization of the optimization domain.

Recently, in view of the high computational complexity of (4) (which is due to the non-analytical form of the objective function), Chen et al. (2017) have proposed a new hyperparameter tuning scheme that essentially solves the same problem but over a small, finite, one-dimensional set in the (γ, C) space. Their method has the following two steps in order: First, for the σ parameter, they interpret it as the local scale of the training data and compute it directly as the average distance of all the training points to their kth nearest neighbors in the same class (for some fixed integer $k \geq 1$). This directly exploits the intrinsic data geometry and is much more efficient than conducting grid search in the γ space. Afterwards (for the fixed σ), they choose an elbow C value from a finite grid of the C domain which only achieves "nearly" the highest validation accuracy. Overall, such a procedure approximately solves the problem in (4), but it yields bigger margins between the different classes (and hence better generalizability of the model) and is computationally much more tractable. Numerical experiments conducted in Chen et al. (2017) seem to demonstrate competitive performance of this new approach in terms of both accuracy and speed.

Despite the advance in efficient hyperparameter tuning for the GSVM classifier, a fundamental limitation of the model in (2) is its use of a single scaling parameter

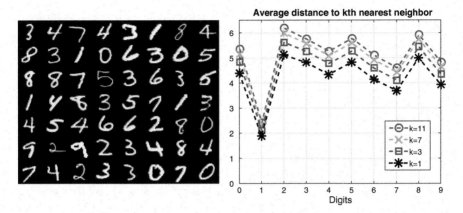

Fig. 1 Left: A small random sample of the MNIST Handwritten Digits (LeCun and Cortes 1998); Right: Average distance (within each digit class) between the different images and their kth nearest neighbors in the same class. We see that the digit 1 class seems to have a smaller local scale than the other digits

σ which essentially assumes that the training data has a uniform scale everywhere. However, in many cases, the data often exhibits different scales in different locations so that a global σ may be unable to fully capture their geometric complexity (see Fig. 1 for an example). As a result, multiscale kernels which combine together several single-scale kernels, each constructed with a distinct value of σ, have been developed in the literature (Kingsbury et al. 2015; Phienthrakul and Kijsirikul 2010; Gonen and Alpaydin 2011; Bao et al. 2017). Apparently, this introduces even more parameters to the GSVM learning problem and consequently further complicates the hyperparameter tuning task. To bypass this difficulty, the method in Kingsbury et al. (2015) simply uses a fixed geometric sequence of σ values which hopefully cover the various scales of the training data, and the work in Bao et al. (2017) focused on the simple case of two scaling parameters $\sigma_1 < \sigma_2$ (along with the tradeoff parameter C) and performed grid search in a three-dimensional space. These methods clearly ignore the actual data geometry and as a result, the training of the GSVM model may be inefficient or inaccurate.

In this paper we extend the work by Chen et al. (2017), which learns a global scale from the training data, to the setting of multiscale GSVM classification by directly learning the scales of the training data at the class level and using them in two different ways to form multiscale kernels: (1) we follow the framework laid out in Gonen and Alpaydin (2011) to build multiple kernels, one with each classwise scale, and then combine them together into a single kernel; (2) we adapt the self-tuning spectral clustering technique (Zelnik-Manor and Perona 2004) for our setting by using all the classwise scales simultaneously to build exactly one kernel matrix. We expect such learned kernels to represent the different scales within the training data (at the class level). Meanwhile, our methods avoid intensive grid-search procedures for tuning the scaling parameters and thus are very efficient.

The rest of the paper is organized as follows. In Sect. 2, we present our methodology for constructing multiscale kernels. Numerical experiments are conducted in Sect. 3 to test the performance of the proposed algorithms. Finally, in Sect. 4, we conclude the paper while pointing out some future directions.

2 Methodology

Assume a set of training examples $x_1, \ldots, x_n \in \mathbb{R}^d$ that are divided into c disjoint classes with index sets $I_1 \cup \cdots \cup I_c = \{1, \ldots, n\}$ and sizes $n_1 + \cdots + n_c = n$. For each $1 \leq i \leq n$, let $\ell(i)$ represent the label of the class containing x_i. We fix a positive integer k representing the number of nearest neighbors to be examined for each x_i. In particular, we denote the kth nearest neighbor (kNN) of x_i in its own class by $x_i^{(k)}$. The distance between x_i and $x_i^{(k)}$ is denoted as $d_i^{(k)} = \|x_i - x_i^{(k)}\|$, which is an estimate of the local scale around x_i: In high density regions such distances tend to be small while in high scatter regions they are much bigger. Collectively, these distances indicate the local geometry of the training set.

Chen et al. (2017) assumed that the training data has a uniform scale everywhere and consequently estimated that scale by averaging all the local scales:

$$\widehat{\sigma} = \frac{1}{n} \sum_{i=1}^{n} d_i^{(k)}. \tag{5}$$

As we pointed out earlier, a single scaling parameter may not fully capture the overall geometric complexity of the data which often displays heterogeneous local scales. Here, we extend the work in Chen et al. (2017) by requiring the training data to have a uniform scale only within each class (between the different classes, their scales may be different). In such cases, it is natural to average the local scales within each class to calculate the classwise scales:

$$\widehat{\sigma}_\ell = \frac{1}{n_\ell} \sum_{i \in I_\ell} d_i^{(k)}, \quad 1 \leq \ell \leq c. \tag{6}$$

Below we present two different ways to construct multiscale Gaussian kernels for SVM classification, by following the procedures in Gonen and Alpaydin (2011) and Zelnik-Manor and Perona (2004), respectively.

2.1 Multiple-Kernel Learning Method

The multiple-kernel learning research (Gonen and Alpaydin 2011) starts by learning several kernel matrices $K_1, \ldots, K_g \in \mathbb{R}^{n \times n}$ from different sources. For example, they may correspond to different kernel types (such as linear, polynomial, and RBF), or different similarity measures (e.g., Gaussian and cosine), or the same kernel function with different parameter settings. Afterwards, those kernels are combined into a single kernel matrix under a pre-selected combination rule f:

$$\mathbf{K} = f(\mathbf{K}_1, \ldots, \mathbf{K}_g). \tag{7}$$

In this paper we consider the simplest combination rule—matrix addition and denote the resulting kernel by $\mathbf{K}^{(\text{sum})}$:

$$\mathbf{K}^{(\text{sum})} = f(\mathbf{K}_1, \ldots, \mathbf{K}_g) = \mathbf{K}_1 + \cdots + \mathbf{K}_g. \tag{8}$$

It is easy to see that $\mathbf{K}^{(\text{sum})}$ is still symmetric and positive semidefinite, thus being a valid kernel. For other combination rules such as the Hadamard product, we refer the reader to Gonen and Alpaydin (2011). Such a synthetic kernel is expected to possess the advantages of individual kernels, and thus capable of handling certain heterogeneous data sets.

In our setting, since we focus on the Gaussian kernel with the Euclidean distance, we will only vary the values of the σ parameter in (3) to generate multiple kernel matrices. Specifically, letting σ be each of the classwise scales $\widehat{\sigma}_\ell$, $1 \leq \ell \leq c$, we obtain a total of c kernel matrices $\{\mathbf{K}_\ell\}$:

$$\mathbf{K}_\ell(i, j) = \kappa_{\widehat{\sigma}_\ell}(\mathbf{x}_i, \mathbf{x}_j), \quad 1 \leq i, j \leq n, \tag{9}$$

each adapted to the scale of a distinct training class. We then use (8) to combine the $\mathbf{K}_\ell$'s into a single kernel matrix for SVM classification.

Remark When test data $\mathbf{t}_1, \ldots, \mathbf{t}_m \in \mathbb{R}^d$ is involved, a second matrix linking the training and test data needs to be built. One can use the same procedure for constructing the synthetic kernel on the training data to construct this link matrix

$$\mathbf{L} = \mathbf{L}_1 + \cdots + \mathbf{L}_c, \tag{10}$$

where for each $\ell = 1, \ldots, c$, the matrix $\mathbf{L}_\ell$ is defined as follows:

$$\mathbf{L}_\ell(i, j) = \kappa_{\widehat{\sigma}_\ell}(\mathbf{x}_i, \mathbf{t}_j), \quad 1 \leq i \leq n, 1 \leq j \leq m. \tag{11}$$

2.2 Self-Tuning Kernel Learning Method

Self-tuning spectral clustering is an unsupervised learning approach introduced by Zelnik-Manor and Perona (2004) to cluster data that has multiple scales. Specifically, given data $\mathbf{x}_1, \ldots, \mathbf{x}_n \in \mathbb{R}^d$, they start by computing an $n \times n$ affinity matrix that adapt to the local scales of the data as follows:

$$\mathbf{A}_{ij} = e^{-\|\mathbf{x}_i - \mathbf{x}_j\|^2/(2\sigma_i\sigma_j)}, \quad 1 \leq i, j \leq n, \tag{12}$$

where σ_i, σ_j represent the distances of the two points $\mathbf{x}_i, \mathbf{x}_j$ to their respective kNN's in the entire data set. Such a similarity matrix has been experimentally shown to improve the vanilla spectral clustering algorithm by Ng et al. (2001) in the setting of multiscale data.

Here, we adapt the self-tuning idea for the supervised setting by constructing a Classwise Self-Tuning (CST) kernel matrix on the training data from its classwise scales $\{\widehat{\sigma}_\ell\}$:

$$\mathbf{K}_{ij}^{(\mathrm{cst})} = e^{-\|\mathbf{x}_i - \mathbf{x}_j\|^2/(2\widehat{\sigma}_{\ell(i)}\widehat{\sigma}_{\ell(j)})}, \quad 1 \leq i, j \leq n. \tag{13}$$

That is, each training example uses the scale of its own class (instead of the global scale) for better adaptivity. We also expect such a kernel to be more robust than that directly built on individual scales $\{d_i^{(k)}\}$, as in (12). In Fig. 2 we compare such a kernel matrix with the single-scale kernel matrix on a toy data set.

To classify new data $\mathbf{t}_1, \ldots, \mathbf{t}_m \in \mathbb{R}^d$ with the correspondingly trained GSVM model, we need to build a similar link matrix from the training data to the test data by following the parameter setting of the CST kernel:

$$\mathbf{L}_{ij}^{(\mathrm{cst})} = e^{-\|\mathbf{x}_i - \mathbf{t}_j\|^2/(2\widehat{\sigma}_{\ell(i)}\widehat{\sigma}_{\ell^*(j)})}, \quad 1 \leq i \leq n, 1 \leq j \leq m, \tag{14}$$

where $\ell^*(j)$ represents the ground-truth label of $\mathbf{t}_j$ which unfortunately is unknown at the time of classification. We propose two strategies to get over this obstacle.

- Method 1: **Global scaling**. Set

$$\widehat{\sigma}_{\ell^*(j)} = \widehat{\sigma}, \quad \forall j = 1, \ldots, m. \tag{15}$$

That is, we use the global scale $\widehat{\sigma}$ (5), which is a weighted average of the classwise scales,

$$\widehat{\sigma} = \frac{1}{n} \sum_{\ell=1}^{c} n_\ell \widehat{\sigma}_\ell \tag{16}$$

to approximate the unknown class scale, yielding:

$$\mathbf{L}_{ij}^{(\mathrm{cst})} = e^{-\|\mathbf{x}_i - \mathbf{t}_j\|^2/(2\widehat{\sigma}_{\ell(i)}\widehat{\sigma})}, \quad 1 \leq i \leq n, 1 \leq j \leq m \tag{17}$$

- Method 2: **kNN scaling**. Set

$$\widehat{\sigma}_{\ell^*(j)} = \widehat{d}_j^{(k)}, \tag{18}$$

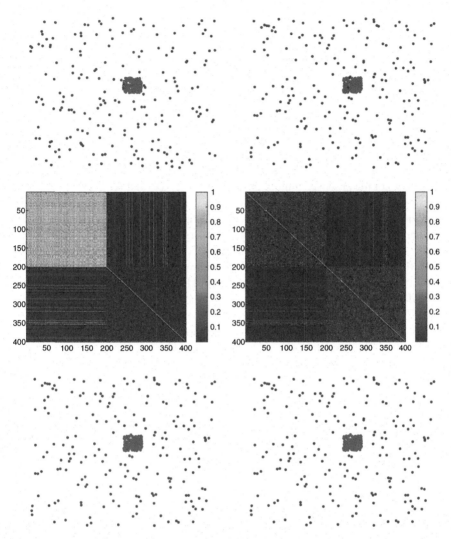

Fig. 2 Demonstration of the single-scale and classwise self-tuning kernels on a toy data set with multiple scales. Top row: training data (left) and test data (right), each having 800 points with equal-sized classes. Middle row: the single-scale RBF kernel matrix (left) and classwise self-tuning kernel (right), both built on the training data (which have been sorted according to the two classes). Bottom row: classification results on the test data corresponding to the two kernels (in same order). We can see that the self-tuning kernel has more balanced blocks and also performs better than the single-scale kernel along the boundary between the two classes

the distance from $\mathbf{t}_j$ to its kNN in the whole training set. We expect it to reasonably approximate the local scale of the ground-truth class in most cases.

We denote the combinations of the CST kernel with the two scaling techniques (for building the link matrices) as CST+global and CST+kNN, respectively.

3   Experiments

In this section we examine the performance (in terms of classification accuracy and CPU time) of the proposed multiscale kernels, $\mathbf{K}^{(\mathrm{sum})}$ (8) and $\mathbf{K}^{(\mathrm{cst})}$ (13), relative to the single-scale kernel (3) whose σ parameter is to be tuned in two different ways: grid search (Hsu and Lin 2002) and (deterministic) kNN (Chen et al. 2017).

We implemented the corresponding classifiers in MATLAB based on the LIB-SVM software (Chang and Lin 2011) which directly performs the grid-search GSVM classification. For this method, we followed the recommendations in Hsu and Lin (2002) to rescale all features to the [0, 1] range and focus on the following candidate values for γ and C, respectively: $\gamma \in \{2^{-10}, 2^{-9}, \ldots, 2^4\}$ and $C \in \{2^{-2}, 2^{-1}, \ldots, 2^{12}\}$. For any training data, the maximizer of the CV accuracy function in (4) restricted to this grid was used to retrain the model for classifying the test data. The number of CV folds was fixed to 10.

When testing the other methods, which all directly infer the scaling parameter(s) from training data based on a kNN procedure, we did not rescale the data to the [0, 1] range (this is not needed and also meant to preserve the original scales of the data). We fixed $k = 7$ for those methods and used MATLAB's *knnsearch* function to perform kNN search tasks. We then computed the kernel and link matrices accordingly for each method and provided them to LIBSVM under the custom kernel option. To tune the C parameter, we conducted grid search in the same candidate set (as above) for each of these methods, including the single-scale, kNN method (Chen et al. 2017) (in order to have a fair comparison between the different kernels). Similarly, we fixed the number of CV folds to 10 in these cases.

We chose nine benchmark data sets from the LIBSVM website[1]: *astroparticle*, *bioinformatics*, *dna*, *madelon*, *pendigits*, *satimage*, *splice*, *usps*, and *vowels*. They were also used in Chen et al. (2017) and some of them were used in Hsu and Lin (2002). Additionally, we include in our study the toy data in Fig. 2 and the digits "1" and "7" in the MNIST data set (LeCun and Cortes 1998), denoted *mnist17*. Summary information of the eleven data sets is displayed in Table 1. For the data sets with more than two classes, we adopted the one-vs-one multiclass extension of the binary GSVM classifier. All the experiments were conducted on a Lenovo Yoga 2 Pro laptop with 8 GB memory and 2.40 GHz CPU.

---

[1] https://www.csie.ntu.edu.tw/~cjlin/libsvmtools/datasets/.

Table 1 Summary information of the data sets used in the experiments of this section

Data sets	# classes	# dims	# train	# test
Astroparticle	2	4	3089	4000
Bioinformatics	2	21	1243	41
DNA	3	180	2000	1186
Madelon	2	500	2000	600
mnist17	2	784	13, 007	2163
Pendigits	10	16	7494	3498
Satimage	6	36	4435	2000
Splice	2	60	1000	2175
toydata	2	2	800	800
usps	10	256	7291	2007
Vowels	11	10	528	462

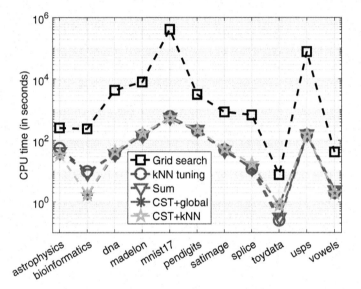

Fig. 3 CPU time needed by each method on each data set

The CPU time used by each of the different methods is shown in Fig. 3. Meanwhile, the test accuracy rates are shown in Table 2 and also displayed in Fig. 4 for easy graphical comparison. The following observations are at hand:

- **CPU time**. Grid search is computationally the most expensive method to run, often taking one to two magnitude more time, while the other methods are all very close except in two cases. This clearly shows the significant speed advantage of the kNN-based kernels.
- **Classification accuracy**. Numerically, grid-search GSVM achieved the highest accuracy rates on six data sets, but the lead over the second best method in each case is at most 0.54%. In particular, its improvement over the single-scale kNN-GSVM is under 1% in all of the six cases. Also, grid-search GSVM performed

Table 2 Test accuracy rates (%) obtained by each kernel method on the data sets in Table 1 (the highest rate for each data set has been highlighted in bold)

| | Single-scale GSVM | | Multiscale GSVM | | |
	Grid search	kNN tuning	Sum	CST+global	CST+kNN
Astroparticle	96.58	**96.83**	96.63	96.70	96.25
Bioinformatics	82.93	87.80	**90.24**	21.95	39.02
DNA	**95.45**	95.19	95.19	95.03	95.03
Madelon	59.33	65.83	65.83	66.00	**66.67**
mnist17	**99.72**	99.68	**99.72**	85.58	97.97
Pendigits	**98.48**	97.51	97.60	97.83	97.94
Satimage	**91.10**	90.45	90.65	89.85	88.60
Splice	**90.11**	89.61	89.56	88.41	88.64
toydata	99.00	99.00	99.25	**99.50**	**99.50**
usps	**95.32**	**95.32**	95.17	82.66	92.28
Vowels	60.82	68.61	**69.26**	67.32	66.67

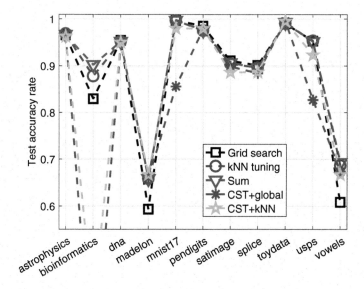

Fig. 4 Test accuracy rate achieved by each method on each data set

quite poorly on three data sets: *bioinformatics*, *madelon*, and *vowels*. The single-scale kNN-GSVM classifier achieved nearly the best accuracy in all but one case (*bioinformatics*), which shows that it is a fairly competitive method. Among the three multiscale kernels, the sum kernel consistently achieved the best (or nearly the best) accuracy rates. It also outperformed the single-scale kNN-GSVM on six data sets (including *mnist17*), demonstrating the advantage of the multiscale kernel. Finally, between the two scaling methods for the CST kernel, kNN scaling seems to be more superior than global scaling overall.

4 Conclusions and Future Work

In this paper we presented two simple yet effective ways of constructing efficient multiscale Gaussian-kernel SVM classifiers by learning the scales of the training data at the class level and using them to build multiscale kernels by following the multiple-kernel learning research (Gonen and Alpaydin 2011) and the self-tuning spectral clustering work (Zelnik-Manor and Perona 2004). Such kernels are expected to adapt better than the single-scale kernel to the geometry of the data. Numerical experiments demonstrated that they can improve the accuracy of the single-scale kernels in some cases, while running very fast. We plan to continue the work along two directions: (1) we will further study the CST kernel (13) to understand its theoretical properties and find better ways to set the scaling parameter for the test data; (2) we will apply our proposed multiscale GSVM classifiers to more challenging data sets and see how they perform in those cases.

Acknowledgement G. Chen was supported by the Simons Foundation Collaboration Grant for Mathematicians.

References

Bao, J., Chen, Y., Yu, L., Chen, C.: A multi-scale kernel learning method and its application in image classification. Neurocomputing **257**, 16–23 (2017)

Bergstra, J., Bengio, Y.: Random search for hyper-parameter optimization. J. Mach. Learn. Res. **13**, 281–305 (2012)

Boser, B., Guyon, I., Vapnik, V.: A training algorithm for optimal margin classifiers. In: Proceedings of the 5th Annual ACM Workshop on Computational Learning Theory, pp. 144–152 (1992)

Chang, C.-C., Lin, C.-J.: LIBSVM: A library for support vector machines. ACM Trans. Intell. Syst. Technol. **2**(27), 1–27 (2011). http://www.csie.ntu.edu.tw/~cjlin/libsvm

Chen, G., Florero-Salinas, W., Li, D.: Simple, fast and accurate hyper-parameter tuning in Gaussian-Kernel SVM. In: Proceedings of International Joint Conference on Neural Networks (2017). https://doi.org/10.1109/IJCNN.2017.7965875

Cortes, C., Vapnik, V.: Support-vector networks. Mach. Learn. **20**(3), 273–297 (1995)

Gonen, M., Alpaydin, E.: Multiple kernel learning algorithms. J. Mach. Learn. Res. **12**, 2211–2268 (2011)

Hsu, C.-W., Lin, C.-J.: A comparison of methods for multiclass support vector machines. IEEE Trans. Neural Netw. **13**(2), 415–425 (2002)

LeCun, Y., Cortes, C. (1998). The MNIST handwritten digits. http://yann.lecun.com/exdb/mnist/

Lin, S.-W., Lee, Z.-J., Chen, S.-C., Tseng, T.-Y.: Parameter determination of support vector machine and feature selection using simulated annealing approach. Appl. Soft Comput. **8**(4), 1505–1512 (2008)

Kingsbury, N., Tay, D., Palaniswami, M.: Multi-scale kernel methods for classification. In: IEEE Workshop on Machine Learning for Signal Processing (2015). https://doi.org/10.1109/MLSP.2005.1532872

Ng, A., Jordan, M., Weiss, Y.: On spectral clustering: analysis and an algorithm. In: Advances in Neural Information Processing Systems, vol. 14, pp. 849–856 (2001)

Phienthrakul, T., Kijsirikul, B.: Evolutionary strategies for hyperparameters of support vector machines based on multi-scale radial basis function kernels. Soft Comput. **14**(7), 681–699 (2010)

Snoek, J., Larochelle, H., Adams, R.: Practical Bayesian optimization of machine learning algorithms. In: Advances in neural information processing systems, vol. 25, pp. 2951–2959 (2012)

Zelnik-Manor, L., Perona, P.: Self-tuning spectral clustering. In: Advances in Neural Information Processing Systems, vol. 17, pp. 1601–1608 (2004)

Random Forests Followed by Computed ABC Analysis as a Feature Selection Method for Machine Learning in Biomedical Data

Jörn Lötsch and Alfred Ultsch

Abstract **Background:** Data from biomedical measurements usually include many parameters (variables/features). To reduce efforts of data acquisition or to enhance comprehension, a feature selection method is proposed that combines the ranking of the relative importance of each parameter in random forests classifiers with an item categorization provided by computed ABC analysis.

Data: The input data space, comprising an example subset of plasma concentrations of $d = 23$ different lipid markers of various classes, acquired in Parkinson patients and healthy subjects ($n = 100$ each).

Methods: Random forest classifiers were constructed with various different scenarios of the number of trees and the number of features in each tree. The relative importance of each feature calculated by the classifier was submitted to computed ABC analysis, a categorization technique for skewed distributions to identify the most important feature subset "A," i.e., a reduced-set containing the important few items.

Results: Using different parameters for the algorithms, the classification performance of all reduced-set random forest classifiers was almost as good as that of a random forest classifier using the full set of $d = 23$ lipid markers; all reaching 95% or better classification accuracy. When including additional "nonsense" features consisting of concentration data permutated across the subject groups, these features were never found in the ABC set "A." The obtained features sets provided better classifiers than those obtained using classical regression methods.

Conclusions: Random forests plus computed ABC analysis provided a feature selection without the necessity to predefine the number of features. A substantial

J. Lötsch (✉)
Institute of Clinical Pharmacology, Goethe – University, Frankfurt am Main, Germany

Fraunhofer Institute for Molecular Biology and Applied Ecology IME, Frankfurt am Main, Germany
e-mail: j.loetsch@em.uni-frankfurt.de

A. Ultsch
Databionics Research Group, University of Marburg, Marburg, Germany

© Springer Nature Singapore Pte Ltd. 2020   57
T. Imaizumi et al. (eds.), *Advanced Studies in Classification and Data Science*,
Studies in Classification, Data Analysis, and Knowledge Organization,
https://doi.org/10.1007/978-981-15-3311-2_5

reduction of the number of features, following the "80/20 rule," was obtained. The classifiers using the A-class performed better than with a regression-based feature selection and were (nearly) as good as using the complete feature set. The obtained small feature sets are also well suited for domain experts' interpretation.

1 Introduction

Data acquisition for biomedical research becomes increasingly complex (McDermott et al. 2013; Rinaldi 2011) due to the rising molecular and clinical knowledge of disease pathomechanisms and technological advances in laboratory equipment and computer science. Such typically high-dimensional data enables data driven research approaches (Lötsch and Geisslinger 2010; Breiman 2001). These are facilitated by developments in data science as a rapidly growing interdisciplinary research area that deals with the problem-oriented processing of large amounts of (complex) data with the aim to discover and process knowledge (President's Information Technology Advisory 2005; James et al. 2013; Dhar 2013).

Among key technologies for complex data evaluation figures machine learning (Murphy 2012), which in its supervised form aims at identifying an intelligent algorithm that maps the input features X comprising vectors $x_i = <x_1, \cdots, x_d>$ with $d > 0$ different parameters (features) acquired from $n > 0$ cases to the output space Y comprising $y_i \in C = \{1, \cdots, c\}$ of c possible classes in the data space, $D = \{(x_i, y_i) | x_i \in X, y_i \in Y, i = 1, \ldots, n\}$. Among reasons to reduce the number of available features are (1) enhancement of understanding of the biological mechanisms in knowledge discovery (Miller 1956), (2) reducing the costs and efforts of data acquisition, and (3) reducing the number of analyzed parameters to increase statistical power and to avoid that it exceeds the number of assessed cases.

Reduction of the number of parameters is obtained by applying methods of feature selection (Saeys et al. 2007). Here a feature selection method is proposed that is based on random forests combined with computed ABC analysis. Specifically, implementations of random forest classifiers (Breiman 2001) often output quantitative measures of the importance of each feature for the overall classification performance. This provides a basis for feature selection, for which a precise statistical limit of the most informative subset is provided by categorization techniques such as computed ABC analysis (Ultsch and Lötsch 2015).

2 Methods

2.1 Biomedical Data Set

A data set suitable for the present assessment of the utility of random forests and computed ABC analysis based feature selection for biomedical problems was available as plasma concentrations of $d = 23$ lipid markers assayed in probes drawn from Parkinson patients and healthy controls. Plasma concentrations had been analyzed using liquid chromatography-electrospray ionization-tandem mass spectrometry (LC-ESI-MS/MS) as described elsewhere (Zschiebsch et al. 2016; Sisignano et al. 2013). The selection included endocannabinoids (AEA, OEA), lysophosphatidic acids (LPA16:0, LPA18:1, LPA18:2, LPA18:3, LPA20:4), ceramides (Cer16:0, Cer18:0, Cer20:0, Cer24:0, Cer24:1, GluCerC16:0, GluCerC24:1, LacCerC16:0, LacCerC24:0, LacCerC24:1; Cer = ceramide, GluCer = glucosylceramide, LacCer = lactosylceramide), and sphingolipids (sphinganine, sphingosine, S1P, SA1P C16Sphinganine, C18Sphinganine, C24Sphinganine, C24:1Sphinganine). The assessment of lipid markers in the context of Parkinson's disease is based on evidence of an involvement of lipid regulation (Pisani et al. 2005; Pyszko and Strosznajder 2014; Mielke et al. 2013; Li et al. 2015; Xing et al. 2016; France-Lanord et al. 1997; Boutin et al. 2016).

2.2 Data Preprocessing

Data were analyzed using the software environments R (version 3.4.0 for Linux; http://CRAN.R-project.org/, R Development Core Team 2008) and SPSS (version 24 for Linux, IBM SPSS Statistics, Chicago, USA) on an Intel Xeon® computer running on Ubuntu Linux 16.04.3 64-bit.

From the originally enrolled $n = 128$ patients and $n = 350$ controls (Ethics approval and informed written consent obtained), BMI and sex matched samples of $n = 100$ subjects per group were drawn. Data preprocessing included (1) log transformation, (2) age correction, (3) uniform scaling, and (4) imputation of missing data. Specifically, (1) as quantile-quantile plots predominantly pointed log-normal distributions of the data, which is in line with general observations in blood-derived concentrations (Lacey et al. 1997), data was zero invariant log-transformed, except for the two endocannabinoids (AE and OEA) for which the plots suggested to prefer the original linear scaling. Subsequently, (2) the influences of age on the lipid marker plasma concentrations (Parkinson patients being older than controls with age $= 69 \pm 8.2$ versus 26.9 ± 6.6 years, respectively) were reduced by applying corrections based on robust linear regression using the Levenberg–Marquardt nonlinear least-squares algorithm implemented in the R library "minpack.lm" (https://cran.r-project.org/package=minpack.lm, Elzhov et al. 2016). To obtain (3) a uniform scaling of all lipid marker plasma concentrations data were transformed into percentages (Milligan and Cooper 1988), i.e., into the interval [0,100]. Finally, a single missing data point was imputed via the k nearest neighbor method with $k = 3$ (Altman 1992) using weighted average and Euclidean distance, as implemented in the "DMwR" R library (https://cran.r-project.org/package=DMwR, Torgo 2010).

2.3   Random Forest and Computed ABC Analysis Based Feature Selection

For the 1000 repeated experiments a training **data set** of 44% ($d = 10$), test data set of 22% ($d = 5$), and a validation data set of 33% ($d = 8$) were created using class-proportional bootstrap resampling from the training data set (Efron and Tibshirani 1995). Sampling was performed using the R library "sampling" (https://cran.r-project.org/package=sampling, Tillé and Matei 2016).

For the training data in each of the 1000 experiments, random forests were created, sized between 500 and 2000 trees and each tree containing $e \leq d$ features randomly drawn from the $d = 23$ lipid marker plasma concentration vectors (Fig. 1).

The classification error rate on the test data sets was monitored for each run and the median and non-parametric 95% confidence intervals (2.5th to 97.5th percentiles of the parameter values obtained in the 1000 Bootstrap resampling runs). The influence of the number of features included in each tree on the resulting selection of features was tested by rerunning the 1000 experiments with nine different ranges for the number of features chosen from the feature space of $d = 23$ different lipid markers to be included per tree, consisting of (1) sqrt(d), which is the standard procedure implemented in the R library "randomForest" (https://cran.r-project.org/package=randomForest, Liaw and Wiener 2002), (2) 0.5*sqrt(d), (3) 2*sqrt(d), (4) 0.5*sqrt(d)–2*sqrt(d), (5) 0.5*sqrt(d)–sqrt(d), (6) sqrt(d)–2*sqrt(d), (7) 1–2*sqrt(d), (8) 0.5*sqrt(d)–d, and (9) 1–d, i.e., one to all lipid markers were allowed to be included in the trees. The experiments were performed modifying the "mtry" parameter of the above-mentioned R library.

For each of the 1000 random forest classifier the influence of the tree count on the classification outcome was submitted to computed ABC analysis (Ultsch and Lötsch 2015). This is a categorization technique for the selection of a most important subset among a larger set of items and it was chosen since it fitted to the basic requirements of feature selection using filtering techniques (Saeys et al. 2007), i.e., it does easily scale to very high-dimensional data sets, it is computationally simple and fast, and independent of the classification algorithm. ABC analysis aims at dividing a set of data into three disjoint subsets called "A," "B," and "C." Set "A" should contain the "important few," i.e., those elements that allow obtaining a maximum of yield with a minimal effort (Pareto 1909; Juran 1975). Set "B" comprises those elements where an increase in effort is proportional to the increase in yield. In contrast, set "C" contains the "trivial many," i.e., those elements with which the yield can only be achieved with an over-proportionally large additional effort (Pareto 1909; Juran 1975). The final size of the feature set was equal to the most frequent size of set "A" in the 1000 runs. The final members of the feature set were chosen in decreasing order of their appearances in ABC set "A" among the 1000 runs. These calculations were done using our R package "ABCanalysis" (http://cran.r-project.org/package=ABCanalysis, Ultsch and Lötsch 2015). A random forest classifier on the reduced feature set using only the most frequent number of features in set "A" was then tested in the 1000 runs.

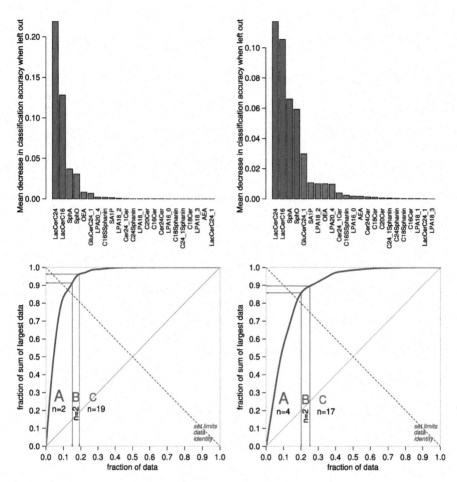

Fig. 1 Importance of single features when many or few features have been allowed for tree building. The bar plot at the **upper left side** shows the features' importance resulting tree building allowed selecting $2*\text{sqrt}(d) = 10$ features from the set of $d = 23$ lipid markers, which had the effect that a few features were recognized to possess a particularly high importance for the classification. On **the upper right side**, tree building was allowed selecting $\text{sqrt}(d) = 5$ features, among which the relative importance of the features was less different. In the subsequent computed ABC analyses (**bottom**), the skewness of the distribution of feature importance was directly related to the size of set size "A" (blue bars). That is, computed ABC analysis is an item selection procedure aiming at identification of most profitable items from a larger list of items. The ABC plot (blue line) shows the cumulative distribution function of the mean decreases in accuracy, along with the identity distribution, $x_i = $ constant (magenta line, i.e., each feature contributes similarly to the classification accuracy (for further details about computed ABC analysis, see Ultsch and Lötsch 2015)

2.4 Comparative Assessment of Classifier Performance

As feature selection, based on random forest and computed ABC analysis, was in the focus of the present assessments, the classifiers using different feature sets were built using the same procedure chosen to be random forests with standard parameters. Specifically, a subsymbolic classifier was generated by means of random forest machine learning using the number of lipid markers in ABC set "A" resulting from the different variants of the feature selection procedure. A further random forest based classifier was created using the complete set of $d = 23$ lipid markers. Finally, for comparison feature selection was done using a classical logistic regression approach as implemented in the SPSS software package (version 24 for Linux, IBM SPSS Statistics, Chicago, USA). Variables were included "stepwise forward" into the logistic regression and the likelihood ratio was used as statistical criterion.

Classifiers were optimized using the test data set. The performances of all classifiers were assessed on the validation data set drawn at the start of the data analysis. Measures of performance (Altman and Bland 1994) included: sensitivity, specificity, precision, recall, prevalence, detection rate, and balanced accuracy (Table 1). The 95% confidence intervals of the performance test parameters were obtained as the 2.5th and 97.5th percentiles of the results of 1000 runs on Bootstrap resampled data.

2.5 Feature Selection Using an Extended Data Set with Features Unrelated to Classes

An extended data set with $d = 46$ features was created by doubling each feature, however, using random permutation of the original biomarker concentrations thereby destroying their possible association with the disease status. Thus, the extended data set comprised $d = 23$ original plus $d = 23$ randomly permuted lipid marker concentrations. Feature selection was repeated using this extended control data set. The expectation was that the feature selection method avoided the "nonsense," i.e., the permutated features.

3 Results

3.1 Classification Performance Using the Full and Reduced Feature Sets

The set of plasma concentration data of several different lipid markers provided a suitable basis for the creation of a diagnostic biomarker with a high accuracy of > 95% (Table 1). The presently proposed feature selection method provided a

Table 1 Classification performance measures of a random forest classifier created following different variants of feature selection

	No feature selection	Sqrt(d)	0.5*sqrt(d)	2*sqrt(d)	0.5*sqrt(d) -2*sqrt(d)	0.5*sqrt(d) -sqrt(d)	sqrt(d) -2*sqrt(d)	1 -2*sqrt(d)	0.5*sqrt(d) -d	1-d 1-d	Logistic regression
Range of features for RF	23	5	2	10	2–10	2–5	5–10	1–10	5–23	1–23	–
Sensitivity	98.5 (98.5–98.5)	95.7 (98.5–98.5)	97.1 (98.4–98.5)	95.6 (97–98.5)	95.6 (97–98.5)	97.1 (97.1–98.5)	95.6 (97–98.5)	95.6 (97–98.5)	95.6 (97–98.5)	95.6 (97–98.5)	94.2 (96.9–98.4)
Precision	98.5 (98.5–98.5)	98.5 (98.5–98.5)	98.5 (98.5–98.5)	97 (97–98.5)	97 (97–98.5)	98.5 (97–98.5)	97 (97–98.5)	97 (97–98.5)	97 (97–98.5)	97 (97–98.5)	98.5 (97–98.5)
Recall	98.5 (98.5–98.5)	95.7 (95.7–97.1)	97.1 (95.7–98.5)	95.6 (95.6–95.7)	95.6 (95.6–95.7)	97.1 (95.7–98.5)	95.6 (95.6–95.7)	95.6 (95.6–95.7)	95.6 (95.6–95.7)	95.6 (95.6–95.7)	94.2 (92.9–94.3)
Prevalence	50 (50–50)	51.5 (50.7–51.5)	50.7 (50–51.5)	50.7 (50.7–51.5)	50.7 (50.7–51.5)	50.7 (50–51.5)	50.7 (50.7–51.5)	50.7 (50.7–51.5)	50.7 (50.7–51.5)	50.7 (50.7–51.5)	52.2 (51.5–53)
Detection rate	49.3 (49.3–49.3)	49.3 (49.3–49.3)	49.3 (49.2–49.3)	48.5 (48.5–49.3)	48.5 (48.5–49.3)	49.3 (48.5–49.3)	48.5 (48.5–49.3)	48.5 (48.5–49.3)	48.5 (48.5–49.3)	48.5 (48.5–49.3)	49.3 (48.5–49.3)
Balanced accuracy	98.5 (98.5–98.5)	97.1 (97.1–97.8)	97.8 (97.1–98.5)	96.3 (96.3–97.1)	96.3 (96.3–97.1)	97.8 (97.1–98.5)	96.3 (96.3–97.1)	96.3 (96.3–97.1)	96.3 (96.3–97.1)	96.3 (96.3–97.1)	95.7 (94.9–96.4)
Features in set "A"	23	4	5	3	3	5	3	3	3	3	3
All	X										
LacCerC16		X	X	X	X	X	X	X	X	X	X
LacCerC24		X	X	X	X	X	X	X	X	X	X
GluCerC24.1			X			X					
Sph0		X	X			X					
SphA		X	X	X	X	X	X	X	X	X	
OEA											X

The different random forest implementations of feature selection mainly differed with respect to the number of features taken from 1 to d features, with the random forest default being proposed at sqrt(d). Numbers represent the medians, with 95% confidence interval obtained during 1000 Bootstrap resampling runs of the random forest classification applied on the test data set substantial reduction of the necessary lipid marker analyses while the classification performance of the reduced-set random forest classifiers was almost as good as that of a random forest classifier using the full set of $d = 23$ lipid markers; all reaching 95% or better classification accuracy. Specifically, as desired the random forest based quantification of the importance of each candidate feature for the classification success with subsequent separation of the most relevant features using computed ABC analysis reduced the original set of $d = 23$ features to 3–5 single markers.

3.2   Performance of Different Parameter Implementation in Random Forests Feature Selection

The results of the random forest analyses were only weakly dependent on the number of trees planted in the random forests. That is, the classification error rate quickly decreased to a minimum that was reached already with approximately 100 trees. Importantly, more trees did not influence the result, which provides a robust parameter setting as a too large number would merely cost computational power and time but not jeopardize the quality of the result.

In contrast to the weak importance of the number of trees, the number of features included in each of the random trees directly influenced the selection of the features resulting from the subsequent analyses (Fig. 1). That is, with the inclusion of an increasing number of features, a few highly relevant features had increasingly often the chance to be part of a tree, which resulted in a highly skewed distribution of feature importance. The skewness was less pronounced when only a few randomly selected features were allowed for tree construction, which increased the chance of the selection of different features as most important among a specific subgroup of features allowed for tree building. The skewness of the distribution of the feature importance was directly related to the size of ABC set "A", hence, the number of features selected by the presently proposed method.

3.3   Reassessment Using Features Unrelated to the Classes

The extended data set included in addition to the original lipid marker concentrations as found in either Parkinson patients of healthy subjects a set of concentration data permutated across the subject groups. During the 1000 runs on resampled data as described above, a dummy, i.e., permutated feature was never found in the ABC set "A." Hence, inclusion of nonsense features did not reduce the classification performance of the reduced feature sets selected using the presently proposed method.

4   Discussion

Computed ABC analysis provides a method to select the most informative variables. ABC analysis is a categorization technique for (positively) skewed distributions such as different importance of features for the classification success. Classical ABC analysis is a heuristic method to identify in a set of positive parameters the "important few," i.e., a subset as small as possible containing the largest values. By contrast, computed ABC analysis uses an optimization of cost (i.e., number of parameters) versus yield (i.e., sum of selected parameters) applying a mathematically precise definition of optimality in cost versus effort. Thus, it replaces heuristics by an algorithmically precise calculation of the three sets (Ultsch and Lötsch 2015). In addition to the present biomedical application, computed ABC analysis has also been used in economic sciences, for example, in business process management (Iovanella 2017) or bankruptcy prediction (Pawelek et al. 2017).

The present experiments using feature selection based on the feature importance measure estimated in the random forest analysis provided the following main results. Firstly, the method substantially reduced the number of features included in final classifiers, which was paid with only a minimum reduction in classification performance. Secondly, random forests qualify as a feature selection procedure and slightly outperformed a classical approach of, e.g., logistic regression. Thirdly, the method was robust against the inclusion of nonsense, i.e., permuted features, which were never selected. Fourthly, for random forest feature based selection, the number of trees in the forest is of minor importance provided that minimum number, presently 100, has been used. More trees did not improve the classification performance; however, they also did not worsen it. Therefore, a too large number can be safely used at the cost of computational power. Fifthly, the number of features allowed to be selected in the trees provides a tuning parameter of the size of ABC set "A." However, the best performance of the resulting classifier was obtained when allowing rather few features for tree building, with the simplest variant being $0.5*sqrt(d)$. This had the effect of avoiding that very few features gain relative high importance reducing the set size of "A," which in turn increased the number of features in the final classifier. An alternative feature ranking criterion is the Gini impurity (https://en.wikipedia.org/wiki/Decision_tree_learning#Gini_impurity); preliminary assessments indicated a similar utility as the decrease in classification accuracy criterion (details not provided). It should be noted that this type of feature selection is not readily suited for uncertain data such as the so-called Japanese Vowels data set (Zhang et al. 2017).

While with the present data set, all scenarios provided good classification performances, differences nevertheless existed (Fig. 2), with trees containing a slightly larger feature set chosen from the $d = 23$ lipid markers providing slightly better random forest classifiers as assessed on the test data set. The classification performance set was obtained when allowing a small set of feature to be included in single tree building, sized $0.5*sqrt(d)$. This resulted in a set of $d = 5$ lipid markers that provided the comparatively best classification. By contrast, the classical

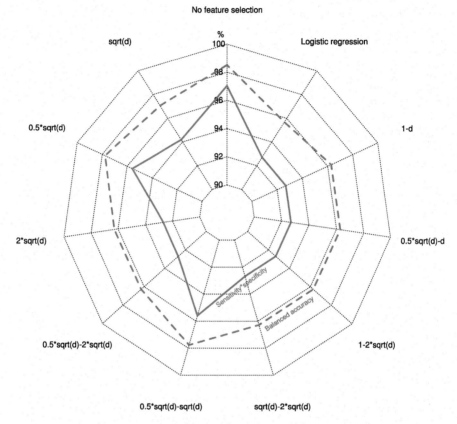

Fig. 2 Radar plot of the product (red) and mean (green) of classification sensitivity and specificity of a random forest classifier created following feature selection based on random forests followed by computed ABC analysis, using different computational parameters for random forests (complete performance measures shown in Table 1). The mean of sensitivity and specificity corresponds to the balanced classification accuracy. The different random forest implementations of feature selection mainly differed with respect to the number of features, d, taken into the random forests, ranging from 1 to d features (d in the ranges around the radar plot denoting the total count of available candidate features)

approach of logistic regression provided a different feature set, which resulted in a comparatively lower classification performance. Of note, the differences were narrow with overlapping confidence intervals (Table 1).

With $d = 3 - 5$ features selected from a candidate feature set of $d = 23$ features, which corresponds to 13–21.7% of the original set size, and considering that the best performing feature set had a size of $d = 5$ (21.7%), the proposed method meets the so-called 80/20 rule (Pareto 1909; Juran 1975), i.e., it is compatible with more general concepts used in the search for a minimum possible effort that gives the maximum yield, which often converge toward the effect that with 20% of the effort 80% of all yield can be obtained. This may substantially reduce efforts to acquire the biomedical data necessary to apply a complex diagnostic tool. A smaller feature list is also better accessible to biomedical interpretation than a more complex subsymbolic classifier. The five features are supported by reports of a particular importance of ceramides and shingosides in Parkinson disease (Pyszko and Strosznajder 2014; Mielke et al. 2013; Xing et al. 2016; France-Lanord et al. 1997).

5 Conclusions

Employing random forests classifier as supervised machine learning provides a ranking of available features according to their importance for successful classification. The application of a feature selection technique in form of computed ABC analysis, proved suitable for creating classifiers from high-dimensional biomedical (laboratory) data that are accessible to topical expert interpretation. Random forests plus computed ABC analysis provide feature selection without the necessity to predefine the number of features that are taken to further analytical steps. The number of selected features can be (slightly) tuned by adapting the computation parameters of mathematically precise definition of optimality in cost vs effort, the random forest analysis. The feature selection method is based on a mathematically precise definition of optimality in cost (= number of features) versus the effort (= sum of selected parameters) (Ultsch and Lötsch 2015). This accommodates the intuitive perception of a substantial and effective feature selection and the obtained small feature set is readily accessible to domain expert's interpretation (Miller 1956). The classifier obtained performed better than an alternative classifier for which the features had been selected using classical regression, and it was (almost) as good as a classifier using the full feature set. Thus, random forests followed by computed ABC analysis may be applied as feature selection method for machine learning on biomedical problems.

Acknowledgments Landesoffensive zur Entwicklung wissenschaftlich—ökonomischer Exzellenz (LOEWE), LOEWE-Zentrum für Translationale Medizin und Pharmakologie (JL) and European Union Seventh Framework Programme (FP7/2007–2013) under grant agreement no. 602919 (JL, GLORIA).

References

Altman, N.S.: An introduction to kernel and nearest-neighbor nonparametric regression. Am. Stat. **46**, 175–185 (1992)
Altman, D.G., Bland, J.M.: Diagnostic tests. 1: sensitivity and specificity. Br. Med. J. **308**, 1552 (1994)
Boutin, M., Sun, Y., Shacka, J.J., Auray-Blais, C.: Tandem mass spectrometry multiplex analysis of glucosylceramide and galactosylceramide isoforms in brain tissues at different stages of Parkinson disease. Anal. Chem. **88**, 1856–1863 (2016)

Breiman, L.: Statistical modeling: the two cultures (with comments and a rejoinder by the author). Stat. Sci. **16**, 199–231 (2001)

Breiman, L.: Random forests. Mach. Learn. **45**, 5–32 (2001)

Dhar, V.: Data science and prediction. Commun. ACM **56**, 64–73 (2013)

Efron, B., Tibshirani, R.J.: An introduction to the bootstrap. Chapman and Hall, San Francisco (1995)

Elzhov, T.V., Mullen, K.M., Spiess, A.-N., Bolker, B.: minpack.lm: R Interface to the Levenberg-Marquardt Nonlinear Least-Squares Algorithm Found in MINPACK, Plus Support for Bounds (2016)

France-Lanord, V., Brugg, B., Michel, P.P., Agid, Y., Ruberg, M.: Mitochondrial free radical signal in ceramide-dependent apoptosis: a putative mechanism for neuronal death in Parkinson's disease. J. Neurochem. **69**, 1612–1621 (1997)

Iovanella, A.: Vital Few e Trivial Many. L'ubiquitá della legge di Pareto e le sue applicazioni nel Business Process Management, pp. 10–13. Il Punto Scientifico (2017)

James, G., Witten, D., Hastie, T., Tibshirani, R.: An Introduction to Statistical Learning. Springer, New York (2013)

Juran, J.M.: The non-Pareto principle; Mea culpa. Qual. Prog. **8**, 8–9 (1975)

Lacey, L.F., Keene, O.N., Pritchard, J.F., Bye, A.: Common noncompartmental pharmacokinetic variables: are they normally or log-normally distributed? J. Biopharm. Stat. **7**, 171–178 (1997)

Li, Z., Zhang, J., Sun, H.: Increased plasma levels of phospholipid in Parkinson's disease with mild cognitive impairment. J. Clin. Neurosci. **22**, 1268–1271 (2015)

Liaw, A., Wiener, M.: Classification and regression by randomForest. R News **2**, 18–22 (2002)

Lötsch, J., Geisslinger, G.: Bedside-to-bench pharmacology: a complementary concept to translational pharmacology. Clin. Pharmacol. Ther. **87**, 647–649 (2010)

McDermott, J.E., Wang, J., Mitchell, H., Webb-Robertson, B.-J., Hafen, R., Ramey, J., Rodland, K.D.: Challenges in biomarker discovery: combining expert insights with statistical analysis of complex omics data. Expert Opin. Med. Diagn. **7**, 37–51 (2013)

Mielke, M.M., Maetzler, W., Haughey, N.J., Bandaru, V.V., Savica, R., Deuschle, C., Gasser, T., Hauser, A.K., Graber-Sultan, S., Schleicher, E., Berg, D., Liepelt-Scarfone, I.: Plasma ceramide and glucosylceramide metabolism is altered in sporadic Parkinson's disease and associated with cognitive impairment: a pilot study. PLoS One **8**, e73094 (2013)

Miller, G.A.: The magical number seven plus or minus two: some limits on our capacity for processing information. Psychol. Rev. **63**, 81–97 (1956)

Milligan, G.W., Cooper, M.C.: A study of standardization of variables in cluster analysis. J. Classif. **5**, 181–204 (1988)

Murphy, K.P.: Machine Learning: A Probabilistic Perspective. The MIT Press, Cambridge (2012)

Pareto, V.: Manuale di economia politica, Milan: Societá editrice libraria, revised and translated into French as Manuel d'économie politique. Giard et Briére, Paris (1909)

Pawelek, B., Pociecha, J., Baryla, M.: Analysis in corporate bankruptcy prediction. Conference of the International Federation of Classification Societies, Tokyo, pp. 215 (2017)

Pisani, A., Fezza, F., Galati, S., Battista, N., Napolitano, S., Finazzi-Agro, A., Bernardi, G., Brusa, L., Pierantozzi, M., Stanzione, P., Maccarrone, M.: High endogenous cannabinoid levels in the cerebrospinal fluid of untreated Parkinson's disease patients. Ann. Neurol. **57**, 777–779 (2005)

President's Information Technology Advisory, C.: Report to the President: Computational Science: Ensuring America's Competitiveness (2005)

Pyszko, J., Strosznajder, J.B.: Sphingosine kinase 1 and sphingosine-1-phosphate in oxidative stress evoked by 1-methyl-4-phenylpyridinium (MPP+) in human dopaminergic neuronal cells. Mol. Neurobiol. **50**, 38–48 (2014)

R Development Core Team: R: A Language and Environment for Statistical Computing. Vienna (2008)

Rinaldi, A.: Teaming up for biomarker future: many problems still hinder the use of biomarkers in clinical practice, but new public–private partnerships could improve the situation. EMBO Rep. **12**, 500–504 (2011)

Saeys, Y., Inza, I., Larranaga, P.: A review of feature selection techniques in bioinformatics. Bioinformatics **23**, 2507–2517 (2007)

Sisignano, M., Angioni, C., Ferreiros, N., Schuh, C.D., Suo, J., Schreiber, Y., Dawes, J.M., Antunes-Martins, A., Bennett, D.L., McMahon, S.B., Geisslinger, G., Scholich, K.: Synthesis of lipid mediators during UVB-induced inflammatory hyperalgesia in rats and mice. PLoS One **8**, e81228 (2013)

Tillé, Y., Matei, A.: Sampling: Survey Sampling (2016)

Torgo, L.: Data Mining with R: Learning with Case Studies. Chapman & Hall/CRC, Boca Raton (2010)

Ultsch, A., Lötsch, J.: Computed ABC analysis for rational selection of most informative variables in multivariate data. PLoS One **10**, e0129767 (2015)

Xing, Y., Tang, Y., Zhao, L., Wang, Q., Qin, W., Ji, X., Zhang, J., Jia, J.: Associations between plasma ceramides and cognitive and neuropsychiatric manifestations in Parkinson's disease dementia. J. Neurol. Sci. **370**, 82–87 (2016)

Zhang, X., Sun, D., Li, Y., Liu, H., Liang, W.: A Novel Extreme Learning Machine-Based Classification Algorithm for Uncertain Data, pp. 176–188. Springer International Publishing, Berlin (2017)

Zschiebsch, K., Fischer, C., Pickert, G., Haeussler, A., Radeke, H., Grosch, S., Ferreiros, N., Geisslinger, G., Werner, E.R., Tegeder, I.: Tetrahydrobiopterin attenuates DSS-evoked colitis in mice by rebalancing redox and lipid signaling. J Crohns Colitis **10**, 965–978 (2016)

Non-hierarchical Clustering for Large Data Without Recalculating Cluster Center

Atsuho Nakayama and Deguchi Shinji

Abstract The k-means non-hierarchical clustering is one of the most widely used methods to partition a dataset into groups of patterns. An advantage of the k-means calculation process is that it is simple and convenient. However, the k-means method converges to one of many local minima, in which the final result depends on the initial starting points. Thus, it is necessary to duplicate the analysis using a different initial value. To obtain the optimum solution for each pair of initial values, the algorithm repeatedly calculates the cluster centers to minimize the average squared distance between objects in the same cluster, so the computational resources required can be costly. Thus, it is important to decrease the number of iterations to reduce processing costs. To solve this problem, we performed k-means analysis without recalculation of the full dataset using the number and the centers of clusters obtained from the analysis of sampling data. The proposed method can be used to simplify and reduce calculation costs by removing the cluster center recalculation step.

1 Introduction

Methods of summarizing and extracting information are often applied to large multivariate datasets, as it can be difficult to understand the relationships among variables or objects. This issue has gained attention due to the large amounts of data that are now electronically collected and gathered. Classification involves the investigation of sets of objects to establish whether these object sets can be summarized into a small number of classes of similar objects (Gordon 1999).

A. Nakayama (✉)
Tokyo Metropolitan University, Hachioji-shi, Japan
e-mail: atsuho@tmu.ac.jp

D. Shinji
DATAEXPLORING, Musashino-shi, Japan
e-mail: shinji.deguchi@dataexploring.com

© Springer Nature Singapore Pte Ltd. 2020
T. Imaizumi et al. (eds.), *Advanced Studies in Classification and Data Science*,
Studies in Classification, Data Analysis, and Knowledge Organization,
https://doi.org/10.1007/978-981-15-3311-2_6

Classification methods are used in many disciplines. The relevant information for extraction depends on the nature of the investigation and the questions of interest. Very large datasets tend to be subjected to less elaborate analysis than smaller datasets. Several clustering algorithms have been proposed; they can be classified into hierarchical and partitioning clustering algorithms (Gordon 1999). Hierarchical algorithms decompose data consisting of n objects into several levels of nested segments, as represented by a dendrogram. Partitioning algorithms construct a single partition of data, consisting of n objects and k clusters, such that the objects in a cluster are more similar than objects in different clusters. The k-means algorithm is one of the most widely used methods to partition a dataset into groups of patterns. The k-means algorithm (MacQueen 1967) is built upon four basic operations: selection of the initial k means for k clusters, calculation of the dissimilarity between an object and the mean of a cluster, allocation of an object to the cluster whose mean is nearest to the object, and recalculation of the mean of a cluster from the objects allocated, such as to minimize intracluster dissimilarity. Except for the first operation, these operations are performed repeatedly until the algorithm converges. The selection of the initial k means may be conducted in a random manner or according to specific heuristics. The k-means algorithm aims at minimizing the following cost function:

$$\sum_{j=1}^{k}\sum_{i=1}^{n} \| x_i^{(j)} - c_j \|^2 \tag{1}$$

where$\| x_i^{(j)} - c_j \|$ is the chosen distance measure between a data point $x_i^{(j)}$ and the cluster center c_j, is an indicator of the distance of the n data points from their respective cluster centers. An advantage of the k-means calculation process is that it is simple and convenient.

However, the k-means method converges to one of many local minima, in which the final result depends on the initial starting points. Thus, it is necessary to duplicate the analysis using a different initial value. To obtain the optimum solution for each pair of initial values, the algorithm repeatedly calculates the cluster centers to minimize the average squared distance between objects in the same cluster, so the computational resources required can be costly. Thus, it is important to decrease the number of iterations to reduce processing costs. To solve this problem, previous studies have proposed algorithms to refine the initial cluster center values by selecting subsamples from the dataset (e.g., Bradley and Fayyad 1998; Fahim et al. 2009).

Fahim et al. (2009) introduced an efficient method to obtain good initial starting points for k-means computation. The dataset is first divided into several blocks. The k-means algorithm is then applied to each block independently, to produce the k centers for each block, as described below. Thus, if the dataset is partitioned into m blocks, then the compressed data will contain $m \times k$ objects. The k-means is then applied to the compressed data to produce k centroids, in which the resulting k centroids are the initial starting points for the k-means algorithm that works on the full dataset. Thus, the cluster center is recalculated using the k-means approach; however, the recalculation step can involve high processing costs, especially for large datasets. Therefore, the present study proposes a method in which recalculation of the cluster center is not required.

2   The Method

The proposed method performs clustering without recalculating the cluster centers in the full dataset, by simplifying the calculations. Here, the Ward method is applied to a compressed dataset to obtain k centroids. The results of this method visually show the grouping process between objects, using a dendrogram; using this approach, it is possible to gain some idea of the suitable number of classes for grouping data. In clustering large datasets, the k-means algorithm is much faster than the hierarchical clustering algorithm, whose general computational complexity is $O(n^2)$ (Murtagh 1992). The hierarchical clustering method is usually applied to small datasets, due to the low efficiency of the approach with larger datasets.

A description of the steps involved in our proposed approach is given below. First, m subsamples are selected from the full dataset. The k-means algorithm is applied to each subsample independently, to produce k centers for each subsample. The size of k is set to be larger than the expected size for the full dataset. Thus, the compressed dataset will contain $m \times k$ objects. Each subsample size is smaller than the full dataset size, such that the compressed dataset is much smaller than the full dataset. The Ward method using a hierarchical clustering algorithm is then applied to the compressed dataset to determine the centers of the full dataset, using the mean value of the variables as the initial starting points of the full dataset. Each case is distributed to the clusters whose cluster center is in closest proximity; this eliminates cluster center recalculation. Distortion takes a set of k estimates of the means and computes the sum of the squared distances of each data point to its nearest mean; this step provides a measure of the degree-of-fit of a set of clusters to the dataset. When creating the cluster center of the original data from multiple sampled data, the proposed method uses hierarchical cluster analysis, as opposed to the k-means method. Notably, it is possible to determine how many clusters to adopt by visually checking the linkage of multiple sampled data between objects when classifying the full dataset.

The advantages of our approach are that cluster centers can be calculated from sample data and are uniquely determined in the proposed method. Recalculation of cluster centers is not required, so the calculation time and computational cost are reduced.

3 The Analysis

We analyzed the customer purchase history data "True Data" at a drug store; these data were provided by True Data, Inc. The customer purchase history data contained 232,030,245 records. Among them, 4,671,856 records with both attribute information and purchasing information were used for this analysis. In the product category, Level 4 (6 digits) of the Japan Item Code File Service/Integrated Flexible Database (JICFS) classification was used, with 629 categories of food and daily necessities. Matrices of individuals (members) and product categories with purchase amounts were calculated based on customers' purchase history data. Multiple sampling data were sufficiently small such that clustering and results analysis could be performed easily. Twenty sampling data were obtained by extracting members based on a systematic sampling process. The sample size n consisted of 2000 individuals and 629 categories. Multiple sampling data were analyzed using the ordinary k-means method; a larger number of clusters than the expected number of true clusters was applied. The number of clusters was set to 20. Thus, 20 sampling data consisting of 2000 individuals and 629 product categories were analyzed using the ordinary k-means method. The compressed data to calculate the initial cluster centers in the full dataset consisted of 400 objects (20 sampling data × 20 clusters). The Ward method of hierarchical cluster analysis was performed on the compressed dataset (size: 400 × 629). The initial cluster centers for the full dataset were calculated using the analysis results of the Ward method. We attempted to obtain an initial solution mechanically. Eleven clusters were adopted based on the dendrogram projections. Using the proposed method, we can interpretatively calculate the number of clusters and cluster average expected from sampling data to obtain the initial solution.

The mean of the purchase amount of all variables for the 11 clusters was calculated as the initial cluster center for the full dataset. Figure 1 shows the mean of each variable of 11 clusters obtained from the analysis of the Ward method. These mean values are the initial cluster centers for the full dataset. In clusters 1 and 2, the means of the purchase amount of various products were slightly higher. Cluster 1 consisted of customers who purchased various products. Cluster 2 consisted of customers who purchased various items and more food than other customers. In clusters 3–11, the means of purchase amounts of specific products were high. Clusters 3 and 11 consisted of customers who frequently purchased "Health food"; the number of product categories in health foods was 190,201 in Level 4 (6 digits) of the JICFS. Cluster 4 included customers who frequently purchased "Coke products"; the number of Coke product categories was 140,307 in JICFS Level 4 (6 digits). Cluster 5 consisted of customers who frequently purchased "Yogurt"; the number of yogurt product categories was 130,205 in JICFS Level 4 (6 digits). Cluster 6 was comprised of customers who frequently purchased "Cat food"; the number of cat food product categories was 262,201 in JICFS Level 4 (6 digits). Cluster 7 consisted of customers who frequently purchased "Baby food"; the number of baby food product categories was 190,103 in JICFS Level

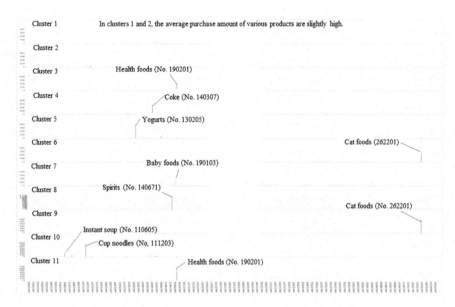

Fig. 1 Mean of each variable of 11 clusters obtained using the Ward method

4 (6 digits). Cluster 8 included customers who frequently purchased "Spirits"; the number of spirit product categories was 140,671 in JICFS Level 4 (6 digits). Cluster 9 consisted of customers who frequently purchased "Cat food"; the number of cat food product categories was 262,201 in Level-4 (6 digits) JICFS. Cluster 10 represented customers who frequently purchased "Instant soup and cup noodles"; the numbers of instant soup and cup noodles product categories were 110,605 and 111,203, respectively, in JICFS Level 4 (6 digits). Clusters 6 and 9 have a similar tendency to buy "Cat food" often, but purchasing trends in categories other than Cat food differ.

We used the purchase amount means of all categories as the initial starting points of the 11 cluster centers in the full dataset. Each case, nearly 4,600,000 in total, was distributed to clusters having the closest cluster center. Although the initial solution was 11 clusters, nearest data did not exist at the center of three clusters. As a result, eight clusters contained many cases; however, three clusters (clusters 6, 7, and 9) were empty. Figure 2 shows the mean of each variable of the final eight clusters obtained from the proposed method. Cluster 1 consisted of customers who purchased various products. In clusters 2–5, 8, 10, and 11, the means of purchase amounts of specific products were high. Cluster 2 consisted of customers who frequently purchased "Biscuits and Cookies"; the number of biscuit and cookie product categories was 130,127 in JICFS Level 4 (6 digits). Cluster 3 was comprised of customers who frequently purchased "Baby food"; the number of baby food product categories was 190,103 in Level 4 (6 digits) of the JICFS. Cluster 4 consisted of customers who frequently purchased "Coke products"; the number of Coke product categories was 140,301 in JICFS Level 4 (6 digits). Cluster 5 included

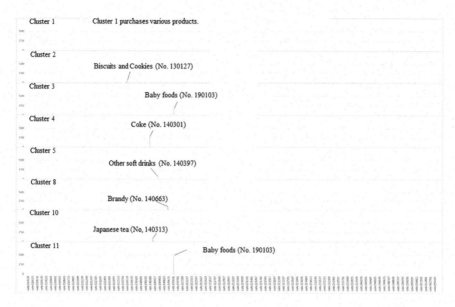

Fig. 2 Mean of each variable of the final eight clusters obtained from the proposed method

customers who frequently purchased "Other soft drinks"; the number of other soft drink product categories was 140,397 in JICFS Level 4 (6 digits). Cluster 8 consisted of customers who frequently purchased "Brandy"; the number of brandy product categories was 140,663 in JICFS Level 4 (6 digits). Cluster 10 was comprised of customers who frequently purchased "Japanese tea"; the number of Japanese tea product categories was 140,313 in JICFS Level 4 (6 digits). Cluster 11 consisted of customers who frequently purchased "Baby food"; the number of baby food product categories was 190,103 in Level 4 (6 digits) of the JICFS. Clusters 3 and 11 have a similar tendency to buy "Baby food" often, but purchasing trends in categories other than "Baby food" differ.

4   Conclusion

We performed k-means analysis without recalculation of the full dataset using the numbers and centers of clusters obtained from the analysis of sampling data. Table 1 lists the final eight clusters. The features of the initial and final clusters had similar tendencies. The proposed method can be used to simplify and reduce calculation costs by removing the cluster center recalculation step. We selected several subsamples from the full dataset and applied the k-means algorithm with a larger number of clusters than expected on each subsample independently, to obtain the centers for each subsample. The centers of each cluster obtained by analyzing a compressed dataset were analyzed using the Ward method, and the

Table 1 Sizes and features of the final clusters

	Final cluster size	Feature of final cluster	Feature of initial cluster
Cluster 1	4, 668, 454	Various product	Various product
Cluster 2	277	Biscuits and cookies	Various product (a little more food is purchased than the others)
Cluster 3	1735	Baby foods	Health foods
Cluster 4	97	Coke	Coke
Cluster 5	453	Other soft drinks	Yogurts
Cluster 8	4	Brandy	Spirits
Cluster 10	747	Japanese tea	Instant soup and cup noodles
Cluster 11	89	Baby foods	Health foods

Table 2 Distances from the center of the cluster to objects and the corresponding standard deviation

	Distances from center of cluster to objects	Standard deviation
Cluster 1	319	56, 530
Cluster 2	175	191
Cluster 3	1408	25, 052
Cluster 4	3502	4879
Cluster 5	378	1700
Cluster 8	551	216
Cluster 10	621	463
Cluster 11	11, 476	20, 416

numbers of clusters and the centers of the clusters for analysis of the full dataset were determined from the results. The proposed method achieved good results.

We mechanically adopted 11 clusters based on the dendrogram, because many variables were used in this analysis. However, some clusters were not chosen as final clusters, e.g., initial clusters 6, 7, and 9. To be selected as a final cluster, there must be an object close to the center of the initial cluster. There were also clusters that did not fit well, such as clusters 4 and 11, in which the means of the distance from the center of the cluster to individual objects and the associated standard deviations were large (Table 2). To solve these problems, the numbers of clusters and the cluster average expected from sampling data must be calculated on the basis of statistical indicators. The motivation for this study was to reduce the calculation cost and time by creating initial values from subsamples and classifying all data based on these initial values. Therefore, the validity of the initial value obtained by the proposed method and the numbers of clusters adopted were not assessed. More research will be needed in the future to examine whether the optimal cluster numbers and centers can be determined by the initial value obtained from the proposed method using statistical indicators such as the Elbow method, the Silhouette method, and gap statistics.

Acknowledgments This work was supported by a Grant-in-Aid for Scientific Research (C) (No. 16K00052) from the Japan Society for the Promotion of Science. We wish to thank True Data Inc., for allowing us to make use of the customer purchase history data "True Data"

References

Bradley, P.S., Fayyad U.M.: Refining initial points for k-means clustering. In: Shavlik, J. (ed.) Proceedings of the 15th International Conference on Machine Learning (ICML98), pp. 91–99. Morgan Kaufmann, San Francisco (1998)

Fahim, A.M., Salem, A.M., Torkey, F.A., Saake G., Ramadan, M.A.: An efficient k-means with good initial starting points. Georgian Electron. Sci. Comput. Sci. Telecommun. **2**(19), 47–57 (2009)

Gordon, A.D.: Classification, 2nd. edn. Chapman & Hall, Boca Raton (1999)

MacQueen, J.: Some methods for classification and analysis of multivariate observations. In: Le Cam, L.M., Neyman, J. (eds.) Proceedings of the 5th Berkeley Symposium on Mathematical Statistics and Probability, vol. 1, pp. 281–297. University of California Press, Berkeley (1967)

Murtagh, F.: Comments on "Parallel algorithms for hierarchical clustering and cluster validity". IEEE Trans. Pattern Anal. Mach. Intell. **14**(10), 1056–1057 (1992)

Supervised Nested Algorithm for Classification Based on K-Means

Luciano Nieddu and Donatella Vicari

Abstract The aim of this paper is to present an extension of the k-means algorithm based on the idea of recursive partitioning that can be used as a classification algorithm in the case of supervised classification. Some of the most robust techniques for supervised classification are those based on classification trees that make no assumptions on the parametric distribution of the data and are based on recursively partitioning the feature space into homogeneous subsets of units according to the class the entities belong to. One of the shortcomings of these approaches is that the recursive partitioning of the data, i.e. the growing of the tree, is achieved considering only one variable at a time and, although it makes the tree pretty simple in terms of determining rules to obtain a classification, it also makes them hard to interpret. Building on these ideas we carry the integration of parametric model into trees one step further and propose a supervised classification algorithm based on the k-means routine that sequentially splits the data according to the whole feature vector. Results from applications to simulated data are shown to address the potentiality of the proposed method in different conditions.

1 Introduction

The problem of supervised classification has been gathering lots of interest since the seminal work of Sir R. Fisher (1936). Given a set of vectors $\mathbf{x}_i$, $i = 1, \ldots, n$ spanning a feature space $\mathcal{X}$, consisting of elements that have been previously classified by an expert or in any other way deemed appropriate into a set of mutually disjunctive classes K, the problem of classification consists in training a classifier

L. Nieddu (✉)
UNINT-Universitá degli Studi Internazionali di Roma, Rome, Italy
e-mail: l.nieddu@unint.eu

D. Vicari
Dipartimento di Scienze Statistiche, Sapienza – Universitá di Roma, Rome, Italy
e-mail: donatella.vicari@uniroma1.it

© Springer Nature Singapore Pte Ltd. 2020
T. Imaizumi et al. (eds.), *Advanced Studies in Classification and Data Science*,
Studies in Classification, Data Analysis, and Knowledge Organization,
https://doi.org/10.1007/978-981-15-3311-2_7

in order to classify new units of unknown class. The classifier is usually a function $\psi : \mathcal{X} \rightarrow K$ which should mimic the behaviour of the expert and the problem is then of getting an estimate of $\psi(\cdot)$ based on the sample at hand.

The dataset of previously classified vectors is known as training set (Watanabe 1985). If the classification of the elements of the training set is certain and not affected by error, then the problem is known as recognition with perfect supervisor, otherwise it is known as recognition with imperfect supervisor (Katre and Krishnan 1989).

In this paper we will be dealing with recognition with perfect supervisor but we will show some insights that should allow the proposed approach to be extended to the problem of recognition with imperfect supervisor.

When the training set is composed of elements of unknown classes then the problem becomes of recognition without supervisor and is usually tackled using clustering or finite mixture models (Fraley and Raftery 2002; Celeux 2007). Recently some attempts have also been carried out to extend the finite mixture model semi-parametric approach to supervised classification (see Hastie and Tibshirani 1996; Nieddu and Vitiello 2013 for instance). Nonetheless this paper will focus only on the problem of recognition with perfect supervisor.

In general, to build a classification rule, given a dataset composed of a set of measurements on n objects, three main steps are required regardless if the problem is with or without supervisor:

1. a pre-processing is applied to the data in order to make them comparable and to try to filter-out the part of information which is not directly related to the classification problem;
2. feature selection and/or extraction is applied to the data in order to reduce the dimension of the space embedding the dataset, retaining as much as possible of the information related to the class while trying to avoid the curse of dimensionality (Bellman 1957). The features selected/extracted should be invariant to incidental changes that are not related to the classification problem, e.g. the features selected in a face recognition problem should be invariant to changes in environmental factors (e.g. lighting) and sensing configuration (e.g. camera placement) (Claudio et al. 2008; Ye et al. 2016);
3. the pattern recognition algorithm is trained on a subset of the dataset (training set) and its performance is evaluated using a cross-validation scheme. The latter is valid only for supervised classification (see Watanabe 1985; Duda and Hart 1973). Estimating the classifier on a subset and testing its performance on an independent subset is a mandatory requirement since each classification rule would be optimal for the dataset it has been trained on, therefore having an independent dataset to verify the classifier is the only way to get an unbiased estimate of the performance of the classifier. The performance of the classifier on the training set is a biased estimate of the real performance and is usually referred to as *apparent recognition rate* (Mclachlan 2004).

Over the years a plethora of approaches have been suggested to handle the problem of supervised classification (see for instance Jain et al. (2000); Nieddu and Patrizi (2000) among many others). It is fair to assume that no algorithm is able to achieve the best performance on any classification problem without any prior assumptions on the data. This has been mathematically clarified in 1996 with the proof of the "no free lunch theorem" (NFL) (Wolpert 1996).

Therefore robustness of a classification algorithm is a key requirement when no specific assumption can be made on the data. Some of the most robust techniques for supervised classification are nonparametric approaches based on classification trees (Breiman 1984; Quinlan 1986) that make no specific parametric assumption on the distribution of the data. These approaches work by recursively partitioning the feature space $\mathscr{X}$ into subsets of elements that are homogeneous according to the class the elements belong to. Usually the Gini index or entropy are used as a measure of homogeneity.

The tree growing process employs a greedy approach (Curtis 2003) to determine the best partition on the data splitting the data on the account of just one variable at a time.

This approach has been recently further developed incorporating simple parametric models into the terminal nodes of the tree. Research in this direction was motivated by the fact that a constant value in the terminal node tends to produce large and thus hard to interpret trees (Chan and Loh 2004).

One of the major shortcomings of partitioning algorithms is that the recursive partitioning of the data, i.e. the growing of the tree, is achieved considering only one variable at a time and, although it makes the trees pretty simple to read, it also may render them hard to interpret (for instance, some variables might show up more than once at various levels in the tree path).

Building on these ideas, we carry the integration of parametric models into trees one step further and propose a supervised classification algorithm where the recursive partitioning of the feature space $\mathscr{X}$ is based on the whole feature vector. The proposed methodology is based on the k-means algorithm and although it has been tested only on supervised classification problems we believe that it can be extended to recognition with imperfect supervisor.

The outline of the paper is as follows: in Sect. 2 the proposed methodology will be presented and an algorithm will be outlined, in Sect. 3 a simulation study will be outlined to test the performance of the proposed methodology with respect to classification trees which use a similar approach to classification. In Sect. 4 finally some conclusions and future extensions will be drawn.

2 The Algorithm

The algorithm presented in this paper is a supervised classification algorithm, i.e. a dataset of elements with known classes is supposed to be available. The aim of this algorithm is to find subclasses in the dataset which can be used to classify new vectors of unknown classes. Starting from the available partition on K known classes the objective is to find a finer partition in subclasses consistent with the original partition that can be used to better classify new entities. This is achieved by recursively partitioning the training set.

The performance of the algorithm is assessed via cross-validation (Watanabe 1985; Mclachlan 2004).

Given a training set of n pattern vectors in $\mathbb{R}^p$, let us assume a partition defined on the dataset, i.e. each pattern vector is assigned to one and only one of K known classes. Let us assume a Euclidean norm defined on the dataset and let ψ be a function from $\mathbb{R}^p$ onto the set $\mathscr{C} = \{1, 2, \ldots, K\}$ which maps each pattern vector $\mathbf{x}_j$, $j = 1, \ldots, n$ into the class $c \in \mathscr{C}$ that it belongs to.

The proposed algorithm works as follows:

- compute the barycentre of each class and compute the distance of each vector from each barycentre;
- if each vector in the training set is closer to the barycentre of its class the algorithm stops, otherwise there will be a nonempty set $\mathscr{M}$ of pattern vectors which belong to a class and are closer to a barycentre of a different class. In $\mathscr{M}$ select the pattern vector $\mathbf{x}_w$ that is farthest from the barycentre of its class. This pattern vector will be used as a seed for a new barycentre for class $\psi(\mathbf{x}_w)$;
- a k-means algorithm (MacQueen 1967) will then be performed for all the pattern vectors in class $\psi(\mathbf{x}_w)$ using, as starting points, the set of barycentres for class $\psi(\mathbf{x}_w)$ and the vector $\mathbf{x}_w$. For ease of exposition let us assume that there are m barycentres in class $\psi(\mathbf{x}_w)$. Once the k-means has been performed the set of barycentres for class $\psi(\mathbf{x}_w)$ will be composed of $m+1$ elements. The barycentres at the new iterations need not be computed for all classes, but only for class $\psi(\mathbf{x}_w)$, since the barycentres for the other classes have remained unchanged. In the following step the distance of each pattern vector from all the barycentres is computed anew, and so is the set $\mathscr{M}$ (see Fig. 1);
- if $\mathscr{M}$ is not empty then the pattern vector in $\mathscr{M}$ which is farthest from a barycentre of its own class is once again selected to serve as a seed for a new barycentre. This procedure iterates until the set $\mathscr{M}$ is empty.

The algorithm is depicted in pseudo-code in Fig. 1.

Upon convergence, the algorithm yields a set of barycentres which, in the worst case, are in a number equal to the number of elements in the training set and which has a lower bound in the number of classes.

It is worth noticing that if the partition defined on the dataset is consistent with the features considered, i.e. if the pattern vectors are linearly separable, then the algorithm generates a number of barycentres equal to the number of classes. On the other hand, if the classes in the dataset are not linearly separable, then the algorithm continues splitting the classes until the subclasses obtained are linearly separable. It is obvious that it can continue splitting until all the subclasses are composed of only one vector (singleton). It must be stressed that it will not converge if two vectors in the training set belong to different classes and are represented by the same pattern vector. Nonetheless if two elements in the training set belong to different classes but show the same pattern vector, either the classification is wrong or the set of measurement is not sufficient to discriminate properly between the two classes. The

Step1   **Let**
- $\mathbf{x}_j, j = 1,\dots,n$ be the pattern vectors in the training set
- $\mathbf{B}_0$ be the set of K initial barycentres $\mathbf{b}_i, i = 1,\dots,K$

Step2

Compute the distances of each $\mathbf{x}_j$ from all the $\mathbf{b}_i \in \mathbf{B}_0$

Let $\mathcal{M}$ be the set of $\mathbf{x}_w$ that are closer to a barycentre of a class different from their own.

$t \leftarrow 0$

Step3   **while** $\mathcal{M} \neq \emptyset$
- **Let** $\mathbf{x}_s, \in \mathcal{M}$ be the vector with the greatest distance from its own barycentre.
- $c \leftarrow \psi(\mathbf{x}_s)$
- **Let** $\mathbf{B}_{t+1} \leftarrow \mathbf{B}_t \cup \mathbf{x}_s$
- for all the elements of class c perform a k-means routine using as starting points the barycentres of $\mathbf{B}_{t+1}$ that belong to class c
- $t \leftarrow t + 1$
- **Compute** the distances of each $\mathbf{x}_j$ from all the $\mathbf{b}_i \in \mathbf{B}_t$
- **Let** $\mathcal{M}$ be the set of $\mathbf{x}_w$ that are closer to a barycentre of a class different from their own.

end

Fig. 1 Algorithm in meta-language

latter problem can be easily overcome increasing the dimension of the vector space gathering more information on the objects we are trying to classify.

Upon convergence, the sets of barycentres can be used to classify new elements (query points) assigning the new element to the class of the barycentre it is closest to. It should be stressed that if elements from the training set are used as query points, then the algorithm always classify them correctly because, once converged, all pattern vectors in the training set are closer to a centroid of their own class, i.e. if the class of the closest barycentre is used to classify new elements, then the apparent error rate for the proposed method is zero (perfect recognition).

The algorithm can be generalized allowing for impurity in the result, i.e. the recursive partitioning of the feature space can be performed until the percentage of elements that are closer to a barycentre of another class has decreased under a certain threshold which can be set to a value different from zero. This can be helpful when the training set has been classified with error (imperfect supervisor): in this case allowing for impurity in the subclasses can prevent the algorithm from overfitting the data.

3   Simulation Study

The proposed algorithm has been tested on a simulated dataset. The dataset has been
generated adding multivariate Gaussian noise to a set of 9 barycentres on $\mathbb{R}^2$.

First the coordinates of the barycentres have been randomly generated according
to a continuous uniform random variable ($\mathbf{B}_j \sim U[-5; 5]^2$, $j = 1, \ldots, 9$) and
then r random points in $\mathbb{R}^2$ have been drawn from a Gaussian process centred in
each barycentre.

Each barycentre and its associated points have then been randomly assigned to
one of $K = 2$ classes according to Bernoulli process with probability $\theta = 0.5$. This
should ensure variable shapes and sizes of the two classes.

In order to test the performance of the algorithm on a dataset with various error
levels, the covariance matrix for the Gaussian noise has been chosen as

$$\Sigma = \gamma \begin{bmatrix} 1 & 0.7 \\ 0.7 & 1 \end{bmatrix} \quad \text{with} \quad \gamma \in \{0.05, 0.10, 0.15, 0.20, 0.25, 0.30, 0.35, 0.40, 0.50\}$$

The number of points for each barycentre has been chosen in $r \in \{5, 10, 15, 20,$
$30, 40\}$ for total sample sizes of $\{45, 90, 135, 180, 270, 360\}$ elements, respectively.

For ease of exposition in Fig. 2 the three steps of the generation of a dataset have
been displayed.

It should be noted that the random assignment of the centroids and their
associated points to one of two classes allows to mimic the pattern of real datasets
with different sizes and correlation levels between variables, simulating various
shapes for each of the two classes and allowing the classes to present a "tangled"
appearance.

The performance of the proposed technique has been assessed using correct
recognition rates i.e. the number of correctly classified units divided by the total
number of units (Watanabe 1985). Since the correct recognition rate determined
on the same set used for training would result in an inflated estimate (apparent
classification rate) a cross-validation framework has been used. More formally
the performance of the proposed technique has estimated via leave one out cross-
validation (Sammut and Webb 2010) and it has been compared with the results
of classification trees on the same datasets. Therefore at each iteration the same
element is singled out for testing and the two techniques are both trained on the
remaining $n - 1$ points and tested on the performance obtained on the singled out
unit. This process has been repeated n times and the average correct recognition for
each configuration of the simulation parameters has been computed.

In Fig. 3 the correct recognition rates for the proposed technique and the
analogous results for the classification trees obtained using the rpart () package
in R have been depicted.

The proposed method almost always performs better than classification trees.
The recognition rates of the proposed method are always greater than the random
recognition rate which, for a two-class problem, is 0.5. The same cannot be said

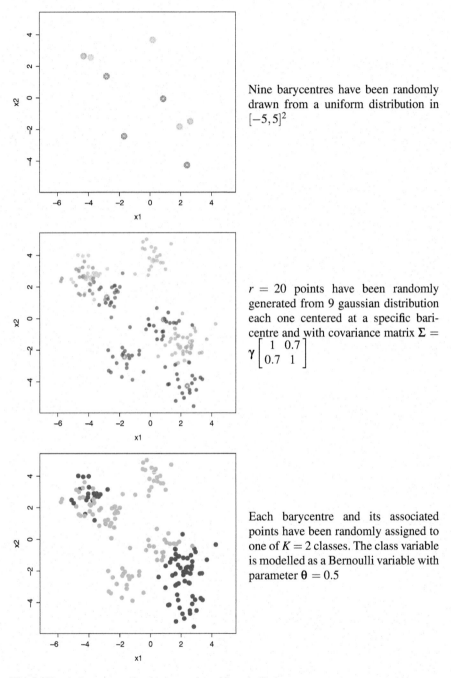

Nine barycentres have been randomly drawn from a uniform distribution in $[-5,5]^2$

$r = 20$ points have been randomly generated from 9 gaussian distribution each one centered at a specific baricentre and with covariance matrix $\Sigma = \gamma \begin{bmatrix} 1 & 0.7 \\ 0.7 & 1 \end{bmatrix}$

Each barycentre and its associated points have been randomly assigned to one of $K = 2$ classes. The class variable is modelled as a Bernoulli variable with parameter $\theta = 0.5$

Fig. 2 Simulation process for the generation of "tangled" classes

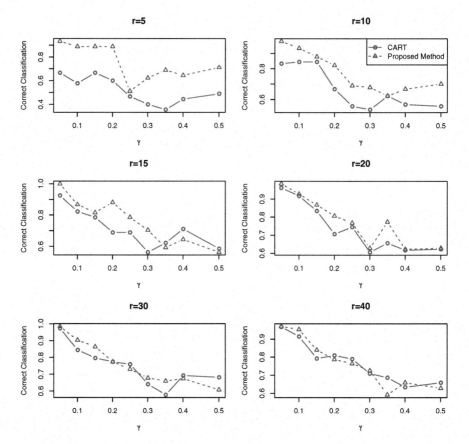

Fig. 3 Recognition rates for the proposed method (dashed blue plot) and for classification trees (continuous red plot) for different sample sizes and varying error levels (γ)

about the classification trees, that, for $r = 5$ yield a recognition rate lower than 0.5 as γ increases (plot on the top left corner of Fig. 3).

Out of the 6×9 cells of the simulation setup (9 levels of noise and 6 different number of points for each barycentre) the proposed methodology obtained a better recognition rate than CART 45 times.

It must be stressed that, although these results can be readily compared since they are based on the same sample and show a better performance of the proposed method, a more extensive simulation study needs to be carried out in order to get a better understanding of the performance of the proposed technique in more different conditions with various shapes and sizes of the datasets.

In Table 1 the average recognition rates and their standard deviations have been displayed. Averages have been calculated over different values of γ (noise size). As the sample size increases both techniques tend to achieve a better average performance with comparable standard deviations. The proposed methodology, on

Table 1 Simulation results: average correct recognition rates and associated standard deviations for the proposed methodology and CARTs over varying number of points in each barycentre

r	Proposed method[a]	SD	CART[a]	SD
5	0.7531	0.1419	0.5185	0.1073
10	0.7741	0.1232	0.6691	0.1271
15	0.7613	0.1377	0.7103	0.1109
20	0.7778	0.1260	0.7401	0.1268
30	0.7504	0.1277	0.7437	0.1126
40	0.7589	0.1302	0.7586	0.1163

[a] Average performance over various levels of noise for different values of points in each original barycentre (r)

average, outperforms classification trees although the performances tend to be very similar with a very large number of points for each barycentre.

It should be noted that since the simulation was carried in $\mathbb{R}^2$ as the number of points and the noise size for each barycentre increase the two classes tend to overlap since each barycentre is associated randomly with one of the two classes (see Fig. 2) and therefore it is possible that the proposed method starts to overfit the data.

4 Conclusions

The idea for a new classification algorithm that builds on the concept of model based recursive partitioning (Zeileis et al. 2008) has been presented in this paper. The classification results on the small simulation study are promising when compared with the performance of classification trees on the same simulated data. CARTs have been chosen as a comparison since they apply a very similar approach to classification. The idea of trying to find a finer partition on the training set that is coherent with the original classes but that allows for a better classification of new entities is at the basis of the proposed method and is applicable to many real data problems where the number of features that have been collected to represent a class are only a part of those that could perfectly separate the classes and therefore the projection of a possible well separated dataset on the observed feature space ends up in a class structure that is not easily separable.

The algorithm has already been tested on its form without allowing impurity, i.e. assuming that the training set has been classified without error (perfect supervisor). The same approach can be applied in the case of imperfect supervisor, allowing for impurity in the training. The possibility to allow for impurity can prevent the proposed method from overfitting possible noise in the data that could result in unnecessary barycentres from the training set.

The version with impurity is currently being tested and the results are promising.

References

Bellman, R.: Dynamic Programming, 1st edn. Princeton University Press, Princeton (1957)

Breiman, L., Friedman, J., Olshen, R., Stone, C.: Classification and Regression Trees. Wadsworth and Brooks, Monterey (1984)

Celeux, G.: Mixture Models for Classification, pp. 3–14. Springer, Berlin (2007)

Chan, K.Y., Loh, W.Y.: Lotus: an algorithm for building accurate and comprehensible logistic regression trees. Journal of Computational and Graphical Statistics 13(4), 826–852 (2004)

Cifarelli, C., Manfredi, G., Nieddu, L.: Statistical Face Recognition via a K-Means Iterative Algorithm. IEEE, Piscataway (2008)

Curtis, S.A.: The classification of greedy algorithms. Sci. Comput. Program. 49(1), 125–157 (2003)

Duda, R.O., Hart, P.E.: Pattern Classification and Scene Analysis. A Wiley Interscience Publication, Wiley (1973)

Fisher, R.A.: The use of multiple measurements in taxonomic problems. Ann. Eugen. 7(7), 179–188 (1936)

Fraley, C., Raftery, A.E.: Model-based clustering, discriminant analysis, and density estimation. J. Am. Stat. Assoc. 97, 611–631 (2002)

Hastie, T., Tibshirani, R.: Discriminant analysis by Gaussian mixtures. J. R. Stat. Soc. Ser. B 58, 155–176 (1996)

Jain, A.K., Duin, R.P.W., Mao, J.: Statistical pattern recognition: a review. IEEE Trans. Pattern Anal. Mach. Intell. 22(1), 4–37 (2000)

Katre, U.A., Krishnan, T.: Pattern recognition with an imperfect supervisor. Pattern Recogn. 22(4), 423–431 (1989)

MacQueen, J.B.: Some methods for classification and analysis of multivariate observations. In: Le Cam, L.M., Neyman, J. (eds.) Proceedings of the Fifth Berkeley Symposium on Mathematical Statistics and Probability, vol. 1, pp. 281–297. University of California Press, California (1967)

Mclachlan, G.J.: Discriminant Analysis and Statistical Pattern Recognition (Wiley Series in Probability and Statistics). Wiley-Interscience, Hoboken (2004)

Nieddu, L., Patrizi, G.: Formal methods in pattern recognition: a review. Eur. J. Oper. Res. 120(3), 459–495 (2000)

Nieddu, L., Vitiello, C.: A proposal for a semiparametric classification method with prior information. RIEDS - Rivista Italiana di Economia, Demografia e Statistica - Italian Review of Economics, Demography and Statistics 67(3–4), 176–183 (2013)

Quinlan, J.R.: Induction of decision trees. Mach. Learn. 1(1), 81–106 (1986)

Sammut, C., Webb, G.I. (eds.) Leave-One-Out Cross-Validation, pp. 600–601. Springer, Boston (2010)

Watanabe, S.: Pattern Recognition: Human and Mechanical. Wiley, New York (1985)

Wolpert, D.H.: The lack of a priori distinctions between learning algorithms. Neural Comput. 8(7), 1341–1390 (1996)

Ye, H., Shao, W., Wang, H., Ma, J., Wang, L., Zheng, Y., Xue, X.: Face recognition via active annotation and learning. In: Proceedings of the 2016 ACM on Multimedia Conference, MM '16, pp. 1058–1062. ACM, New York (2016)

Zeileis, A., Hothorn, T., Hornik, K.: Model-based recursive partitioning. J. Comput. Graph. Stat. 17(2), 492–514 (2008)

Using Classification of Regions Based on the Complexity of the Global Progress Indices for Supporting Development in Competitiveness

Éva Sándor-Kriszt and Anita Csesznák

Abstract On the methodological difficulties of creating social measurement indicators. This paper focuses on the three pillars of the Regional Competitiveness Index with a special emphasis on higher education. The research used data gathered from the 262 regions of the European Union on higher education and lifelong learning, labour market efficiency and market size. Using cluster analysis, the 262 regions were arranged into five clusters.

1 Introduction

The purpose is to bridge existing geographical distances between a small East-Central European country, Hungary and the member countries of the International Federation of Classification Societies analysing the methodology of global progress indices in the process.

This is a great opportunity for economists from a small country to show their findings in Tokyo. Although Hungary is very far from Asia, our aims in higher education are common: to improve academic standards, to foster internationalization and to meet the new requirements triggered by globalization.

This paper gives an overview of the applicability of recently published global indices for measuring social, environmental and economic progress at different regional levels. The critical investigation is based on determining the relevance and comparability of the referred data sources and aggregation techniques used for composing complex indicators. Advantages and barriers for appropriate geographical/regional classification in context of measuring achievement of globally recognized sustainable development goals, are presented and prioritized. Classification of sustainability measures related to the development of new business and

É. Sándor-Kriszt (✉) · A. Csesznák
Budapest Business School, Budapest, Hungary
e-mail: Kriszt.Eva@uni-bge.hu; OroszneCsesznak.Anita@uni-bge.hu

© Springer Nature Singapore Pte Ltd. 2020
T. Imaizumi et al. (eds.), *Advanced Studies in Classification and Data Science*,
Studies in Classification, Data Analysis, and Knowledge Organization,
https://doi.org/10.1007/978-981-15-3311-2_8

non-profit models based on shared values provides new tools for evaluation and interpretation of both temporary and longer term regional differences in support of global and local development policies. In this way, the current and desirable states of the key drivers for the development of human well being, like innovation in higher education systems, can be better identified and explained for the policy-makers, especially in Central-Eastern Europe, where dramatic changes of current "business as usual" strategies and models and the proper and fast adoption of globally successful innovative solutions are necessary at all overarching macro, medium and micro levels within the common universe of society, environment and economy.

2   Statistical Literacy as a Precondition of Globalization

The basic skills required to do any research work in any field include linguistic skills—researchers need to understand research literature published in English; intercultural skills—research work entails cooperation between specialists coming from diverse cultural backgrounds; IT skills and last, but not least, statistical skills.

It is easy to see that to be able to carry out research work one must work with data and information. Today's question is not how or where to acquire the necessary information. To be able to process data into information and use that in a meaningful way, it is imperative that one should acquire statistical literacy. This means being skilled in reading, selecting, identifying, analysing and interpreting information. The authors of this paper have also been challenged by this problem: it was easy to gather data about the regions of the European Union, but less so to select the necessary indicators and to construct the Regional Competitiveness Index.

Selecting the right indicators poses problems: economic and social processes do not lend themselves to easy measurability.

3   Problems with the Measurability of Social Processes

Economic indicators can be and are calculated for countries, regions, industries, etc., but unfortunately they are not exact measures as compared with the ones used in natural sciences. Some of these indicators are volatile—the prices of securities are continuously changing on the Stock Exchange, and so are the values of commodities and services. Most economic indicators are usually based on uncertain data. This means that GDP or GNP figures, inflation rates and unemployment statistics are more or less estimates.

Ranking is also an accepted way to evaluate performance. It is used in all walks of life—from singing competitions to measuring creditworthiness.

Ranking universities is also done this way. Surveys are carried out asking employers or alumni how satisfied they are with a particular institute of higher education.

However, in the case of universities, it is possible to make a more precise evaluation of their performance, which can be done using a set of indicators and statistics.

Unfortunately, even these indicators and statistics do not provide an exact measure of the performance of universities. There are about 18,000 institutions of higher education in the world so to be able to evaluate the performance of a particular university, we need to examine its own mission statement and the demand for its alumni on the labour market among other things. Such complex phenomena cannot be presented using only one indicator; rather, a whole system of indicators or a complex indicator ought to be created and used.

4 An Overview of the Regional Competitiveness Index

(RCI, Pillars, Index Construction)

To improve the understanding of territorial competitiveness at the regional level, the European Commission has developed the Regional Competitiveness Index—RCI—that shows the strengths and weaknesses of each of the EU NUTS 2 regions. We are going to discuss the Index on the basis of EU Regional Competitiveness Index RCI 2013 in JRC Scientific and Policy Reports Angrist and Pischke (2015).

4.1 Changes in the RCI Over Time

The development of the index started in 2008 and has built on the methodology developed by the World Economic Forum for the Global Competitiveness Index. It covers a wide range of issues related to territorial competitiveness including innovation, quality of institutions, infrastructure and measures of health and human capital.

In 2010, a joint project led to the publication of the RCI 2010, the first composite in Europe aiming at mapping the economic performance and competitiveness of regions. Results showed great variation within each country, with under-competitive regions scattered all around strong regions. In this respect, the national level is not assumed to make the real difference in terms of competitiveness.

The project provided a method to benchmark regional competitiveness and to identify the key factors which would allow a low competitive region to catch-up. RCI can be considered as an overall but synthetic picture of regional competitiveness.

RCI 2013 is the second edition of the index and includes updated and more data together with method refinements.

RCI can provide a guide to what each region should focus on, taking into account its specific situation and its overall level of development. In this perspective, RCI may play a critical role in the debate on the future of cohesion policy.

RCI 2013 was based on a set of 80 candidate indicators of which 73 have been eventually included in the index.

The results demonstrate that territorial competitiveness in the EU has a strong regional dimension, which national level analysis does not properly capture in the EU. The gap and variation in regional competitiveness should stimulate a debate as to what extent these gaps are harmful for their national competitiveness and to what extent the internal variation can be remediated.

4.2 Composition of the Current RCI

RCI 2013 has basically the same framework and structure as the 2010 edition and includes most recent data for all the indicators. As for the previous version, the index is based on eleven pillars describing both inputs and outputs of territorial competitiveness, grouped into three sets describing basic, efficiency and innovative factors of competitiveness.

The basic pillars represent the basic drivers of all economies. They include (1) Quality of Institutions, (2) Macro-economic Stability, (3) Infrastructure, (4) Health and the (5) Quality of Primary and Secondary Education. These pillars are most important for less developed regions.

The efficiency pillars are (6) Higher Education and Lifelong Learning, (7) Labour Market Efficiency and (8) Market Size.

The innovation pillars, which are particularly important for the most advanced regional economies, include (9) Technological Readiness, (10) Business Sophistication and (11) Innovation. This group plays a more important role for intermediate and especially for highly developed regions. Overall, the RCI framework is designed to capture short- and long-term capabilities of the regions.

When the Index was constructed, some regions were merged with surrounding ones to correct for commuting patterns following the new OECD-EC city definition. For example, more capital regions were merged with their surrounding regions: Wien (AT), Brussels (BE), Praha (CZ), Berlin (DE), Amsterdam (NL) and London (UK).

Candidate indicators are mainly selected from Eurostat with some additional official sources, such as the World Economic Forum. These indicators have been standardized and, using Principal Component Analysis, artificial variables have been created.

A score is computed for each pillar as simple average of the z-score standardized and/or transformed indicators. Sub-indexes for the basic, efficiency and innovation groups of pillars are computed as arithmetic means of pillar scores. The overall RCI score is the result of a weighted aggregation of the three sub-indexes. We standardized the published data for cluster analysis and calculated the Z-scores. We used SPSS software to do the calculations.

We have examined the Efficiency sub-index with a view to formulating homogenous groups from the regions. The Pillars included in the Efficiency sub-index are in Table 1.

Table 1 Name and description of indicators

Name	Description	Pillar
Population 25–64 with higher education	Population aged 25–64 with higher educational attainment (ISCED5_6) % of total population of age group	Higher education
Lifelong learning	Participation of adults aged 25–64 in education and training, % of population aged 25–64	Higher education
Accessibility to universities	Population living at more than 60 min from the nearest university, % of total population	Higher education
Employment rate (excluding agriculture)	% of population 15–64 years	Labour market
Long-term unemployment	% of labour force	Labour market
Unemployment	% of active population	Labour market
Labour productivity	GDP/person employed in industry and services (€) Index, EU27 = 100	Labour market
Gender balance unemployment	ABSOLUTE difference between female and male unemployment rates	Labour market
Female unemployment	% of female unemployed	Labour market
Share of population aged 15–24 not in education, employment or training (NEET)	% of population aged 15–24	Labour market
Disposable income per capita	Gross adjusted disposable household income in PPCS per capita index	Market size
Potential GDP in PPS	Potential market size expressed in GDP (pps), index EU28 = 100	Market size
Potential POP	Potential market size expressed in population, index EU28 = 100	Market size

5 The Results of the Cluster Analysis

5.1 The Characteristics of Clusters

After standardization, we have used the published Efficiency pillar scores (Higher Education and Lifelong Learning; Labour Market Efficiency and Market Size) for a cluster analysis. Data were not fully available for 4 out of the 262 regions so those four regions have been left out of the cluster analysis.

The first step in a cluster analysis is to determine the number of clusters. In order to do that we used hierarchical clustering, the end result of which gave a recommendation of one of 5-7-10 clusters. For reasons of manageability and having more balanced cluster sizes, we have decided to use five clusters. We have calculated EESS%—one of the cluster quality coefficient measures recommended by Vargha et al. (2016)—for all the three cluster structures and it showed the highest value (75.7%) at the 5-cluster solution.

We have performed our cluster analysis using K-means clustering. The cluster method used was between-groups linkage with squared Euclidean distance.

Cluster 4 in Table 2 shows the regions that have the most beneficial position within the five clusters as they have the highest scores in Higher Education, Labour Market and Market Size. These are the regions that have the largest market potential and can be regarded as well positioned in the other variables as well. All in all, 51 regions make up this cluster.

Cluster 5 includes the regions that have a better-than-average Higher Education score (0.68786) and Labour market (0.79666) and around average Market Size (0.07966) scores. These regions are in a similar position as the ones in Cluster 4, but the size of the market is smaller, around average size. The number of regions in this cluster is 70.

Regions of a worse than average position belong to the other three clusters. The best of these regions can be found in Cluster 3 (71 regions) with all the three factors being slightly under-average.

The members of Cluster 1 (22 regions) and Cluster 2 (44 regions) are very much below average as regards all the three factors. The biggest difference between Cluster 1 and Cluster 2 can be seen in the Labour Market factor. The members of Cluster 1 are in the worst position since they have the lowest scores in both the Higher Education (−1, 44557) and Labour Market (−1, 97820) factors.

Variance analysis proves that in the case of clusters, there is a significant difference in the expected values of variables. Table 3 shows that the Market Size

Table 2 Final cluster centres

	1	2	3	4	5
HE scores	−1.44557	−1.10819	−0.25790	0.99458	0.68786
Labour market	−1.97820	−0.71304	−0.16317	0.76643	0.79666
Market size	−0.94794	−1.14655	−0.14239	1.53838	0.07966

Table 3 ANOVA table

Factors		Sum of squares	Degree of freedom	Mean square	F	Significant
HE	Between groups	188.300	4	47.075	173.362	0.000
	Within groups	68.700	253	0.272		
	Total	257.000	257			
Labour market	Between groups	184.466	4	46.116	200.502	0.000
	Within groups	58.191	253	0.230		
	Total	242.657	257			
Market size	Between groups	200.164	4	50.041	222.247	0.000
	Within groups	56.965	253	0.225		
	Total	257.129	257			

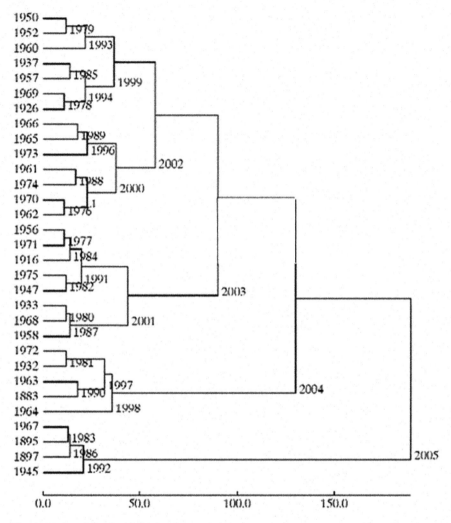

Fig. 1 Market size values for the regions

variable has the largest F-value, which is to say that this factor has the most important role in distinguishing clusters.

We have used Fig. 1 to show the Market Size scores of individual regions. The various geometric shapes indicate which cluster the particular region belongs to. The graph highlights the fact that members in Clusters 4 and 5 are in the best position as they have the highest value Market Size factor. It is also interesting to note that regions with similar Market Size values may belong to different clusters as the value of their other factors may be very different.

5.2   Homogeneity

We have examined how many regions in every country belong to a cluster. Belgium, for example, has six regions in Cluster 4, and two in Cluster 5, which indicates the country's very good position in the examined areas. Greece, on the other hand, has twelve regions in Cluster 1 and only one in Cluster 3, which is a clear sign of Greece's bad position in comparison with other European countries. Hungary is also having problems: four of its regions are in Cluster 2 and three in Cluster 3, which places the country near the bottom of the list (Table 4).

Table 4 Number of national regions in the individual clusters

Country	Cluster 1 Worst position	Cluster 2 Smallest market size	Cluster 3 Close to average	Cluster 4 Best position	Cluster 5 Good pos. with smaller market	Total
Austria					8	8
Belgium			1	6	2	9
Bulgaria	1	5				6
Cyprus			1			1
Czech Republic			6		1	7
Germany			1	16	20	37
Denmark					5	5
Estonia		1				1
Spain	5	2	9	1	2	19
Finland					4	4
France		1	18	1	2	22
Greece	12		1			13
Croatia		2				2
Hungary		4	3			7
Ireland			2			2
Italy	4	2	15			21
Lithuania		1				1
Luxembourg				1		1
Latvia		1				1
Malta		1				1
Netherlands				8	3	11
Poland		11	5			16
Portugal		4	3			7
Romania		7			1	8
Sweden					8	8
Slovenia			1		1	2
Slovakia		2	1		1	4
United Kingdom			4	18	12	34
Total	22	44	71	51	70	258

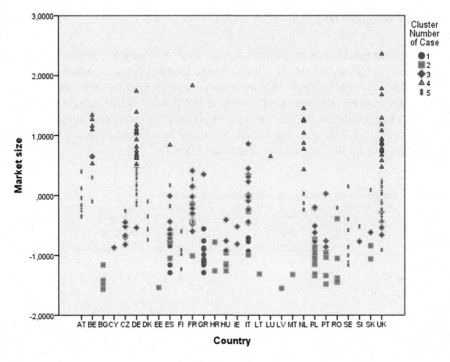

Fig. 2 Market size by region and country

From the point of view of cluster analysis, countries with all their regions falling into one category are considered homogenous.

Countries with up to three regions are considered small. Examples are Estonia, Cyprus, Croatia, Ireland, Lithuania, Luxembourg, Latvia, Malta and Slovenia, and only Croatia (2 regions) may be called homogenous. Slovenia's two regions are not in the same cluster: one is in Cluster 3 (average position), the other is in Cluster 5 (good position).

Countries with 4–10 regions are medium sized. Austria, Denmark, Finland and Sweden are homogenous as there is no significant difference between their regions. In other medium-sized countries differences can be observed between the regions, for example, Portugal's four regions are in Cluster 2 and three in the around-average Cluster 3, which is very similar to Hungary's situation.

Among countries with 10–20 regions, Spain can be considered rather heterogeneous: at least one of Spain's 19 regions appears in all the clusters.

Countries with more than 20 regions (Germany, France, the UK, Italy) are considered large. France is the only one of these countries that has regions in as many as four clusters, the regions of the others fall into three clusters.

We have shown the position of the various regions of each country according to the factors of Market Size, Labour Market and Higher Education (Fig. 2).

To which cluster a region belongs is determined by the market size factor. By far the largest market is the UK00 (Bedfordshire and Hertfordshire + Essex + Inner London + Outer London). The graph shows which clusters the regions of a country belong to and the clusters are represented by the same shapes as in Fig. 1.

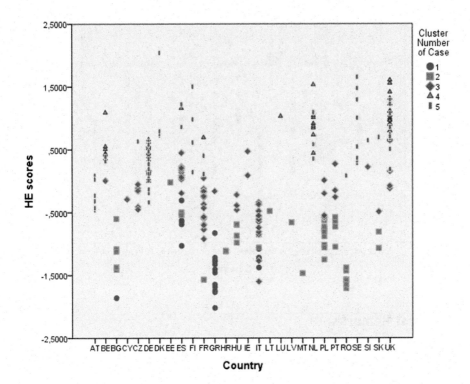

In the case of the Higher Education factor the position of a region in the cluster structure is a bit less unambiguous. (From the point of view of clustering this factor has the least important role as its F-value in Table 3 has the lowest value.) We can see regions which have similar values but belong to different clusters, the reason for which is that we have used not one but three factors creating the clusters. Denmark, for example, has been categorized as a homogenous country; however, if we consider only one factor, namely Higher Education, we can observe quite significant differences between the regions. If we consider all the factors, however, the differences between the regions are not sizeable enough to place them in different clusters.

Examining the Labour Market factor, similar conclusions can be drawn. It is easy to see in which countries there are significant differences between regions concerning either factor values or belonging to a particular cluster (e.g. Spain or Italy).

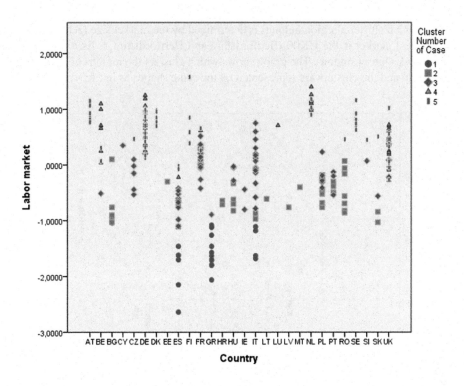

6   Final Remarks

The results demonstrate that territorial effectivity in the EU has a strong regional dimension. Regions within a country may show greater differences in development than the countries they belong to. In this way the regions of a country may fall in different clusters. This information can be highly important when the distribution of various EU development funds is considered.

As we have shown, the Market Size factor has the most important role in differentiating clusters—and Higher Education the least important. That needs to be taken into consideration when assessing the development potential of a particular region.

As we have seen, there are regionally homogenous and heterogeneous countries. Austria, Denmark, Finland and Sweden can be considered homogenous: their regions are all in Cluster 5, obviously, they are all very highly developed economically.

A further question to examine may be whether there is a significant difference in competitiveness between regions belonging to different clusters. That is to say, can we draw conclusions as to the competitiveness of a region on the basis of just a few factors?

Another question is which factor has the most important role in measuring competitiveness. As various factors may be more or less important in making up competitiveness, we may need to find the most relevant factors.

References

Angrist, J.D., Pischke, J.-S.: Mastering Metrics, the Path from Cause to Effect. Princeton University Press, Princeton (2015). ISBN: 978-0-691-15283-7

Vargha, A., Bergman, L.R., Takács, S.: Performing cluster analysis within a person-oriented context: Some methods for evaluating the quality of cluster solutions. J. Person-Oriented Res. **2**(1–2), 78–86 (2016). https://www.researchgate.net/profile/Andras_Vargha/publication/301622365_Performing_Cluster_Analysis_Within_a_Person-Oriented_Context_Some_Methods_for_Evaluating_the_Quality_of_Cluster_Solutions/linksV57233ac808aef9c00b810aba/Performing-Cluster-Analysis-Within-a-Person-Oriented-Context-Some-Methods-for-Evaluating-the-Quality-of-Cluster-Solutions.pdf

Estimation Methods Based on Weighting Clusters

Roland Szilágyi, Beatrix Varga, and Renáta Géczi-Papp

Abstract Research based on samples and their conclusions play an increasing role in making business decisions and also in creating information. The study gives an overview of the different estimation methods that can be used for the description of socio-economic relations. In addition, we searched for solution variants for handling of mistakes, based on different distributions. Identifying tendencies plays a significant role in eliminating the bias caused by failed assumptions. It is worth examining the differences between tendencies of different groups. In order to achieve a successful procedure the samples should be grouped based on variables which are in stochastic relation with the examined criterion. The tendencies must be examined with the help of simulations and modeled according to this. That was the reason why we created an estimating model of weighted tendencies, in which the estimated values of the above-mentioned tendencies were defined as average estimated values by weighting the explanatory features of functions.

1 Introduction

Due to alterations in market structure, the rapid changes in the environment and the development of information technology, the time necessary for preparing a decision has significantly decreased. The decision-makers of business life can only compete effectively with both time and competitors if they constantly improve the techniques used during the preparation of decisions. There is a wide choice of quantitative analytic methods for the inspection of the characteristic features of individuals. However, choosing and applying the adequate quantitative method is not the only key to success. The other important factor must be the reliability of the data used to obtain information. It is known from the specialized literature that

R. Szilágyi (✉) · B. Varga · R. Géczi-Papp
University of Miskolc, Institute of Economic Theory and Methodology, Miskolc-Egyetemváros, Hungary
e-mail: strolsz@uni-miskolc.hu; stbea@uni-miskolc.hu; stgpren@uni-miskolc.hu

© Springer Nature Singapore Pte Ltd. 2020
T. Imaizumi et al. (eds.), *Advanced Studies in Classification and Data Science*,
Studies in Classification, Data Analysis, and Knowledge Organization,
https://doi.org/10.1007/978-981-15-3311-2_9

valid results, those producing and publishing relatively few errors resulting from surveys based on sampling, can only be guaranteed if there are adequate databases. It is not uncommon to use auxiliary information, which is mainly used to determine the sampling error, to increase the response ratio, and to reveal bias (see Estevao and Särndal 2002; Roy 2006). The auxiliary information necessary for correcting the results of surveys based on sampling is widely dispersed and can be found in the data collections of different offices and organizations (though it may not always be accessible). In practice, however, companies and researchers often do not have access to the necessary auxiliary information for financial or technical reasons, so they use as much internal information as possible. For this reason, other methods are used that are not able to meet the general estimation criteria, like computational cost, unbiasedness, efficiency, or asymptotic properties (Kennedy 1998). The aim of the research is to work out an analytic system which enables minimization of mistakes and the reduction of bias. The method was created in a generalized way, so that this process could be of help to both individual researchers in the social sciences and to research organizations. It can be utilized not only on a macro-level but even on the level of small-size enterprises. Such enterprises typically do not have access to error calculating software and algorithms, unlike the representatives of official statistical organizations. Without adequate methods they are not able to produce quality information. The methodological results of this empirical research aim at filling this gap.

During the analysis, we offer a solution for those who do not have the possibility of using auxiliary information. For this purpose, partial samples were created from the available population data by simple random sampling. For each subsample, several independent cluster analyses were performed, and the results of the cluster analyses obtained were weighted according to their homogeneity using the weight system we developed. Finally, we calculated the results of the different cluster procedures to the extent of the weighting, thus improving the estimation procedure. The structure of the study is as follows: the second section describes the related literature; the third section details the database and the created estimation procedure; in the fourth section empirical results are presented; the study ends with the conclusions in the fifth section.

2 Related Literature

The literature related to the statistical inference and sampling is quite wide. In practice, it is rare to know the population data, because of financial, methodological, and other reasons. Companies and researchers therefore use different sampling methods in many cases and try to correctly (or incorrectly) describe the characteristics of the population on the basis of samples. Since the paper's goal is the introduction of a new estimation process, the different sampling techniques are not presented. Many estimation procedures are known, and because of the IT development their application is becoming simpler. Here we introduce basic estimation procedures closely related to the purpose of the study. The described methods are valid for a finite population. The following notations were used in the equations: N—population size; n—sample size; $(y_1, y_2, \ldots, y_N)$ vector is a realization of N independent, randomly distributed random variable; μ—expected value of the population; σ^2—population variance.

2.1 Independent, Identically Distributed Random Variables, Simple Sampling Without Replacement

If we have independent, identically distributed random variables (μ, σ_y^2) and the sampling method is simple, without replacement, then the sample mean is an unbiased estimate of the expected value of the population. In this case, the following formula can be used (Fuller 2009):

This means that

$$E\{(\sigma_{\bar{y}}^2 - \mu)^2\} = n^{-1}\sigma_y^2$$

is the best estimation, which minimizes the mean square error (MSE).

2.2 Stratified Sample

The characteristics of the population do not change in the case of stratified sample. (The description of stratification procedures exceeds the scope of this study.) Basically, it is advisable to use stratified sampling for a heterogeneous population. For the same sample size, with appropriate stratification, the expected value and the standard deviation are obtained with a smaller error than for simple sampling (Bunce et al. 1983). The following formula contains an estimate in the case of stratified sampling where H represents the number of groups (Fuller 2009):

$$\bar{y}_{st} = \sum_{h=1}^{H} W_h \bar{y}_h, \text{ where } W_h = N^{-1} N_h$$

3 Estimation Methods Based on Weighting Clusters

In cases where no auxiliary information is available, the researcher may choose from two options: use internal (sample) information or apply typical trends of clusters and project the pattern to the entire population. In the present study, the latter option was selected.

3.1 Dataset

The database of the Household Budget Survey (HBS) in 2005 was used for the analysis, provided by the Hungarian Central Statistical Office, revised by Eurostat. The size of the population is 9058, with 42 different stock and flow variables. Subsequently, 15 more variables were created based on existing variables (per capita or per employee). There are both quantitative and qualitative; categorical and continuous; nominal, ordinal and ratio scales variables. This variety ensures that the analysis is implemented as fully as possible. Observation units of the database were included in the analysis without weighting, as individuals. Since the data comes from a population survey, many variables can be considered with high missing values. During the selection of the variables included in the analysis, the ratio of the missing values also played an important role. In the specification of model the continuous variables with few missing values were preferred. We proposed to estimate the household's average net income (net income means the total income from all sources including non-monetary components minus income taxes) in the population. For clustering we used the following variables:

- Age (in completed years) of reference person;
- Useful living area in (m^2) (principal residence);
- Number of cars;
- Total consumption expenditure;
- Consumption on food and non-alcoholic beverages;
- Consumption on clothing and footwear;
- Consumption on housing, water, electricity, gas, and other fuels;
- Consumption on furnishings, household equipment, and routine maintenance of the house;
- Consumption on communication;
- Sum of household members.

We took 10 different samples, each consisting of 900 households, according to the rules of simple random sampling. With the help of these samples we made a total of 100 classifications where different numbers of clusters were created. In order to ensure comparability, 16 different stratified samples were used to evaluate the results where the stratifying variables were the region, density, and sex.

3.2 Methodology

In cases where it is not possible to take a stratified sample, we need to estimate the parameters of the heterogeneous population based on a sample. The estimation method based on weighting clusters is capable of drawing conclusions, even in the absence of auxiliary information. Simulation studies performed with the HBS database proved to be useful in three ways. First of all, because the original database provided the population's expected value—estimating this effectively and accurately was the fundamental aim. Secondly, the procedure takes advantage of cluster analysis. Finally, the results of classification are used not only partially, but for describing the entire database. The stages of the weighted estimation based on cluster analysis are the following:

1. Creating subsamples, using simple random sampling.
2. For each subsample, making different cluster analyses using different variables for the classification in each case.
3. Ranking the solutions of classifications based on their homogeneity.
4. Creating a suitable system of weights.
5. Estimating points.
6. Estimating confidence intervals.

3.2.1 Cluster Analysis

It should be taken into account that cluster analysis is primarily an exploratory method and that there is no single best solution. The development of clusters depends on the chosen methods, distance calculation methods, and variables involved in the analysis (Hair et al. 2010). From among the agglomerative hierarchical classification methods, the Ward method was chosen. In this method, the cluster to be merged is the one which will minimize the increase in within-group variance (Ward 1963). During the formation of groups, general statistical knowledge has to be kept in mind. We know that having too few groups does not assist in effective analysis, while the formation of too many groups leads to the results of analysis of data without grouping. The clusters were created according to the following criteria: the grouping variables are uncorrelated; the database does not contain outliers; the grouping variables should be continuous, and categorical variables should be avoided. However, in order to illustrate the effects of ignoring the criteria, some clusters were allowed to keep outliers and to include categorical variables. If we measure n observations in criteria/variables m, this can be described with a matrix of $n \times m$:

$$X = [x_{ij}] = \begin{bmatrix} x_{11} & x_{12} & x_{1m} \\ x_{21} & x_{22} & x_{2m} \\ x_{n1} & x_{n2} & x_{nm} \end{bmatrix},$$

where $i = 1, 2, \ldots, n$ and $j = 1, 2, \ldots, m$. Each row of the matrix represents an observation; x_{ij} is the value of variable j for observation i.

In this way the clusters are

$$G = \{G_1, G_2, \ldots, G_k, \ldots, G_l\},$$

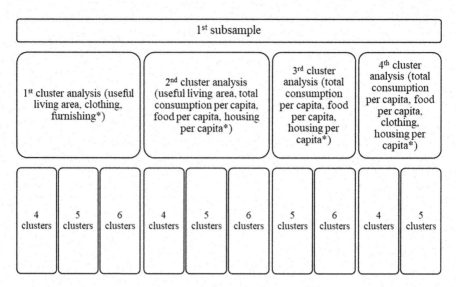

Fig. 1 The result of the different cluster analysis in the first subsample. *Clusters were created based on the variables in parentheses. Source: Own compilation

where k is the id of the cluster (where: $k = 1, 2, \ldots, l$) and l means the total number of clusters, n_k is the number of observations/elements of the kth cluster (where $n_k = 1, 2, \ldots, u$). By clustering, we mean the division of the set X into subgroups to which the following properties must be fulfilled: for every $x_i : x_i \in G_k$ i.e. all elements belong to a cluster; $G_k \cap G_l = \emptyset$ if $k \neq l$, i.e. all elements belong to only one cluster, and clusters are mutually disjoint sets; $n_k \neq 0$, i.e. there are no clusters having no elements; $\cup_{k=1}^{l} G_k = X$ together the clusters cover all elements. During the simulation, ten subsample and several different cluster analyses was performed (according to the cluster procedure assumptions) using different groups of variables as clustering criteria. At the end of the process, 100 classifications were finally obtained from the 10 subsamples. Each of the classification procedures divided the samples into 4 to 6 clusters, depending on the result of cluster analyses (Varga and Szilágyi 2011). The analyses were carried out using the SPSS (Statistical Package for the Social Sciences) software. Figure 1 illustrates the process and Table 1 helps to understand the process.

3.2.2 Ranking Classifications

The determinant of the covariance matrix of the grouping variable is the generalized variance of the given cluster: $\det(C_k)$. This is used to describe the homogeneity of the clusters. The larger the generalized variance, the more heterogeneous the clusters are. Cluster analysis can be considered successful when the homogeneity within a group is high and the distance among groups is also high. In this paper we introduce

Table 1 Number of cluster analyses by subsamples

Subsample	Number of classifications
1	10
2	14
3	10
4	7
5	9
6	10
7	10
8	10
9	10
10	10
Total	100

Source: Own compilation

an indicator that can characterize the homogeneity of the clusters, which helps in ranking the clusters. For this indicator we use ω.

$$\omega_k = 1 - \frac{\det(C_k)}{\sum_{k=1}^{l} \det(C_k)}.$$

A low value of ω indicates that the cluster is more heterogenous and a higher value indicates more homogeneity. The advantage is that ω it is independent of the measurement and scale of original variables; it is a relative indicator which makes it easier to compare the homogeneity of clusters. Based on the above formula, the clusters became comparable for each of the 100 classifications. Later on this feature is used in the weight system design. Evaluation of homogeneity allows us to take the "quality" of the classifications into account during the weighting process so that it is incorporated in the estimation process.

3.2.3 Developing a Weight System

The design of the weight system has a great importance in the estimation process. If weights are poorly defined, results will be biased. In order to benefit from the cluster analysis in the estimation process, it is important to consider the cluster properties. Determination of the weight system was based on the following criteria: higher weight was given to clusters that had more elements and that were more homogeneous; clusters with higher dispersion in the target variable (estimated variable) were also given higher weight. The formula based on these considerations is the following:

$$W_k = \frac{n_k \cdot s_k^2 \cdot \omega_k}{\sum_{k=1}^{l} n_k \cdot s_k^2 \cdot \omega_k},$$

where $\sum_{k=1}^{l} W_k = 1$; s_k is the variance of the target variable (estimated variable) in the kth cluster.

4 Result

In the study 10 randomly selected random samples of 900 elements were generated from the 9058 item database of the Hungarian Central Statistical Office's Household Budget Survey. Based on these samples, a total of 100 different cluster classifications were made; in each case the number of clusters was between 4 and 6. Point estimates were prepared based on the methodology described in the previous section. In order to illustrate the consequences of ignoring the assumptions, so-called test clusters were also included among the clusters (raw cluster based on Fig. 2). In these cases, the expectations concerning outliers or the type of variables included in the survey (avoiding categorical variables) are not met. As we assumed would happen, estimates from these clusters significantly differed from the expected value. From the population data, it was also possible to determine the expected value (mean) of the population. In the present case, the value of the net annual income in the population is HUF 2,114,213. In addition to the estimation based on weighting clusters, simple point estimates for clusters based on simple random samples were also determined to ensure comparability. Since the main purpose of the method was to provide an alternative when stratified sampling is not possible, it is also advisable to compare the results with the stratified samples. There were 16 different stratifications based on different criteria and we got 900-element samples. From these samples we estimated the expected net income. We illustrate the results graphically, as this helps to understand and process the data better (Kovács and Kriszt Sándorné 2016). We can compare the results of the estimation based on four different aspects altogether. The results are summarized in Fig. 2.

Figure 2 shows the differences of the point estimates and the expected value for annual household net income in Hungary. As you can see, estimates are randomly dispersed around the expected value. We did not find systematic bias. The unusual clusters in raw cluster based category can be traced back to different causes: the involved variables have a significant amount of missing data; outlier values were not filtered out during the analysis; categorical and qualitative variables were used in the model. Compared to estimation of simple random sampling, it can be seen that the range of the method now introduced is smaller. However, the results are less asymmetric than in the case of simple random samples. Compared to stratified sampling, the range of estimation based on clustering is larger. Based on the results,

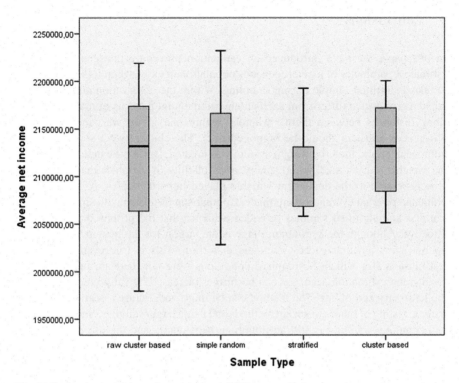

Fig. 2 Comparison of point estimates in case of different methods

the method offers an effective, unbiased alternative in cases where no stratified sample is available to perform the analysis on. There are several advantages to our method that make it suitable for situations such as corporate data analysis. When the method is used with a simple random sample, it acts as a kind of post-stratification. The bias of outliers is decreased. Information that is internal to the sample is used effectively, and external information is not necessary. Just like any analytical method, the new procedure also has limitations. This method is suitable for a relatively large sample size and when categorical variables are not used. A complete data set is also advantageous, since a missing item leads to the entire element being filtered out, which reduces sample size. Improving the method (maybe with the contribution of institutions of higher education outside Hungary) can provide wider use. International cooperation provides great opportunities and can bring significant results.

5 Conclusions

In this paper we have introduced an estimation procedure that can be used to estimate a parameter of a heterogeneous population in cases when it is not possible to take a stratified sample during sampling. Within the observation units, we have created groups that differ from each another in the joint patterns of many variables (long distances between them), whereas within one group we cannot find big differences (clusters should be homogeneous). The clusters were sorted by their homogeneity and then the weight system was created. During the determination of the weights, cluster size, homogeneity, and volatility of variables that were more closely related to the dependent variable played important roles. As a final step, point and interval estimates were made. Comparison of results with simple random samples and stratified samples provided evidence that the proposed procedure is effective. Although the estimation range is the largest for the new method, values do not show asymmetry or systematic distortion. This is true even though the test clusters (for which the required conditions were not met) are present in the results; the estimation accuracy can be further increased by fully complying with the necessary conditions. The method can be improved. Future research directions include testing of other clustering methods and using larger samples. Simple random and stratified sampling is still an important benchmark, and the use of additional sampling methods is not justified, as our experience is that more complicated methods are not used in corporate research. The estimation method based on weighting clusters can serve as an effective alternative for companies when it is not possible to include auxiliary information or stratified sampling.

Acknowledgments Renáta Géczi-Papp would like to thank the Pallas Athéné Domus Scientiae Foundation for its support.

References

Bunce, R.G.H., Barr, C.J., Whittaker, H.A.: A stratification system for ecological sampling. In: Fuller, R.M. (ed.) Ecological Mapping from Ground, Air and Space, pp. 39–46. Institute of Terrestrial Ecology, Cambridge (1983)

Estevao, V.M., Särndal, C.E.: The ten cases of auxiliary information for calibration in two-phase sampling. J. Off. Stat. **18**(2), 233–255 (2002)

Fuller, W.A.: Sampling Statistics. John Wiley & Sons Inc., Hoboken (2009)

Hair, J.F., Black, W.C., Babin, B.J., Anderson, R.E.: Multivariate Data Analysis, 7th edn. Pearson Prentice Hall, Upper Saddle River (2010)

Kennedy, P.: A Guide to Econometrics. The MIT Press, Cambridge (1998)

Kovács, P., Kriszt Sándorné, É.: Hungarian visualization tools on official social-economic data in classes. In: Engel, J. (ed.) Promoting Understanding of Statistics About Society: Engaging Social Data with Technology: Teaching and Visualizing Aggregated Data and Microdata, Berlin, pp. 1–6 (2016)

Roy, D., Safiquzzaman, M.D.: Variance estimation by Jackknife method under two-phase complex survey design. J. Off. Stat. **22**(1), 35–51 (2006)

Varga, B., Szilágyi, R.: Quantitative Information Forming Methods (2011). https://www.tankonyvtar.hu/hu/tartalom/tamop425/0049_08_quantitative_information_forming_methods/6126/index.html

Ward, J.H.: Hierarchical grouping to optimize an objective function. J. Am. Stat. Assoc. **58**(301), 236–244 (1963)

Part II
Multidimensional Analysis and Visualization

Five Strategies for Accommodating Overdispersion in Simple Correspondence Analysis

Eric J. Beh and Rosaria Lombardo

Abstract Traditionally, simple correspondence analysis applied to a two-way contingency table is performed by decomposing a matrix of standardised residuals using singular value decomposition where the sum-of-squares of these residuals gives Pearson's chi-squared statistic. Such residuals, which are treated as being asymptotically normally distributed, arise by assuming that the cell frequencies of the table are Poisson random variables so that their expectation and variance are equivalent. However there is clear evidence in the statistics literature that suggests that the variance of these residuals is less than one. Thus, we observe overdispersion in the table. Various strategies can be undertaken to study, and deal with, overdispersion. In this paper we shall briefly review five possible strategies. Although we conceed that the purpose of this paper is not to provide a comprehensive examination of their utility—future investigations of their properties will confirm any further benefits in their use.

1 Introduction

Correspondence analysis is a very popular approach for visualising the association between two or more categorical variables. There is a plethora of books, articles and reports that describe the practical and technical features of correspondence analysis. One may consider, for example, the texts of Beh and Lombardo (2014), Greenacre (1984) and Lebart et al. (1984) for a comprehensive description of the

E. J. Beh (✉)
School of Mathematical & Physical Sciences, University of Newcastle, Callaghan, NSW, Australia
e-mail: eric.beh@newcastle.edu.au

R. Lombardo
Department of Economics, University of Campania "Vanvitelli", via Gran Priorato di Malta, Capua (CE), Italy
e-mail: rosaria.lombardo@unicampania.it

© Springer Nature Singapore Pte Ltd. 2020   117
T. Imaizumi et al. (eds.), *Advanced Studies in Classification and Data Science*,
Studies in Classification, Data Analysis, and Knowledge Organization,
https://doi.org/10.1007/978-981-15-3311-2_10

range of issues that a correspondence analyst is confronted with. There are a number of starting points from which one may consider correspondence analysis but the most common approach involves the singular value decomposition (SVD) of the standardisation of the cell frequencies of a contingency table. The standardisation rests on the assumption that the cell frequencies are random variables from a Poisson distribution so that their expectation and variance are identical. However, in many situations, such an equality is rarely observed. Instead, the data may either exhibit characteristics that are consistent with overdispersion (variance > expectation) or, less likely, underdispersion (variance < expectation). These features are not confined to just the Poisson distribution, but also apply to all single parameter distributions including the binomial and exponential distributions (Cox 1983). The possibility of over/underdispersion in a contingency table has implications for how one performs simple correspondence analysis. As we shall discuss, overdispersion is much more of a concern than underdispersion for contingency table analysis and so we shall confine our attention to overdispersed cross-classified categorical data. In doing so, we will briefly explore five strategies for accommodating overdispersion in a contingency table. This shall be achieved by first providing a brief overview of some of the fundamental features of simple correspondence analysis; see Sect. 2. Section 3 provides a glimpse at the problem of overdispersion in contingency tables as described by Agresti (2013) and Haberman (1973). The five strategies that we consider here can be divided between those involving variants of the Poisson distribution (which accommodate for overdispersion by the presence of an additional, dispersion, parameter—see Sect. 4) and variance stabilising approaches (Sect. 5). The variance stabilising approaches we discuss include the SVD of what we term the *Freeman–Tukey* residual (see Beh et al. 2018 for a comprehensive discussion of this approach) and the adjusted (Pearson) residual proposed by Haberman (1973); a comprehensive description of this particular approach to accommodating overdispersion in the correspondence analysis of a two-way table was given by Beh (2012).

We must acknowledge that the discussion of overdispersion for the Poisson distribution is not new. For example, Greenwood and Yule (1920) and Satterthwaite (1942) provide early discussions of this issue which led to an evolution of numerous variants and extensions of the Poisson distribution. What has not been adequately discussed, however, is how one may take advantage of these variants and undertake a correspondence analysis of a two-way contingency table when the cross-classified data exhibit overdispersion. There has been limited attention paid to this aspect of correspondence analysis and the purpose of this paper is NOT to provide a comprehensive discussion of overdispersion in correspondence analysis. Nor is our intention to definitively state a preference of one strategy over another. Instead, the aim of this paper is to discuss possible generalisations and alternatives of the Poisson distribution to accommodate overdispersion in contingency tables. A detailed examination of each of these strategies will be left for future consideration.

2   A Brief Overview of Correspondence Analysis

2.1   Notation

Suppose we have two categorical variables that are cross-classified to form an $I \times J$ two-way contingency table, $\mathbf{N}$, where the (i, j)th cell entry is denoted by n_{ij} for $i = 1, 2, \ldots, I$ and $j = 1, 2, \ldots, J$. Let the grand total of $\mathbf{N}$ be n and let the matrix of relative frequencies be $\mathbf{P}$ so that the (i, j)th cell entry is $p_{ij} = n_{ij}/n$ where $\sum_{i=1}^{I} \sum_{j=1}^{J} p_{ij} = 1$. Define the ith row marginal proportion by $p_{i\bullet} = \sum_{j=1}^{J} p_{ij}$. Similarly, define the jth column marginal proportion as $p_{\bullet j} = \sum_{i=1}^{I} p_{ij}$.

To determine whether there exists a statistically significant association between the row and column variables, one may calculate Pearson's chi-squared statistic

$$X^2 = \sum_{i=1}^{I} \sum_{j=1}^{J} \frac{\left(n_{ij} - n_{i\bullet}n_{\bullet j}/n\right)^2}{n_{i\bullet}n_{\bullet j}/n} = n \sum_{i=1}^{I} \sum_{j=1}^{J} \frac{\left(p_{ij} - p_{i\bullet}p_{\bullet j}\right)^2}{p_{i\bullet}p_{\bullet j}}.$$

Given a level of significance, α, the statistical significance of the association may be tested by comparing X^2 with the $1 - \alpha$ percentile of the chi-squared distribution with $(I - 1)(J - 1)$ degrees of freedom.

2.2   Pearson's Residuals

When it is found that a statistically significant association exists between two categorical variables, a visual inspection of this association can be made using correspondence analysis; see, for example, Beh and Lombardo (2014), Greenacre (1984) and Lebart et al. (1984). Central to correspondence analysis is the partition of Pearson's chi-squared statistic. This is achieved by applying the singular value decomposition (SVD) to Pearson's residuals, r_{ij}^P, such that

$$r_{ij}^P = \frac{p_{ij} - p_{i\bullet}p_{\bullet j}}{\sqrt{p_{i\bullet}p_{\bullet j}}} = \sum_{m=1}^{M} a_{im}\lambda_m b_{jm},$$

where $M = \min(I, J) - 1$. Here a_{im} is the mth element of the singular vector associated with the ith row category. Similarly, b_{jm} is the mth element of the singular vector associate with the jth column category. These elements have the property

$$\sum_{i=1}^{I} a_{im}a_{im'} = \begin{cases} 1 & m = m' \\ 0 & m \neq m' \end{cases} \qquad \sum_{j=1}^{J} b_{jm}b_{jm'} = \begin{cases} 1 & m = m' \\ 0 & m \neq m' \end{cases}.$$

The values of a_{im} and b_{jm} are commonly viewed as scores that provide insight into those rows and columns that play an important role in defining the association structure between the variables. They are also sometimes used as *standard* coordinates for visualising the association. Also, λ_m is the mth singular value of the Pearson residuals so that

$$\lambda_m = \sum_{i=1}^{I} \sum_{j=1}^{J} p_{ij} a_{im} b_{jm},$$

where $1 > \lambda_1 > \lambda_2 > \ldots, > \lambda_M > 0$. Through the SVD of r_{ij}^P, a visualisation of the association between the row and column categories of a two-way contingency table can be made by considering either a traditional correspondence plot or a biplot. For a traditional correspondence plot, the ith row and jth column categories are depicted using the *principal* coordinates along the mth dimension by

$$f_{im} = \frac{a_{im}}{\sqrt{p_{i\bullet}}}\lambda_m \qquad g_{jm} = \frac{b_{jm}}{\sqrt{p_{\bullet j}}}\lambda_m,$$

respectively. Biplot coordinates can be depicted by plotting the principal coordinates for one variable and standard coordinates for the other variable. There has been plenty of discussion of the use of plotting schemes, including biplots, in the correspondence analysis literature; see Beh and Lombardo (2014, Section 4.5) for more details.

An important property of these features is that they are directly linked to Pearson's chi-squared statistic such that

$$\frac{X^2}{n} = \sum_{m=1}^{M} \lambda_m^2 = \sum_{i=1}^{I}\sum_{m=1}^{M} p_{i\bullet} f_{im}^2 = \sum_{j=1}^{J}\sum_{m=1}^{M} p_{\bullet j} g_{jm}^2.$$

From this classical approach to correspondence analysis, the (i, j)th cell proportion can be exactly reconstituted from the association model

$$p_{ij(M)}^P = p_{i\bullet} p_{\bullet j}\left(1 + \sum_{m=1}^{M} \breve{a}_{im} \lambda_m \breve{b}_{jm}\right),$$

where the inclusion of the subscript (M) indicates that the model accounts for the association reflected in an M-dimensional correspondence plot. For this model, $\breve{a}_{im} = a_{im}/\sqrt{p_{i\bullet}}$ and $\breve{b}_{jm} = b_{jm}/\sqrt{p_{\bullet j}}$ so that

$$\sum_{i=1}^{I} p_{i\bullet}\breve{a}_{im}\breve{a}_{im'} = \begin{cases} 1 & m = m' \\ 0 & m \neq m' \end{cases} \qquad \sum_{j=1}^{J} p_{\bullet j}\breve{b}_{jm}\breve{b}_{jm'} = \begin{cases} 1 & m = m' \\ 0 & m \neq m' \end{cases}. \qquad (1)$$

Therefore, simple correspondence analysis may be performed using a generalised SVD (GSVD) of the matrix of Pearson's ratio's whose (i, j) element is $p_{ij} / (p_{i \bullet} p_{\bullet j}) - 1$ using these constraints; see, for example, Beh (2004) and Beh and Lombardo (2014, p. 125) for more details.

3 Dealing with Overdispersion

Despite the popularity of Pearson's chi-squared statistic, the problem is that the Pearson residuals, r_{ij}^P, are based on the assumption that the cell frequencies, n_{ij}, are random variables from a Poisson distribution with expectation $np_{i \bullet} p_{\bullet j}$. This implies that the mean and variance of n_{ij} are assumed to be identical, which may not be valid. In fact, Agresti (2013, p. 80) points out that the asymptotic variance of $\sqrt{n} r_{ij}$ "is less than 1.0, averaging $[(I-1)(J-1)]/(\text{number of cells})$"—or $(1 - 1/I)(1 - 1/J)$. Also, it was pointed out by Haberman (1973) that, under independence, the maximum likelihood estimate (using the Poisson distribution, for example) of the variance of $\sqrt{n} r_{ij}^P$ is $(1 - p_{i \bullet})(1 - p_{\bullet j}) < 1.0$. Both results imply that the variance of the residuals exceeds its expectation and so the residuals, and hence n_{ij}, exhibit overdispersion when it is treated as a Poisson random variable. Therefore, performing correspondence analysis using r_{ij}^P may, in some cases, not be the most appropriate means of studying the association structure between the variables. By adopting the strategy used by Efron (1992), one may easily detect overdispersion in cross-classified data by plotting the standard deviation of each cell against the Pearson residual, r_{ij}^P. Due to space restrictions here we confine our discussion to introducing some of the technical aspects of various strategies for dealing with overdispersion in the correspondence analysis of a two-way contingency table. However, for a more visual perspective of this issue, one may consult, for example, Beh and Lombardo (2014, Sections 9.2 & 9.3) and Beh et al. (2018) who provided a practical demonstration of the role of correspondence analysis and overdispersion in contingency tables.

4 Strategies Based on Variants of the Poisson Distribution

4.1 A General Standardised Residual

There are a range of alternative strategies that one can take to deal with overdispersion. Suppose we define γ to be the dispersion parameter associated with a particular variant of the Poisson distribution so that $E(n_{ij}; \gamma)$ and $Var(n_{ij}; \gamma)$ is the expectation and variance of n_{ij}, respectively, given the dispersion parameter. So, a general standardised residual of n_{ij} is

$$r_{ij}^G (\gamma) = \frac{n_{ij} - E\left(n_{ij}; \gamma\right)}{\sqrt{\text{Var}\left(n_{ij}; \gamma\right)}}.$$

Therefore a general approach to performing correspondence analysis of a contingency table that consists of overdispersed data is to apply an SVD to the matrix with elements $r_{ij}^G (\gamma)$ whose generic element is

$$r_{ij}^G (\gamma) = \sum_{m=1}^{M} \tilde{a}_{im} \tilde{\lambda}_m \tilde{b}_{jm},$$

where

$$\sum_{i=1}^{I} \tilde{a}_{im} \tilde{a}_{im'} = \begin{cases} 1 & m = m' \\ 0 & m \neq m' \end{cases} \qquad \sum_{j=1}^{J} \tilde{b}_{jm} \tilde{b}_{jm'} = \begin{cases} 1 & m = m' \\ 0 & m \neq m' \end{cases}. \qquad (2)$$

The choice of γ that is made to reflect the overdispersion in a contingency table depends on the choice of which variant of the Poisson distribution is considered. We now discuss a few such variants and other variance stabilising options that the correspondence analyst can use. Since further investigation of each strategy described below, and the properties of their γ parameter, needs to be undertaken, we shall include a subscript on each γ.

4.2 Generalised Poisson Distribution

One strategy that can be adopted when performing a simple correspondence analysis on overdispersed data is to assume that the cell frequencies are random variables from the generalised Poisson distribution with a mean and variance of

$$E\left(n_{ij}; \gamma_1\right) = \frac{n p_{i\bullet} p_{\bullet j}}{1 - \gamma_1} \qquad \text{Var}\left(n_{ij}; \gamma_1\right) = \frac{n p_{i\bullet} p_{\bullet j}}{(1 - \gamma_1)^3},$$

respectively; see Consul (1989), Consul and Jain (1973) and Tripathi and Gupta (1984) for more details on this distribution. Note that when the dispersion parameter is $\gamma_1 = 0$, $E\left(n_{ij}; \gamma_1 = 0\right) = \text{Var}\left(n_{ij}; \gamma_1 = 0\right)$ so that there is zero dispersion. Conversely, $\gamma_1 > 0$ and $\gamma_1 < 0$ reflect overdispersion and underdispersion, respectively.

By considering such an amendment to the mean and variance of the Poisson distribution, correspondence analysis of overdispersed cross-classified data may be performed by considering the SVD of the (generalised) standardised residual

$$r_{ij}^{GP}(\gamma) = \frac{p_{ij} - p_{i\bullet}p_{\bullet j}/(1-\gamma_1)}{\sqrt{p_{i\bullet}p_{\bullet j}/(1-\gamma_1)^3}} = \sum_{m=1}^{M} \tilde{a}_{im}\tilde{\lambda}_m \tilde{b}_{jm}. \tag{3}$$

Note that for zero dispersion, $\gamma_1 = 0$, the left hand side of (3) simplifies to the standardised residual r_{ij}^{P}.

By applying an SVD to the matrix of elements consisting of r_{ij}^{GP}, the (i, j)th cell proportion can be exactly reconstituted using the model

$$p_{ij(M)}^{GP} = \frac{p_{i\bullet}p_{\bullet j}}{1-\gamma_1} + \sqrt{\frac{p_{i\bullet}p_{\bullet j}}{(1-\gamma_1)^3}} \sum_{m=1}^{M} \tilde{a}_{im}\tilde{\lambda}_m \tilde{b}_{jm}$$

$$= \frac{p_{i\bullet}p_{\bullet j}}{1-\gamma_1}\left(1 + \frac{1}{\sqrt{1-\gamma_1}} \sum_{m=1}^{M} \breve{a}_{im}\tilde{\lambda}_m \breve{b}_{jm}\right),$$

where $\breve{a}_{im}$ and $\breve{b}_{jm}$ are constrained by (1). Note that, as expected when $\gamma_1 = 0$, $p_{ij(M)}^{GP} = p_{ij(M)}^{P}$. Therefore, by constraining $\breve{a}_{im}$ and $\breve{b}_{jm}$ by (1), correspondence analysis may be equivalently performed by applying a GSVD to the matrix with elements $\sqrt{1-\gamma_1}\,p_{ij}/(p_{i\bullet}p_{\bullet j}) - 1/\sqrt{1-\gamma_1}$. Note that, again, when there is zero dispersion (so that $\gamma_1 = 0$) this results in the GSVD of Pearson's ratio.

4.3 Negative Binomial Distribution

Another strategy that can be adopted to deal with overdispersed cross-classified data is to consider that n_{ij} is a random variable from a negative binomial distribution (Fisher 1941). By doing so, the expectation and variance of n_{ij} is

$$E\left(n_{ij}; \gamma_2\right) = np_{i\bullet}p_{\bullet j} \qquad Var\left(n_{ij}; \gamma_2\right) = np_{i\bullet}p_{\bullet j} + \frac{\left(np_{i\bullet}p_{\bullet j}\right)^2}{\gamma_2},$$

respectively. Overdispersion arises for all $\gamma_2 > 0$, although as $\gamma_2 \to \infty$, $Var\left(n_{ij}; \gamma_2\right) \to E\left(n_{ij}; \gamma_2\right)$. In this case, correspondence analysis may be performed by first considering the SVD of the standardised n_{ij} under this distribution such that

$$r_{ij}^{NB}(\gamma_2) = \frac{p_{ij} - p_{i\bullet}p_{\bullet j}}{\sqrt{p_{i\bullet}p_{\bullet j} + (n/\gamma_2)\left(p_{i\bullet}p_{\bullet j}\right)^2}} = \sum_{m=1}^{M} \tilde{a}_{im}\tilde{\lambda}_m \tilde{b}_{jm}.$$

By applying an SVD to the matrix consisting of r_{ij}^{NB}, the (i, j)th cell proportion can be exactly reconstituted using the model

$$p_{ij(M)}^{NB} = \frac{p_{i\bullet}p_{\bullet j}}{p_{i*}p_{*j}} \left(1 + \sqrt{1 + \frac{np_{i\bullet}p_{\bullet j}}{\gamma_2}} \sum_{m=1}^{M} \breve{a}_{im} \tilde{\lambda}_m \breve{b}_{jm} \right)$$

where $\breve{a}_{im}$ and $\breve{b}_{jm}$ are constrained by (1). Note that, as $\gamma_2 \to \infty$, $p_{ij(M)}^{NB} \to p_{ij(M)}^{P}$.

4.4 Conway–Maxwell Poisson Distribution

Another variant of the Poisson distribution that accommodates for dispersion is the Conway–Maxwell Poisson (CMP) distribution; see Conway and Maxwell (1962). For such a distribution, there is no closed form for the expectation and variance of n_{ij}; however, they may be approximated (Sellers et al. 2012) by

$$\mathrm{E}\left(n_{ij}; \gamma_3\right) \approx \left(np_{i\bullet}p_{\bullet j}\right)^{1/\gamma_3} - \frac{\gamma_3 - 1}{2\gamma_3} \qquad \mathrm{Var}\left(n_{ij}; \gamma_3\right) \approx \frac{\left(np_{i\bullet}p_{\bullet j}\right)^{1/\gamma_3}}{\gamma_3},$$

where $\gamma_3 > 1$ reflects underdispersion in the contingency table and $0 < \gamma_3 < 1$ reflects overdispersion. Note that when $\gamma_3 = 1$, there is zero dispersion since $\mathrm{E}\left(n_{ij}; \gamma_3 = 1\right) = \mathrm{Var}\left(n_{ij}; \gamma_3 = 1\right)$. It was shown by Shmueli et al. (2005) that such approximations work well when $\gamma_3 \leq 1$ or when $np_{i\bullet}p_{\bullet j} > 10^{\gamma_3}$. In this case, correspondence analysis of overdispersed cross-classified data may be performed by considering the SVD of a matrix whose general element is the residual

$$r_{ij}^{CM}(\gamma_3) = \frac{p_{ij} - n^{(1-\gamma_3)/\gamma_3}\left(p_{i\bullet}p_{\bullet j}\right)^{1/\gamma_3} + (\gamma_3 - 1)/(2n\gamma_3)}{\sqrt{n^{(1-\gamma_3)/\gamma_3}\left(p_{i\bullet}p_{\bullet j}\right)^{1/\gamma_3}/\gamma_3}} = \sum_{m=1}^{M} \tilde{a}_{im}\tilde{\lambda}_m \tilde{b}_{jm}.$$

$$(4)$$

Note that when there is zero dispersion, $\gamma_3 = 1$, the left hand side of (4) simplifies to r_{ij}^{P}. Otherwise, one cannot escape from the standardised residual being dependent on the sample size, n.

5 Variance Stabilising Strategies

Strategies to accommodate for overdispersion of the cell frequencies of a contingency table are not just confined to considering variants of the Poisson distribution. Here we consider two relatively common approaches that can be used to stabilise the variance of the Poisson distribution—by incorporating the variance structure of r_{ij}^P described by Haberman (1973) and by using the Freeman–Tukey transformation.

5.1 Adjusted Standardised Residual

One strategy that may be adopted to accommodate the presence of overdispersion in a contingency table is to consider the adjusted Pearson residuals proposed by Haberman (1973)

$$r_{ij}^A = \frac{p_{ij} - p_{i\bullet}p_{\bullet j}}{\sqrt{p_{i\bullet}p_{\bullet j}\left(1 - p_{i\bullet}\right)\left(1 - p_{\bullet j}\right)}} .$$

One advantage of considering r_{ij}^A is that, unlike r_{ij}^P, $\sqrt{n}r_{ij}^A$ is a standard normally distributed random variable. In fact, Beh (2012) showed how correspondence analysis can be performed through the SVD of the matrix whose general element is r_{ij}^A. For example, it was shown that the (i, j)th cell proportion can be exactly reconstituted by

$$p_{ij(M)}^A = p_{i\bullet}p_{\bullet j}\left(1 + \sum_{m=1}^{M} \breve{a}_{im}\tilde{\lambda}_m \breve{b}_{jm}\right),$$

where $\breve{a}_{im}$ and $\breve{b}_{jm}$ are constrained by

$$\sum_{i=1}^{I}\left(\frac{p_{i\bullet}}{1 - p_{i\bullet}}\right)\breve{a}_{im}\breve{a}_{im'} = \begin{cases} 1 & m = m' \\ 0 & m \neq m' \end{cases}$$

$$\sum_{j=1}^{J}\left(\frac{p_{\bullet j}}{1 - p_{\bullet j}}\right)\breve{b}_{jm}\breve{b}_{jm'} = \begin{cases} 1 & m = m' \\ 0 & m \neq m' \end{cases} .$$

Therefore, with these orthogonal constraints, a correspondence analysis may be equivalently performed by applying a GSVD to the matrix whose elements are Pearson's ratio, $p_{ij}/\left(p_{i\bullet}p_{\bullet j}\right) - 1$.

5.2 Freeman–Tukey Residual

Since overdispersion arises because the variance of n_{ij} is larger than its expectation, a popular method used to stabilise its variance is to consider the transformation proposed by Freeman and Tukey (1950) who, adapting their notation for a two-way contingency table, showed that

$$\sqrt{n_{ij}} + \sqrt{n_{ij} + 1}$$

has a mean and variance of $\sqrt{4np_{i\bullet}p_{\bullet j} + 1}$ and 1, respectively. Thus, Freeman and Tukey (1950) proposed

$$\tilde{T}^2 = \sum_{i=1}^{I}\sum_{j=1}^{J}\left(\sqrt{n_{ij}} + \sqrt{n_{ij} + 1} - \sqrt{4np_{i\bullet}p_{\bullet j} + 1}\right)^2$$

as an alternative to Pearson's chi-squared statistic. Note that $\tilde{T}^2$ may be alternatively, but equivalently, expressed as

$$\tilde{T}^2 = n\sum_{i=1}^{I}\sum_{j=1}^{J}\left(\sqrt{p_{ij}} + \sqrt{p_{ij} + \frac{1}{n}} - \sqrt{4p_{i\bullet}p_{\bullet j} + \frac{1}{n}}\right)^2.$$

For large sample sizes, $\tilde{T}^2$ may be approximated by eliminating $1/n$. Doing so gives the *Freeman–Tukey* statistic

$$\check{T}^2 = 4n\sum_{i=1}^{I}\sum_{j=1}^{J}\left(\sqrt{p_{ij}} - \sqrt{p_{i\bullet}p_{\bullet j}}\right)^2.$$

From the perspective of correspondence analysis of a contingency table, Beh and Lombardo (2014) introduced the idea that overdispersed cross-classified data can be treated by applying an SVD to the *Freeman–Tukey residual* such that

$$r_{ij}^{FT} = \sqrt{p_{ij}} + \sqrt{p_{ij} + \frac{1}{n}} - \sqrt{4p_{i\bullet}p_{\bullet j} + \frac{1}{n}} = \sum_{m=1}^{M}\tilde{a}_{im}\tilde{\lambda}_m\tilde{b}_{jm}. \tag{5}$$

This strategy was extensively elaborated upon in the context of studying sparse archaeological data by Beh et al. (2018). Here $\tilde{\lambda}_m$ is the mth singular value of the $I \times J$ matrix of Freeman–Tukey residuals whose (i, j)th element is r_{ij}^{FT} and $\tilde{b}_{jm}$ are subject to the constraints given by (2). When n is large, one may consider the SVD of r_{ij}^T such that

$$r_{ij}^T = 2\left(\sqrt{p_{ij}} - \sqrt{p_{i\bullet}p_{\bullet j}}\right) = \sum_{m=1}^{M}\tilde{a}_{im}\lambda_m\tilde{b}_{jm}. \tag{6}$$

For many practical situations where n is large, choosing between (5) or (6) will have a minimal effect on the numerical summaries when performing a correspondence analysis. Nor will there be any considerable difference in the configuration of points when comparing their correspondence plots. Refer to Beh et al. (2018) for more details on this approach, including specific comments on the association model

$$p_{ij(M)}^T = p_{i\bullet}p_{\bullet j}\left(1 + \frac{1}{2\sqrt{p_{i\bullet}p_{\bullet j}}}\sum_{m=1}^{M}\tilde{a}_{im}\lambda_m\tilde{b}_{jm}\right)^2$$

which, unlike $p_{ij(M)}^P$, guarantees that any reconstitution of the cell frequencies (or their marginal frequencies) for any $M < \min(I, J) - 1$ will be non-negative.

6 Discussion

The aim of this paper is to propose a variety of strategies that one may adopt for dealing with overdispersion in cross-classified data when performing correspondence analysis. In all of the cases described above, a visualisation of the association can be undertaken by either constructing a traditional correspondence plot or a biplot. For a correspondence plot, the coordinate along the mth axis for the ith row, and jth column, is

$$f_{im} = \frac{\tilde{a}_{im}\tilde{\lambda}_m}{\sqrt{p_{i\bullet}}} \qquad g_{jm} = \frac{\tilde{b}_{jm}\tilde{\lambda}_m}{\sqrt{p_{\bullet j}}},$$

respectively, where $\tilde{a}_{im}$ and $\tilde{b}_{jm}$ are constrained by (2). For all of the strategies outlined here, the total inertia of the contingency table can still be quantified through the sum-of-squares of the residuals, and with the orthogonality constraints (2), can be expressed as

$$\frac{\tilde{X}^2}{n} = \sum_{m=1}^{M}\tilde{\lambda}_m^2 = \sum_{i=1}^{I}\sum_{m=1}^{M}p_{i\bullet}f_{im}^2 = \sum_{j=1}^{J}\sum_{m=1}^{M}p_{\bullet j}g_{jm}^2.$$

However, unlike the classical approach to correspondence analysis, such a measure will not always be equal to Pearson's chi-squared statistic.

In all cases, one may also construct a biplot to visualise the association. If we consider a more general set of coordinates such that

$$\tilde{f}_{im} = \frac{\tilde{a}_{im}\tilde{\lambda}_m^{\beta}}{\sqrt{p_{i\bullet}}} \qquad \tilde{g}_{jm} = \frac{\tilde{b}_{jm}\tilde{\lambda}_m^{1-\beta}}{\sqrt{p_{\bullet j}}},$$

then there are two types of biplots that are typically considered, namely the *row isometric* biplot where $\beta = 1$ and the *column isometric* biplot where $\beta = 0$; see Beh and Lombardo (2014, Section 4.5.3), and the references mentioned therein, for more details on the numerous biplot displays when performing correspondence analysis.

When reflecting upon the benefits of the strategies outlined in this paper, it is important to keep in mind some of the key features that make correspondence analysis very versatile at visualising association between categorical variables. We have alluded to some of the features associated with the models obtained from each of the strategies, but perhaps of paramount importance is the interpretation of the distance between row, and column, points. When an SVD of the matrix of r_{ij}^P residuals is performed, it is well established that the (squared) distance between two row (or column) points whose coordinate is defined by f_{im} (or g_{jm}) is Euclidean. For the row categories, these Euclidean distances reflect the chi-squared distance between two centred row profiles, $p_{ij}/p_{i\bullet} - p_{\bullet j}$ and $p_{i'j}/p_{i'\bullet} - p_{\bullet j}$, say. When an SVD is applied to the matrix of Freeman–Tukey residuals, r_{ij}^T, the distance between two row (or column) points in the correspondence plot is measured using the Hellinger distance (Beh et al. 2018) and reflects the difference between the square-root of two centred profiles; in the case of the rows, such profiles are defined by $\sqrt{p_{ij}/p_{i\bullet}} - \sqrt{p_{\bullet j}}$. In all other cases described in this paper, further investigation needs to be undertaken to more fully understand the interpretation of distances between two row, or column, points and how they relate to comparing the centred row, and column, profiles. The issue of assessing the distance between a row point and a column point is a long, and contentious, one. One may refer to, for example, Beh et al. (2018, Section 4.6.3) (and the references within this section) and Nishisato and Clavel (2003) for more of a discussion of this issue.

While this paper introduces some possible options for accommodating overdispersion in cross-classified data when performing correspondence analysis, it certainly does not provide a comprehensive overview of the issues. There are certainly various other issues that require deeper investigation to more fully appreciate the utility of these strategies.

References

Agresti, A.: Categorical Data Analysis, 3rd edn. Wiley, New York (2013)

Beh, E.J.: Simple correspondence analysis: a bibliographic review. Int. Stat. Rev. **72**, 257–284 (2004)

Beh, E.J.: Simple correspondence analysis using adjusted residuals. J. Stat. Plann. Infer. **142**, 965–973 (2012)

Beh, E.J., Lombardo, R.: Correspondence Analysis: Theory, Practice and New Strategies. Wiley, Chichester (2014)

Beh, E.J., Lombardo, R., Alberti, G.: Correspondence analysis and the Freeman-Tukey statistic: a study of archaeological data. Comput. Stat. Data Anal. **128**, 73–86 (2018)

Consul, P.C.: Generalized Poisson Distributions: Properties and Applications. Dekker, New York (1989)

Consul, P.C., Jain, G.C.: A generalization of the Poisson distribution. Technometrics, **15**, 791–799 (1973)

Conway, R.W., Maxwell, W.L.: A queuing model with state dependent service rates. J. Ind. Eng. **12**, 132–136 (1962)

Cox, D.R.: Some remarks on overdispersion. Biometrika **70**, 269–274 (1983)

Efron, B.: Overdispersion estimates based on the method of asymmetric maximum likelihood. J. Am. Stat. Assoc. **87**, 98–107 (1992)

Fisher, R.A.: The negative binomial distribution. Ann. Eugenic. **11**, 182–187 (1941)

Freeman, M.F., Tukey, J.W.: Transformations related to the angular and square root. Ann. Math. Stat. **21**, 607–611 (1950)

Greenacre, M.J.: Theory and Applications of Correspondence Analysis. Academic, London (1984)

Greenwood, M., Yule, G.U.: An inquiry into the nature of frequency distributions representative of multiple happenings with particular reference to the occurrence of multiple attacks of disease or of repeated accidents. J. Roy. Stat. Soc. **83**, 255–279 (1920)

Haberman, S.: The analysis of residuals in cross-classified tables. Biometrics **75**, 457–467 (1973)

Lebart, L., Morineau, A., Warwick, K.M.:Multivariate Descriptive Statistical Analysis. Wiley, New York (1984)

Nishisato, S., Clavel, J.G.: A note on between-set distances in dual scaling and correspondence analysis. Behaviormetrika **30**, 87–98 (2003)

Satterthwaite, F.E.: Generalized Poisson distribution. Ann. Math. Stat. **13**, 410–417 (1942)

Sellers, K.F., Borle, S., Schmueli, G.: The COM-Poisson model for count data: a survey of methods and applications. Appl. Stoch. Model. Bus. Ind. **28**, 104–116 (2012)

Shmueli, G., Minka, T.P., Kadane, J.B., Borle, S., Boatwright, P.: A useful distribution for fitting discrete data: revival of the Conway-Maxwell-Poisson distribution. Appl. Stat. **54**, 127–142 (2005)

Tripathi, R.C., Gupta, R.C.: Statistical inference regarding the generalized Poisson distribution. Sankhya (Ser. B) **46**, 166–173 (1984)

From Joint Graphical Display to Bi-Modal Clustering: [2] Dual Space Versus Total Space

José G. Clavel and Shizuhiko Nishisato

Abstract There are two ways to use dual scaling results of the condensed response-pattern matrix, one in terms of dual space and the other in total space. The main task of this paper is to explore the differences in dealing with dual space and total space, and it is hoped that the study will offer some insights into the characteristics of these two types of space for cluster analysis. Bi-modal clustering is defined as a method of cluster analysis to identify only between-set clusters (i.e., clusters consisting of both rows and columns of the contingency table), and the surest way to find such clusters is to use the between-set distance matrix as an input for clustering. Since this input matrix is typically rectangular, most traditional and currently popular methods of cluster analysis cannot handle rectangular distance matrices. Thus, the current study has chosen an intuitive method, called exploratory clustering, as a start. Since there are two ways to compute the between-set distance matrix, namely in dual space and total space, the current study aims at demonstrating the differences in clusters formed by two approaches. After examining the results, the study concludes that bi-modal clustering should be carried out in dual space. This conclusion is discussed in the current paper, and it is hoped that our conclusion can be supported by researchers. This conclusion also means that the initial object of total information analysis or comprehensive dual scaling by Nishisato and Clavel can be justified. This connection is also discussed in the current paper.

J. G. Clavel (✉)
Universidad de Murcia, Murcia, Spain
e-mail: jjgarvel@um.es

S. Nishisato
University of Toronto, Toronto, ON, Canada
e-mail: shizuhiko.nishisato@utoronto.ca

© Springer Nature Singapore Pte Ltd. 2020
T. Imaizumi et al. (eds.), *Advanced Studies in Classification and Data Science*,
Studies in Classification, Data Analysis, and Knowledge Organization,
https://doi.org/10.1007/978-981-15-3311-2_11

131

1 Introduction

The main purpose of this paper is to present numerical examples to explore the differences in the distance matrices in dual space and total space of the same data. Under these two situations, the between-set distance matrices will be calculated and then subjected to exploratory clustering with flexible filters to examine the differences in cluster compositions. We aim to find clues for justifying the use of dual space in bi-modal clustering through numerical demonstrations. This choice of the space is primarily the problem of whether we should use the same multidimensional space to define the distance between two variables, or allow the situation in which one spans, for example, in 2-dimensional space and the other in 4-dimensional space. We hope that numerical examples will justify the use of dual space for bi-modal clustering.

As mentioned in the paper [1] (Nishisato in this volume), dual space and total space are identical when the number of rows and that of column of the contingency table are equal. When they are not equal, total space is larger in dimensionality than dual space. In computing the distance between a row and a column of the contingency table, we have two points of view: one is to define the distance between the two variables in the common space (e.g., two sets of color chips which vary in hue and brightness), here referred to as dual space, and the other view is to define the distance between variables which span different numbers of dimensions (e.g., one set of color chips that vary in hue and brightness, and the other set of chips which vary in hue, brightness, and saturation). The second case requires space which accommodate not only the common space of hue and brightness, but also an additional dimension associated with saturation, and the space for the second case is called total space.

These two views appear to be equally reasonable, namely, one is the view to define the distance between two variables over the common attributes, and the other is to define the distance between two variables which possess different numbers of attributes. In quantification research, we do not deal with such distinct attributes as hue, brightness, and saturation, but attributes are deduced from the common set of variables which contribute to particular components (e.g., introvert-extravert dimension from a set of common personality traits). The views of using dual space and total space seem to be equally reasonable to define distances between two objects, but our numerical examples will show quite convincingly that dual space is more appropriate for bi-modal clustering than total space.

2 Exploratory Clustering in Dual Space and Total Space

The main role of the paper is to provide a number of numerical examples of exploratory clustering, carried out in both dual space and total space to offer a glimpse of differences between the two approaches. It should be noted that this problem of two alternatives has not arisen in the traditional quantification theory, and the main reason for the rise of this new problem is due to the fact that quantification analysis is expanded to between-set analysis from the traditional within-set analysis.

2.1 Example 1: Rorschach Responses Under Different Moods

Garmize and Rycklak (1964) collected data with the focus on eleven Rorschach inkblot responses (e.g., bat, blood, butterfly, . . .) under six types of experimentally induced moods: fear, anger, depression, ambition, security, and love. The results are as summarized in the contingency table (Table 1).

In the current paper, this contingency is then converted into the condensed form of response-pattern table (Nishisato 1980, 2019). Since this is one of the key procedures, we need to explain this process clear enough to understand. The original data are too large to explain the procedure in detail. Therefore, for this demonstration of constructing a condensed form of response-pattern table, let us use a subset of the above table as shown below.

The Table 2 can be expressed as a condensed form of the response-pattern table, that is, a summarized form of the incidence table, as shown in Table 3. In case it is difficult to follow this part of conversion, please refer to Nishisato (1980, 2019).

Table 1 Rorschach responses and induced moods

| Responses | Induced moods | | | | | |
	Fear	Anger	Depression	Ambition	Security	Love
Bat	33	10	18	1	2	6
Blood	10	5	2	1	0	0
Butterfly	0	2	1	26	5	18
Cave	7	0	13	1	4	2
Clouds	2	9	30	4	1	6
Fire	5	9	1	2	1	1
Fur	0	3	4	5	5	21
Mask	3	2	6	2	2	3
Mountains	2	1	4	1	18	2
Rocks	0	4	2	1	2	2
Smoke	1	6	1	0	1	0

Table 2 Subset of the Rorschach data

Responses	Fear	Anger	Depression	Ambition	Security	Love
Bat	33	10	18	1	2	6
Butterfly	0	2	1	26	5	18
Mountains	2	1	4	1	18	2

Table 3 Condensed response-pattern table

Bat	Butterfly	Mountains	Fear	Anger	Depression	Ambition	Security	Love
33	0	0	33	0	0	0	0	0
10	0	0	0	10	0	0	0	0
18	0	0	0	0	18	0	0	0
1	0	0	0	0	0	1	0	0
2	0	0	0	0	0	0	2	0
6	0	0	0	0	0	0	0	6
0	2	0	0	2	0	0	0	0
0	1	0	0	0	1	0	0	0
0	26	0	0	0	0	26	0	0
0	5	0	0	0	0	0	5	0
0	18	0	0	0	0	0	0	18
0	0	2	2	0	0	0	0	0
0	0	1	0	1	0	0	0	0
0	0	4	0	0	4	0	0	0
0	0	1	0	0	0	1	0	0
0	0	18	0	0	0	0	18	0
0	0	2	0	0	0	0	0	2

The rows of this new table consist of combinations of the row and column variables, thus the total number of rows of this table corresponds to the total number of non-zero elements of the original 3×6 contingency table, that is, 17. The columns of this table consist of the row variables (3) and the column variables (6) of the original contingency table. Notice that the rows and the columns of the contingency table are now placed in the columns of the condensed response-pattern table.

Recall (1) that if we subject the original 3×6 contingency table, we obtain two components, but (2) that these two components require 4-dimensional space because of the discrepancy between row and column space. In contrast, if we subject the response-pattern table to dual scaling, we obtain seven components. Out of the seven components, four components constitute dual space, that is, the space shared by both sets of variables. How to identify these four components is explained in Nishisato (2019).

These four components and the remaining three components comprise total space. To make this distinction clear, let us show the principal coordinates of the seven components (see Table 4). As (Nishisato 1980) indicated, the average eigenvalue is 0.5. Notice that the average of the four components which constitute dual space also has the average eigenvalue of 0.5. In addition, the remaining three eigenvalues are each equal to 0.5. Notice, further, that the three Rorschach inkblots have principal coordinates in these three dimensions all zero, indicating that these inkblots have no contributions to the three components. In other words, the extra three components have nothing to do with the Rorschach inkblots. To show that those four components for dual space are correct can be numerically demonstrated

Table 4 Principal coordinates[a] of condensed response-pattern Table 3

	Comp. 1	Comp. 2	Comp. 3	Comp. 4	Comp. 5	Comp. 6	Comp. 7
Bat	−0.924	−0.391	0.000	0.000	0.000	−0.189	−0.317
Butterfly	1.194	−0.485	0.000	0.000	0.000	−0.234	0.410
Mountains	0.093	1.878	0.000	0.000	0.000	0.907	0.032
Fear	−1.097	−0.420	−0.555	−0.775	−0.903	0.203	0.377
Anger	−0.659	−0.371	3.121	−0.086	0.371	0.179	0.226
Depression	−0.830	0.000	−0.588	1.831	1.026	0.000	0.285
Ambition	1.367	−0.640	0.058	0.771	−1.080	0.309	−0.469
Security	0.293	1.968	0.062	−0.232	−0.259	−0.951	−0.101
Love	0.786	−0.453	−0.415	−1.141	1.534	0.219	−0.270
Eigenvalue	0.8947	0.8108	0.500	0.500	0.500	0.1892	0.1053

[a]Component 1, 2, 6 and 7 constitute the dual space. See Nishisato (2019)

by the results that the between-set (rows and columns of the contingency table) distances calculated from the coordinates of the four components are exactly the same as the between-set distances calculated from the formulation of total information analysis TIA (Nishisato and Clavel 2010), that is, the distances from two dimensions with space discrepancies taken into consideration. This in turn means that TIA offers analysis in dual space, not in total space. The extra three components in total space are solely to accommodate the variations of the moods, not explained by the Rorschach inkblot responses. So, in summary, the space common for both Rorschach responses and induced moods is called dual space, which is in the current example 4-dimensional, while the entire space for the columns of the response-pattern table is called total space, which is in this example 7-dimensional. Another way of interpreting dual space is that it is the space in which the column space in the current example is projected onto the row space, hence it is the space in which the relations between the row and the column variables are maximized, and this is another way of looking at dual space. With this much introduction, let us get back to the original data in Table 1, and look at the between-set (Rorschach-by-moods) distance matrices obtained in total space (Table 5) and dual space (Table 6).

Let us now subject these two distance matrices to clustering with filters, a procedure in which all distances greater than the threshold filter are discarded and the remaining distances are used to form clusters. The threshold filters we used are 10, 20, 30, 40, 50, and above 50 percentiles. These choices are quite arbitrary, and the study is indeed only exploratory. Table 7 shows a number of interesting differences with respect to the threshold values and two types of space, dual and total.

Before we look at the clusters of the variables, let us keep in mind that altogether 17 variables (11 Rorschach responses and 6 induced moods) are considered for exploratory clustering. In Table 7, clusters are indicated by alphabets (upper case letters for total space, lower case letters for dual space). Let us summarize our

Table 5 Between-set distances in total space

	Fear	Anger	Depression	Ambition	Security	Love
Bat	2.28	3.11	2.63	3.53	3.57	3.09
Blood	4.29	4.69	4.72	5.09	5.23	4.96
Butt	3.46	3.57	3.25	2.56	3.65	2.88
Cave	3.91	4.40	3.58	4.46	4.31	4.18
Cloud	3.40	3.31	2.44	3.63	3.82	3.30
Fire	4.54	4.28	4.66	4.91	5.05	4.80
Fur	3.80	3.83	3.51	3.84	3.89	2.90
Mask	4.76	4.92	4.52	5.00	5.05	4.77
Mount	4.11	4.29	3.90	4.41	3.14	4.13
Rock	6.04	5.74	5.81	6.12	6.03	5.89
Smoke	6.50	5.98	6.42	6.77	6.67	6.60

Table 6 Between-set distances in dual space

	Fear	Anger	Depression	Ambition	Security	Love
Bat	1.90	2.85	2.31	3.30	3.34	2.82
Blood	2.22	2.92	2.96	3.53	3.72	3.34
Butt	3.45	3.55	3.23	2.54	3.63	2.86
Cave	2.55	3.26	2.02	3.33	3.13	2.95
Cloud	3.21	3.12	2.16	3.45	3.65	3.10
Fire	3.07	2.69	3.25	3.61	3.78	3.45
Fur	3.75	3.78	3.46	3.79	3.85	2.84
Mask	2.22	2.55	1.64	2.71	2.79	2.24
Mount	4.02	4.21	3.81	4.33	3.02	4.04
Rock	3.25	2.64	2.79	3.39	3.23	2.95
Smoke	4.47	3.67	4.36	4.85	4.71	4.62

observations on clustering of 17 variables in terms of the number of clusters as functions of threshold percentiles and two different types of space, dual and total:

- As a demonstration, we used only up to 50 percentile point, which turned out to be too small to include all 17 variables into some clusters, namely, smoke, mask, and rock are absent in total space, and smoke is absent in dual space. This raises a question if one should increase the threshold value until all the variables are included in at least one cluster.
- Fixing the 10% threshold, we have four clusters in total space [Bat; Fear], [Cloud, Bat; Depression], [Butterfly; Ambition], [Fur, Butterfly, Bat; Love] and only two in dual space [Bat, Blood, Mask; Fear], [Cave, Cloud, Mask; Depression, Love]. In both spaces, the moods Depression, Fear, and Ambition and the responses Bat, and Cloud are clustered, but Butterfly only appears in total space, and Mask only is clustered in dual space. Selecting 30% as threshold for the filtering, that is, discarding all distances greater than the 30-percentile point and identifying the clusters from the remaining distances, we find five clusters in both spaces

Table 7 Clusters formed by different threshold percentile points

Filter threshold	Clusters	Responses	Moods
Clusters in total space			
10%	A	Bat	Fear
	B	Cloud, Bat	Depression
	C	Butterfly	Ambition
	D	Fur, Butterfly, Bat	Love
20%	A	Bat	Fear, Anger
	B	Cloud, Bat, Butterfly	Depression
	C	Butterfly	Ambition
	D	Fur, Butterfly, Bat	Love
	E	Mountain	Security
30%	A	Bat, Cloud	Fear, Anger
	B	Cloud, Bat, Butterfly, Cave	Depression
	C	Butterfly	Ambition
	D	Fur, Butterfly, Bat, Cloud	Love
	E	Mountain	Security
40%	A	Bat, Cloud	Fear, Anger
	B	Cloud, Bat, Butterfly, Cave, Fur	Depression
	C	Butterfly	Ambition
	D	Fur, Butterfly, Bat, Cloud	Love
	E	Mountain	Security
50%	A	Bat, Cloud, Fire	Fear, Anger
	B	Cloud, Bat, Butterfly, Cave, Fur	Depression
	C	Butterfly	Ambition
	D	Fur, Butterfly, Bat, Cloud	Love
	E	Mountain	Security
Clusters in dual space			
10%	a	Bat, Blood, Mask	Fear
	b	Cave, Cloud, Mask	Depression, Love
20%	a	Bat, Blood, Mask	Fear
	b	Cave, Cloud, Mask	Depression, Love
	c	Butterfly	Ambition
	d	Fire, Rock, Mask	Anger
30%	a	Bat, Blood, Mask	Fear
	b	Cave, Cloud, Mask, Fur	Depression, Love
	c	Butterfly	Ambition
	d	Fire, Rock, Mask	Anger
	e	Mask	Security
40%	a	Bat, Blood, Mask, Fire	Fear
	b	Cave, Cloud, Mask, Fur, Rock	Depression, Love
	c	Butterfly	Ambition
	d	Fire, Rock, Mask	Anger
	e	Mask, Mountain	Security

(continued)

Table 7 (continued)

Filter threshold	Clusters	Responses	Moods
50%	a	Bat, Blood, Mask, Fire	Fear
	b	Cave, Cloud, Mask, Fur, Rock	Depression, Love
	c	Butterfly	Ambition
	d	Fire, Rock, Mask	Anger
	e	Mask, Mountain, Cave, Rock	Security

including all the moods. In total space: [Bar, Cloud; Fear, Anger], [Cloud, Bat, Butterfly, Cave; Depression], [Butterfly; Ambition], [Fur, Butterfly, Bat, Cloud; Love], [Mountain; Security]. In dual space: [Bat, Blood, Mask; Fear], [Cave, Cloud, Mask, Fur; Depression, Love], [Butterfly; Ambition], [Fire, Rock, Mask; Anger], [Mask; Security]. Although some clusters are similar, see cluster "C" in total space and cluster "c" in dual space, both containing [Butterfly; Ambition], there are important differences. For example, in total space, the response Butterfly appears clustered with the moods Depression, and Love, but in dual space, it is related only with Ambition. Other clear difference is the response Mask that forms overlapping clusters in dual space: Mask belongs to four out of five clusters and it is related with all the moods but Ambition. In total space, Mask has not appeared yet. Using the 50% threshold the response Fire, that was clustered in several groups in dual space, finally appears clustered with Fear and Anger in total space, meaning that its distance in total space was so big that they were discarded with filters below 50-percentile. At 50% there are some coincidences between spaces: [Mountain; Security], [Bat; Fear], [Fur, Cloud; Love], [Fire; Anger]... but the differences show clearly that we should be careful clustering between-set distances.

- As a rule of thumb, as we go up with the threshold values from 10% on, there seems a definite tendency for clustering in dual space to show more inclusive and gradual changes as opposed to clustering in total space. This can be interpreted as a reflection in which variations in space orthogonal to the common space are affecting the outcomes of clustering in total space.
- If we are to seek a stable set of clusters, it looks as though the clusters [Bat, Blood, Mask; Fear], [Cave, Cloud, Mask; Depression, Love], [Fire, Rock, Mask; Anger], [Mask; Security], obtained with the filters 20 and 30% in dual space are the solution. However, "Smoke" is not included in this exploratory clustering.
- As we see it clearly in the exploratory clustering outcomes in total space and dual space are quite different. As anticipated, dual space is where the quantification is carried out in such a way that the row variables and the column variables are maximally correlated. This observation is very important for dual space to be preferred to total space for exploratory clustering.

Due to the limited space, we cannot discuss some details of the exploratory clustering differences between total space and dual space. However, without even looking at numerical results, one can advance the view that dual space should be used for bi-modal clustering since dual space is where row variables and

Table 8 Between-set distances in total space

	Fear	Anger	Depression	Ambition	Security	Love
Bat	2.02	2.85	2.42	3.28	3.49	3.15
Blood	3.77	4.18	4.25	4.6	4.86	4.71
Butt	3.14	3.27	3.03	2.26	3.5	2.72
Cloud	3.08	3.01	2.11	3.35	3.69	3.29
Fire	4.03	3.78	4.22	4.44	4.68	4.54
Mask	4.24	4.41	4.02	4.51	4.66	4.44
Mount	3.68	3.89	3.54	4.02	2.64	3.98
Smoke	5.77	5.25	5.73	6.06	6.04	6.08

column variables are most highly correlated. As is stated above, this aspect was demonstrated in the results that all row and column variables are better presented in clusters than in total space. In contrast, total space involves some independent contributions of the larger set of variables, and this condition is totally against the fundamental purpose of bi-modal clustering, that is, to find clusters of row variables and column variables.

2.2 Example 2: A Subset of Rorschach Responses Under Different Moods

The above example is a case in which the discrepancy in dimensionality between dual space and total space is relatively large, that is, 10 dimensions of dual space and 15 dimensions of total space. Let us now see a case in which the dimensions of two kinds of space are a little closer than the previous case, and see what will happen to the clustering differences. As an example, let us choose eight Rorschach inkblots (Bat, Blood, Butterfly, Cloud, Fire, Mask, Mountain, and Smoke) and the six moods. Then, dual space is 10-dimensional (5 dimensions times 2), and total space is 12-dimensional $(8 - 1 + 6 - 1)$ so that the difference is 2.

The two distance matrices in dual space and total space are given in Tables 8 and 9. In this comparison, dual space and total space yield the following clusters as shown in Table 10.

Table 9 Between-set distances in dual space

	Fear	Anger	Depression	Ambition	Security	Love
Bat	1.92	2.78	2.34	3.23	3.43	3.09
Blood	3.44	3.88	3.96	4.33	4.61	4.44
Butt	3.13	3.26	3.01	2.24	3.49	2.7
Cloud	3.06	2.99	2.09	3.33	3.68	3.27
Fire	2.94	2.59	3.19	3.48	3.79	3.61
Mask	2.58	2.86	2.22	3.02	3.23	2.9
Mount	3.68	3.88	3.53	4.02	2.63	3.97
Smoke	4.67	4.01	4.63	5.03	5.00	5.05

Table 10 Clusters formed by different threshold percentile point

Filter threshold	Clusters	Responses	Moods
Clusters in total Space			
10%	A	Bat	Fear
	B	Cloud, Bat	Depression
	C	Butterfly	Ambition
	D	Mountain	Security
20%	A	Bat	Fear, Anger
	B	Cloud, Bat	Depression, Anger
	C	Butterfly	Ambition, Depression
	D	Mountain	Security
	E	Butterfly	Love
30%	A	Bat	Fear, Anger
	B	Cloud, Bat	Depression, Anger
	C	Butterfly	Ambition, Depression
	D	Mountain	Security
	E	Butterfly, Bat, Cloud	Love
40%	A	Bat	Fear, Anger
	B	Cloud, Bat	Depression, Anger
	C	Butterfly	Ambition, Depression
	D	Mountain	Security, Depression
	E	Butterfly, Bat, Cloud	Love
50%	A	Bat, Blood	Fear, Anger
	B	Cloud, Bat	Depression, Anger
	C	Butterfly	Ambition, Depression
	D	Mountain	Security, Depression
	E	Butterfly, Bat, Cloud	Love
Clusters in dual space			
10%	a	Bat	Fear
	b	Mask, Cloud, Bat	Depression
	c	Butterfly	Ambition
20%	a	Bat, Mask	Fear
	b	Mask, Cloud, Bat	Depression
	c	Butterfly	Ambition, Love
	d	Fire, Bat	Anger
	e	Mountain	Security
30%	a	Bat, Mask	Fear
	b	Mask, Cloud, Bat, Butterfly	Depression
	c	Butterfly	Ambition, Love
	d	Fire, Bat, Mask, Cloud	Anger
	e	Mountain	Security

(continued)

Table 10 (continued)

Filter threshold	Clusters	Responses	Moods
40%	a	Bat, Mask	Fear
	b	Mask, Cloud, Bat, Butterfly	Depression
	c	Butterfly, Bat	Ambition, Love
	d	Fire, Bat, Mask, Cloud	Anger
	e	Mountain	Security
50%	a	Bat, Mask	Fear
	b	Mask, Cloud, Bat, Butterfly	Depression
	c	Butterfly, Bat	Ambition, Love
	d	Fire, Bat, Mask, Cloud	Anger
	e	Mountain, Mask	Security

Let us again summarize the main outcomes of exploratory clustering.

- Again in this case too, we can see that more variables are involved in clustering in dual space than in total space.
- Without specifying detailed comparisons, the results of exploratory clustering in total space are closer to those of dual space than the above case in which the dimensionality of the two spaces is wider than the current case.
- Even at the 50% threshold value, Mask and Smoke are not included in total space clustering, and Blood and Smoke are not included in dual space.
- In total space, a strange combination of Depression and Ambition appears, while in dual space, Ambition appears with Love and Depression is a sole existence, which common sense would understand.
- Another strange result in total space is the combination of Depression and Security—why are they clustered together? In dual space Security is a lone existence.

These are only some observations, and what we can conclude are: first we must deal with the question of whether or not we should seek the threshold value in such a way that all variables are included in some clusters; second we should use dual space to capture the essence of the between-set relations; thirdly, the use of total space for exploratory clustering appears to bring in information which is not relevant to the objective of bi-modal clustering.

3 Concluding Remarks

Following the thorough historical background for this giant leap in quantification theory as presented in the first part of the paper by Nishisato, this paper provided numerical examples to draw some conclusions on the new approach to quantification theory. The main new developments are twofold: the strategic shift from popular, but problematic, joint graphical display to exploratory clustering of the quantification results—we observed that the correct way for joint graphical display requires doubling the multidimensional space, which makes graphical displays even more difficult than what is done in practice today; the second, but equally important, point is to apply cluster analysis to the between-set distance matrix, unlike the traditional within-set analysis, which makes it certain that the resultant clusters are bi-modal, that is, clusters consisting of both row variables and column variables of the contingency table. Under these objectives, we went on a step further to quantify the condensed response-pattern table, generated by the contingency table, which led to the distinction between dual space and total space. Noting that a typical between-set distance matrix is rectangular, we applied Nishisato's exploratory clustering method (Nishisato 2014) to the between-set distance matrices calculated in dual space and total space. The between-set distance matrix calculated in dual space is identical to the between-set distance matrix obtained by total information analysis (Nishisato and Clavel 2010). Theoretically, at the onset of the study, we arrived at the conclusion that cluster analysis of the between-set distance matrix in dual space should be preferred to that in total space. This was partly supported by the fact that the between-set distance matrix calculated in dual space is identical to the between-set distance matrix used in total information analysis TIA, which is the traditional extension of dual scaling to the doubled space. We should further note that bi-modal clustering is to identify clusters consisting of both row variables and columns variables of the contingency table, and that it is natural to consider the joint space of both row and column variables, which is dual space. Nevertheless, the current study also investigated total space for the reason that we need total space to reproduce the response-pattern table, generated from the contingency table.

We used the exploratory clustering method with flexible filters as proposed by Nishisato (2014) for the purpose to demonstrate analysis of between-set distance matrices. As is obvious, this clustering method is totally descriptive and exploratory, and requires much more elaborations before it can be regarded as one of the routine methods of cluster analysis. It is interesting, though, that the method does not depend on any optimization algorithm nor a cluster model, and as such it has also revealed the "truth" of data structure, more concretely that there were a number of overlapping clusters in data. This suggests an interesting problem that most cluster analysis methods which are currently in use impose the condition that the clusters are non-overlapping. Our numerical results strongly suggest that this condition of non-overlapping clusters may be too restrictive to understand data in hand. For the flexible filter method, the most urgent problem is how to define the most reasonable threshold value. This question requires much more theoretical and empirical work in response to such a question as whether or not all variables in the data set should be clustered into at least one cluster. As demonstrated in the current study, such an inclusive threshold value will undoubtedly result in many overlapping clusters.

The main contributions of the current two-part research [1] and [2] are: development of an alternative approach to joint graphical display, an implementation of analysis in doubled Euclidean space, analysis of between-set distance matrices, use of a cluster analysis method which can handle rectangular input matrices, and separation of dual space and total space for data analysis. The main task for the future research will be on the development of objective criteria for determining an optimal threshold percentile point for the flexible clustering procedure. As was demonstrated in the current study, it is very likely that real data would favor overlapping clusters, rather than forced non-overlapping clusters as typically done today.

References

Garmize, L.M., Rycklak, J.F.: Role-play validation of socio-cultural theory of symbolism. J. Consult. Psychol. (28), 107–115 (1964)

Nishisato, S.: Analysis of Categorical Data: Dual Scaling and Its Applications. The University of Toronto Press, Toronto (1980)

Nishisato, S.: Structural representation of categorical data and cluster analysis through filters. In: Gaul, W., Geyer-Schulz, A., Baba, Y., Okada, A. (eds.) German-Japanese Interchange of Data Analysis Results. pp. 81–90. Springer International Publishing, Cham (2014)

Nishisato, S.: Reminiscence: Quantification theory and graphs. Theory Appl. Data Anal. 8(1), 47–57 (2019) (in Japanese)

Nishisato, S., Clavel, J.G.: Total information analysis: comprehensive dual scaling. Behaviometrika 37(1), 15–32 (2010)

Linear Time Visualization and Search in Big Data Using Pixellated Factor Space Mapping

Fionn Murtagh

Abstract It is demonstrated how linear computational time and storage efficient approaches can be adopted when analysing very large data sets. More importantly, interpretation is aided and furthermore, basic processing is easily supported. Such basic processing can be the use of supplementary, i.e. contextual, elements, or particular associations. Furthermore, pixellated grid cell contents can be utilized as a basic form of imposed clustering. For a given resolution level, here related to an associated m-adic (m here is a non-prime integer) or p-adic (p is prime) number system encoding, such pixellated mapping results in partitioning. The association of a range of m-adic and p-adic representations leads naturally to an imposed hierarchical clustering, with partition levels corresponding to the m-adic-based and p-adic-based representations and displays. In these clustering embedding and imposed cluster structures, some analytical visualization and search applications are described.

1 Introduction

Often in data analytics the objective is to have clusters formed and studied, but even then, some open questions may need to be: to define the dissimilarity or distance measure to use; then to define the cluster optimization criterion, or the hierarchical agglomerative clustering criterion. This provides both motivation and justification for the following.

Our approach here is to assume a factor or principal component space, thoroughly taking semantic relationships into account, and that is endowed with the Euclidean metric. For original data that is comprised of categorical (qualitative) and quantitative values, Correspondence Analysis is most suitable. Since the factor space is constructed through eigenvalue, eigenvector decomposition of the source data, it

F. Murtagh (✉)
University of Huddersfield, Huddersfield, UK
e-mail: fmurtagh@acm.org

© Springer Nature Singapore Pte Ltd. 2020
T. Imaizumi et al. (eds.), *Advanced Studies in Classification and Data Science*,
Studies in Classification, Data Analysis, and Knowledge Organization,
https://doi.org/10.1007/978-981-15-3311-2_12

145

follows that if the number or rows, $n \gg m$, the latter here being the number of columns, then the computational requirement is for $O(m^3)$ processing time. This is likely to be achievable in practice.

A particular benefit of Correspondence Analysis is its suitability for carrying out an orthonormal mapping, or scaling, of power law distributed data. Power law distributed data are found in many domains. Correspondence factor analysis provides a latent semantic or principal axes mapping. Cf. Murtagh (2017b).

The case study used in this paper comes from the data studied in Murtagh (2016). It is a large set of Twitter, social media, data. There are 880,664 retained tweets and 481 retained dates. Our major aims here are (1) to simplify, in an easily interpretable way, the output, biplot or factor space planar plot, display; (2) to use an image-like approach to displaying such a plot; and (3) to relate our data to forms of data encoding that are other than real-valued and that are complementary to real-valued data. In Sect. 3.1, a first stage of agglomerative hierarchical clustering is at issue, but this is contiguity- or adjacency-constrained. While the latter terms are relevant, it is better expressed as being sequence-constrained where that constraint applies to what is to be clustered (to be seen later, this applies to row or column sets of grid cells).

Our priority is to consider here the two-dimensional, principal factor plane. This is produced from 880,664 Twitter tweets, sometimes also termed micro-blogs. In Fig. 1, rows, here tweets, and columns, here attributes that can be selected words in the tweets, and perhaps other tweet properties, these here are mapped into the principal plane of factors 1 and 2, so terms are used such as planar projection or mapping, and also being termed a biplot, with the mapping too of the three references in these tweets to the Cannes film festival. Used here are "C", "c" and

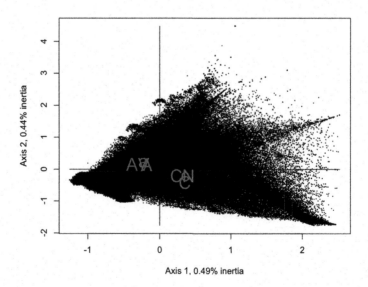

Fig. 1 880,664 Twitter tweets projected on the principal factor, i.e. principal axis plane, with attributes projected

"CN" for different use of upper can lower case in these references to the Cannes film festival. Also in this mapping, there are references to the Avignon theatre festival, using labels "A", "a" and "AV", relating to the user of upper can lower case letters in the tweets.

Our first task is to have an analogy of a two-dimensional histogram of what is displayed in the two previous figures. This can be advanced towards an image representation. Pixellation of a spatial domain can be displayed using heatmaps or false colour coding, implying the predominant visualization role for such display.

2 Algorithm for Determining and Contextual Displaying of a Two-Dimensional Histogram

Pixellating the data is equivalent to hashing, and for an image representation, the viewpoint is employed, of having this considered as a two-dimensional (2D) histogram.

2.1 Pixellating the Mapped Data Point Cloud

Firstly, given that the factor space is endowed with the Euclidean distance, the coordinates of the projected cloud of points can be rescaled. One reason for this is to have a standardized algorithmic processing approach. If there were to be particular topological patterns (e.g. the horseshoe effect, curvilinear pattern related to cluster ordering through embeddedness) or other targeted cluster morphologies (shapes, whether in two dimensions or in the full dimensional space), the analytics would focus on such patterns. Here, algorithmically, the general and generic objective is carried out by mapping the coordinates of the projected data cloud, in the factor plane, onto coordinates that are in the (closed lower bound, open upper bound) interval [0, 1). This mapping results from the interval of minimum value, maximum value, on each coordinate.

This then allows pixellation of the rescaled, unit square area containing the mapped cloud of points. This could also be generalized to unit volumes if more than two factors are at issue. Pixellation, i.e. imposing a regular grid, can be displayed as a grid structure. Furthermore, the pixel value is to be defined by the frequency of occurrence of points mapped into the given pixel area. To be both precise about what is displayed and computational and storage efficient, the frequency of occurrence values, comprising our pixel values here, will be displayed.

Such objectives in data analysis are largely consistent with the computational and interpretational advantages and benefits that are described in Murtagh et al. (2000). First, we may note that the heatmap display provides colour coding of the data matrix values. This is based on permutation of rows and columns with adjacency that is compatible with the hierarchical clustering of both rows and columns. Such visualization of hierarchical clustering is at issue in Zhang et al. (2017).

Now, a similar view of analysis might start with a very large data array and decide that a convenient and computationally efficient analysis process is as follows. Firstly, have the rows and columns permuted so that there is at least some relevance of the proximity taken into account by the permutation. In Murtagh et al. (2000), this was based on principal factor projections. Then, secondly, the row and column permuted input data array is considered such that all array values are the pixel values of an image. So, the input data array is thus considered as an image. Then, computationally efficient processing can be undertaken on the image representation of the input data array. Such processing is quite likely to involve a wavelet transform of the image, with filtering carried out in wavelet transform space. This provides a manner of determining clusters, and all is very relevant when visualization is a desirable outcome.

An initial process here of the analysis is the carrying out of the factor space mapping, endowed with the Euclidean metric, and that can be computationally efficient if, for example, having n rows, and m columns, then eigen-reduction that determines the factor space mapping, when $m \ll n$, is computationally, $O(m^3)$. This can be quite efficient, assuming that $m \ll n$. Then, all of the work here is, analogously, to map large data arrays into images, here directly based on the Euclidean metric endowed factor space.

Sample R code used for pixellation is available at this website, http://www.multiresolutions.com/strule/papers.

3   Visualization Through 2D Histogram Representation

Following Figs. 1 and 2 shows how we can have displays that are both informative and also that provide alternative display capability for very large numbers of projection locations, i.e. mapped rows or columns, observations and attributes.

In Fig. 2, it seems visually that the high frequency value in the grid cell that has a projection greater than 4 on axis 2, and greater than 1 on axis 1, is unacceptably high in value. This grid cell has 13,333 projected points. In fact, we verified that there are 13,333 overlapping points here. (Their values on factor 2 are all: 4.476693).

A heatmap may be derived from this representation, that is grid-based, and can be characterized as a two-dimensional histogram. The heatmap, being false colour coding of the data matrix being hierarchically clustered, in regard to both the row set and the column set, is used in Zhang et al. (2017). Such a heatmap display has a somewhat different objective, relative to what will now follow. Our later aim is to have the data structured to support search and retrieval.

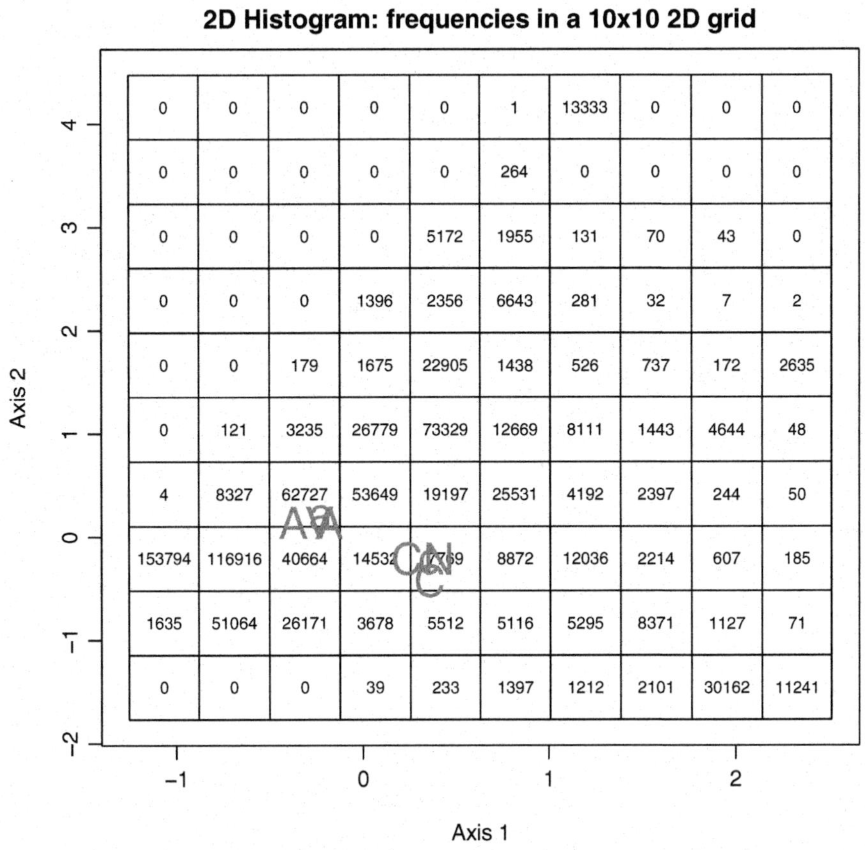

2D Histogram: frequencies in a 10x10 2D grid

0	0	0	0	0	1	13333	0	0	0
0	0	0	0	0	264	0	0	0	0
0	0	0	0	5172	1955	131	70	43	0
0	0	0	1396	2356	6643	281	32	7	2
0	0	179	1675	22905	1438	526	737	172	2635
0	121	3235	26779	73329	12669	8111	1443	4644	48
4	8327	62727	53649	19197	25531	4192	2397	244	50
153794	116916	40664	14532	7769	8872	12036	2214	607	185
1635	51064	26171	3678	5512	5116	5295	8371	1127	71
0	0	0	39	233	1397	1212	2101	30162	11241

Axis 2

Axis 1

Fig. 2 A 10×10 grid display with supplementary elements projected. These projections, associated with tweet content, are related to the Cannes Film festival and the Avignon Theatre festival

3.1 Varying Number Theoretic Representations of the Data: m-Ary and p-Adic Representations

Having pixellated the projected cloud of points, this is based on associating each projected point with its grid cell. Integral to this is that the grid cell containing the projected point becomes an expression that labels each of its grid cell members. This is analogous to the projected point being a member of a cluster, and perhaps it is also analogous to the projected point being conceptually characterized by the superset of projected points associated with the grid cell. Since the grid cells are defined by default as decimal numbers, that will also be termed here, 10-ary, i.e. m-ary with $m = 10$, we will next consider alternative number theoretic representations.

With p being a prime number and m being a non-prime integer, we will consider the best-fitting grid cell mapping representations that are derived from the decimal or $m = 10$, m-ary, representation. We considered: $m = 9$, $m = 8$, $p = 7$, $m = 6$, $p = 5$, $m = 4$, $p = 3$, $p = 2$, number representations.

Considerable background discussion on p-adic number systems, and their role in various domains, is in Murtagh (2017a). Specifically using the algorithm now to be described, for closest fit of one number theoretic system to another, this is described in Murtagh (2016).

Algorithmically to move from an $m = 10$ m-ary representation to an $m = 9$ m-ary representation, we take the definition of the grid cells from their projections on factor 1 and on factor 2. We first consider factor 1, i.e. axis 1, with identical reasoning applied to factor 2, i.e. axis 2. Let v be the constant interval between grid lines. For $m = 10$, we have the axis 1 values as follows: $0, 0 + v, 0 + 2v, 0 + 3v, \ldots, 0 + 9v$. We now want to form grid cell boundaries for $m = 9$, so that on axis 1, we will have: $0, 0 + v', 0 + 2v', 0 + 3v', \ldots, 0 + 0 + 8v'$.

Because we want a closest fit by the 9-ary representation to the 10-ary representation, the former is based on the latter. We take the least difference between the total sum of successive grid bins. In effect, then, we merge these two successive grid bins. By relabelling higher valued grid bin sequence numbers, this directly provides us with a 9-ary representation.

The same approach is used for factor 2, i.e. axis 2. Then, we can proceed through further stages to find a best-fit 8-ary representation to the just determined 9-ary representation. The following stage is to find a best-fit 7-ary representation to the just determined 8-ary representation. This continues stagewise until we have a 2-ary representation, i.e. a binary representation of the grid cell axis 1 and axis 2 projections.

Further study of pixellation was for a 9×9 grid display, an 8×8 grid display, a 7×7 grid display, and so, continuing to a 2×2 grid display. The latter is associated with a binary encoding of the grid cell boundary coordinates.

Thus, to state what is at issue here, it is visualization through 2D (two-dimensional) histogram display, possibly accompanying m-adic and p-adic representation, in our mapped or represented data.

We may wish to determine a rather good display of the supplementary elements relative to particular grid cells.

There can be a display of local densities using the projected elements. Taking previous outputs, Fig. 3 displays local densities of the tweets, provided by the 10×10 grid. The grid cells can be considered as three-dimensional histogram bins. Then in Fig. 4, the supplementary variables are projected, relating to the words used for the Cannes Film Festival and the Avignon Theatre Festival. At issue here is largely the display.

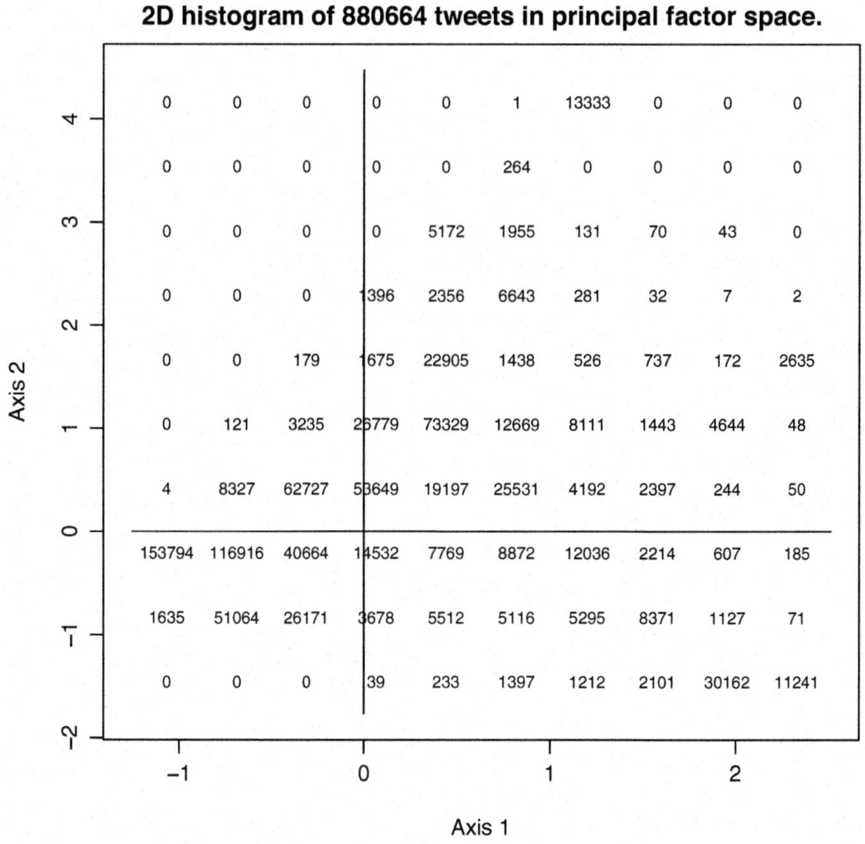

2D histogram of 880664 tweets in principal factor space.

0	0	0	0	0	1	13333	0	0	0
0	0	0	0	0	264	0	0	0	0
0	0	0	0	5172	1955	131	70	43	0
0	0	0	396	2356	6643	281	32	7	2
0	0	179	675	22905	1438	526	737	172	2635
0	121	3235	26779	73329	12669	8111	1443	4644	48
4	8327	62727	53649	19197	25531	4192	2397	244	50
153794	116916	40664	14532	7769	8872	12036	2214	607	185
1635	51064	26171	3678	5512	5116	5295	8371	1127	71
0	0	0	39	233	1397	1212	2101	30162	11241

Axis 2 (vertical): 4, 3, 2, 1, 0, −1, −2

Axis 1 (horizontal): −1, 0, 1, 2

Fig. 3 The grid structure, used for the two-dimensional histogram binning, is not displayed but the histogram bin values alone

4 Hashing and Binning for Nearest Neighbour Searching

In Murtagh (1993), a framework for all that is at issue here was described with relevance in the fields of information retrieval and kindred areas related to search and retrieval. For nearest neighbour searching, what is discussed initially is: hashing, or simple binning or bucketing. The grid structure cell constitutes the way forward for searching in other neighbours of the query point that have also been mapped into the one grid cell. Some consideration may need to be given to one or more adjacent grid cells if the query point is closer to a grid cell boundary, than it is to any potential nearest neighbour in the given grid cell. With uniformly distributed data, here in the two-dimensional context, then it is noted how constant time, i.e. $O(1)$ time, is the expectation (statistical first-order moment, mean) of the computational time. A proof of this is in Bentley et al. (1980).

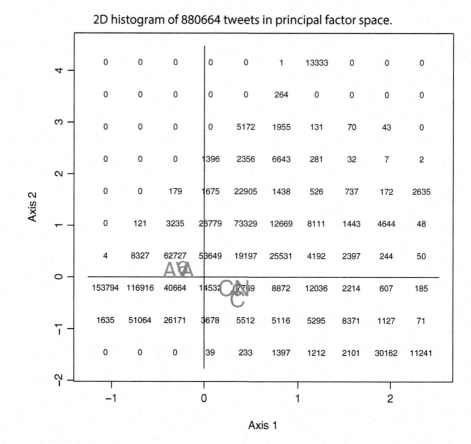

2D histogram of 880664 tweets in principal factor space.

Fig. 4 As previous figure, Fig. 3, with supplementary elements projected. These projections, associated with tweet content, are related to the Cannes Film festival and the Avignon Theatre festival

It is to be noted, Murtagh (1993, p. 33), that when searching requires use of adjacent multidimensional grid cells, by design hypercube in their hypervolume, then this implies a computational limitation that increases with dimensionality. In Murtagh (1993), for dimensionality reduction that supports nearest neighbour searching, reference is made to principal components analysis, non-metric multidimensional scaling, and self-organizing feature maps. In our work here, Correspondence Analysis provides semantic mapping, i.e. the factor that is, in effect, the principal component space, with a unified scale for both observations and attributes.

Extending this approach to higher dimensional spaces, there is the k-d (k-dimensional, for integer, k) tree, or multidimensional binary search tree. This is a balanced tree, by design, representation of the multidimensional point cloud. Stepwise binarization of the data is carried out using the median projection on each axis. However, the computational complexity of requiring the checking of adjacent clusters that are, by design, hyperrectangular has the following consequence: computationally such an approach needs to be limited in the dimensionality. Up to dimensionality of 8 is reported on in the literature.

Also discussed in Murtagh (1993) are bounds on the nearest neighbour distance, given a candidate nearest neighbour; other such bounding using metric properties, especially the triangular inequality; branch and bound is the at issue in such methodology; for high dimensional, sparse, binary data, and where binary here represents presence or absence values in, e.g., keyword-based document data (as an example, there could be 10,000 documents, crossed by the presence or absence of 10,000 keywords). For the high dimensional, sparse, binary data, such data have always been traditional in information retrieval, and what is used is the mapping of documents to keywords, and the inverted file that is the mapping of keywords to documents. Cluster-based retrieval can extend some of these approaches. Further discussion in Murtagh (1993) is for range searching using the quadtree, for two-dimensional images, the octree, for three-dimensional data cubes (just one example here is a three-dimensional image volume), and a quadtree implementation on a sphere, for spherical data. The latter is relevant for remotely senses earth data, and for cosmological data. Range searching involves moving beyond location-oriented search.

5   Multidimensional Baire Distance

An open issue, motivated by this work, is to aim at having a multidimensional Baire distance. This could be based on the following. Take a full factor space, perhaps with 5 factors retained (as it the default in the FactoMineR package in R), and for such labels here as C, A, etc., looking at grid binned factor pairwise (biplots) supplementary mapping. This is to see what grid cells are relevant for the supplementary elements. But, based on this approach, we may have supplementary rows or individuals, that with Twitter data, is to then do digit-wise mapping of tweets against the selected supplementary elements.

In general, related to such a multidimensional Baire distance is the Baire distance formulated for multi-channel data, i.e. hyperspectral images, and used for machine learning (Support Vector Machine, supervised classification) in Bradley and Braun (2015).

6   Conclusion

A central theme of this work can be expressed as follows: performing data mapping that results in a domain-relevant data encoding. One aim of the work here has been to benefit from just how very evident it is, that human visual-based information, and both measure and approximated data, become very efficient as well as effective. At issue are: image, display, and biplot. Further practical benefits are demonstrated in Murtagh et al. (2000), by representing the data to be analysed as an image, and thereby carrying out wavelet transform-based filtering, and object detection, and so on.

Informally expressed, therefore, we may state that this work is in relation to the visualization of data that accompanies having the data verbalized: see Blasius and Greenacre (2014) for this phrasing. The application objectives cover data mining (as contrasted with supervised classification and mainstream data mining), data analytics (encompassing both processing and output), and inductive reasoning (what the analysis achieves). Furthermore, the computational complexity of the processing and all of the implementation is implicit in this work.

Largely, the terms used here, pixellation and a 2D histogram, are identical. Search in Big Data is the basis for matching, such as in nearest neighbour matching, and associated or relevant data querying. Although not always done, from a mathematical sciences point of view, well-based, past work ought to be cited. In Murtagh (1985), chapter 2 entitled "Fast nearest neighbour searching" covers multidimensional binary search tree structuring of the data and hashing for nearest neighbour or best match searching, accompanied by implementation optimization, and with twenty-two references on these, and directly related, themes.

All that is at issue here is open to the possibility of implementation using distributed computing. For future research, there can be applications that are well associated with this work, and a further objective can well be to have distributed computing implementations set up and having them functioning operationally.

References

Bentley, J.L., Weide, B.W., Yao, A.C.: Optimal expected time algorithms for closest point problems. ACM Trans. Math. Softw. **6**, 563–580 (1980)

Blasius, J., Greenacre, M. (eds.): Visualization and Verbalization of Data. Chapman and Hall/CRC Press, London/Boca Raton (2014)

Bradley, P.E., Braun, A.C.: Finding the asymptotically optimal Baire distance for multi-channel data. Appl. Math. **6**(3), 484–495 (2015)

Murtagh, F.: Multidimensional Clustering Algorithms. Physica-Verlag, Vienna-Würzburg (1985)

Murtagh, F.: Search algorithms for numeric and quantitative data. In: Heck, A., Murtagh, F. (eds.) Intelligent Information Retrieval: The Case of Astronomy and Related Space Sciences, chap. 4, pp. 29–48. Kluwer, Dordrecht (1993)

Murtagh, F.: Sparse p-Adic data coding for computationally efficient and effective big data analytics. p-Adic Number. Ultrametric Anal. Appl., **8**(3), 236–247 (2016). https://arxiv.org/abs/1604.06961

Murtagh, F.: Semantic mapping: Towards contextual and trend analysis of behaviours and practices. In: Balog, K., Cappellato, L., Ferro, N., MacDonald, C. (eds.) Working Notes of CLEF 2016 - Conference and Labs of the Evaluation Forum, Évora, 5–8 September, 2016, pp. 1207–1225 (2016). http://ceur-ws.org/Vol-1609/16091207.pdf

Murtagh, F.: Data Science Foundations: Geometry and Topology of Complex Hierarchic Systems and Big Data Analytics. Chapman and Hall, CRC Press, New York/Boca Raton (2017a)

Murtagh, F.: Big data scaling through metric mapping: Exploiting the remarkable simplicity of very high dimensional spaces using correspondence analysis. In: Palumbo, R. et al. (eds.) Data Science, Studies in Classification, Data Analysis, and Knowledge Organization, pp. 279–290. Springer, Berlin (2017b)

Murtagh, F., Starck, J.-L., Berry, M.W.: Overcoming the curse of dimensionality in clustering by means of the wavelet transform. Comput. J. **43**(2), 107–120 (2000)

Zhang, Z., Murtagh, F., Van Poucke, S., Lin, S., Lan, P.: Hierarchical cluster analysis in clinical research with heterogeneous study population: highlighting its visualization with R. Ann. Trans. Medicine **5**(4), (2017). http://atm.amegroups.com/article/view/13789/pdf

From Joint Graphical Display to Bi-Modal Clustering: [1] A Giant Leap in Quantification Theory

Shizuhiko Nishisato

Abstract Joint graphical display is popular in correspondence analysis as a tool to illustrate the results of quantification analysis, known by many names such as dual scaling, correspondence analysis, homogeneity analysis, and optimal scaling. Most methods of joint graphical display, however, are at best rough approximations to the analytical results. The present study starts with an extensive historical overview of joint graphical display as opposed to algebraic analysis without graphs to identify some perennial problems surrounding graphical display in quantification theory. This review leads to the conclusion that a better alternative to joint graphical display may be a special form of cluster analysis, called bi-modal clustering, defined later. The current study, therefore, marks the beginning of replacing the long-standing graphical approach with cluster analysis. This transition also implies a giant leap in quantification theory from the traditional heavily weighted within-set analysis to between-set analysis. Recently, some studies have been carried out in the new direction, and this paper is devoted to the historical background for this new approach. Some controversial views will be presented here, but it is hoped that the paper will lay a foundation for the future direction of quantification theory. The paper by Clavel and Nishisato, which accompanies this paper, will show some preliminary numerical results with the hope that future problems of this new horizon will be identified.

1 To Begin With

In the history of quantification theory (see, for example, Nishisato 1980; Greenacre 1984; Gifi 1990; Beh and Lombardo 2014), we can identify two extreme positions of emphasis, one "no emphasis" on joint graphical display of quantification results and the other "strong emphasis." During the past 50 years or so, joint graphical

S. Nishisato (✉)
University of Toronto, Toronto, ON, Canada
e-mail: shizuhiko.nishisato@utoronto.ca

© Springer Nature Singapore Pte Ltd. 2020
T. Imaizumi et al. (eds.), *Advanced Studies in Classification and Data Science*,
Studies in Classification, Data Analysis, and Knowledge Organization,
https://doi.org/10.1007/978-981-15-3311-2_13

display of rows and columns of the contingency table has become the mainstream of quantification theory, and the algebraic approach without graphical display has been pushed away into oblivion. The current study is an attempt to remedy this situation to resuscitate the non-graphical algebraic approach as a better strategy for meaningful multidimensional data analysis than the approach of joint graphical display, plagued with theoretical justifications. In this new context, the traditional within-set analysis of quantification theory is extended to between-set analysis. This is a tall order, but it is important to realize that the current within-set analysis, coupled with joint graphical display, does not offer a good approximation to the exact representation of data structure, and therefore should be used only with serious cautions that Lebart et al. (1977) insightfully pointed out 43 years ago.

2 Bock–Nishisato Versus Benzécri–Greenacre Streams

In the history of quantification, we have seen two contrasting teams of supervisor (mentor)–student relations:

Bock–Nishisato Team Bock (1960) published a famous article on optimal scaling and supervised Ph.D. thesis on "minimum entropy clustering of test items" (Nishisato 1966). In turn, Nishisato (1980) published a book on dual scaling, the name coined by him for optimal scaling.

Benzécri–Greenacre Team Benzécri et al. (1973) published the bible of French "analyse des correspondances" and his student Greenacre (1984) published a very popular book on correspondence analysis in English.

As mentioned above, the two teams created contrasting approaches: the Bock–Nishisato stream did not emphasize joint graphical display at all, while the Benzécri–Greenacre stream emphasized joint graphical display to the extent that correspondence analysis has become almost synonymous to graphical analysis.

3 Early Publications in English

Historically, the book by Nishisato (1980) appears to be one of the first English books on quantification theory, yet, probably due to the popularity of joint graphical analysis in recent years, some claim that Greenacre (1984) is the first English book. This discrepancy may have arisen because (Nishisato 1980) discussed "dual scaling" which he coined, while (Greenacre 1984) presented the now popular "correspondence analysis." (Note: the story behind various names for quantification theory is a topic for another paper). It should be remembered, however, that Nishisato (1980) covered not only simple correspondence analysis (dual scaling of contingency tables) and multiple correspondence analysis (dual scaling of multiple-choice data) but also dual scaling of paired comparison and rank-order data, dual scaling of multidimensional tables, analysis of variance of categorical data (i.e., data with a designed structure for the row, a forerunner of multi-way analysis by Nishisato and Lawrence (1989), or that of conditional principal component analysis by Takane and Shibayama (1991)), order constraints, and missing responses, the only intentional omission being graphical display. This omission of graphical display may be another reason why Greenacre (1984) is considered to be the first English book. There is also a historical background that researchers at the Soviet Academy of Sciences chose Nishisato (1980) for Russian translation and that the contract was signed between the University of Toronto Press, Canada and the Finansy I Statistika Publishing House in Moscow, USSR. B. Mirkin and S. Adamov completed the Russian translation in 1990, just before the collapse of the USSR. This historical event, however, led to the closure of the publisher, and in spite of the efforts for finding funds through 1994, the publication of the Russian translation of Nishisato (1980) was never realized.

Please note that the above discussion should not give a wrong impression that researchers in English speaking community emerged after the Hayashi School in Japan (e.g., Hayashi 1950) or the Benzécri School in France (e.g., Benzécri et al. 1973). On the contrary, quantification theory was well known in the English community and researchers were well versed in the developments before Hayashi or Benzécri school. See, for example, Richardson and Kuder 1933; Hirschfeld 1935; Horst 1935; Fisher 1940, 1948; Guttman 1941, 1946; Maung 1941; Mosier 1946; Hayashi 1950, 1952, 1954, 1964, 1967, 1968; Williams 1952; Bock 1956, 1960; Lord 1958; Torgerson 1958; Baker 1960; Slater 1960; Kendall and Stuart 1961; Bradley et al. 1962; Lingoes 1964, 1968; Shiba 1965; McKeon 1966; Whittaker 1967; McDonald 1968; Benzécri 1969; Gabriel 1971, 1972; Nishisato 1971, 1972, 1973, 1976, 1978, 1979; Nishisato and Inukai 1972; de Leeuw 1973; Hill 1973, 1974; Nishisato and Arri 1975; Nishisato and Leong 1975; Teil 1975; de Leeuw et al. 1976; Healy and Goldstein 1976; Young et al. 1976; Gauch et al. 1977; Greenacre and Degos 1977, Lebart et al. 1977; Takane et al. 1977; Greenacre 1978; Iwatsubo 1978; Lang 1978; Tanaka 1978; Van Rijckevorsel and de Leeuw 1978; Young et al. 1978; McDonald et al. 1979.

Many of these are cited in Nishisato (1980). In addition, a book by Gauch (1982) was published 2 years before (Greenacre 1984), and we also have Lebart et al. (1984). In the late 1970s and early 1980s, there was an impressive surge of publications by such Dutch researchers as de Leeuw, van de Geer, van Ricjckevorsel, Heiser, Meulman, van der Burg, van der Heijden, ter Braak, ten Berge, Israls, Sikkel, Stoops, Kroonenberg, Mooijart, Bethlehem, Bettonvil, and Wansbeek (Note: due to the space, their publications are not included in References). Their work was eventually culminated into an outstanding book (Gifi 1990). In the 1980s, we saw enhanced and excellent communications among researchers from different parts of the world. Noting that recent researchers tend to cite only relatively new publications, one should remember that many papers on quantification theory were published before 1984 in English. There were also many publications in French and Japanese as well in early days.

4 Principal Coordinates and Standard Coordinates

To identify the basic aspect of Euclidean graphs, let us consider continuous variables. Suppose we have a set of standardized continuous variables, which span two-dimensional space. Then, if we draw a two-dimensional graph, all the variables are located somewhere on the circle of radius 1 (see Fig. 1a); similarly, if a set of standardized variables span three-dimensional space, each variable is located at a distance of 1 from the origin on the sphere (Fig. 1b). These configurations, whether a circle or a sphere, remain the same, irrespective of the eigenvalues. The eigenvalues are associated with the distributions of the data points on the circle or the sphere: if the first eigenvalue is much larger than the second eigenvalue in the two-dimensional data, most variables would be close to the both ends of the first (horizontal) axis, while only a few would be located close to the both ends of the second (vertical axis) (Fig. 1c). No matter what, these positions of the variables are at the distance of 1 from the origin. The corresponding coordinates of the variables are called **principal coordinates** (or any orthogonal transformations of principal coordinates will do). In data analysis, we also use **standard coordinates**, which are normed to the same constant for all the axes, irrespective of the eigenvalues. Thus, in the two-dimensional case, corresponding to Fig. 1c, those points lying along the vertical axis (the smaller eigenvalue) are located much further than 1 from the origin and those points lying along the horizontal axis (the larger eigenvalue) are located much closer than 1 from the origin. Therefore, one can conclude that the data points on the standard coordinates can no longer depict the original structure of data, or rather their locations have nothing to do with the geometry of the data. Note therefore that only the principal coordinates provide the configuration of the data. This point was stressed in Nishisato (1996, 2016). We should also note that the principal coordinates are the square root of the eigenvalue times the standard coordinates, and that the square root of the eigenvalue is the singular value.

[a]   [b]   [c]

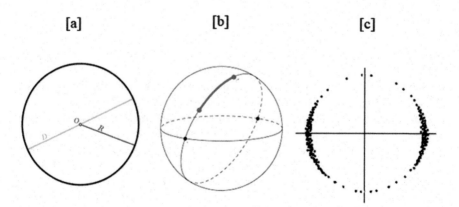

Fig. 1 Circle, sphere of radius 1, and 2-dimensional configuration of data

5 Joint Graphical Display as a Problematic Approach

In quantification of the standardized contingency table $\mathbf{C} = (c_{ij})$, we use singular-value decomposition, $\mathbf{C} = \mathbf{Y} \Lambda \mathbf{X}'$, or more precisely as

$$c_{ij} = \frac{c_{i.}c_{.j}}{c_{..}}[1 + \rho_1 y_{i1}x_{1j} + \rho_2 y_{i2}x_{j2} + \cdot + \rho_K y_{iK}x_{jK}$$

where ρ_k is the k-th singular value, Λ is the diagonal matrix of singular values, K is the rank of $\mathbf{C} - 1$, y_{ik} is the i-th element of the left-singular vector k, x_{kj} is the j-th element of the right-singular vector k, $c_{i.}$ is the i-th row marginal of $\mathbf{C}$, $c_{.j}$ is the j-th column marginal of $\mathbf{C}$, and $c_{..}$ is the total frequency of $\mathbf{C}$. In this expansion, principal coordinates for rows and columns are $\rho_k y_{ik}$ and $\rho_k x_{kj}$, standard coordinates are y_{ik} and x_{kj}. Popular joint graphical methods which are currently used are as follows:

Symmetric Plot (Also Called French Plot) We plot $\rho_k y_{ik}$ and $\rho_k x_{kj}$ in the same graph, but these variates do not span the same space. Thus, if we want to be exact, a unidimensional symmetric graph, for example, is a summary of a two-dimensional graph, the reason why Lebart et al. (1977) warned not to calculate the distance of two between-set points in a symmetric graph. The fact that the graph uses principal coordinates is correct, but the graph in reality needs doubled dimensions because the axis for rows and the axis for columns are separated by an angle (θ) equal to the arccosine of the singular value, that is, $\theta = \arccos(\rho)$. Thus, under the symmetric scaling, a four-dimensional configuration, for example, is presented by a two-dimensional graph, corresponding to the first two largest singular values, thus making the graph at best a rough approximation to the correct graph.

Non-symmetric Graph We plot ρy_{ik} and x_{kj} or y_{ik} and ρx_{kj}. This graph is sometimes considered logically correct since this involves projection of one variate onto the space of the other variable, placing both on the same space. Note that ρ is known in dual scaling as a projection operator. Although the idea behind non-symmetric graph may appear logically correct, we should know that ρy_{ik}, for example, is the projection of the standard coordinate onto the standard space of the other variable. As previously noted, standard coordinates do not reflect information in data. Therefore, the projection of standard coordinates of one set of variables onto the standard space of the other set of variables cannot properly represent a graph of data. Thus, unlike the general belief, a non-symmetric graph is not logical and it is not a graph of the data structure at all.

Biplot (See Gabriel 1971) The singular decomposition of $\mathbf{Y} \Lambda \mathbf{X}'$ is expressed as $\mathbf{Y} \Lambda^\alpha \Lambda^\beta \mathbf{X}'$, where $\alpha + \beta = 1$. In the context of joint graphical display, when $\alpha = 1$, or when $\beta = 1$, this decomposition provides non-symmetric joint graphical display, that is, the one we have rejected above. There is no form for symmetric graphical display because symmetric decomposition means that $\alpha = \beta = 0.5$, and we should note that the square root of the singular value is not a projection operator (recall that the singular value is the projection operator). For the correct graphical representation, we must have that $\alpha = 1, \beta = 1$ and must use doubled space. In biplot, however, doubling the space is not a considered operation.

So, none of these popular joint graphical methods would provide a correct Euclidean representation of data. In this regard, the strong emphasis on joint graphical display by the Benzécri–Greenacre stream faces a serious drawback against the popularity that it has so far enjoyed. But, because of the established popularity of joint graphical display, it would be difficult for most researchers to give up joint graphical display. Our view, however, is that researchers should be warned about its inherent problems and pose for a moment of reflection. If one must choose between symmetric scaling and non-symmetric scaling, choose symmetric scaling or French plot, and remind yourself of the warning by Lebart et al. (1977). See also an interesting debate on the Carroll–Green–Schaffer scaling (Caroll et al. 1986; Greenacre 1989).

6 Total Information Analysis or Comprehensive Dual Scaling

Nishisato and Clavel (2010) proposed total information analysis (TIA) or comprehensive dual scaling (CDS). The basic idea is to extend the traditional within-set analysis in quantification theory to both within-set and between-set analyses. The procedure is to extract all components from the contingency table by dual scaling, and then using all components we calculate the within-row distance matrix $\mathbf{D_{yy}}$, the between-row-column distance matrix $\mathbf{D_{yx}}$, the between-column-row distance matrix $\mathbf{D_{xy}}$, and the within-column distance matrix$\mathbf{D_{xx}}$, resulting in the super-distance matrix $\mathbf{D}$, consisting of these four sub-matrices. It is known that $\mathbf{D}$ spans the space with doubled dimensions of the original contingency table, thus no more problem of discrepancy between row space and column space. This doubled space is called dual space. Because their analysis deals with doubled multidimensional space, which is likely to make correct joint graphs impractical, they suggest subjecting $\mathbf{D}$ to cluster analysis.

Clavel and Nishisato (2020) then discovered that cluster analysis of the super-distance matrix often yields within-set clusters (clusters consisting of only row variables or only column variables), which defeats the original purpose of quantification of the contingency tables, that is, to find relations between rows and columns. Their conclusion then is to subject the between-set distance matrix $\mathbf{D_{yx}}$ to cluster analysis, which guarantees finding between-set clusters. This is called **bi-modal** clustering. Notice that almost all varieties of clustering methods deal with only square distance matrices, hence fail to deal with typically rectangular between-set distance matrices, they recommended clustering with the p-percentile filter, or more generally clustering with flexible filters (Nishisato 2014), which simply screen out relatively large distances to identify clusters from the remaining small distances. We will look at some applications of this simple procedure in the paper by Clavel and Nishisato.

7 Contingency Table and Response-Pattern Table

Relations between the contingency table and the corresponding response-pattern table are thoroughly described in Nishisato (1980), where he showed that the response-pattern table, derived from the contingency table, requires twice the dimensions for the contingency table when the number of rows, m, and that of columns, n, are equal, and more than twice the dimensions for the contingency table when $m \neq n$. Nishisato (2019) has identified two ways to calculate the between-set distance matrix from a given contingency table, one in dual space and the other in total space. This distinction will be demonstrated in this section.

Let us use an example, which is a subset of the (Garmize and Rychlak 1964) data on Rorschach inkblots and induced moods as shown in Table 1, the corresponding condensed form of the response-pattern matrix (Table 2), and the

Table 1 Rorschach responses and induced mood

Responses	Fear	Anger	Depression	Love	Ambition	Security
Bat	33	10	18	1	2	6
Butterfly	0	2	1	26	5	18
Mountains	2	1	4	1	18	2

Contingency table
Notes: Selection from Garmize and Rychlak (1964, Table 1, p. 111)

Table 2 Rorschach responses and induced mood

Bat	Butterfly	Mountains	Fear	Anger	Depression	Love	Ambition	Security
33	0	0	33	0	0	0	0	0
10	0	0	0	10	0	0	0	0
18	0	0	0	0	18	0	0	0
1	0	0	0	0	0	1	0	0
2	0	0	0	0	0	0	2	0
6	0	0	0	0	0	0	0	6
0	2	0	0	2	0	0	0	0
0	1	0	0	0	1	0	0	0
0	26	0	0	0	0	26	0	0
0	5	0	0	0	0	0	5	0
0	18	0	0	0	0	0	0	18
0	0	2	2	0	0	0	0	0
0	0	1	0	1	0	0	0	0
0	0	4	0	0	4	0	0	0
0	0	1	0	0	0	1	0	0
0	0	18	0	0	0	0	18	0
0	0	2	0	0	0	0	0	2

Response-Pattern table
Notes: Selection from Garmize and Rychlak (1964, Table 1, p. 111)

Table 3 Principal coordinates of variables in dual space and total space

Variables	1	2	6	7	3	4	5
Bat	−0.92	−0.39	−0.19	−0.32	0.00	0.00	0.00
Butterfly	1.19	−0.49	−0.23	0.41	0.00	0.00	0.00
Mountains	0.09	1.88	0.91	0.03	0.00	0.00	0.00
Fear	−1.10	−0.42	0.20	0.38	−0.01	−1.25	0.29
Anger	−0.66	−0.37	0.18	0.23	2.74	1.23	0.78
Depression	−0.83	−0.00	0.00	0.28	−1.48	1.75	−0.43
Ambition	1.37	−0.64	0.31	−0.47	−0.09	0.11	1.23
Security	0.29	1.97	−0.95	−0.10	0.20	−0.29	0.16
Love	0.79	−0.45	0.22	−0.27	−0.14	−0.31	−1.88
Dual space	***	***	***	***	−	−	−
Total space	***	***	***	***	***	***	***

Notes: *** indicates a check mark

principal coordinates of both Rorschach inkblots and induced moods in dual space (components 1, 2, 6, and 7) and total space (all seven components) (Table 3). As discussed by Nishisato (2019), dual space is the space common for both rows and columns of the contingency table, and total space consists of seven components. In the current example, the extra components are for the induced moods (components 3, 4, and 5) where the corresponding principal components for the Rorschach inkblots are all zero, that is, no more variations left among the Rorschach inkblots.

Dual space is the space in which all relations between rows and columns are accommodated, thus, consisting of components 1, 2, 6, and 7 (see Nishisato 2019, how these components are identified), while total space is that accommodates these joint contributions and the unique contributions of induced moods in the current example.

7.1 Justification of Total Space

First of all, note that the seven components are needed to reproduce the response-pattern table. Secondly, consider the situation in which we are interested in the distance judgments between two sets of color chips, where one set consists of chips which vary in hue and brightness and the other set which consists of color chips that vary in hue, brightness, and saturations. If we assume that the first set spans two dimensions (hue, brightness) and the second set three dimensions (hue, brightness, and saturation), it is quite legitimate for us to calculate distances between two sets of these color chips, that is, distances between variables in different dimensions. This is the case in which one must consider total space for between-set distances.

Which space we should use for distance calculation, dual space or total space, is something that remains unsettled at this stage of research, and the choice is likely to rest on the opinion of the researchers, and therefore this matter of choice will be left for future investigation.

8   Bi-Modal Clustering in Dual Space and Total Space

With this much background information, the following paper by Clavel and Nishisato will be devoted to numerical illustrations of bi-modal clustering of the between-set distance matrix and total space. Because of the philosophical problem of which space to analyze, the current study will remain exploratory, and its main goal is to provide some numerical results which can be used for further discussion of analysis in dual space versus total space. It is hoped that the information provided by the two papers will offer strong enough evidence in support of bi-modal clustering as a method preferred to the currently popular joint graphical analysis.

Acknowledgement The author wishes to express his heart-felt thanks to Prof. José Garcia Clavel of Universidad de Murcia, Spain for his great assistance and helpful comments.

References

Baker, F.G.: Quantifying qualitative variables by the method of reciprocal averages. Occasional Paper No. 7, Laboratory of Experimental Design, Department of Educational Psychology, University of Wisconsin (1960)

Beh, E.J., Lombardo, R.: Correspondence Analysis: Theory, Practice and New Strategies. Wiley, Hoboken (2014)

Benzécri, J.-P.: Statistical analysis as a tool to make patterns emerge from data. In: Watanabe, S. (ed.) Methodologies of Pattern Recognition, pp. 35–74. Academic, New York (1969)

Benzécri, J.-P., et al.: L'analyse des données: II. L'analyse des Correspondances. Dunod, Paris (1973)

Bock, R.D.: The selection of judges for preference testing. Psychometrika **21**, 349–366 (1956)

Bock, R.D.: Methods and Applications of Optimal Scaling. The University of North Carolina Psychometric Laboratory Research Memorandum No.25 (1960)

Bradley, R.A., Katti, S.K., Coons, I.J.: Optimal scaling for ordered categories. Psychometrika **27**, 355–374 (1962)

Caroll, J.D., Green, P.E., Schaffer, C.M.: Interpoint distance comparisons in correspondence analysis. J. Market. Res. **23**, 271–280 (1986)

Clavel, J.G., Nishisato, S.: From joint graphical display to bi-modal clustering. In: Imaizumi, T. et al (eds.) Dual space versus total space. Advanced Studies in Classification and Data Science. Springer Nature Singapore 16 (2020)

de Leeuw, J.: Canonical Analysis of Categorical Data. Psychological Institute, Leiden University (1973)

de Leeuw, J., Young, F.W., Takane, Y.: Additive structure in qualitative data: an alternating least squares method with optimal scaling features. Psychometrika **41**, 471–504 (1976)

Fisher, R.A.: The precision of discriminant functions. Ann. Eugenic. **10**, 422–429 (1940)

Fisher, R.A.: Statistical Methods for Research Workers, 10th edn. Oliver and Boyd, London (1948)

Gabriel, K.R.: Biplot graphic display of matrices with application to principal component analysis. Biometrika **58**, 453–467 (1971)

Gabriel, K.R.: Analysis of meteorological data by means of canonical decomposition and biplots. J. Appl. Meteorol. **11**, 1071–1077 (1972)

Garmize, L.M., Rychlak, J.F.: Role-play validation of a sociocultural theory of symbolism. J. Consult. Psychol. **28**, 107–115 (1964)

Gauch, Jr., H.G.: Multivariate Analysis in Community Ecology. Cambridge University Press, Cambridge (1982)

Gauch, Jr., H.G., Whittaker, R.H., Wentworth, T.R.: A comparative study of reciprocal averaging and other ordination techniques. J. Ecol. **65**, 157–174 (1977)

Gifi, A.: Nonlinear Multivariate Analysis. Wiley, New York (1990)

Greenacre, M.J.: Some objective methods of graphical display of a data matrix. Special Report, Department of Statistics and Operations Research, University of South Africa (1978)

Greenacre, M.J.: Theory and Applications of Correspondence Analysis. Academic, London (1984)

Greenacre, M.J.: The Carroll-Green-Schaffer scaling in correspondence analysis: a theoretical and empirical appraisal. J. Market. Res. **26**, 358–365 (1989)

Greenacre, M.J., Degos, L.: Correspondence analysis of HLA gene frequency data from 125 population samples. Amer. J. Hum. Genet. **29**, 60–75 (1977)

Guttman, L.: The quantification of a class of attributes: A theory and method of scale construction. In: The Committee on Social Adjustment (ed.) The Predication of Personal Adjustment, pp. 319–348. Social Research Council, Swindon (1941)

Guttman, L.: An approach for quantifying paired comparisons and rank order. Ann. Math. Stat. **17**, 144–163 (1946)

Hayashi, C.: On the quantification of qualitative data from the mathematico-statistical point of view. Ann. Inst. Stat. Math. **2**, 35–47 (1950)

Hayashi, C.: On the prediction of phenomena from qualitative data and the quantification of qualitative data from the mathematico-statistical point of view. Ann. Inst. Stat. Math. **3**, 69–98 (1952)

Hayashi, C.: Multidimensional quantification with the applications to analysis of social phenomena. Ann. Inst. Stat. Math. **5**, 121–143 (1954)

Hayashi, C.: Multidimensional quantification of the data obtained by the method of paired comparison. Ann. Inst. Stat. Math. **16**, 231–245 (1964)

Hayashi, C.: Note on multidimensional quantification of data obtained by paired comparison. Ann. Inst. Stat. Math. **19**, 363–365 (1967)

Hayashi, C.: One dimensional quantification and multidimensional quantification. Ann. Jpn. Associat. Philos. Sci. **3**, 115–120 (1968)

Healy, M.J.R., Goldstein, H.: An approach to the scaling of categorical attributes. Biometrika **63**, 219–229 (1976)

Hill, M.O.: Reciprocal averaging: an eigenvector method of ordination. J. Ecol. **61**, 237–249 (1973)

Hill, M.O.: Correspondence analysis: a neglected multivariate method. Appl. Stat. **23**, 340–354 (1974)

Hirschfeld, H.O.: A connection between correlation and contingency. In: Mathematical Proceedings of the Cambridge Philosophical Society, vol. 31, pp. 520–524. (1935)

Horst, P.: Measuring complex attitudes. J. Soc. Psychol. **6**, 367–374 (1935)

Iwatsubo, S.: An optimal scoring method for detecting clusters and interpretations from multi-way qualitative data. Behaviormetrika **5**, 1–22 (1978)

Kendall, M.G., Stuart, A.: The Advanced Theory of Statistics, vol. 2. Griffin, London (1961)

Lang, C.: Factorial correspondence analysis of oligochaeta communities according to eutrophication level. Hydrobiologia **57**, 241–247 (1978)

Lebart, L., Morineau, A., Tabard, N.: Techniques de la Description Statistique: Méthodes et Logiciels pour l'Analyse des Grands Tableaux. Dunod, Paris (1977)

Lebart, L., Morineau, A., Warwick, K.M.: Multivariate Descriptive Statistical Analysis. Wiley, England (1984)

Lingoes, J.C.: Simultaneous linear regression: an IBM 7090 program for analyzing metric/nonmetric or linear/nonlinear data. Behav. Sci. **9**, 87–88 (1964)

Lingoes, J.C.: The multivariate analysis of qualitative data. Mult. Behav. Res. **3**, 61–94 (1968)

Lord, F.M.: Some relations between Guttman's principal components of scale analysis and other psychometric theory. Psychometrika **23**, 291–296 (1958)

Maung, K.: Measurement of association in contingency table with special reference to the pigmentation of hair and eye colours of Scottish school children. Ann. Eugenic. **11**, 189–223 (1941)

McDonald, R.P.: A unified treatment of the weighting problem. Psychometrika **33**, 351–381 (1968)

McDonald, R.P., Torii, Y., Nishisato, S.: Some results on proper eigenvalues and eigenvectors with applications to scaling. Psychometrika **44**, 211–227 (1979)

McKeon, J.J.: Canonical analysis: some relations between canonical correlations, factor analysis, discriminant function analysis, and scaling theory. Psychometric Monograph No. 13 (1966)

Mosier, C.I.: Machine methods in scaling by reciprocal averages. In: Proceedings, Research Forum, pp. 35–39. International Business Corporation, Endicath (1946)

Nishisato, S.: Minimum entropy clustering of test items. The University of North Carolina at Chapel Hill Ph.D. Thesis (1966)

Nishisato, S.: Analysis of variance through optimal scaling. In: Proceedings of the First Canadian Conference on Applied Statistics, pp. 306–316. Sir George Williams University Press, Montreal (1971)

Nishisato, S.: Optimal scaling and its generalizations. I. Methods. Measurement and Evaluation of Categorical Data (MECD) Technical Report 1. The Ontario Institute for Studies in Education (OISE), Toronto (1972)

Nishisato, S.: Optimal scaling and its generalizations. I. Applications. MECD Technical Report 2, OISE, Toronto (1973)

Nishisato, S.: Optimal scaling as applied to different forms of data. MECD Technical Report 4, OISE, Toronto (1976)

Nishisato, S.: Optimal scaling of paired comparison and rank order data: an alternative to Guttman's formulation. Psychometrika **43**, 263–271 (1978)

Nishisato, S.: Dual scaling and its variants. In: Traub, R.E. (ed.) Analysis of Test Data, pp. 1–12. Jossey-Bass, San Francisco (1979).

Nishisato, S.: Analysis of Categorical Data: Dual Scaling and Its Applications. University of Toronto Press, Toronto (1980)

Nishisato, S.: Gleaning in the field of dual scaling. Psychometrika **61**, 559–599 (1996)

Nishisato, S.: Quantification theory: Reminiscence and a step forward. In: Gaul, W., Geyer-Schultz, A., Schmidt-Thiéme, Kunze, J. (eds.) Challenges at the Interface of Data Analysis, Computer Science, and Optimization, pp. 109–119. Springer, Berlin (2014)

Nishisato, S.: Multidimensional joint graphical display of symmetric analysis: Back to the basics. In: Vander Ark, L.A., Bold, D.M., Wang, W.C., Douglas, J.A., Wiberg, M. (eds.) Quantitative Psychology Research, 291–298. Springer, Berlin (2016)

Nishisato, S.: Notes on the Carroll-Green-Schaffer scaling and joint Graphical display. (2017, under review for publication)

Nishisato, S.: Reminiscence: Quantification theory and graphs. Theory Appl. Data Anal. **8**(1), 47–57 (2019) (in Japanese)

Nishisato, S., Arri, P.S.: Nonlinear programming approach to optimal scaling of partially ordered categories. Psychometrika **40**, 525–548 (1975)

Nishisato, S., Clavel, J.G.: Total information analysis: Comprehensive dual scaling. Behaviormetrika **37**, 15–32 (2010)

Nishisato, S., Inukai, Y.: Partially optimal scaling of items with ordered categories. Jpn. Psychol. Res. textbf14, 109–119 (1972)

Nishisato, S., Lawrence, R.D.: Dual scaling of multiway data matrices: Several variants. In: Copi, R., Bolasco, S. (eds.) Multiway Data Analysis, pp. 317–326. Elsevier, Amsterdam (1989)

Nishisato, S., Leong, K.S.: OPSCAL: A Fortran IV program for analysis of qualitative data by optimal scaling. MECD Technical Report 3, Toronto (1975)

168   S. Nishisato

Richardson, M., Kuder, G.F.: Making a r5ating scale that measures. Personnel J. **12**, 36–40 (1933)
Shiba, S.: A method for scoring multicategory items. Jpn. Psychol. Res. **7**, 75–79 (1965)
Slater, P.: The analysis of personal preferences. Br. J. Stat. Psychol. **3**, 119–135 (1960)
Takane, Y., Shibayama, T.: Principal component analysis with external information on both subjects and variables. Psychometrika **56**, 97–120 (1991)
Takane, Y., Young, F.W., de Leeuw, J.: Nonmetric individual differences multidimensional scaling: an alternating least squares method with optimal scaling features. Psychometrika **42**, 7–67 (1977)
Tanaka, Y.: Some generalized method of optimal scaling and their asymptotic theories: the case of multiple-response-multiple factors. Ann. Inst. Stat. Math. **30**, 329–348 (1978)
Teil, H.: Correspondence factor analysis: an outline of its method. Math. Geol. **7**, 3–12 (1975)
Torgerson, W.S.: Theory and Methods of Scaling. Wiley, New York (1958)
Van Rijckevorsel, J., de Leeuw, J.: An outline of HOMALS-1. Department of Data Theory, Faculty of Social Sciences, University of Leiden (1978)
Whittaker, R.H.: Gradient analysis of vegetation. Biol. Rev. **42**, 207–264 (1967)
Williams, E.J.: Use of scores for the analysis of association in contingency tables. Biometrika **39**, 274–280 (1952)
Young, F.W., de Leeuw, J., Takane, Y.: Regression with qualitative and quantitative variables: an alternating least squares with optimal scaling features. Psychometrika **41**, 505–529 (1976)
Young, F.W., Takane, Y, de Leeuw, J.: The principal components of mixed measurement level multivariate data: an alternating least squares method with optimal scaling features. Psychometrika **43**, 279–281 (1978)

External Logistic Biplots for Mixed Types of Data

José L. Vicente-Villardón and Julio C. Hernández-Sánchez

Abstract A simultaneous representation of individuals and variables in a data matrix is called a biplot. When variables are binary, nominal, or ordinal, a classical linear biplot representation is not adequate. Recently, biplots for categorical data-based logistic response models have been proposed. The coordinates of individuals and variables are computed to have logistic responses along the biplot dimensions. The methods are related to logistic regression in the same way as classical biplot analysis (CBA) is related to linear regression, thus are referred as logistic biplot (LB). Most of the estimation methods are developed for matrices in which the number of individuals is much higher than the number of variables. When the number of variables is high, external logistic biplots can be used; row coordinates are obtained by principal coordinates analysis and then logistic regression is fitted to obtain the variables representation. In this work, external logistic biplots for binary data are extended to nominal and ordinal data using parametric and nonparametric logistic fits and then combined in a single representation.

1 Introduction

A simultaneous representation of individuals and variables in a data matrix is called a biplot. Biplots were proposed in Gabriel (1971) originally for continuous data. More information about biplots can be found in Gower and Hand (1995) or Gower et al. (2011). For binary, nominal, or ordinal variables, a classical linear biplot

J. L. Vicente-Villardón (✉)
Dpto. de Estadística, Universidad de Salamanca, Salamanca, Spain
e-mail: villardon@usal.es

J. C. Hernández-Sánchez
Instituto Nacional de Estadística, Madrid, Spain
e-mail: juliocesar.hernandez.sanchez@ine.es

© Springer Nature Singapore Pte Ltd. 2020   169
T. Imaizumi et al. (eds.), *Advanced Studies in Classification and Data Science*,
Studies in Classification, Data Analysis, and Knowledge Organization,
https://doi.org/10.1007/978-981-15-3311-2_14

representation is not adequate. A popular alternative for categorical variables is the *Gifi System for Descriptive Multivariate Analysis* (Gifi 1990, Michailidis and de Leeuw 1998). The methods are based on the optimal scaling of categorical variables using alternating least squares procedures. Gower and Hand (1995) or Gower et al. (2011) also study some properties of the procedures based on quantification in relations to the associated biplot representations. Other generalizations of the main idea are, for example, the non-linear biplots (Gower and Harding 1988) or the generalized biplot (Gower 1992) in which some pseudo-samples are projected onto a principal coordinates analysis (Gower 1966, Gower 1968) to obtain a non-linear trajectory that can be interpreted as a biplot, the introduction of positive definite metrics for rows and columns (Greenacre 1984, Vicente-Villardon 1992), or the symmetrical representation of rows and columns (Galindo 1986).

A newer trend is developing similar methods based on logistic rather than linear relations among observed and combined variables. Principal components for binary data based on logistic responses were proposed in (De Leeuw 2006), a biplot version (Logistic Biplot) by Vicente-Villardón et al. (2006), later extended in Demey et al. (2008) to matrices with more variables than individuals (external logistic biplot). Extension of the logistic biplots to nominal data can be found in Hernández-Sánchez and Vicente-Villardón (2017) and to ordinal data in Vicente-Villardóon and Henández-Sánchez (2014).

In this paper, we consider external biplots for mixtures of different data types, continuous, binary, nominal, or ordinal. In Sect. 2.1, we present a general definition of a biplot, wide enough to accommodate the interpretations for different types of data. In Sect. 2.2, we describe principal coordinates analysis on the Gower's distances that will be used to obtain the euclidean map representing the rows. Section 2.2 describes how to project continuous variables onto the euclidean map, Sect. 2.3 has the procedure for binary variables, and Sects. 2.4 and 2.5 for the nominal and ordinal cases, respectively. In Sect. 3, we illustrate the methods with some examples.

2   Biplots for Mixed Types of Data

We think of a biplot, in a very general way, as a joint representation in a scattergram of the rows and the columns of a data matrix. Given a data matrix $\mathbf{X}_{I \times J}$ contains the measures of J variables of mixed types on I individuals, an S-dimensional biplot is a joint representation of the rows and columns of the matrix, usually with a set of points $\mathbf{A}_{I \times S}$ for the rows and a set of points (or vectors) $\mathbf{B}_{J \times S}$ for the columns. Similarities and differences among rows (individuals) are interpreted as distances while correlations or similarities among columns (variables) have different interpretations depending on the kind of data and the kind of biplot used. Joint interpretation of rows and columns is made using inner products (projection of the row points onto the column vectors) in the linear, binary, and ordinal cases, and using distances among row and column points in the nominal case; for the latter, there is also a division of the row space into prediction regions for each category of the nominal variables.

Here, we use an external version of the biplot where the scores for individuals will be calculated using principal coordinates analysis (PCoA) (Gower 1966) or classical multidimensional scaling. The scores are represented as points on a euclidean map (scattergram). This technique tends to cluster together individuals with similar profiles. The external biplot enables the variables to be represented on the euclidean map and then to look for the characteristics responsible for the configuration. In the following sections, we describe how to project and interpret each type of variable (continuous, binary, nominal, or ordinal).

2.1 Principal Coordinates Analysis and Gower's Similarity Coefficient

PCoA is concerned with the problem of constructing a configuration $\mathbf{A}$ of I points in a low dimensional euclidean space in such a way that the distance between any two points of the configuration approximates, as closely as possible, a dissimilarity (δ_{ij}) between the individuals represented by these points. The objective is then to find a configuration $\mathbf{A}$ in an S-dimensional euclidean space, whose inter-point distance matrix $\mathbf{D} = (d_{ij}) = \sqrt{(\mathbf{a}_i - \mathbf{a}_j)'(\mathbf{a}_i - \mathbf{a}_j)}$ is as close as possible to the observed dissimilarity $\Delta = (\delta_{ij})$ matrix. When the observed dissimilarities/distances are *euclidean*, it is possible to find an exact configuration in $I - 1$ dimensions. A lower dimensional approximation can be obtained projecting onto the first S principal coordinates (usually $S = 2$). The theoretical considerations and demonstrations of the method can be found, for example, in Mardia et al. (1979). We will calculate the dissimilarities among the individuals using the Gower's coefficient (Gower 1971) for mixed types of data with the addition in Podani (1999) to cope with ordinal variables. We do not describe these methods in more detail here because they are very well known to most researchers.

2.2 Representation of Continuous Variables

Let $\mathbf{X}_{I \times J}$ be a data matrix that contains the measures of J continuous variables on I individuals, and consider the following S-dimensional reduced rank model

$$\mathbf{X} = \mathbf{1}_I \mathbf{b}_0' + \mathbf{A}\mathbf{B}' + \mathbf{E} \tag{1}$$

where $\mathbf{b}_0'$ is a vector of constants, usually the column means ($\mathbf{b}_0 = \bar{\mathbf{x}} = \frac{1}{I}\mathbf{X}'\mathbf{1}_I$), $\mathbf{A}$ and $\mathbf{B}$ are matrices of rank S with I and J rows, respectively, and $\mathbf{E}$ is an $I \times J$ matrix of errors or residuals. The expected values of $\mathbf{X}$ can be written as

$$E[\mathbf{X}] = \mathbf{1}_I \mathbf{b}_0' + \mathbf{A}\mathbf{B}', \tag{2}$$

or

$$\hat{\mathbf{X}} = E\left[\mathbf{X} - \mathbf{1}_I \mathbf{b}_0'\right] = E\left[\tilde{\mathbf{X}}\right] = \mathbf{AB}' \tag{3}$$

That approximation $\hat{\mathbf{X}}$ of the centered data matrix $\tilde{\mathbf{X}}$ is called a biplot (Gabriel 1971) because it can be used to simultaneously plot the individuals and variables in a reduced rank subspace, using the rows $\mathbf{a}'_1, \ldots, \mathbf{a}'_I$ of $\mathbf{A}$ and the rows $\mathbf{b}'_1, \ldots, \mathbf{b}'_J$ of $\mathbf{B}$ as markers or coordinates on the low rank subspace. The inner product $\hat{x}_{ij} = \mathbf{a}'_i \mathbf{b}_j$ approximates the element $\tilde{x}_{ij}$. Figure 1 shows the geometry of the biplot. In summary, the expected values on the original data matrix are obtained on the biplot using a simple scalar product, that is, projecting the point $\mathbf{a}_i$ onto the direction defined by $\mathbf{b}_j$ (Fig. 1a). This is why row markers are usually represented as points and column markers as vectors or directions (also called biplot axis in Gower and Hand 1995). Then the set of points predicting a particular value of one variable are on a straight line perpendicular to the direction representing the variable (Fig. 1b).

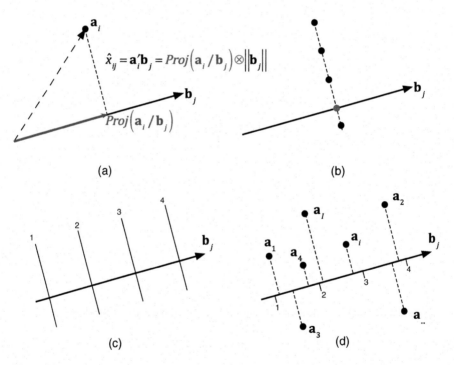

Fig. 1 Biplot approximation: (**a**) Inner product of the row and column markers. (**b**) The set of points predicting the same value are all on a straight line perpendicular to the direction defined by the column marker $\mathbf{b}_j$. (**c**) Points predicting different values are on parallel lines. (**d**) The variable direction can be supplemented with scales to visually obtain the prediction

Graded scales on the direction of the variable can be added to guide the interpretation of the projections (Fig. 1c and d). The calculations for obtaining the scale markers are simple. To find the marker for a fixed value μ, we look for the point (x, y) that predicts μ and is on the direction of $\mathbf{b}_j$, i.e., on the line joining the points $(0, 0)$ and $\mathbf{b}_j = (b_{j1}, b_{j2})$, that is $y = \frac{b_{j2}}{b_{j1}}x$. The prediction also verifies $\mu = b_{j1}x + b_{j2}y$. Then, we obtain

$$x = \frac{\mu\, b_{j1}}{b_{j1}^2 + b_{j2}^2}; \qquad y = \frac{\mu\, b_{j2}}{b_{j1}^2 + b_{j2}^2} \tag{4}$$

Because we assume that the columns are centered to have zero mean, the point that predicts 0 for all the variables is the origin, that is, it predicts the mean of each variable. It is convenient to label the graded scales with the initial rather than the transformed values. For example, if the variable is centered, we could use $\mu + \bar{x}_i$ as the label of the point μ. When the variables have different measurement units is convenient to standardize each column individually dividing by the standard deviation, in that case the label should be $\mu s_i + \bar{x}_i$. An algorithm to find *pretty* values for the scales may be needed.

If we consider the row markers $\mathbf{A}$ as fixed and the data matrix previously centered, the column markers can be computed by regression through the origin:

$$\mathbf{B}' = (\mathbf{A}'\mathbf{A})^{-1}\mathbf{A}'\tilde{\mathbf{X}}. \tag{5}$$

The global goodness of fit is the amount of variability accounted by the prediction, that is,

$$\rho^2 = tr(\hat{\mathbf{X}}'\hat{\mathbf{X}})/tr(\tilde{\mathbf{X}}'\tilde{\mathbf{X}}) \tag{6}$$

Even for cases in which we obtain a good global fit, this does not imply obtaining a good fit for each row or each column of the original data matrix. The goodness of fit for each column is

$$\rho_j^2 = diag(\hat{\mathbf{X}}'\hat{\mathbf{X}}) \div diag(\tilde{\mathbf{X}}'\tilde{\mathbf{X}}) \tag{7}$$

where $\div$ means the element by element operation. ρ_j^2 is like the *R-squared* of the regression of each column of $\mathbf{X}$ on $\mathbf{A}$. We call that quantity *quality of representation* of the variable in analogy with the terminology of correspondence analysis (Benzecri 1973, Greenacre 1984). In (Gardner-Lubbe et al. 2008), this quantity is called *predictiveness* of the column. The measures are used to identify which variables are most related to the representation. Only when the predictiveness of a variable is reasonable, the approximation of its values is interpretable.

2.3 Representation of Binary Variables

The logistic biplot for binary data was proposed in Vicente-Villardón et al. (2006) and later extended in Demey et al. (2008) to its external version. Let **X** be an $I \times J$ binary data matrix containing the measures of J binary characters on I individuals. A binary matrix can be understood as a matrix of observed probabilities. Let $\hat{p}_{ij} = E(x_{ij})$ be the expected probability that the character j be present at individual i. A logistic bilinear model can be written as

$$\hat{p}_{ij} = \frac{e^{(b_{j0} + \sum_k b_{jk} a_{ik})}}{1 + e^{(b_{j0} + \sum_k b_{jk} a_{ik})}} \tag{8}$$

where a_{ik} and b_{jk}, $(i = 1, \ldots, I; j = 1, \ldots, J; k = 1, \ldots, S)$ are the model parameters that will be used as row and column markers, respectively. The model is a generalized bilinear model having the *logit* as link function.

$$\text{logit}(\hat{p}_{ij}) = b_{j0} + \sum_{k=1}^{q} b_{jk} a_{ik} = b_{j0} + \mathbf{a}'_i \mathbf{b}_j$$

In matrix form,

$$\text{logit}(\mathbf{P}) = \mathbf{1}_n \mathbf{b}'_0 + \mathbf{AB}' \tag{9}$$

So, we have a biplot in *logit* scale except for the vector of constants. Except for the constant, the geometry of the biplot is the same of the linear case to predict the logits and then, to predict the probabilities. The intercept has to be fitted because it is not possible to previously center the binary data (see Fig. 2).

The calculations for obtaining the scale markers are simple, is the same as the linear case but keeping the intercept. To find the marker for a fixed probability p, we look for the point (x, y) that predicts p and is on the biplot axis, i.e., on the line joining the points $(0, 0)$ and $\beta_j = (b_{j1}, b_{j2})$, that is

$$y = \frac{b_{j2}}{b_{j1}} x$$

The prediction verifies

$$\text{logit}(p) = b_{j0} + b_{j1} x + b_{j2} y$$

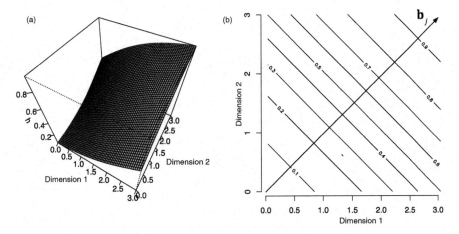

Fig. 2 Geometry of the binary logistic biplot: (**a**) Response surface of a binary logistic regression. (**b**) Level curves of the response predicting different probabilities and the direction to represent on the biplot

We obtain

$$x = \frac{(\text{logit}(p) - b_{j0})b_{j1}}{b_{j1}^2 + b_{j2}^2}; \, y = \frac{(\text{logit}(p) - b_{j0})b_{j2}}{b_{j1}^2 + b_{j2}^2}$$

For example, the point on axis β_j predicting 0.5 (logit(0.5)=0) is

$$x = \frac{-b_{j0}b_{j1}}{b_{j1}^2 + b_{j2}^2}; \, y = \frac{-b_{j0}b_{j2}}{b_{j1}^2 + b_{j2}^2}$$

If the intercept is not fitted, the point predicting 0.5 is always the origin and the goodness of fit is much lower. To simplify the graph, we represent the segment (or arrow) joining the points predicting 0.5 and 0.75. The line perpendicular to the segment at the 0.5 point divides the plot in two regions, one predicting presence and the other absence of the variable.

If we consider the row markers **A** as fixed, obtained from a principal coordinates analysis as before, the column markers **b′₀** and **B** can be computed by standard logistic regression (probably penalized if presences and absences are completely separated on the scattergram).

Predictiveness of the variables is now measured using, for example, the Nagelkerke pseudo R-squared, AIC, BIC, or any fit measure traditionally used in logistic regression. The percent of correct classifications (assigning presence when the probability is higher than 0.5) is also a good measure of fit.

2.4   Representation of Nominal Variables

Let $\mathbf{X}$ be a data matrix containing the values of J nominal variables, each with K_j $(j = 1, \ldots, J)$ categories, for I individuals. The last (or the first) category of each variable will be used as a baseline. Let $\pi_{ij(k)}$ denote the expected probability that the category k of variable j be present at individual i. As before, we suppose that we already have a euclidean map to represent the individuals and we want to obtain the representation of a nominal variable. We will use a multinomial logistic model in which the probabilities are obtained as:

$$\pi_{ij(k)} = \frac{e^{b_{j(k)0} + \sum_{s=1}^{S} b_{j(k)s} a_{is}}}{\sum_{l=1}^{K_j} e^{b_{j(l)0} + \sum_{s=1}^{S} b_{j(l)s} a_{is}}}, (k = 1, \ldots, K_j). \tag{10}$$

Using the last category as a baseline in order to make the model identifiable, the parameter for that category is restricted to be 0, i.e., $b_{j(K_j)0} = b_{j(K_j)s} = 0$, $(j = 1, \ldots, J; \quad s = 1, \ldots, S)$. The model can be rewritten as:

$$\pi_{ij(k)} = \frac{e^{b_{j(k)0} + \sum_{s=1}^{S} b_{j(k)s} a_{is}}}{1 + \sum_{l=1}^{K_j-1} e^{b_{j(l)0} + \sum_{s=1}^{S} b_{j(l)s} a_{is}}}, (k = 1, \ldots, K_j - 1). \tag{11}$$

The model is fitted using standard nominal logistic regressions. It is easy to see that the log-odds of each response (relative to the last category) follow a bi-linear model:

$$\log \left(\frac{\pi_{ij(k)}}{\pi_{ij(K_j)}} \right) = b_{j(k)0} + \sum_{s=1}^{S} b_{j(k)s} a_{is} = b_{j(k)0} + \mathbf{a}'_i \mathbf{b}_{j(k)},$$

where a_{is} and $b_{j(k)s}$ $(i = 1, \ldots, I; \quad j = 1, \ldots, J; \quad k = 1, \ldots, K_j - 1; \quad s = 1, \ldots, S)$ are the model parameters. Although the biplot for the log-odds may be useful, it would be more interpretable in terms of predicted probabilities and categories.

The points predicting different probabilities are no longer on parallel straight lines (see Fig. 3); this means that predictions on the logistic biplot are not made in the same way as in the linear biplots, the surfaces now define prediction regions for each category as shown in the graph. An algorithm to obtain the prediction regions and the complete geometry of the biplot can be found in Hernández-Sánchez and Vicente-Villardón (2017). When the number of categorical variables is high, it would be very difficult to visualize all the prediction regions. Using methods of *Computational Geometry*, the tessellation (prediction regions) obtained from the

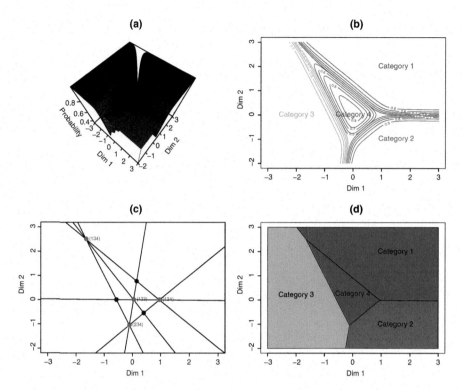

Fig. 3 Nominal logistic biplot: (**a**) Response surface of a nominal logistic regression. (**b**) Level curves of the response predicting 0.5 for each category. (**c**) Lines for the comparison of each pair of categories. (**d**) Prediction regions for each category

multinomial logistic regression on the principal coordinates can be approximated by a Voronoi diagram and then a set of generators of the diagram obtained (see Fig. 4). The generators have the role of *category points*. The interpretation is quite simple, the prediction for an individual is the category corresponding to the nearest *category point*.

Predictiveness of the variables is now measured using any fit measure traditionally used in multinomial logistic regression.

2.5 Representation of Ordinal Variables

Let **X** be a data matrix containing the measures of I individuals on J ordinal variables with K_j, $(j = 1, \ldots, J)$ ordered categories each.

Let $\pi^*_{ij(k)} = P(x_{ij} \leq k)$ be the cumulative probability that individual i has a value lower than k on the $j - th$ ordinal variable, and let $\pi_{ij(k)} = P(x_{ij} = k)$

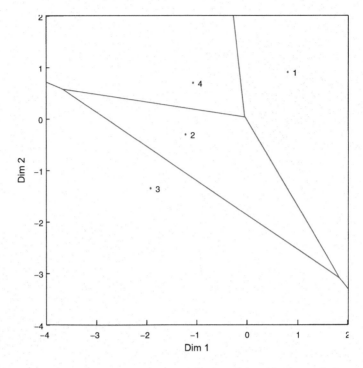

Fig. 4 Category points (generators of the Voronoi diagram) induced by the prediction regions

the (expected) probability that individual i takes the k-th value on the j-th ordinal variable. Then $\pi^*_{ij(K_j)} = P(x_{ij} = K_j) = 1$ and $\pi_{ij(k)} = \pi^*_{ij(k)} - \pi^*_{ij(k-1)}$ (with $\pi^*_{ij(0)} = 0$). An S-dimensional ordinal logistic model for the cumulative probabilities can be written for $(1 \leq k \leq K_j - 1)$ as

$$\pi^*_{ij(k)} = \frac{1}{1 + e^{-\left(d_{jk} + \sum_{s=1}^{S} a_{is} b_{js}\right)}} = \frac{1}{1 + e^{-(d_{jk} + \mathbf{a}'_i \mathbf{b}_j)}} \tag{12}$$

where $\mathbf{a}_i = (a_{i1}, \ldots, a_{iS})'$ is the vector of latent trait scores for the $i-th$ individual and d_{jk} and $\mathbf{b}_j = (b_{j1}, \ldots, b_{jS})'$ the parameters for each item or variable.

Observe that we have defined a set of binary logistic models, one for each category, where there is a different intercept for each but a common set of slopes for all.

The $\mathbf{b}_j$ parameters can also be represented on the graph as directions on the scores space that best predict probabilities and they are used to help in searching for the variables or items responsible for the configuration of the individuals.

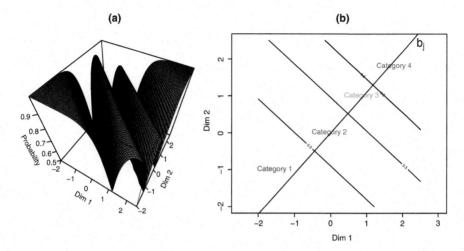

Fig. 5 Ordinal logistic biplot: (**a**) Response surfaces of an ordinal logistic regression. (**b**) Prediction regions and direction to represent on the biplot

In logit scale, the model is

$$\text{logit}(\pi^*_{ij\,(k)}) = d_{j(k)} + \sum_{s=1}^{S} a_{is}b_{js} = d_{j(k)} + \mathbf{a}'_i\mathbf{b}_j, \tag{13}$$

with $k = 1, \ldots, K_j - 1$. That defines a binary logistic biplot for the cumulative categories. From a practical point of view, we are interested mainly in the regions that predict each category, that is, the set of points whose expected probabilities are higher in each category. Those regions are separated by parallel straight lines, all perpendicular to the direction of $\mathbf{b}_j$ (See Fig. 5). The details can be found in Vicente-Villardóon and Henández-Sánchez (2014).

As in the binary case, predictiveness of the variables is now measured using, for example, the Nagelkerke pseudo *R-squared*, AIC, BIC, or any fit measure traditionally used in ordinal logistic regression.

3 Illustrative Example

As illustration, we have used some classical data set available as an example in R (*mtcars*). The source of the data is Henderson and Velleman (1981) and contains information about 32 cars on 11 variables.

- **mpg** Miles/(US) gallon
- **cyl** Number of cylinders (ordinal)
- **disp** Displacement (cu.in.)

Table 1 Predictiveness of
the continuous variables

mpg	disp	hp	drat	wt	qsec	carb
0.593	0.737	0.629	0.727	0.570	0.672	0.349

- **hp** Gross horsepower
- **drat** Rear axle ratio
- **wt** Weight (1000 lbs)
- **qsec** 1/4 mile time
- **vs** V/S
- **am** Transmission (0 = automatic, 1 = manual)
- **gear** Number of forward gears (nominal in the example)
- **carb** Number of carburetors.

The euclidean configuration of the individuals (cars) has been obtained using principal coordinates from a dissimilarity matrix calculated with the Gower's similarity coefficient for mixed types of data. The first two principal dimensions account for 80.2% of the variability, thus reflecting quite well the actual similarities and differences among the car models. All the variables have been projected on the graph in order to interpret the characteristics of the groups (clusters) of cars. Continuous, binary, and ordinal variables are represented as straight lines with the adequate graded scales and nominal variables with prediction regions.

There are three main clusters on the graph, coincident with the number of gears of the car, i.e., the nominal variable *gear* has a perfect fit. Most of the cars in the red cluster with 4 gears has 4 cylinders (category 1 of *cyl)*), the cars in the cluster with 5 gears 6 cylinders and 5 gears is associated to 8 cylinders. The Nagelkerke pseudo R-squared for the variable *cyl* is 0.72, i.e., there is a strong relation of the variable with the principal coordinates solution.

The variable *am* has also a perfect classification of the cars. All the cars in cluster 3 have automatic transmission and cars in cluster 5 and most of cluster 3 have manual transmission. The pseudo R-squared is 0.92.

The variable *vs* has a slightly lower percent of correct classification (84.37) with a pseudo R-squared of 0.82.

The predictiveness for the continuous variables is in Table 1 and is lower than the ones for the discrete variables. *mpg* and *disp* have inverse high correlation, *mpg* is higher in cluster 4 and *disp* in cluster 3. Cluster 5 has intermediate values for both variables. *drat* and *wt* have also inverse correlation, the first is higher in clusters 4 and 5, while the second is higher in cluster 3. A similar interpretation could be done for the rest of the variables (Fig. 6).

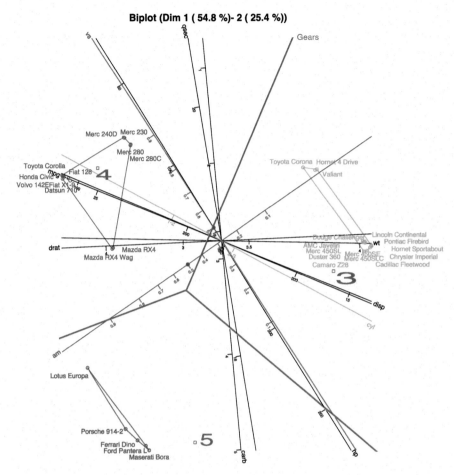

Fig. 6 Principal coordinates analysis of the *mtcars* data with the continuous and the binary variables projected onto the euclidean map (biplot)

4 Software Note

All the procedures explained in this paper can be calculated using the R (R Core Team 2019) packages: *MultbiplotR* (Vicente-Villardon 2019), *NominalLogisticBiplot* (Hernandez-Sanchez and Vicente-Villardon 2014), and *OrdinalLogisticBiplot* (Hernandez-Sanchez and Vicente-Villardon 2014).

References

Benzecri, J.P.: L'analyse des donnees, vol. 2, p. l. Dunod, Paris (1973)

De Leeuw, J.: Principal component analysis of binary data by iterated singular value decomposition. Comput. Stat. Data Anal. **50**(1), 21–39 (2006)

Demey, J.R., Vicente-Villardón, J.L., Galindo-Villardón, M.P., Zambrano, A.Y.: Identifying molecular markers associated with classification of genotypes by External Logistic Biplots. Bioinformatics **24**(24), 2832–2838 (2008)

Gabriel, K.R.: The biplot graphic display of matrices with application to principal component analysis. Biometrika **58**(3), 453–467 (1971)

Galindo, M.P.: Una alternativa de representacion simultanea: HJ-Biplot. Questiio **10**(1), 13–23 (1986)

Gardner-Lubbe, S., Le Roux, N.J., Gower, J.C.: Measures of fit in principal component and canonical variate analyses. J. Appl. Stat. **35**(9), 947–965 (2008)

Gifi, A.: Nonlinear Multivariate Analysis. Wiley, New York (1990)

Gower, J.C.: Some distance properties of latent root and vector methods used in multivariate analysis. Biometrika **53** (3–4), 325–338 (1966)

Gower, J.C.: Adding a point to vector diagrams in multivariate analysis. Biometrika **55**(3), 582–585 (1968)

Gower, J.C.: A general coefficient of similarity and some of its properties. Biometrics **27**(4), 857–871 (1971)

Gower, J.C.: Generalized biplots. Biometrika **79**(3), 475–493 (1992)

Gower, J.C., Hand, D.J.: Biplots: Monographs on Statistics and Applied Probability, vol. 54. Chapman and Hall, London (1995)

Gower, J.C., Harding, S.A.: Nonlinear biplots. Biometrika **75** (3), 445–455 (1988)

Gower, J.C., Lubbe, S.G., Le Roux, N.J.: Understanding Biplots. Wiley, Hoboken (2011)

Greenacre, M.J.: Theory and Applications of Correspondence Analysis. Academic, Cambridge (1984)

Henderson, H.V., Velleman, P.F.: Building multiple regression models interactively. Biometrics **37**(2), 391–411 (1981)

Hernandez-Sanchez, J.C., Vicente-Villardon, J.L.: NominalLogisticBiplot: Nominal Logistic Biplots in R. Salamanca (2014). Available via https://cran.r-project.org/package=NominalLogisticBiplot

Hernandez-Sanchez, J.C., Vicente-Villardon, J.L.: OrdinalLogisticBiplot: Biplot representations of ordinal variables. Salamanca (2014). Available via https://cran.r-project.org/package=OrdinalLogisticBiplot

Hernández-Sánchez, J.C., Vicente-Villardón, J.L.: Logistic biplot for nominal data. Adv. Data Anal. Classif. **11** (2) 307–326 (2017)

Mardia, K.V., Kent, J.T., Bibby, J.M.: Multivariate Analysis. Probability and Mathematical Statistics. Academic Press, London (1979)

Michailidis, G., de Leeuw, J.: The Gifi system of descriptive multivariate analysis. Stat. Sci. **13**, 307–336 (1998)

Podani, J.: Extending Gower's general coefficient of similarity to ordinal characters. Taxon **48**, 331–340 (1999)

R Core Team: R: A language and environment for statistical computing. R Foundation for Statistical Computing, Vienna (2019). Available via https://www.R-project.org/

Vicente-Villardon, J.L.: Una alternativa a los métodos factoriales clásicos basada en una generalización de los metodos biplot. An alternative to the classical factor methods based on a generalization of biplot methods (Doctoral dissertation, MS Thesis. Universidad de Salamanca. Spain) (1992).

Vicente-Villardon, J.L.: MultBiplotR: Multivariate Analysis using Biplots. R Package Version 19.11.19 (2019). Available via http://biplot.usal.es/multbiplot/multbiplot-in-r/

Vicente-Villardón, J.L., Galindo-Villardón, M.P., Blazquez-Zaballos, A.: Logistic biplots. In: Multiple Correspondence Analysis and Related Methods, pp. 503–521. Chapman and Hall, London (2006)

Vicente-Villardóon, J.L., Henández-Sánchez, J.C.: Logistic Biplots for Ordinal Data with an Application to Job Satisfaction of Doctorate Degree Holders in Spain (2014). Preprint arXiv:1405.0294

Part III
Statistical Methods

Functional Clustering Approach for Analysis of Concentration

Yifan Chen, Yuriko Komiya, Hiroyuki Minami, and Masahiro Mizuta

Abstract In this research, we study characteristics of concentration variation with functional data analysis. Functional data analysis, which combines traditional data analysis with the characteristics of functions, is suitable to analyze changing trend of observed data with the utility of derivatives. We applied functional clustering analysis to milk dataset, which contains milk concentration, sum of rain, average speed of wind, and average temperature, to reveal the relationship between milk concentration and other variables. We get eight dendrograms in functional clustering analysis. Cophenetic correlation coefficient is used to measure the similarities among the dendrograms. Multidimensional scaling is used to visualize the dissimilarities of the dendrograms clearly. As a result, we find that the trend of milk concentration has relation to those of sum of rain and of average speed of wind.

1 Introduction

Concentration variation happens everywhere, such as medicine aspect and dietary aspect. It has a great meaning in our daily life; therefore, revelation of the affecting factors of concentration variation is important and necessary. The purpose of this research is to represent an application of functional data analysis to solve real problems like "what are affecting factors to concentration variation."

Functional data analysis is an analysis when data are functions proposed by Ransay and Silverman (1982). It is a powerful approach in the biological and pharmaceutical fields. There are many methods in functional data analysis that we may see in the traditional data analysis as well, including functional clustering and functional multidimensional scaling (MDS) (Ransay and Silverman 2005).

Y. Chen
Graduate School of Information Science and Technology, Hokkaido University, Hokkaido, Japan

Y. Komiya · H. Minami · M. Mizuta (✉)
Information Initiative Center, Hokkaido University, Hokkaido, Japan
e-mail: komiya@iic.hokudai.ac.jp; min@iic.hokudai.ac.jp; mizuta@iic.hokudai.ac.jp

© Springer Nature Singapore Pte Ltd. 2020
T. Imaizumi et al. (eds.), *Advanced Studies in Classification and Data Science*,
Studies in Classification, Data Analysis, and Knowledge Organization,
https://doi.org/10.1007/978-981-15-3311-2_15

In this research, we propose a method to analyze relationship between concentration and affecting factors. Meanwhile we also use functional data analysis to explain the relationship between the changing trends of concentration and those of affecting factors. Finally, to get a better interpretation, we show some examples of execution of the proposed method.

2 Proposed Method

Before the introduction of our method, some notations should be claimed. We assume there are time dependent multivariate objects as described below:

$$y_{ij}, x_{ij}^{(1)}, x_{ij}^{(2)}, \ldots, x_{ij}^{(p)} \qquad i = 1, \ldots, n; j = 1, \ldots, m \qquad (1)$$

where p is the number of variables, n is the number of objects, m is the number of period, and $x_{ij}^{(p)}$ refers to the i-th object at j-th period for the r-th variable. In this research, we assume y_{ij} as an object of concentration, and $x_{ij}^{(1)}, x_{ij}^{(2)}, \ldots, x_{ij}^{(p)}$ as the affecting factors at the j-th period for i-th object.

We describe the proposed method as follows:

STEP 1:	Functionalization
STEP 2:	Functional hierarchical clustering
STEP 3:	Measure of similarity between dendrograms
STEP 4:	Visualization

2.1 Functionalization

Functionalization is a step of smoothing. In the step, basis functions are chosen to make data into functions, which we can calculate by any argument t, with basis function system. It is a linear combination of basis functions $\Phi_k = (\phi_1(t), \ldots, \phi_k(t))^T$ and coefficients that are independent from each other. There are several options, such as Legendre polynomials and Fourier basis system.

Legendre polynomials system is a linear combination of Legendre polynomials functions, which are the solutions to Legendre's differential equation with order n:

$$P_n(t) = \frac{1}{2^n n!} \frac{d^n}{dt^n} (t^2 - 1)^n, \qquad (2)$$

therefore, the basis function can be written as

$$\Phi(t) = (P_0(t), P_1(t), \ldots, P_k(t))^T.$$

Fourier basis system is a linear combination as well, but this time it is with the Fourier series, which is:

$$c_0 + c_1 \sin(wt) + c_2 \cos(wt) + c_3 \sin(2wt) + c_4 \cos(2wt) + \cdots \tag{3}$$

and its corresponding basis function can be written as

$$\Phi(t) = \left(\frac{1}{\sqrt{2\pi}}, \frac{1}{\sqrt{\pi}} \sin(wt), \frac{1}{\sqrt{\pi}} \cos(wt), \frac{1}{\sqrt{\pi}} \sin(2wt), \frac{1}{\sqrt{\pi}} \cos(2wt), \ldots \right)^{\mathrm{T}}.$$

Since basis functions are independent, basis function system can be an orthonormal system. Thus, we assume basis functions are orthonormal,

$$\int \phi_k(t)^{\mathrm{T}} \phi_s(t) dt = \begin{cases} 1 & \text{if } s = k \\ 0 & \text{if } s \neq k \end{cases}. \tag{4}$$

We can choose the kind of basis function system according to the characteristic of dataset. In this research, the corresponding coefficients $c_{ki}^{(r)}$ and c_{ki} are calculated by minimizing the least square criterions,

$$\mathrm{SMSSE}(\mathbf{c}_i^{(r)} | x_{i1}^{(r)}, \ldots, x_{im}^{(r)}) = \sum_{j=1}^{m} \left(x_{ij}^{(r)} - \sum_{k=1}^{K} c_{ki}^{(r)} \phi_k(t_j) \right)^2, \tag{5}$$

$$\mathrm{SMSSE}(\mathbf{c}_i | y_{i1}, \ldots, y_{im}) = \sum_{j=1}^{m} \left(y_{ij} - \sum_{k=1}^{K} c_{ki} \phi_k(t_j) \right)^2, \tag{6}$$

where $x_{ij}^{(r)}$ and y_{ij} are the observed values, $\sum_{k=1}^{K} c_{ki}^{(r)} \phi_k(t_j)$ and $\sum_{k=1}^{K} c_{ki} \phi_k(t_j)$ are the corresponding functions.

Then, for each i and r, $\left\{ x_{ij}^{(r)}, j = 1, \ldots, m \right\}$ can be approximated well arbitrarily with a linear combination of K basis function (Ransay and Silverman 2005). The result of functionalization of the original data are

$$x_i^{(r)}(t) = \sum_{k=1}^{K} c_{ki}^{(r)} \phi_k(t) = \mathbf{c}_i^{(r)\mathrm{T}} \Phi(t), \tag{7}$$

$$y_i(t) = \sum_{k=1}^{K} c_{ki} \phi_k(t) = \mathbf{c}_i^{\mathrm{T}} \Phi(t), \tag{8}$$

where $\mathbf{c}_i^{(r)} = (c_{1i}^{(r)}, c_{2i}^{(r)}, \ldots, c_{Ki}^{(r)})^{\mathrm{T}}$ and $\mathbf{c}_i = (c_{1i}, c_{2i}, \ldots, c_{Ki})^{\mathrm{T}}$ correspondingly.

With the functionalization, data have the characteristics of function, which means that we can take derivatives and find the integration of those dataset. More precisely, we can explain changing trends with the derivative of the functions.

To find the effects of changing trends of data, we focus on the derivatives of basis functions. So we calculated their derivatives of functions as $\frac{d}{dt} y_i(t)$, $\frac{d}{dt} x_i^{(1)}(t)$, ..., $\frac{d}{dt} x_i^{(p)}(t)$ from $y_i(t)$, $x_i^{(1)}(t)$, ..., $x_i^{(p)}(t)$ correspondingly.

2.2 Functional Hierarchical Clustering

In the step of functional hierarchical clustering, functional clustering should be used to get an interpretation of the relationship between observed objects.

Functional clustering is used not only in the functions but also in the derivatives of the functions, because both original functions and their derivative functions may be useful for analysis of concentration changes.

First, dissimilarity should be calculated. Dissimilarity between two functions $f(t)$, $g(t)$ is defined by

$$\|f - g\|_1 = \int |f(t) - g(t)| dt \tag{9}$$

or

$$\|f - g\|_2 = \int (f(t) - g(t))^2 dt. \tag{10}$$

Then, we carry out functional hierarchical clustering to functions $\{y_i(t); i = 1, \ldots, n\}$, $\{\frac{d}{dt} y_i(t); i = 1, \ldots, n\}$, $\{x_i^{(r)}(t); i = 1, \ldots, n, r = 1, \ldots, p\}$, $\{\frac{d}{dt} x_i^{(r)}; i = 1, \ldots, n, r = 1, \ldots, p\}$. We can get $2p + 2$ dendrograms.

2.3 Measure of Similarity Between Dendrograms and Visualization

After functional hierarchical clustering, a criterion should be used to evaluate similarities among dendrograms. Cophenetic correlation coefficient is used in this research. It is the height between two nodes in the dendrograms (Saraçli1 et al. 2013) and usually considered as a measure to judge how well dendrograms can fit the original distances.

More precisely speaking, cophenetic distance between two observations that have been clustered represents the intergroup dissimilarity at which the two observations are first combined into a single cluster. That means cophenetic distance is the height of point where two elements first intersect as a union.

Cophenetic correlation coefficient varies from 0 to 1. When two kinds of distance fit in 100%, it becomes 1. When two kinds of distance do not fit each other at all, it moves to 0.

In this research, we use it as a measure of similarity of two classifications or two dendrograms. We denote i_a and i_b as the i-th elements of the dendrogram_a and dendrogam_b, respectively. Also $D_a(i_a, i_b)$, $D_b(i_a, i_b)$ are denoted as the cophenetic distance between objects i_a and i_b in dendrogram_a and dendrogam_b, respectively, and D_a, D_b as the average cophenetic distance in dendrogram_a and dendrogam_b, respectively. Cophenetic correlation coefficient is defined as below:

$$C_{ab} = \frac{\sum_{i_a=1}^{n} \sum_{i_b=1}^{m} (D_a(i_a, i_b) - D_a)(D_b(i_a, i_b) - D_b)}{\sqrt{\sum_{i_a=1}^{n} \sum_{i_b=1}^{m} (D_a(i_a, i_b) - D_a)} \sqrt{\sum_{i_1=a}^{n} \sum_{i_b=1}^{m} (D_b(i_a, i_b) - D_b)}}.$$

(11)

If the outcome is close to 1, the correlation between two dendrogram is strong. If it is moving to 0, the correlation between two dendrograms goes weak. Then we apply MDS to the cophenetic correlation coefficients

3 Analysis of Concentration Data

The proposed method is applied to an actual dataset below:

$$\left\{ \left(y_{ij}, \ x_{ij}^{(1)}, \ x_{ij}^{(2)}, \ x_{ij}^{(3)} \right); i = 1, \dots, n, j = 1, \dots, m \right\},$$

where y_{ij} is the concentration value, $x_{ij}^{(1)}$ is the sum of rain in one month, $x_{ij}^{(2)}$ is the average speed of wind in one month, and $x_{ij}^{(3)}$ is the average temperature in one month for the i-th place and the j-th period. The number of data from different places is eighteen ($n = 18$). All those data are observed every two months, and continued in one year ($m = 6$).

In this analysis, the concentration dataset is made of same kind of food but from different places. The concentration varies from places to places in different time. The main characteristics of corresponding places are also obtained as the observed dataset in sum of rain, average speed of wind, and average temperature in one month. Therefore, the goal is to find out whether the factors are effective to the changing by using the proposed method above.

In functionalization of this experiment, Fourier basis system resulting from Fourier series is used, for the dataset is periodic and Fourier series is suitable. Fourier basis function system below is chosen with five basis functions, here is the outcome of functionalization on concentration:

$$x^{(r)}(t) = a_{i0}^{(r)} + a_{i1}^{(r)}\sin(t) + a_{i2}^{(r)}\cos(t) + a_{i3}^{(r)}\sin(2t) + a_{i4}^{(r)}\cos(2t), \quad (12)$$

and the derivatives as follows:

$$x^{(r)'}(t) = a_{i1}^{(r)}\cos(t) - a_{i2}^{(r)}\sin(t) + 2a_{i3}^{(r)}\cos(2t) - 2a_{i4}^{(r)}\sin(2t). \quad (13)$$

Three more variables are obtained as $\frac{d}{dt}x_{ij}^{(1)}(t)$, $\frac{d}{dt}x_{ij}^{(2)}(t)$, $\frac{d}{dt}x_{ij}^{(3)}(t)$, which are representations of changing trends of sum of rain in one month, changing trends of average speed of wind in one month, and changing trends of average temperature in one month correspondingly.

In this case, eight dendrograms can be obtained by L_2-norm. We can see four dendrograms of the outcomes below Figs.1, 2, 3, and 4.

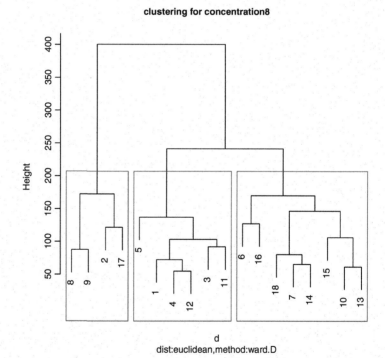

Fig. 1 Clustering for concentration

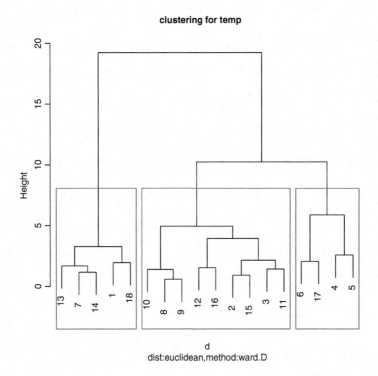

Fig. 2 Clustering for temperature

There are four dendrograms of clustering outcomes from the differential functions below Figs. 5, 6, 7, and 8. In the figures, the number denotes 18 different locations, which means same number corresponds to same place.

Then cophenetic correlation coefficient (2.11) is applied to measure the similarity between dendrograms of concentration and the variables. In this case, "Kendall" method is used instead of "Pearson" method, because coefficient of rank correlation is considered to be a more suitable measure of rank than others in hierarchical clustering.

We summarize the similarities in Table 1, where "concentration_d," "rain_d," "wind_d," and "temperature_d" represent derivatives of concentration, derivatives of sum of rain in one month, derivatives of average speed of wind in one month, and derivatives of average temperature in one month correspondingly.

Multidimensional scaling (Torgerson method) is used to visualize the dissimilarity between dendrograms. We can see from Fig. 9 that it reflects directly the outcome of cophenetic correlation coefficients. It also shows that the dendrogram of the speed of wind has the closest relationship with the dendrogram of concentration, while the dendrograms of the derivatives of sum of rain and the derivatives of average speed of wind in one month are in close relationship with the dendrogram of derivatives

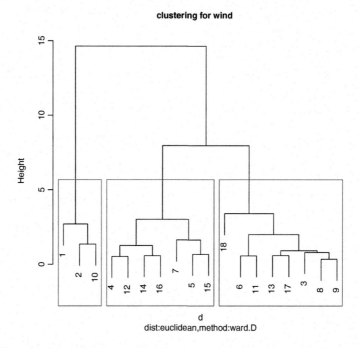

Fig. 3 Clustering for wind

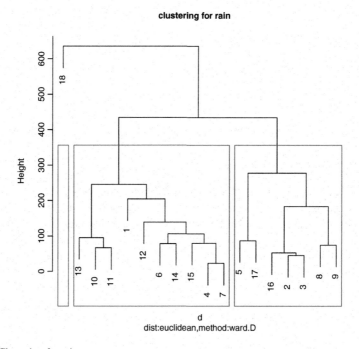

Fig. 4 Clustering for rain

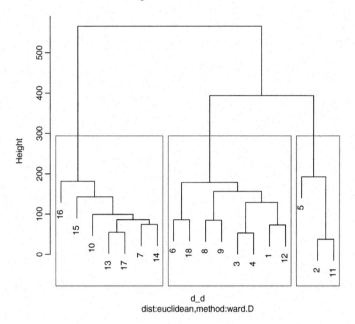

Fig. 5 Clustering for derivatives of concentration

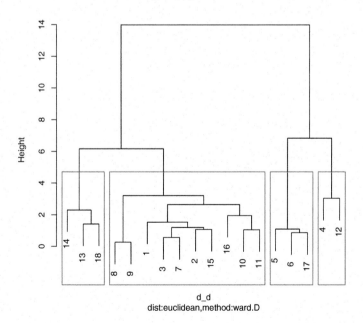

Fig. 6 Clustering for derivatives of temperature

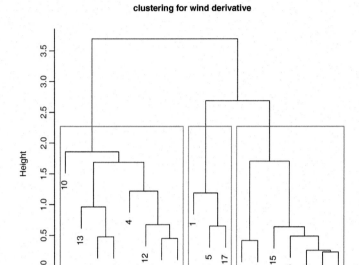

Fig. 7 Clustering for derivatives of wind

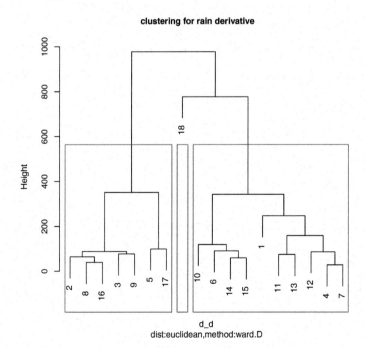

Fig. 8 Clustering for derivatives of rain

Table 1 Similarities between dendrograms

	Concentration	concentration_d	Rain	Wind	Temperature	rain_d	wind_d	temperature_d
Concentration	1.00	0.34	0.39	0.71	0.56	0.44	0.40	0.20
concentration_d	0.34	1.00	0.26	0.35	0.48	0.72	0.60	0.18
Rain	0.39	0.26	1.00	0.44	0.24	0.23	0.43	0.41
Wind	0.71	0.35	0.44	1.00	0.40	0.40	0.42	0.23
Temperature	0.56	0.48	0.24	0.40	1.00	0.55	0.37	0.26
rain_d	0.44	0.72	0.23	0.40	0.55	1.00	0.67	0.13
wind_d	0.39	0.60	0.43	0.42	0.36	0.67	1.00	0.21
temperature_d	0.20	0.18	0.41	0.23	0.26	0.13	0.21	1.00

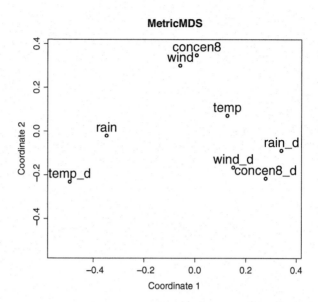

Fig. 9 Multidimensional scaling of variables

of concentration. With the functionalization, factors affecting the changing trends of concentration are shown.

4 Conclusion

Functional data analysis, especially functional clustering, gives us more choices to analyze dataset. Cophenetic correlation coefficient can be used as a measure of degree of fitness of a classification to a set of data to reveal relationship between two dendrograms. Multidimensional scaling is a good visualization to show the dissimilarity between variables in this research.

References

Ransay, J.O., Silverman, B.W.: When the data are functions. Psychometrika **47**, 379–396 (1982)
Ransay, J.O., Silverman, B.W.: Functional Data Analysis, Springer, Berlin (2005)
Saraçli1, S., Doǧa, N., Doǧan, İ.: Comparison of hierarchical cluster analysis methods by cophenetic correlation coefficient. J. Inequal. Appl. **2013**, 203 (2013). https://journalofinequal itiesandapplications.springeropen.com/articles/10.1186/1029-242X-2013-203

Generalized Additive Models for the Detection of Copy Number Variations (CNVs) Using Multi-gene Panel Sequencing Data

Corinna Ernst, Rita K. Schmutzler, and Eric Hahnen

Abstract We present a generalized additive models framework for the detection of germline chromosomal copy number variations from multi-gene panel sequencing data. Mean read abundances along a gene panel target are modelled as a product of two smooth functions, namely a generic background function that contributes to all samples under consideration and a sample-specific smooth function which is used for final copy number variation calling. We validated our approach on 442 germline samples that were sequenced on a customized diagnostic gene panel comprising exons of 49 genes and found that the proposed method outperforms existing approaches both in sensitivity and specificity.

1 Introduction

We present an application of generalized additive models (GAMs) in the area of genomic sequence data analysis, namely the identification of germline chromosomal copy number variations (CNVs) in multi-gene panel data.

Next generation sequencing (NGS) has become an established tool for the investigation and diagnosis of various diseases, e.g., cancer or mental disorders. Targeted sequencing approaches restrict analyses to genomic regions of special interest, e.g., the exome, or, in case of so-called multi-gene panels, to exons of genes known or assumed to be implicated in a special phenotype. Thus, costs, storage requirements, and computation times are further decreased, and panel approaches have become a widely used tool in clinical diagnostics (Antoniadi et al. 2015; Harter et al. 2017; Schreiber et al. 2013). Targeted sequencing data is typically characterized by high sequence read coverage within targeted regions next to a low, partially non-existent overall coverage, as well as strong biases based on local

C. Ernst (✉) · R. K. Schmutzler · E. Hahnen
Center for Familial Breast and Ovarian Cancer, Medical Faculty, University Hospital Cologne,
University of Cologne, Cologne, Germany
e-mail: corinna.ernst@uk-koeln.de; rita.schmutzler@uk-koeln.de; eric.hahnen@uk-koeln.de

© Springer Nature Singapore Pte Ltd. 2020   199
T. Imaizumi et al. (eds.), *Advanced Studies in Classification and Data Science*,
Studies in Classification, Data Analysis, and Knowledge Organization,
https://doi.org/10.1007/978-981-15-3311-2_16

mappability, GC-content, and further factors affecting capture efficiency (Kuilman et al. 2015; Sims et al. 2014). Therefore, multi-gene panels allow for the reliable detection of genetic variants involving only a few nucleotides, namely single nucleotide polymorphisms (SNPs) and short insertion/deletion events (indels), but recent approaches for read depth-based estimation of large genomic rearrangements, so-called CNVs, show notable lacks of accuracy and robustness (Johansson et al. 2016; Povysil et al. 2017). The term CNV describes large-scale chromosomal copy number variations, i.e., gains or losses of >50 bp of genomic DNA (Alkan et al. 2011). CNVs occur frequently in healthy individuals (almost 10% of a typical human genome is supposed to be affected by copy number variation (Zarrei et al. 2015)), but may also be associated with several diseases. For instance, CNVs in *BRCA1* or *BRCA2* are known to confer increased risks of breast and ovarian cancer (Engert et al. 2008).

For the reliable detection of CNVs, complex wet lab experiments are required, such as multiplex ligation-dependent probe amplification (MLPA) (Schouten et al. 2002), fluorescence *in situ* hybridization (FISH), or array comparative genomic hybridization (aCGH) (Oostlander et al. 2004). As these additional analyses are costly and time-consuming, they are usually applied solely to genes known to be frequently affected by CNVs (Johansson et al. 2016), e.g., *BRCA1* and *BRCA2* in case of hereditary breast and ovarian cancer (Harter et al. 2017).

To explore the whole spectrum of genetic variability among the entire set of a panel's sequencing targets, but without the necessity of large-scale laborious wet lab analyses, several computational approaches for CNV detection from targeted sequencing data have been published (Johansson et al. 2016; Povysil et al. 2017; Pugh et al. 2016). All of these methods take single values per sequencing target and sample as input for read depth-based approximation of copy numbers, e.g., read counts per target (Povysil et al. 2017), average depth of read coverage (Johansson et al. 2016), or the fraction of a sample's total coverage per target (Pugh et al. 2016). Doing so, the amount of available information is pruned, and subregions with non-existing coverage, non-linear effects typically occurring at target edges, as well as copy number changes within sequencing targets hinder reliable detection of CNVs.

We propose an approach for CNV detection which is tailored to the challenges of multi-gene panel analysis by investigation of read abundances on all base pair positions along a sequencing target instead of taking solely summed up or averaged read depths per sequencing target into account. Our method relies on the usage of generalized additive model (GAMs), which have recently been shown to comprise a powerful tool for the identification of ChIP-Seq peaks and genomic regions of aberrant DNA methylation (Stricker et al. 2017). We present a GAM which models the mean of observed read counts as a product of two smooth functions, namely a generic background function that contributes to all samples under consideration and a sample-specific smooth function. The latter function is used for final CNV calling. It is assumed to deviate significantly from zero in case a CNV exists. We evaluated our approach on 442 samples containing 17 verified CNVs and found that it outperforms existing state-of-the-art approaches.

2   Methods

We assume mapped sequencing reads of m, $m \geq 30$, samples which are captured on the same gene panel, and to be re-aligned around indels and filtered for duplicates as input. Furthermore, a gene panel specification is required, defining the start and stop positions, as well as the chromosomes, where the sequencing targets are located. Let C be the set of chromosomes containing sequencing targets.

Our approach is divided into three processing steps. In the pre-processing step, sequencing targets are filtered for minimum length and minimum coverage, sample outliers are determined, and inter-sample normalization factors are computed. Then, GAM fitting is applied for sample/target combinations comprising putative CNVs. Finally, fitted GAMs are used for CNV prediction.

2.1   Pre-processing

Pre-processing aims to identify and exclude targets and samples suffering from low sequencing read coverage or aberrant read abundance patterns due to technical errors, as these may cause false positive CNV calls. In order to account for varying sequencing depths among samples, normalization factors are pre-computed.

In a first pre-filter step, sequencing targets covering less than five adjacent base positions are excluded from downstream analysis. In order to reduce artificial correlation between adjacent nucleotides, the number of reads centered per genomic position (henceforth referred to as read count) is used as desired quantity for the estimation of observed read abundances. Thus, input for CNV detection is a matrix $M = (r_{ij}) \in \mathbb{N}_0^{n \times m}$, where r_{ij} is the number of reads in sample j that are centered at DNA sequence position x_i corresponding to the ith entry in M. If $M_{\bullet j}$, the jth column of M, is zero at more than the half of all genomic positions, the corresponding sample is assumed to be a subject of sequencing dropout and is removed from M. Furthermore, sequencing targets t, with

$$\frac{1}{n_t} \left(\sum_{i=1...n_t} \mathbf{1} \left(\sum_{j=1,...,m} r_{ij} = 0 \right) \right) > \frac{1}{3} \tag{1}$$

with n_t denoting the number of positions covered by t and $\mathbf{1}$ denoting the indicator function, are supposed to be insufficiently covered for technical reasons and excluded from M.

A common strategy in existing CNV detection methods is the identification and exclusion of samples that show aberrant read abundance patterns due to technical reasons. This approach aims to reduce the variance of inter-sample read abundances per target in order to prevent false positive CNV calls. For that purpose, a condensed matrix $\overline{M} = (\bar{r}_{tj}) \in \mathbb{Q}_+^{T \times m}$ with $t = 1, \ldots, T$ is generated, where T is the number of sequencing targets under consideration, and values $\bar{r}_{tj}$ are the mean of observed read counts per target t in sample j.

For each sample j^*, the mean of pairwise Pearson correlations

$$\bar{\rho}_{j^*} = \frac{1}{m-1} \sum_{\substack{j=1...m \\ j \neq j^*}} \rho_{\overline{M}_{\bullet j^*}, \overline{M}_{\bullet j}} \tag{2}$$

is used for determination of its read abundance similarity to other samples. Correlation of target-wise read abundances is a widely used measure for the similarity between samples, e.g., panelcn.MOPS (Povysil et al. 2017) uses the correlation of summed read counts per target for that purpose. However, determination of a suitable correlation threshold value t_{corr} that $\bar{\rho}_{j^*}$ has to exceed in order to be further processed is a crucial task, as observed correlations are affected by the overall sequencing quality, the number of input samples, and the panel specification itself. As a data-driven approach, we propose to exclude samples with a mean pairwise Pearson correlation $\bar{\rho}_{j^*} < t_{corr} := Q_1 - 1.5(Q_3 - Q_1)$ with Q_1, respectively Q_3, denoting the first, respectively the third, quartile of the mean pairwise Pearson correlations of all m samples. Usage of interquartile ranges for putative outlier detection is intended to stringently exclude samples that may cause false positive CNV calls, as required in a clinical setting, but our implementation also offers the possibility of setting user-defined constraints.

In order to avoid the exclusion of samples due to low correlation caused by the existence of (large) CNVs, we also compute the mean correlations for putative outlier samples under iterative exclusion of each single chromosome $c \in C$, i.e., exclusion of all sequencing targets that are located on chromosome c. If one or more values of observed mean correlations exceed t_{corr}, the corresponding sample is re-included into downstream analysis.

Comparison of read counts requires normalization of fluctuating overall sequencing depths. A straightforward approach for inter-sample read abundance normalization is division by sample-specific normalization factors s_j. We make use of an inter-sample normalization procedure proposed by Anders and Huber (2010) for the use in differential expression analysis of RNA sequencing experiments (RNA-seq), namely determination of s_j using the median of the ratios of observed read counts per sequencing target via

$$\hat{s}_j = \operatorname*{median}_{\substack{t=1,\dots,T \\ \bar{r}_{tk} \neq 0 \forall j=1,\dots,m}} \frac{\bar{r}_{tj}}{\left(\prod_{k=1}^{m} \bar{r}_{tk} \right)^{1/m}}. \tag{3}$$

We use $\overline{M}$, i.e., the mean of read counts per target, for computation of $\hat{s}_j$ in order to reduce required computational resources and biases towards outstanding long targets.

2.2 A Generalized Additive Model for CNV Detection

We present a GAM which models centered read counts r along a multi-gene panel target using two smooth functions.

Given a set of x_i, $i = 1, \ldots, n$, genomic positions on a single target, the numbers of reads $r_{i\bullet}$ centered at genomic position x_i are assumed to follow a negative binomial distribution with mean μ_i, i.e., $r_{i\bullet} \sim \text{NB}(\mu_i, \Theta)$ and dispersion parameter Θ relating the variance to the mean such that $Var(r_i) = \mu_i + \mu_i^2/\Theta$ (Anders and Huber 2010). A generalized additive model for examination of sample j^*, $1 \le j^* \le m$, on a single target arises via

$$\log(\mu_i) = \log(\hat{s}_j) + f_{all}(x_i) + f_{j^*}(x_i)\mathbf{1}(j = j^*) \tag{4}$$

with $\mathbf{1}(.)$ denoting the indicator function. In summary, the mean of observed read counts over all samples is modelled as a product of two smooth functions, namely $f_{all}(x_i)$ that contributes to all m samples and f_{j^*} that contributes exclusively to the single sample j^* under consideration. f_{j^*} is assumed to deviate significantly from zero in case a CNV exists in sample j^*, and is used for final CNV calling. In concordance with Stricker et al. (2017) second-order, cubic P-splines were chosen, i.e., low rank smoothers using a B-splines basis. Thus, smooth functions f_{all} and f_{j^*} are basically represented by linear combinations of a set of k regularly spaced B-spline basis functions (De Boor 1978). Given a set of $k + 4$ interval boundaries (knots), these basis functions take non-zero values over four adjacent intervals.

GAMs are fitted per target and sample using R's `mgcv` (Wood 2017) package. Smoothing parameter λ as well as dispersion factor Θ are estimated as part of fitting, but k is controlled explicitly by definition of $k := \max(4, \frac{n}{20})$. Fitting the proposed GAM m times for $j^* = 1, \ldots, m$ results in each m fitted smooth functions f_{all} and m fitted sample-specific smooth functions f_j. Furthermore, estimated standard errors per position x_i can be obtained via the `predict.gam` functionality of the `mgcv` package (Wood 2017). See Fig. 1 for visualization of f_{all} and f_j on a sequencing target in *BRCA1* containing a verified single deletion.

The introduced model allows for the examination of read abundances along a single sequencing target instead of simple comparison of isolated values per sample and target such as mean coverage or read counts. But, GAM fitting is time-consuming and computationally expensive, especially if applied for all samples on all panel targets. Thus, a heuristic strategy is applied for the identification of sample/target combinations that are suspected to comprise a CNV. Samples j^* for which

$$0.75 < \frac{r_{ij^*}/\hat{s}_{j^*}}{\frac{1}{m}\sum_{j=1}^{m} r_{ij}/\hat{s}_j} < 1.25 \; \forall \, i \in t \tag{5}$$

holds, are expected to not be affected by copy number variation in the corresponding sequencing target t without further examination. If the heuristic fold change

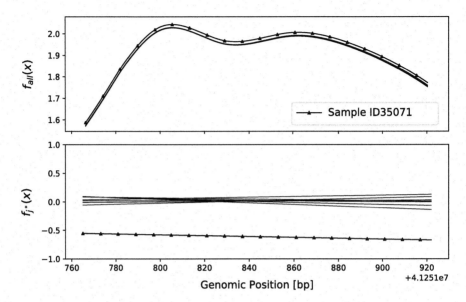

Fig. 1 Fitted smooth functions $f_{all}(x)$ and $f_{j*}(x)$ for ten samples on a sequencing target on chromosome 17 from genomic position 41251765 to 41251922. Sample ID35071 is known to be heterozygously deleted in the shown target. Thus, corresponding function values of $f_{j*}(x)$ deviate noticeably from zero

undercuts 0.75 (putative deletion) or exceeds 1.25 (putative duplication) on at least one target position, GAM fitting is executed for the corresponding sample j^* and under consideration of $n_c - 1$ control samples with highest pairwise Pearson correlations between $\overline{M}_{\bullet j*}$ and $\overline{M}_{\bullet j} \forall j \in \{1 \ldots m | j \neq j^*\}$. n_c is set to 30, or in case $n_c > m$, to the number of samples remaining after outlier exclusion (see Sect. 2.1). Obtained function values of f_{all} and f_j are used for CNV detection in sample j^* exclusively.

As computational complexity of GAM fitting grows polynomially with the number of basis functions k, separate GAM fitting in sequential, overlapping intervals is performed as proposed by Stricker et al. (2017) for sequencing targets comprising more than $L_{max} := 300$ base positions.

2.3 Final CNV Calling

Our aim is the derivation of the copy number state of sample j^* in target t from observed function values $f_j(x)$. Let J^*, $|J^*| = n_c$, be the set of samples that GAMs were fitted for in order to determine the copy number state of sample j^* in target t, and let x_{i*} be an arbitrary base position within t. Furthermore, let X be the values of $\exp(f_j(x_{i*})) \forall j \in J^*$. It is assumed that $f_j(x_{i*})$ will deviate noticeably from zero in case a CNV exists in sample j. Furthermore, due to our GAM definition, $X_j = \exp(f_j(x_{i*}))$ is expected to correspond to a chromosome copy number factor, i.e., to values close to 0.5 in case of a heterozygous deletion, and values close to 1.5 in case of a copy number of three alleles. However, the focus lies on the reliable detection of CNVs rather than the determination of exact copy number states. Consequently, an approach is introduced which focuses mainly on the classification of X_{j*} in the "twilight zone", i.e., values of X_{j*} for which it is hard to decide if aberration from zero might be explained by noisy data or the existence of a CNV. Therefore, sample j is expected to contain a CNV per default if $X_j \geq 2$ (or at least to not represent a proper control sample). $X_j, j \in \{J^*|X_j < 2\}$, are used to optimize log-likelihood function

$$l = \prod_{j \in J} ((1 - p - q)f(X|1, s) + p\ f(X|0.5, s) + q\ f(X|1.5, s))$$

(6)

with $0 \leq p + q \leq 1$,

using the Nelder–Mead method of R's optim utility, where $f(.|\mu, s)$ stands for the probability density function of the normal distribution with mean μ and standard deviation s. See Fig. 2 for some examples of fitted data. Equation (6) represents a mixture model of three normal distributions, each constituting a distinct event of possible copy number states, i.e., deletion ($DEL, \mu = 0.5$), copy number 2 ($CN = 2, \mu = 1$), and duplication ($DUP, \mu = 1.5$). Hence, likelihoods of observed values X are obtained via

$$\mathbb{P}(X|DEL) \sim \mathcal{N}(X|0.5, s)$$

$$\mathbb{P}(X|CN = 2) \sim \mathcal{N}(X|1, s)$$

(7)

$$\mathbb{P}(X|GAIN) \sim \mathcal{N}(X|1.5, s)$$

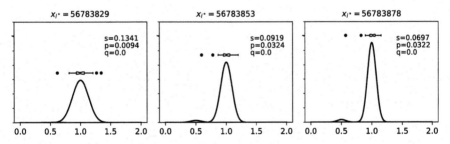

Fig. 2 Boxplots of exponentiated function values $\exp(f_j(x_{i*}))$ and corresponding log-likelihood functions due to (6) for three positions each in target chr17:56783823–56783994 in *RAD51C*. The shown sequencing target is known to contain a CNV, namely a heterozygous deletion. Estimated values of standard deviation s serve as a measure for quality control, i.e., position 56783829 would not be evaluated if $\mathbb{P}(0.75 < x < 1.25|x \sim \mathcal{N}(1, s)) \geq 0.95$ is claimed, resulting in an upper threshold $smax := 0.1276 < 0.1341$

Note, that Eqs. (6) and (7) omit modelling of more extreme events such as $CN = 0$ and $CN > 3$, as these events are expected to occur too seldom to distort the adjusted parameters p, q, s (at least in the use case of examination of non-tumor samples), and are further reduced due to pre-filtering for extreme fold changes, or rather, values of $X_j \geq 2$.

Posterior probabilities of copy number events $\Omega \in \{DEL, \ CN = 2, \ DUP\}$ arise due to Bayes' theorem via

$$\mathbb{P}(\Omega|X) = \frac{\mathbb{P}(X|\Omega)\mathbb{P}(\Omega)}{\mathbb{P}(X)}, \tag{8}$$

whereby $\mathbb{P}(X)$ is constant for all copy number events Ω, and can be neglected. Hence, estimates of conditional probabilities $\mathbb{P}(\Omega|X_j)$ are given by

$$\mathbb{P}(DEL|X_j) = \alpha_{DEL} \ f(X_j|0.5, s)\mathbb{P}(DEL)$$
$$\mathbb{P}(CN=2|X_j) = \alpha_{CN=2} \ f(X_j|1, s)\mathbb{P}(CN=2) \tag{9}$$
$$\mathbb{P}(DUP|X_j) = \alpha_{DUP} \ f(X_j|1.5, s)\mathbb{P}(DUP)$$

with normalization factors α_Ω, $\Omega \in \{DEL, CN=2, DUP\}$, arising from

$$\sum_{\Omega \in \{DEL, CN=2, DUP\}} \mathbb{P}(\Omega|X_j) = 1. \tag{10}$$

Figure 3 shows posterior probabilities $\mathbb{P}(DEL|X)$ and $\mathbb{P}(DUP|X)$ in dependence to X for $\mathbb{P}(DEL) = \mathbb{P}(DUP) = 0.1$ and five different values of standard

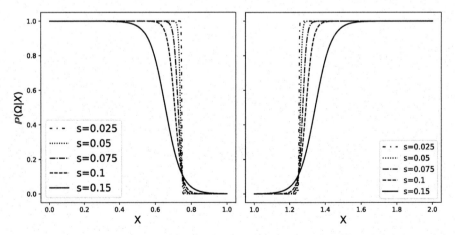

Fig. 3 Posterior probabilities $\mathbb{P}(DEL|X)$ for $\mathbb{P}(DEL) := 0.1$ (left) and $\mathbb{P}(DUP|X)$ for $\mathbb{P}(DUP) := 0.1$ (right) due to (8) and (9) for several values of standard deviation s

deviation s. It can be seen that, at least in case of comparatively small values of s, $\mathbb{P}(DEL|X_j)$ behaves like a slightly smoothed step function. Hence, determination of a threshold for $\mathbb{P}(\Omega|X)$, $\Omega \in \{DEL, DUP\}$ does only affect a small range of possible values X, as long as the chosen threshold is neither close to 0 nor close to 1. In the sense of the proposed Bayesian framework, a CNV should be called if the corresponding conditional probability exceeds the two other values of conditional probabilities. For meaningful, i.e., sufficient small, values of s, $\mathbb{P}(DUP|X)$ can be expected to be negligibly small if $\mathbb{P}(DEL|X) \gg 0$, and vice versa. Hence, it is proposed to call a CNV at position x_{i*} in case $\mathbb{P}(DEL|X) > 0.5$ or $\mathbb{P}(DUP|X) > 0.5$, respectively.

Determination of $\mathbb{P}(DEL)$ and $\mathbb{P}(DUP)$ is crucial, as proper values for coding regions in humans are not available and prior probabilities for the existence of CNVs may be target- and cohort-specific. As Zarrei et al. (2015) reported that almost 10% of a typical human genome is supposed to be affected by CNVs, we propose to set $\mathbb{P}(DEL) := \mathbb{P}(DUP) := 0.1$ to provide an upper bound, at least.

Besides determination of $\mathbb{P}(X|\Omega)$ due to (7), estimated standard errors s do also serve as a quality measure for the purpose of preventing false positive CNV calls. For positions with increased s, values of X resulting in $\mathbb{P}(\Omega|X) > 0.5$ for $\Omega \in \{DEL, DUP\}$ may be more likely caused by variance than by the existence of CNVs. For definition of a suitable upper threshold s_{max}, we propose investigation of the 95% confidence intervals of $\mathcal{N}(1, s)$. Claiming $\mathbb{P}(0.75 < x < 1.25|x \sim \mathcal{N}(1, s)) \geq 0.95$ results in an upper threshold $s_{max} := 0.1276$, which s has to undercut in order to assess the corresponding x_{i*} as an evaluable position.

We have introduced an approach for CNV calling on the basis of values $f_j(x_{i*})$ for a single position i^*. A possible criterion for selection of the most suitable position i^* in a given target is the minimum of the mean standard error of functions $f_{all}(x_i)$ over $i = 1, \ldots, n$. Standard error estimates are returned by the `predict.gam()` utility of `mgcv` on the basis of the Bayesian posterior covariance matrix of the parameters in f_{all} (see Wood 2017 for details). Logically, standard errors reach local maxima at knot locations, where smooth functions are supported by only three splines. Hence, we propose to set the values of i^* to the local minima of the standard errors of f_{all} between two adjacent knot locations (Fig. 4). Such an approach scales the number of evaluated positions by target length and ensures its regularly spacing along the target. In order to determine a copy number state per target, we propose to claim the existence of a CNV in case $\mathbb{P}(DEL|X) > 0.5$ or $\mathbb{P}(DUP|X) > 0.5$, respectively, was observed for at least two adjacent evaluated positions. Given that splines are located at 20 bp intervals, this criterion is justified by the 50 bp minimum length of CNVs.

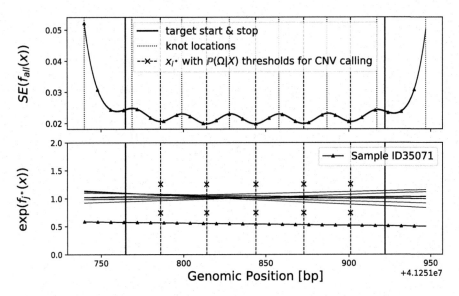

Fig. 4 Estimated standard error of smooth functions $f_{all}(x)$ (above) and exponentiated values of $f_{j*}(x)$ (below) for the fitted GAM also shown in Figure 1. Position-wise CNV calling is performed on positions x_{i*} corresponding to the minima of the mean of $SE(f_{all}(x))$ between two adjacent knot locations

3   Results

We evaluated our approach on 442 germline samples from ovarian cancer patients that were analyzed in the course of the AGO-TR1 study (Harter et al. 2017). Sequencing was performed on HiSeq 2000 using a customized diagnostic gene panel (Agilent) comprising exons of 49 genes known or assumed to be implicated in hereditary cancer. High risk genes *BRCA1* and *BRCA2* were entirely screened for CNVs via MLPA. Thereby, six deletions and two duplications were found in *BRCA1*, whereas no CNVs could be found in *BRCA2*. CNVs in the remaining genes were identified via the SOPHIA DDM® platform (Sophia Genetics), and subsequently confirmed via aCGH or MLPA. See Table 1 for an enumeration of all confirmed CNVs. We compared the sensitivity and specificity of our method against three state-of-the-art approaches for CNV detection on gene panel data, namely panelcn.MOPS (Povysil et al. 2017), CoNVaDING (Johansson et al. 2016), and VisCap (Pugh et al. 2016). All of these tools are read depth-based approaches, i.e., based on the assumption that depth of coverage is correlated with chromosomal copy number, and have been published recently.

Table 1 Verified CNVs in six sequencing runs comprising 442 germline samples

Run	#Sample	CNV	Gene	Chr	Start	Stop	#Targets
1	ID35071	DEL	*BRCA1*	17	41,251,765	41,258,575	4
	ID35073	DEL	*BRCA1*	17	41,197,668	41,197,844	1
	ID35080	DEL	*FANCM*	14	45,667,820	45,669,236	2
	ID36319	DEL	*RAD51C*	17	56,783,823	56,811,608	6
2	ID40100	DEL	*FANCA*	16	89,818,519	89,825,138	2
3	ID41108	DEL	*BRCA1*	17	41,262,525	41,276,138	3
4	ID41396	DEL	*RAD51D*	17	33,427,945	33,428,080	1
	ID41272	DEL	*BRCA1*	17	41,215,323	41,267,821	18
	ID41421	DUP	*XRCC2*	7	152,345,700	152,373,189	3
	ID41477	DEL	*BRCA1*	17	41,219,598	41,276,138	17
5	ID42064	DEL	*BRCA1*	17	41,242,934	41,247,964	3
	ID42068	DUP	*FANCA*	16	89,804,982	89,882,419	43
	ID42070	DUP	*MLH1*	3	37,088,983	37,092,169	4
6	ID42585	DUP	*BRCA1*	17	41,215,323	41,215,993	2
	ID42614	DEL	*RAD51C*	17	56,783,823	56,811,608	6
	ID42618	DUP	*BRCA1*	17	41,251,765	41,251,922	1
	ID42640	DEL	*RAD51C*	17	56,787,193	56,787,376	1

Samples were entirely screened for CNVs in *BRCA1* and *BRCA2* via MLPA. CNVs in additional genes were verified via aCGH or MLPA. Column #Targets refers to the count of involved adjacent sequencing targets

3.1 Determination of Sample Outliers

All approaches under consideration identify samples with aberrant read abundance patterns for prevention of false positives or at least in order to warn of low quality samples. Whereas panelcn.MOPS and CoNVaDING only report putative outliers, VisCap and our approach exclude them entirely from analysis. See Table 2 for an overview of the results of outlier sample detection in our evaluation data set. Our approach, which is based on outlier detection due to the interquartile range of mean pairwise Pearson correlations of observed mean read counts per target (see Sect. 2.1), identified all samples as outliers that were detected by at least one of the other methods, but turned out to be more stringent. Sample ID42068 from run 5 makes an exception as it was excluded by CoNVaDING and VisCap, but re-included by our approach due to its mean correlation to other samples under exclusion of all sequencing targets located in chromosome 16. It can be concluded that its low correlation to other samples is caused by the existence of a large duplication in gene *FANCA* (see Table 1), rather than by an overall deviating read abundance pattern. We evaluated our approach exclusively on the basis of samples that were not identified as outlier by none of CoNVaDING, VisCap, or panelcn.MOPS, but adapted t_{corr} such that the remaining samples were entirely included by our GAM approach. Thus, we set $t_{corr} := 0.99$ for sequencing runs 1 and 3, $t_{corr} := 0.994$ for 4 and 5, and $t_{corr} := 0.9965$ for sequencing run 6.

Table 2 Identification of samples with aberrant read abundance patterns in six sequencing runs by our approach, ConVaDING, VisCap, and panelcn.MOPS

Run	m	t_{corr}	Excluded	Corr	GAM	CoNVaDING	VisCap	pcn.MOPS
1	80	0.9977	ID35093	0.9971	x			
			ID35098	0.9958	x			
			ID36232[a]		x	x	x	x
			ID36307	0.9948	x			
			ID36317	0.9973	x			
			ID36319	0.9975	x			
			ID39328	0.9973	x			
			ID39350	0.9885	x	x		x
2	32	0.9964	ID40083	0.9864	x	x	x	x
			ID40086	0.9939	x	x	x	x
			ID40091	0.9950	x	x	x	
			ID40099[a]		x	x	x	x
3	96	0.9971	ID41110	0.9957	x			
			ID41142	0.9957	x			
			ID41150	0.9968	x			
4	96	0.9960	ID41271	0.9920	x		x	x
			ID41486	0.9944	x			
			ID41507[a]		x		x	x
5	48	0.9957	ID42068	0.9934[b]		x	x	
			ID42078	0.9953	x			
			ID42174[a]		x	x	x	x
			ID42175[a]		x	x	x	x
			ID42176[a]		x	x	x	x
			ID42177[a]		x	x	x	x
			ID42178[a]		x	x	x	x
			ID42180[a]		x	x	x	x
6	90	0.9970	ID42586	0.9954	x		x	
			ID42616	0.9963	x			
			ID42646	0.9905	x	x	x	

The table shows the number of samples per run (m) and the correlation thresholds t_{corr} a sample's mean correlation have to exceed per default to be analyzed by our approach. Samples that were identified as outlier and excluded by our approach (GAM), CoNVaDING, VisCap, or panelcn.MOPS (pcn.MOPS) are characterized by x

[a] Excluded due to low coverage

[b] Re-included into analysis due to mean correlation under exclusion of chromosome 16

Table 3 Evaluation of CNV detection results for 425 samples for our approach (GAM), panelcn.MOPS, CoNVaDING, and VisCap

	GAM	panelcn.MOPS	CoNVaDING	VisCap
Specificity ($N = 21148$)	>**0.9999**	0.9991	>**0.9999**	0.9976
Sensitivity ($N = 71$)	**1.0**	**1.0**	0.8493	**1.0**
MCC	**0.9725**	0.8856	0.8691	0.7663

Specificities were determined on 21,148 sample/target combinations in *BRCA1* and *BRCA2*, whereas sensitivities were determined on 71 sample/target combinations on eight different genes (see Table 1). Maximum values are shown in bold

3.2   CNV Detection Results

We evaluated the specificity of our approach, panelcn.MOPS, CoNVaDING, and VisCap on all sequencing targets in *BRCA1* (24 targets) and *BRCA2* (26 targets), as these genes were entirely screened for CNVs via MLPA (Harter et al. 2017). Given 49 sample/target combinations known to contain a verified CNV in *BRCA1* (see Table 1), and due to the exclusion of 17 out of 442 samples as outliers, this results in a set of 21,201 sample/target combinations. Sensitivity was evaluated on the 74 sequencing targets containing verified CNVs as reported in Table 1, but under exclusion of the large deletion in *FANCA* in sample ID42068 which was excluded from analysis by VisCap (Table 2). Our approach classified 56 sample/target combinations in *BRCA1* and *BRCA2* as not evaluable, as no two adjacent interval positions fulfilled criterion $s < s_{max}$ and show identical position-wise calls Ω. We examined specificities and sensitivities under exclusion of these sample/target combinations resulting in sample sets of size $N = 21148$ for determination of specificities and $N = 71$ for sensitivities. Evaluation results are summarized in Table 3.

As CNVs are rare events, evaluation of accuracies is naturally biased towards the number of true negatives. Therefore, we evaluated Matthews correlation coefficients (MCCs). The introduced GAM approach reached the highest MCC among all approaches, outperforms panelcn.MOPS and Viscap due to specificity, and CoNVaDING due to its sensitivity.

4   Discussion

We introduced generalized additive models as a promising approach for the analysis of CNVs in multi-gene panel data. The proposed method outperforms existing tools due to the investigation of fold changes at several target positions instead of examination of single values per target and sample, such as read counts or mean read coverage. Furthermore, our GAM approach allows for independent quality assessment per sample and sequencing target.

However, besides omitting the estimation of exact chromosomal copy numbers, an obvious drawback of our approach is its computational complexity due to the fitting of n_c GAMs per sample/target combination that is expected to represent a putative CNV. Furthermore, most GAM fittings are applied at low coverage sequencing targets which then subsequently turned out to be not evaluable due to criterion $s > s_{max}$. Clearly, time complexity could be reduced by a parallelized implementation and a strategy for the identification of low quality sample/target combinations previous to GAM fitting. Finally, we have to note that we omit to discuss here the handling of sex chromosomes X and Y which cannot be assumed to appear twice per default.

Acknowledgments We sincerely thank Andreas Beyer, Achim Tresch, and Michael Nothnagel for helpful discussions.

References

Alkan, C., Coe, B.P., Eichler, E.E.: Genome structural variation discovery and genotyping. Nat. Rev. Gen. **12**(5), 363 (2011)

Anders, S., Huber, W.: Differential expression analysis for sequence count data. Genome Biol. **11**(10), R106 (2010)

Antoniadi, T., Buxton, C., Dennis, G., Forrester, N., Smith, D., Lunt, P., Burton-Jones, S.: Application of targeted multi-gene panel testing for the diagnosis of inherited peripheral neuropathy provides a high diagnostic yield with unexpected phenotype-genotype variability. BMC Med. Genet. **16**(1), 84 (2015)

De Boor, C.: A Practical Guide to Splines, vol. 27. Springer, New York (1978)

Engert, S., Wappenschmidt, B., Betz, B., Kast, K., Kutsche, M., Hellebrand, H., Goecke, T.O., Kiechle, M., Niederacher, D., Schmutzler, R.K., et al.: MLPA screening in the BRCA1 gene from 1,506 German hereditary breast cancer cases: novel deletions, frequent involvement of exon 17, and occurrence in single early-onset cases. Human Mutat. **29**(7), 948–958 (2008)

Harter, P., Hauke, J., Heitz, F., Reuss, A., Kommoss, S., Marmé, F., Heimbach, A., Prieske, K., Richters, L., Burges, A., et al.: Prevalence of deleterious germline variants in risk genes including BRCA1/2 in consecutive ovarian cancer patients (AGO-TR-1). PLoS One **12**(10), e0 186043 (2017)

Johansson, L.F., van Dijk, F., de Boer, E.N., van Dijk-Bos, K.K., Jongbloed, J.D., van der Hout, A.H., Westers, H., Sinke, R.J., Swertz, M.A., Sijmons, R.H., et al.: CoNVaDING: single exon variation detection in targeted NGS data. Human Mutat. **37**(5), 457–464 (2016)

Kuilman, T., Velds, A., Kemper, K., Ranzani, M., Bombardelli, L., Hoogstraat, M., Nevedomskaya, E., Xu, G., de Ruiter, J., Lolkema, M.P., et al.: CopywriteR: DNA copy number detection from off-target sequence data. Genome Biol. **16**(1), 49 (2015)

Oostlander, A.E., Meijer, G., Ylstra, B.: Microarray-based comparative genomic hybridization and its applications in human genetics. Clin. Genet. **66**(6), 488–495 (2004)

Povysil, G., Tzika, A., Vogt, J., Haunschmid, V., Messiaen, L., Zschocke, J., Klambauer, G., Hochreiter, S., Wimmer, K.: panelcn.MOPS: copy-number detection in targeted NGS panel data for clinical diagnostics. Hum. Mutat. **38**(7), 889–897 (2017)

Pugh, T.J., Amr, S.S., Bowser, M.J., Gowrisankar, S., Hynes, E., Mahanta, L.M., Rehm, H.L., Funke, B., Lebo, M.S.: VisCap: inference and visualization of germ-line copy-number variants from targeted clinical sequencing data. Genet. Med. **18**(7), 712 (2016)

Schouten, J.P., McElgunn, C.J., Waaijer, R., Zwijnenburg, D., Diepvens, F., Pals, G.: Relative quantification of 40 nucleic acid sequences by multiplex ligation-dependent probe amplification. Nucleic Acids Res. **30**(12), e57–e57 (2002)

Schreiber, M., Dorschner, M., Tsuang, D.: Next-generation sequencing in schizophrenia and other neuropsychiatric disorders. Am. J. Med. Gen. B Neuropsychiatr. Genet. **162**(7), 671–678 (2013)

Sims, D., Sudbery, I., Ilott, N.E., Heger, A., Ponting, C.P.: Sequencing depth and coverage: key considerations in genomic analyses. Nat. Rev. Genet. **15**(2), 121 (2014)

Stricker, G., Engelhardt, A., Schulz, D., Schmid, M., Tresch, A., Gagneur, J.: GenoGAM: genome-wide generalized additive models for ChIP-Seq analysis. Bioinformatics **33**(15), 2258–2265 (2017)

Wood, S.N.: Generalized Additive Models: An Introduction with R. Chapman and Hall/CRC, Boca Raton (2017)

Zarrei, M., MacDonald, J.R., Merico, D., Scherer, S.W.: A copy number variation map of the human genome. Nat. Rev. Genet. **16**(3), 172 (2015)

Variable Selection for Classification of Multivariate Functional Data

Tomasz Górecki, Mirosław Krzyśko, and Waldemar Wołyński

Abstract New variable selection method is considered in the setting of classification with multivariate functional data (Ramsay and Silverman, Functional data analysis, 2005). The variable selection is a dimensionality reduction method which leads to replace the whole vector process, with a low-dimensional vector still giving a comparable classification error. The various classifiers appropriate for functional data are used. The proposed variable selection method is based on functional distance covariance (Székely et al. Ann Appl Stat 3(4):1236–1265, 2009; Stat Probab Lett 82(12):2278–2282, 2012). and is a modification of the procedure given by Kong et al. (Stat Med 34:1708–1720, 2015). The proposed methodology is illustrated on real data example.

1 Introduction

Much attention has been paid in recent years to methods for representing data as functions or curves. Such data are known in the literature as functional data (Ramsay and Silverman 2005; Horváth and Kokoszka 2012). Applications of functional data can be found in various fields, including medicine, economics, meteorology, and many others. In many applications there is a need to use statistical methods for objects characterized by multiple variables observed at many time points (doubly multivariate data). Such data are called multivariate functional data. In this paper we focus on the classification problem for multivariate functional data. In many cases, in the classification procedures, number of predictors p is much greater than the sample size n. It is thus natural to assume that only a small number of predictors are relevant to response Y.

Various basic classification methods have also been adapted to functional data, such as linear discriminant analysis (Hastie et al. 1995), logistic regression (Rossi

T. Górecki (✉) · M. Krzyśko · W. Wołyński
Faculty of Mathematics and Computer Science, Adam Mickiewicz University, Poznań, Poland
e-mail: tomasz.gorecki@amu.edu.pl; mkrzysko@amu.edu.pl; wolynski@amu.edu.pl

© Springer Nature Singapore Pte Ltd. 2020
T. Imaizumi et al. (eds.), *Advanced Studies in Classification and Data Science*,
Studies in Classification, Data Analysis, and Knowledge Organization,
https://doi.org/10.1007/978-981-15-3311-2_17

et al. 2002), penalized optimal scoring (Ando 2009), kNN (Ferraty and Vieu 2003), SVM (Rossi and Villa 2006), and neural networks (Rossi et al. 2005). Moreover, the combining of classifiers has been extended to functional data (Ferraty and Vieu 2009). Górecki et al. (2016) adapted multivariate regression models to the classification of multivariate functional data.

Székely et al. (2007), Székely and Rizzo (2009), Székely and Rizzo (2012, 2013) defined the measures of dependence between random vectors: the distance covariance (dCov) coefficient and the distance correlation (dCor) coefficient. These authors showed that for all random variables with finite first moments, the dCor coefficient generalizes the idea of correlation in two ways. Firstly, this coefficient can be applied when X and Y are of any dimensions and not only for the simple case where $p = q = 1$. Secondly, the dCor coefficient is equal to zero, if and only if there is independence between the random vectors. Indeed, a correlation coefficient measures linear relationships and can be equal to 0 even when the variables are related. Based on the idea of the distance covariance between two random vectors, we introduced the functional distance correlation between two random processes. We select a set of important predictors with large value of functional distance covariance. Our selection procedure is a modification of the procedure given by Kong et al. (2015). Entirely different approach to the variable selection in functional data classification is presented by Berrendero et al. (2016). It is clear that variable selection has, at least, an advantage when compared with other dimension reduction methods (functional principal component analysis (FPCA), see Górecki et al. 2014; Jacques and Preda 2014, functional partial least squares (FPLS) methodology, see Delaigle and Haal 2012, and other methods) based on general projections: the output of any variable selection method is always directly interpretable in terms of the original variables, provided that the required number d of selected variables is not too large.

The rest of this paper is organized as follows. In Sect. 2 we present the classification procedures used through the paper. In Sect. 3 we present the problem of representing functional data by orthonormal basis functions. In Sect. 4, we define a functional distance covariance and distance correlation. In Sect. 5 we propose a variable selection procedure based on the functional distance covariance. In Sect. 6 we illustrate the proposed methodology through a real data example. We conclude in Sect. 7.

2 Classifiers

The classification problem involves determining a procedure by which a given object can be assigned to one of q populations based on observation of p features of that object.

The object being classified can be described by a random pair (X, Y), where $X = (X_1, X_2, \ldots, X_p)' \in \mathbf{R}^p$ and $Y \in \{1, \ldots, q\}$. An automated classifier can be viewed as a method of estimating the posterior probability of membership in groups. For a given X, a reasonable strategy is to assign X to that class with the highest posterior probability. This strategy is called the Bayes' rule classifier.

2.1 Linear and Quadratic Discriminant Classifiers

Now we make the Bayes' rule classifier more specific by the assumption that all multivariate probability densities are multivariate normal having arbitrary mean vectors and a common covariance matrix. We shall call this model the linear discriminant classifier (LDC). Assuming that class-covariance matrices are different, we obtain quadratic discriminant classifier (QDC).

2.2 Naive Bayes Classifier

A naive Bayes classifier is a simple probabilistic classifier based on applying Bayes' theorem with independence assumptions. When dealing with continuous data, a typical assumption is that the continuous values associated with each class are distributed according to a one-dimensional normal distribution or we estimate density by kernel method.

2.3 k-Nearest Neighbor Classifier

Most often we do not have sufficient knowledge of the underlying distributions. One of the important nonparametric classifiers is a k-nearest neighbor classifier (kNN classifier). Objects are assigned to the class having the majority in the k nearest neighbors in the training set.

2.4 Multinomial Logistic Regression

It is a classification method that generalizes logistic regression to multiclass problem using one vs. all approach.

3 Functional Data

We now assume that the object being classified is described by a p-dimensional random process $X = (X_1, X_2, \ldots, X_p)' \in L_2^p(I)$, where $L_2(I)$ is the Hilbert space of square-integrable functions, and $E(X) = 0$.

Moreover, assume that the kth component of the vector X can be represented by a finite number of orthonormal basis functions $\{\varphi_b\}$

$$X_k(t) = \sum_{b=0}^{B_k} \alpha_{kb}\varphi_b(t), \ t \in I, \ k = 1, \ldots, p, \tag{1}$$

where $\alpha_{k0}, \alpha_{k1}, \ldots, \alpha_{kB_k}$ are the unknown coefficients.

Let $\boldsymbol{\alpha} = (\alpha_{10}, \ldots, \alpha_{1B_1}, \ldots, \alpha_{p0}, \ldots, \alpha_{pB_p})'$
and

$$\boldsymbol{\Phi}(t) = \begin{bmatrix} \boldsymbol{\varphi}_1'(t) & \mathbf{0}' & \cdots & \mathbf{0}' \\ \mathbf{0}' & \boldsymbol{\varphi}_2'(t) & \cdots & \mathbf{0}' \\ \cdots & \cdots & \cdots & \cdots \\ \mathbf{0}' & \mathbf{0}' & \cdots & \boldsymbol{\varphi}_p'(t) \end{bmatrix},$$

where $\boldsymbol{\varphi}_k(t) = (\varphi_0(t), \ldots, \varphi_{B_k}(t))', k = 1, \ldots, p$.

Using the above matrix notation, process X can be represented as:

$$X(t) = \boldsymbol{\Phi}(t)\boldsymbol{\alpha}, \tag{2}$$

where $E(\boldsymbol{\alpha}) = \mathbf{0}$. This means that the realizations of a process X are in finite-dimensional subspace of $L_2^p(I)$. We will denote this subspace by $\mathscr{L}_2^p(I)$.

We can estimate the vector $\boldsymbol{\alpha}$ on the basis of n independent realizations $x_1, x_2, \ldots, x_n$ of the random process X (functional data). We will denote this estimator by $\hat{\boldsymbol{\alpha}}$.

Typically data are recorded at discrete moments in time. Let x_{kj} denote an observed value of the feature X_k, $k = 1, 2, \ldots, p$ at the jth time point t_j, where $j = 1, 2, \ldots, J$. Then our data consist of the pJ pairs (t_j, x_{kj}). These discrete data can be smoothed by continuous functions x_k and I is a compact set such that $t_j \in I$, for $j = 1, \ldots, J$.

Details of the process of transformation of discrete data to functional data can be found in Ramsay and Silverman (2005) or in Górecki et al. (2014).

4 Functional Distance Covariance and Distance Correlation

For jointly distributed random process $X \in L_2^p(I)$ and random vector $Y \in \mathbb{R}^q$, let

$$f_{X,Y}(l, m) = \mathrm{E}\{\exp[i < l, X >_p + i < m, Y >_q]\}$$

be the joint characteristic function of (X, Y), where

$$< l, X >_p = \int_I l'(t) X(t) dt$$

and

$$< m, Y >_q = m' Y.$$

Moreover, we define the marginal characteristic functions of X and Y as follows: $f_X(l) = f_{X,Y}(l, 0)$ and $f_Y(m) = f_{X,Y}(0, m)$.

Here, for generality, we assume that $Y \in \mathbb{R}^q$, although the label Y in the classification problem is a random variable, with values in $\{1, \ldots, q\}$. Label Y has to be transformed into the label vector $Y = (Y_1, \ldots, Y_q)'$, where $Y_i = 1$ for $i = 1, \ldots, q$ if X belongs to class i, and 0 otherwise.

Now, let us assume that $X \in \mathcal{L}_2^p(I)$. Then the process X can be represented as:

$$X(t) = \Phi(t)\alpha, \tag{3}$$

where $\alpha \in \mathbb{R}^{K+p}$ and $K = B_1 + \cdots + B_p$.

In this case, we may assume (Ramsay and Silverman 2005) that the vector weight function l and the process X are in the same space, i.e. the function l can be written in the form

$$l(t) = \Phi(t)\lambda, \tag{4}$$

where $\lambda \in \mathbb{R}^{K+p}$.

Hence

$$< l, X >_p = \int_I l'(t) X(t) dt = \lambda' [\int_I \Phi'(t)\Phi(t) dt]\alpha = \lambda'\alpha,$$

where α and λ are vectors occurring in the representations (3) and (4) of process X and function l, and

$$f_{X,Y}(l, m) = \mathrm{E}\{\exp[i\lambda'\alpha + im'Y]\} = f_{\alpha,Y}(\lambda, m),$$

where $f_{\alpha,Y}(\lambda, m)$ is the joint characteristic function of the pair of random vectors (α, Y).

On the basis of the idea of distance covariance between two random vectors (Székely et al. 2007), we can introduce functional distance covariance between random processes X and random vector Y as a nonnegative number $\nu_{X,Y}$ defined by

$$\nu_{X,Y} = \nu_{\alpha,Y},$$

where

$$\nu_{\alpha,Y}^2 = \frac{1}{C_{K+p}C_q} \int_{\mathbb{R}^{K+p+q}} \frac{|f_{\alpha,Y}(\lambda, m) - f_{\alpha}(\lambda) f_Y(m)|^2}{\|\lambda\|_{K+p}^{K+p+1} \|m\|_q^{q+1}} d\lambda dm,$$

and $|z|$ denotes the modulus of $z \in \mathbb{C}$, $\|\lambda\|_{K+p}$, $\|m\|_q$ the standard Euclidean norms on the corresponding spaces V chosen to produce scale free and rotation invariant measure that does not go to zero for dependent random vectors, and

$$C_r = \frac{\pi^{\frac{1}{2}(r+1)}}{\Gamma(\frac{1}{2}(r+1))}$$

is half the surface area of the unit sphere in $\mathbb{R}^{r+1}$.

The functional distance correlation between random vector process X and random vector Y is a nonnegative number defined by

$$\mathscr{R}_{X,Y} = \frac{\nu_{X,Y}}{\sqrt{\nu_{X,X}\nu_{Y,Y}}}$$

if both $\nu_{X,X}$ and $\nu_{Y,Y}$ are strictly positive, and defined to be zero otherwise.

We have $\mathscr{R}_{X,Y} = \mathscr{R}_{\alpha,Y}$ as $\nu_{X,Y} = \nu_{\alpha,Y}$.

For distributions with finite first moments, distance correlation characterizes independence in that $0 \leq \mathscr{R}_{X,Y} \leq 1$ with $\mathscr{R}_{X,Y} = 0$ if and only if X and Y are independent. We can estimate functional distance covariance using data $\{(\hat{\alpha}_1, y_1), \ldots, (\hat{\alpha}_n, y_n)\}$.

Let

$$\bar{\alpha} = \frac{1}{n}\sum_{i=1}^{n} \hat{\alpha}_k, \quad \bar{y} = \frac{1}{n}\sum_{i=1}^{n} \hat{k}_k,$$

$$\tilde{\alpha}_k = \hat{\alpha}_k - \bar{\alpha}, \quad \tilde{y}_k = y_k - \bar{y}, \ k = 1, \ldots, n$$

and

$$A = (a_{kl}), \quad B = (b_{kl}),$$

$$\tilde{A} = (A_{kl}), \quad \tilde{B} = (B_{kl}),$$

where

$$a_{kl} = \|\hat{\boldsymbol{\alpha}}_k - \hat{\boldsymbol{\alpha}}_l\|_{K+p}, \quad b_{kl} = \|\mathbf{y}_k - \mathbf{y}_l\|_q,$$
$$A_{kl} = \|\tilde{\boldsymbol{\alpha}}_k - \tilde{\boldsymbol{\alpha}}_l\|_{K+p}, \quad B_{kl} = \|\tilde{\mathbf{y}}_k - \tilde{\mathbf{y}}_l\|_q, \quad k, l = 1, \ldots, n.$$

Hence

$$\tilde{A} = HAH, \quad \tilde{B} = HBH,$$

where

$$H = I_n - \frac{1}{n}\mathbf{1}_n\mathbf{1}_n'$$

is the centering matrix.

Let $\tilde{A} \circ \tilde{B} = (A_{kl}B_{kl})$ denote the Hadamard product of the matrices $\tilde{A}$ and $\tilde{B}$. Then, on the basis of the result of Székely et al. (2007), we have

$$\hat{v}_{X,Y}^2 = \frac{1}{n^2}\sum_{k,l=1}^n A_{kl}B_{kl}.$$

The sample functional distance correlation is then defined by $\hat{\mathcal{R}}_{X,Y} = \hat{\mathcal{R}}_{\alpha,Y}$, where

$$\hat{\mathcal{R}}_{\alpha,Y} = \frac{\hat{v}_{\alpha,Y}}{\sqrt{\hat{v}_{\alpha,\alpha}\hat{v}_{Y,Y}}}$$

if both $\hat{v}_{\alpha,\alpha}$ and $\hat{v}_{Y,Y}$ are strictly positive, and zero otherwise.

5   Variable Selection Based on the Distance Covariance

In this section we propose the selection procedure built upon the distance covariance. Let $Y = (Y_1, \ldots, Y_q)'$ be the response vector, and $X = (X_1, \ldots, X_p)'$ be the predictor p-dimensional process. Assume that only a small number of predictors are relevant to Y. We select a set of important predictors with large $\hat{\mathcal{R}}_{X,Y} = \hat{\mathcal{R}}_{\alpha,Y}$. We utilize the functional distance covariance because it allows for arbitrary relationship between Y and X, regardless of whether it is linear or nonlinear.

The functional distance covariance also permits univariate and multivariate response. Thus, this distance covariance procedure is completely model-free. Kong et al. (2015) prove the following theorem.

Theorem 1 *Suppose random vectors* $X, Z \in \mathbb{R}^p$ *and* $Y \in \mathbb{R}^q$, *and assume* Z *is independent of* (X, Y), *then*

$$\nu^2_{(X,Z),Y} \le \nu^2_{X,Y}.$$

And a consequence of this theorem is the statement in the next corollary.

Corollary 1 *For the sample distance covariance, if* n *is large enough, we should have*

$$\hat{\nu}^2_{(X,Z),Y} \le \hat{\nu}^2_{X,Y},$$

under the assumption of independence between (X, Y) *and* Z.

We implemented the above theorem as a stopping rule in the selections of responses. The procedure took the following steps:

1. Calculate marginal distance covariances for $X_k, k = 1, \ldots, p$ with the response Y.
2. Rank the variables in decreasing order of the distance covariances. Denote the ordered predictors as $X_{(1)}, X_{(2)}, \ldots, X_{(p)}$. Start with $X_S = \{X_{(1)}\}$.
3. For k from 2 to p, keep adding $X_{(k)}$ to X_S if $\hat{\nu}^2_{X_S,Y}$ does not decrease. Stop otherwise.

6 Real Example

As a real example we used Japanese Vowels data set which is available at UCI Machine Learning Repository (Lichman 2013). Nine male speakers uttered two Japanese vowels /ae/ successively. For each utterance, it was applied $12°$ linear prediction analysis to obtain a discrete-time series with 12 LPC cep-strum coefficients. This means that one utterance by a speaker forms a time series whose length is in the range 7–29 and each point of a time series is of 12 features (12 coefficients). The number of the time series is 640 in total. The samples in this data set are of different lengths. They were extended to the length of the longest sample in the data set (Górecki and Łuczak 2015).

During the smoothing process we used Fourier basis with five components. In the next step we applied the described earlier method of selecting variables (we stopped the procedure if the increase in covariance measure was less than 0.01). In such way we obtained four variables (Fig. 1).

Next, we applied described classifiers to reduced functional data and to full functional data. To estimate the error rate of the classifiers we used tenfold cross-validation method. The results are in Table 1.

We can observe that the error rate increases if we reduce our data set. This behavior is expected. However, the increase seems not too big. Particularly inter-

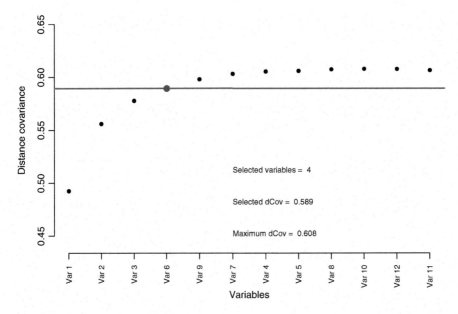

Fig. 1 Variables selection for Japanese Vowels data set

Table 1 Classification accuracy (in %) for Japanese Vowels data set

Classifier	Selected variables (4)	All variables (12)
LDC	93.60	99.37
Logistic regression	91.06	97.97
kNN ($k = 1, \ldots, 8$)	90.94	96.71
Naive Bayes (normal)	90.77	95.50
Naive Bayes (kernel)	90.15	94.34
QDC	89.85	Too small groups

esting is the case of QDC. For this method we do not have enough data to estimate covariance matrices for all groups for full data. When we select only four variables this procedure could be performed. We can also notice that the order of classifiers stays unchanged (the best classifier for full data is LDC, and the same is the best for reduced data).

During the calculations we used R (R Core Team 2017) software and caret (Kuhn 2017), energy (Rizzo and Székely 2016), and fda (Ramsay et al. 2014) packages.

7 Conclusion

The paper introduces variable selection for classification of multivariate functional data. Use of distance covariance as a tool to reduce dimensionality of data set suggests that the technique provides useful results for classification of multivariate functional data. For the analyzed data set only four from twelve variables were included in the final model. We can observe that classification accuracy could drop a little. However, we expect that this drop should be reasonable and in return we could gain a lot of computation time.

In practice, it is important not to depend entirely on variable selection criteria because none of them works well under all conditions. So our approach could be seen as a competitive to another variable selection methods. Additionally, model obtained by the proposed method of variable selection seems comparable with the full model (model without variables reduction). Finally, the researcher needs to evaluate the models using various diagnostic procedures.

References

Ando, T.: Penalized optimal scoring for the classification of multi-dimensional functional data. Stat. Methodol. 6, 565–576 (2009)

Berrendero, J.R., Cuevas, A., Torrecilla, J.L.: Variable selection in functional data classification: a maxima-hunting proposal. Stat. Sin. 26(2), 619–638 (2016)

Delaigle, A., Haal, P.: Methodology and theory for partial least squares applied to functional data. Ann. Stat. 40, 322–352 (2012)

Ferraty, F., Vieu, P.: Curve discrimination. A nonparametric functional approach. Comput. Stat. Data Anal. 44, 161–173 (2003)

Ferraty, F., Vieu, P.: Additive prediction and boosting for functional data. Comput. Stat. Data Anal. 53(4), 1400–1413 (2009)

Górecki, T., Krzyśko, M., Waszak, Ł., Wołyński, W.: Methods of reducing dimension for functional data. Stat. Transition New Series 15, 231–242 (2014)

Górecki, T., Krzyśko, M., Wołyński, W.: Multivariate functional regression analysis with application to classification problems. In: Wilhelm Adalbert, F.X., Kestler Hans, A. (eds.) Analysis of Large and Complex Data, Studies in Classification, Data Analysis, and Knowledge Organization, pp. 173–183. Springer, Berlin (2016)

Górecki, T., Łuczak, M.: Multivariate time series classification with parametric derivative dynamic time warping. Expert Syst. Appl. 42(5), 2305–2312 (2015)

Hastie, T.J., Tibshirani, R.J., Buja, A.: Penalized discriminant analysis. Ann. Stat. 23, 73–102 (1995)

Horváth, L., Kokoszka, P.: Inference for Functional Data with Applications. Springer, New York (2012)

Jacques, J., Preda, C.: Model-based clustering for multivariate functional data. Comput. Stat. Data Anal. 71, 92–106 (2014)

Kong, J., Wang, S., Wahba G.: Using distance covariance for improved variable selection with application to learning genetic risk models. Stat. Med. 34, 1708–1720 (2015)

Kuhn, M., Wing, J., Weston, S., Williams, A., Keefer, C., Engelhardt, A., Cooper, T., Mayer, Z., Kenkel, B.: caret: Classification and Regression Training. R package version 6.0-76 (2017). https://CRAN.R-project.org/package=caret

Lichman, M.: UCI Machine Learning Repository. University of California, School of Information and Computer Science, Irvine (2013). http://archive.ics.uci.edu/ml

R Core Team: R: A language and environment for statistical computing. R Foundation for Statistical Computing, Vienna (2017). https://www.R-project.org/

Ramsay, J.O., Silverman, B.W.: Functional Data Analysis. Springer, New York (2005)

Ramsay, J.O., Wickham, H. Graves, S., Hooker, G.: fda: Functional Data Analysis. R package Version 2.4.4 (2014). https://CRAN.R-project.org/package=fda

Rizzo, M.L., Székely, G.J.: Energy: E-Statistics: Multivariate Inference via the Energy of Data. R Package Version 1.7-0 (2016). https://CRAN.R-project.org/package=energy

Rossi, F., Delannayc, N., Conan-Gueza, B., Verleysenc, M.: Representation of functional data in neural networks. Neurocomputing **64**, 183–210 (2005)

Rossi, F., Villa, N.: Support vector machines for functional data classification. Neural Comput. **69**, 730–742 (2006)

Rossi, N., Wang, X., Ramsay, J.O.: Nonparametric item response function estimates with EM algorithm. J. Educ. Behav. Stat. **27**, 291–317 (2002)

Székely, G.J., Rizzo, M.L., Bakirov, N.K.: Measuring and testing dependence by correlation of distances. Ann. Stat. **35**(6), 2769–2794 (2007)

Székely, G.J., Rizzo, M.L.: Brownian distance covariance. Ann. Stat. **3**(4), 1236–1265 (2009)

Székely, G.J., Rizzo, M.L.: On the uniqueness of distance covariance. Stat. Probab. Lett. **82**(12), 2278–2282 (2012)

Székely, G.J., Rizzo, M.L.: The distance correlation t-test of independence in high dimension. J. Multivar. Anal. **117**, 193–213 (2013)

Initial Value Selection for the Alternating Least Squares Algorithm

Masahiro Kuroda, Yuichi Mori, and Masaya Iizuka

Abstract The alternating least squares (ALS) algorithm is a popular computational algorithm for obtaining least squares solutions minimizing the loss functions in nonlinear multivariate analysis with optimal scaling (NMVA). The ALS algorithm is a simple computational procedure and has a stable convergence property, while the algorithm only guarantees local convergence. In order to avoid finding a local minimum of a loss function, the most commonly used method is to start the ALS algorithm with various random initial values. Such random initialization ALS algorithm tries to find the least squares solution that globally minimizes the loss function. However, the drawback of the random initialization ALS algorithm with multiple runs is to take a huge number of iterations and long computation time. For these problems, we consider initial value selection for selecting an initial value leading to a global minimum of the loss function. The proposed procedure enables efficiently selecting an initial value of the ALS algorithm. Furthermore, we can increase the computation speed when applying the vector ε acceleration for the ALS algorithm to the initial value selection procedure and the least squares estimation in NMVA.

1 Introduction

The alternating least squares (ALS) algorithm is a popular computational algorithm for obtaining least squares solutions minimizing the loss functions in nonlinear multivariate analysis with optimal scaling (NMVA). The ALS algorithm is a simple computational procedure and has a stable convergence property, while the algorithm

M. Kuroda (✉) · Y. Mori
Okayama University of Science, Okayama, Japan
e-mail: kuroda@mgt.ous.ac.jp; mori@mgt.ous.ac.jp

M. Iizuka
Okayama University, Okayama, Japan
e-mail: iizuka@okayama-u.ac.jp

© Springer Nature Singapore Pte Ltd. 2020
T. Imaizumi et al. (eds.), *Advanced Studies in Classification and Data Science*,
Studies in Classification, Data Analysis, and Knowledge Organization,
https://doi.org/10.1007/978-981-15-3311-2_18

only guarantees local convergence. In order to avoid finding a local minimum of a loss function, the most commonly used method is to start the ALS algorithm with various random initial values. Such random initialization ALS algorithm tries to find the least squares solution that globally minimizes the loss function. However, the drawback of the random initialization ALS algorithm with multiple runs is to take a huge number of iterations and long computation time.

For these problems, we consider initial value selection for selecting an initial value leading to a global minimum of the loss function. The initial value selection procedure proposed in this paper is based on the emEM algorithm of Biernacki et al. (2003) that is the EM algorithm including an initial value selection step for Gaussian mixture models. The proposed procedure enables efficiently selecting an initial value of the ALS algorithm. Furthermore, we can increase the computation speed when applying the vector ε acceleration for the ALS algorithm of Kuroda et al. (2011) to the initial value selection procedure and the least squares estimation in NMVA.

The paper is organized as follows. Section 2 gives the ALS algorithm for NMVA. Section 3 describes the ALS algorithm for principal component analysis of categorical data [nonlinear PCA (NPCA)] and proposes the initial value selection procedure of the ALS algorithm. Section 4 introduces the vector ε acceleration for the ALS algorithm of Kuroda et al. (2011) and describes the initial value selection procedure using this acceleration algorithm. Section 5 presents numerical experiments to evaluate the performance of the proposed initial value selection procedures. Section 6 gives our discussion.

2 The ALS Algorithm for NMVA

Let $\mathbf{X} = [\mathbf{x}_1, \ldots, \mathbf{x}_p]$ be a data matrix of n objects by p categorical variables measured with nominal and ordinal scales. Let $\mathbf{x}_j$ of $\mathbf{X}$ be a qualitative vector with K_j categories. To quantify $\mathbf{x}_j$, the vector is coded by using an $n \times K_j$ indicator matrix

$$\mathbf{G}_j = [\mathbf{g}_{j1}, \ldots, \mathbf{g}_{jK_j}] = \begin{bmatrix} g_{j11} & \cdots & g_{j1K_j} \\ \vdots & \ddots & \vdots \\ g_{jn1} & \cdots & g_{jnK_j} \end{bmatrix},$$

where

$$g_{jik} = \begin{cases} 1 & \text{if object } i \text{ belongs to category } k, \\ 0 & \text{if object } i \text{ belongs to some other category } k' (\neq k). \end{cases}$$

In NMVA, $\mathbf{X}$ is quantified using an optimal scaling technique that optimally assigns numerical values to qualitative scales (Young 1981). Then the optimal scaling finds $K_j \times 1$ category quantifications $\mathbf{q}_j$ under the restrictions imposed by the measurement level of variable j and transforms $\mathbf{x}_j$ into an optimally scaled vector $\mathbf{x}_j^* = \mathbf{G}_j \mathbf{q}_j$. There are different ways for quantifying observed data of nominal and ordinal variables:

- For nominal scale data, the quantification is unrestricted. Objects i and $h (\neq i)$ in the same category for variable j obtain the same quantification. Thus, if $x_{ji} = x_{jh}$, then $x_{ji}^* = x_{jh}^*$.
- For ordinal scale data, the quantification is restricted to the order of categories. If observed data x_{ji} and x_{jh} for objects i and h in variable j have order $x_{ji} > x_{jh}$, then quantified data x_{ji}^* and x_{jh}^* have order $x_{ji}^* \geq x_{jh}^*$.

Let $\mathbf{Q} = \{\mathbf{q}_1, \ldots, \mathbf{q}_p\}$ be a set of category quantifications. We denote optimally scaled data

$$\mathbf{X}(\mathbf{Q}) = [\mathbf{G}_1 \mathbf{q}_1, \ldots, \mathbf{G}_p \mathbf{q}_p] = [\mathbf{x}_1^*, \ldots, \mathbf{x}_p^*]$$

as the function of $\mathbf{Q}$. The problem of least squares fitting a model $\mathbf{M}(\boldsymbol{\theta}) = [\mathbf{m}_1, \ldots, \mathbf{m}_p]$ with a set of parameters $\boldsymbol{\theta}$ to $\mathbf{X}(\mathbf{Q})$ can be solved in minimizing the loss function

$$\sigma(\mathbf{X}(\mathbf{Q}), \mathbf{M}(\boldsymbol{\theta})) = \|\mathbf{X}(\mathbf{Q}) - \mathbf{M}(\boldsymbol{\theta})\|_2^2 \tag{1}$$

over $\mathbf{Q}$ and $\boldsymbol{\theta}$, and then we can find the solutions $\hat{\mathbf{Q}}$ and $\hat{\boldsymbol{\theta}}$. Because it is not possible to obtain simultaneously them in the closed-form solutions for this minimization problem, we do not utilize a non-iterative procedure such as the least squares method. The alternating least squares (ALS) algorithm is a possible computational algorithm for the simultaneous estimation of $\mathbf{Q}$ and $\boldsymbol{\theta}$. The algorithm updates each of $\mathbf{Q}$ and $\boldsymbol{\theta}$ in turn, keeping the other fixed.

Let $\mathbf{Q}^{(t)}$ and $\boldsymbol{\theta}^{(t)}$ be the t-th estimates of $\mathbf{Q}$ and $\boldsymbol{\theta}$. Given $\mathbf{X}(\mathbf{Q}^{(0)})$, the ALS algorithm iterates alternatively the following two steps:

Parameter estimation step: Compute $\boldsymbol{\theta}^{(t+1)}$ giving the minimum value of the loss function (1) for fixed $\mathbf{X}(\mathbf{Q}^{(t)})$:

$$\boldsymbol{\theta}^{(t+1)} = \arg\min_{\boldsymbol{\theta}} \sigma(\mathbf{X}(\mathbf{Q}^{(t)}), \mathbf{M}(\boldsymbol{\theta})).$$

Category quantification step: Obtain the least squares estimate $\mathbf{Q}^{(t+1)}$ in minimizing the loss function (1) for fixed $\boldsymbol{\theta}^{(t+1)}$:

$$\mathbf{Q}^{(t+1)} = \arg\min_{\mathbf{Q}} \sigma(\mathbf{X}(\mathbf{Q}), \mathbf{M}(\boldsymbol{\theta}^{(t+1)}))$$

and update $\mathbf{X}(\mathbf{Q}^{(t+1)}) = [\mathbf{G}_1\mathbf{q}_1^{(t+1)}, \ldots, \mathbf{G}_p\mathbf{q}_p^{(t+1)}]$. Check the convergence using

$$\sigma(\mathbf{X}(\mathbf{Q}^{(t)}), \mathbf{M}(\theta^{(t)})) - \sigma(\mathbf{X}(\mathbf{Q}^{(t+1)}), \mathbf{M}(\theta^{(t+1)})) < \delta, \tag{2}$$

where δ is a desired accuracy.

De Leeuw et al. (1976) shows that the ALS algorithm can reduce the value of the function (1) at each iteration and has the monotonic convergence property. If the function (1) is bounded, the function will be locally minimized over the entire set of $\mathbf{Q}$ and θ (Krijnen 2006).

3 Initial Value Selection of the ALS Algorithm for NPCA

3.1 The ALS Algorithm for NPCA

In principal component analysis, the set of parameters is $\theta = \{\mathbf{Z}, \mathbf{A}\}$ and the model is given by

$$\mathbf{M}(\theta) = \mathbf{Z}\mathbf{A}^\top, \tag{3}$$

where $\mathbf{Z} = [\mathbf{Z}_1 \cdots \mathbf{Z}_r]$ is an $n \times r$ matrix of n component scores on r components and $\mathbf{A} = [\mathbf{A}_1 \cdots \mathbf{A}_r]$ is a $p \times r$ matrix of p loadings on the r components. Let $\mathbf{X}$ be an $n \times p$ matrix of categorical data and $\mathbf{X}(\mathbf{Q})$ be the optimally scaled data of $\mathbf{X}$.

For NPCA, the loss function is given by

$$\sigma(\mathbf{X}(\mathbf{Q}), \mathbf{M}(\theta)) = \|\mathbf{X}(\mathbf{Q}) - \mathbf{Z}\mathbf{A}^\top\|_2^2. \tag{4}$$

Then the ALS algorithm finds the least squares solutions $\hat{\mathbf{Q}}$ and $\hat{\theta} = \{\hat{\mathbf{Z}}, \hat{\mathbf{A}}\}$ minimizing the loss function (4). PRINCIPALS of Young et al. (1978) and PRINCALS of Gifi (1990) are the ALS algorithms for solving the least squares minimization problem of the loss function (4). We use PRINCIPALS as the ALS algorithm for NPCA.

Under restrictions

$$\mathbf{X}(\mathbf{Q})^\top \mathbf{1}_n = \mathbf{0}_p \quad \text{and} \quad \text{diag}\left[\frac{\mathbf{X}(\mathbf{Q})^\top\mathbf{X}(\mathbf{Q})}{n}\right] = \mathbf{I}_p, \tag{5}$$

the algorithm alternates between the updates of $\mathbf{Q}$ and θ. In the initialization, the observed categorical data $\mathbf{X}$ may be used as $\mathbf{X}(\mathbf{Q}^{(0)})$. For nominal variables, we initialize $\mathbf{Q}^{(0)}$ with random numbers and determine $\mathbf{X}(\mathbf{Q}^{(0)})$ after standardizing each column of $X(Q^{(0)})$ under the restriction (5). The ALS algorithm iterates the following two steps:

Parameter estimation step: Obtain $A^{(t+1)}$ by solving the eigen-decomposition of $X(Q^{(t)})^\top X(Q^{(t)})/n$ or the singular value decomposition of $X(Q^{(t)})$. Compute $Z^{(t+1)} = X(Q^{(t)})A^{(t+1)}$ and update $M(\theta^{(t+1)}) = Z^{(t+1)}A^{(t+1)\top}$.

Category quantification step: Obtain $X(Q^{(t+1)})$ by separately estimating x_j^* for each variable j. Compute $q_j^{(t+1)}$ for nominal variables as

$$q_j^{(t+1)} = \left(G_j^\top G_j\right)^{-1} G_j^\top m_j^{(t+1)}.$$

Re-compute $q_j^{(t+1)}$ for ordinal variables using the monotone regression (Kruskal 1964). Update $X(Q^{(t+1)}) = [G_1 q_1^{(t+1)}, \ldots, G_p q_p^{(t+1)}]$ under the restriction (5). Check the convergence using the criterion (2).

3.2   Initial Value Selection Procedure of the ALS Algorithm

A numerical experiment illustrates the local convergence of the ALS algorithm for NPCA for $r = 10$ principal components. X is a random data matrix of $n = 50$ objects by $p = 25$ nominal variables with 10 levels each. The random initialization ALS algorithm starting with 100 random initial values of Q runs to find $\hat{Q}$ and $\hat{\theta} = \{\hat{Z}, \hat{A}\}$ and then computes

$$\sigma(X(\hat{Q}), M(\hat{\theta})) = \|X(\hat{Q}) - \hat{Z}\hat{A}^\top\|_2^2.$$

The histogram of Fig. 1 is drawn from 100 values of $\sigma(X(\hat{Q}), M(\hat{\theta}))$. We can see from the histogram that $\sigma(X(Q), M(\theta))$ has several local minimum values and its minimization is deeply related to initial values. In order to obtain $\hat{Q}$ and $\hat{\theta}$ that globally minimize $\sigma(X(Q), M(\theta))$, we commonly perform the random initialization ALS algorithm, such as this illustration. However, this random initialization ALS algorithm takes the large number of iterations and long computation time.

We consider an initial value selection procedure that reduces them and finds initial values of the ALS algorithm leading to the global minimum of $\sigma(X(Q), M(\theta))$. We propose an initial value selection procedure of the ALS algorithm based on the emEM algorithm of Biernacki et al. (2003). The emEM algorithm includes an initialization step that selects an initial value of the EM algorithm for getting the global maximum of the likelihood in Gaussian mixture models.

Our proposed initial value selection procedure has three steps: Random starting, Short running, and Selection. We describe the procedure.

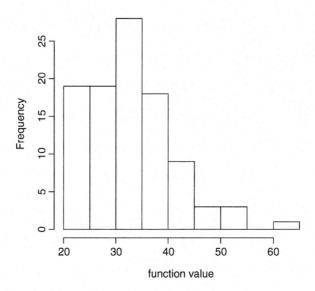

Fig. 1 Histogram of 100 values of $\sigma(\mathbf{X}(\hat{\mathbf{Q}}), \mathbf{M}(\hat{\boldsymbol{\theta}}))$ obtained from the random initialization ALS algorithm

Iterate the Random starting and Short running steps for $h = 1, \ldots, H$:

Random starting step: Generate a random initial value $\mathbf{Q}^{(0,h)}$.

Short running step: Find $\mathbf{Q}^{(t+1,h)}$ and $\boldsymbol{\theta}^{(t+1,h)}$ using the ALS algorithm. Stop ALS iterations starting from $\mathbf{X}(\mathbf{Q}^{(0,h)})$ at the $(t + 1)$-iteration using

$$\text{Condition 0:}\quad \frac{\sigma(\mathbf{X}(\mathbf{Q}^{(t+1,h)}), \mathbf{M}(\boldsymbol{\theta}^{(t+1,h)})) - \sigma(\mathbf{X}(\mathbf{Q}^{(t,h)}), \mathbf{M}(\boldsymbol{\theta}^{(t,h)}))}{\sigma(\mathbf{X}(\mathbf{Q}^{(t+1,h)}), \mathbf{M}(\boldsymbol{\theta}^{(t+1,h)})) - \sigma(\mathbf{X}(\mathbf{Q}^{(0,h)}), \mathbf{M}(\boldsymbol{\theta}^{(0,h)}))} < \delta_{ini}$$

$$\text{or}\quad t > T_{max}, \tag{6}$$

where T_{max} is the prefixed maximum number of iterations.

Obtain the h-th candidate initial value $\mathbf{Q}^{(t_h,h)}$ and compute $\sigma(\mathbf{X}(\mathbf{Q}^{(t_h,h)}), \mathbf{M}(\boldsymbol{\theta}^{(t_h,h)}))$, where t_h is the final number of iterations.

Selection step: From H candidate initial values $\{\mathbf{Q}^{(t_h,h)}\}_{h=1,\ldots,H}$, select the initial value $\mathbf{Q}^{(0)}$ such that

$$\mathbf{Q}^{(0)} = \arg\min\{\sigma(\mathbf{X}(\mathbf{Q}^{(t_h,h)}), \mathbf{M}(\boldsymbol{\theta}^{(t_h,h)}))\}_{h=1,\ldots,H}.$$

The initial value selection procedure does not wait for convergence and stops the ALS iterations using a lax stop condition (6), and thus we expect to reduce both the total number of iterations and total computation time of the ALS algorithm in the Short running step. Given $\mathbf{Q}^{(0)}$ selected by the initial value selection procedure, the

ALS algorithm iterates until convergence and finds $\hat{\mathbf{Q}}$ and $\hat{\theta}$ that globally minimize $\sigma(\mathbf{X}(\mathbf{Q}), \mathbf{M}(\theta))$.

Next, we consider to improve the computation speed of the initial value selection procedure by speeding up the ALS computation in the Short running step.

4 Acceleration of the ALS Algorithm in the Initial Value Selection Procedure

Kuroda et al. (2011) provided the vector ε acceleration for the ALS algorithm that can accelerate the convergence of the ALS algorithm for NPCA and demonstrated that its speed of convergence is significantly faster than that of the ordinary ALS algorithm.

We briefly introduce the vector ε ($v\varepsilon$) algorithm of Wynn (1962) for accelerating the convergence of the ALS algorithm. The $v\varepsilon$ algorithm is utilized to accelerate the convergence of a linear convergent vector sequence. It is well known that the algorithm is very effective when the vector sequence converges slowly. Let $\{\mathbf{Y}^{(t)}\}_{t \geq 0}$ be a linear convergent vector sequence generated by an iterative computational procedure, and let $\{\dot{\mathbf{Y}}^{(t)}\}_{t \geq 0}$ be a faster convergent sequence of $\{\mathbf{Y}^{(t)}\}_{t \geq 0}$ obtained from the $v\varepsilon$ algorithm. We denote $\Delta \mathbf{Y}^{(t)} = \mathbf{Y}^{(t+1)} - \mathbf{Y}^{(t)}$ and define $[\mathbf{Y}]^{-1} = \mathbf{Y}/\|\mathbf{Y}\|_2^2$. Then, the $v\varepsilon$ algorithm enables producing $\{\dot{\mathbf{Y}}^{(t)}\}_{t \geq 0}$ using

$$\dot{\mathbf{Y}}^{(t-1)} = \mathbf{Y}^{(t)} + \left[\left[\Delta \mathbf{Y}^{(t)} \right]^{-1} - \left[\Delta \mathbf{Y}^{(t-1)} \right]^{-1} \right]^{-1}. \tag{7}$$

When $\{\mathbf{Y}^{(t)}\}_{t \geq 0}$ converges to a limit point $\hat{\mathbf{Y}}$ of $\{\mathbf{Y}^{(t)}\}_{t \geq 0}$, the $v\varepsilon$ algorithm generates $\{\dot{\mathbf{Y}}^{(t)}\}_{t \geq 0}$ that converges to $\hat{\mathbf{Y}}$ faster than $\{\mathbf{Y}^{(t)}\}_{t \geq 0}$.

Assume that $\{\mathbf{Q}^{(t)}\}_{t \geq 0}$ generated by the ALS algorithm converges to a limit point $\hat{\mathbf{Q}}$. Then, in order to accelerate the convergence of the ALS algorithm, we apply the $v\varepsilon$ acceleration (7) to $\{\mathbf{Q}^{(t)}\}_{t \geq 0}$ and obtain the faster convergent sequence $\{\dot{\mathbf{Q}}^{(t)}\}_{t \geq 0}$ of $\{\mathbf{Q}^{(t)}\}_{t \geq 0}$. We refer to the $v\varepsilon$ acceleration for the ALS algorithm as the $v\varepsilon$-ALS algorithm. The $v\varepsilon$-ALS algorithm iterates the following two steps:

ALS step: Find $\mathbf{Q}^{(t+1)}$ and $\theta^{(t+1)}$ using the ALS algorithm:

$$\theta^{(t+1)} = \arg\min_{\theta} \sigma(\mathbf{X}(\mathbf{Q}^{(t)}), \mathbf{M}(\theta)),$$

$$\mathbf{Q}^{(t+1)} = \arg\min_{\mathbf{Q}} \sigma(\mathbf{X}(\mathbf{Q}), \mathbf{M}(\theta^{(t+1)})).$$

vε acceleration step: Compute $\dot{\mathbf{Q}}^{(t-1)}$ using $\{\mathbf{Q}^{(t-1)}, \mathbf{Q}^{(t)}, \mathbf{Q}^{(t+1)}\}$ from the vε algorithm:

$$\mathrm{vec}\dot{\mathbf{Q}}^{(t-1)} = \mathrm{vec}\mathbf{Q}^{(t)} + \left[\left[\varDelta\mathrm{vec}\mathbf{Q}^{(t)}\right]^{-1} - \left[\varDelta\mathrm{vec}\mathbf{Q}^{(t-1)}\right]^{-1}\right]^{-1},$$

where $\mathrm{vec}\dot{\mathbf{Q}} = [\dot{\mathbf{q}}_1^\top, \cdots, \dot{\mathbf{q}}_p^\top]^\top$. Check the convergence by

$$\|\mathrm{vec}\dot{\mathbf{Q}}^{(t-1)} - \mathrm{vec}\dot{\mathbf{Q}}^{(t-2)}\|^2 < \delta,$$

where δ is a desired accuracy.

Before starting the iteration of the vε acceleration step, the ALS step runs twice to generate $\{\mathbf{Q}^{(0)}, \mathbf{Q}^{(1)}, \mathbf{Q}^{(2)}\}$. The vε-ALS algorithm finds $\hat{\mathbf{Q}}$ that is the final value of $\{\dot{\mathbf{Q}}^{(t)}\}_{t\geq 0}$ in the vε acceleration step and then computes $\mathbf{X}(\hat{\mathbf{Q}})$. Given $\mathbf{X}(\hat{\mathbf{Q}})$, we can immediately obtain $\hat{\boldsymbol{\theta}}$ from

$$\hat{\boldsymbol{\theta}} = \arg\min_{\boldsymbol{\theta}} \sigma\left(\mathbf{X}(\hat{\mathbf{Q}}), \mathbf{M}(\boldsymbol{\theta})\right)$$

in the ALS step. It is most reasonable and efficient to generate $\{\dot{\mathbf{Q}}^{(t)}\}_{t\geq 0}$ in the vε acceleration step.

Note that $\dot{\mathbf{Q}}^{(t-1)}$ obtained at the t-th iteration of the vε acceleration step is not used as the estimate $\mathbf{Q}^{(t+1)}$ at the $(t+1)$-th iteration of the ALS step. Therefore, the vε-ALS algorithm guarantees the monotonic convergence property of the ALS algorithm and moreover enables the acceleration of convergence of $\{\mathbf{Q}^{(t)}\}_{t\geq 0}$.

In order to speed up the computation of the initial value selection procedure, we use the vε-ALS algorithm in the Short running step.

Iterate the Random starting and Short running steps for $h = 1, \ldots, H$:

Random starting step: Generate a random initial value $\mathbf{Q}^{(0,h)}$.
Short running step: Compute $\mathbf{Q}^{(t+1,h)}$ and $\boldsymbol{\theta}^{(t+1,h)}$ in the ALS step. Obtain $\dot{\mathbf{Q}}^{(t-1,h)}$ in the vε acceleration step. Stop the iterations at the $(t+1)$-th iteration using either

Condition 1: $\dfrac{\|\mathrm{vec}\dot{\mathbf{Q}}^{(t-1,h)} - \mathrm{vec}\dot{\mathbf{Q}}^{(t-2,h)}\|_2^2}{\|\mathrm{vec}\dot{\mathbf{Q}}^{(t-1,h)} - \mathrm{vec}\dot{\mathbf{Q}}^{(0)}\|_2^2} < \delta_{ini}$ or $t > T_{max}$, or

Condition 2: $\|\mathrm{vec}\dot{\mathbf{Q}}^{(t-1,h)} - \mathrm{vec}\dot{\mathbf{Q}}^{(t-2,h)}\|_2^2 < \delta_{ini}$ or $t > T_{max}$,

Obtain the h-th candidate initial value $\dot{\mathbf{Q}}^{(t_h,h)}$ and compute $\sigma(\mathbf{X}(\dot{\mathbf{Q}}^{(t_h,h)})$, $\mathbf{M}(\dot{\boldsymbol{\theta}}^{(t_h,h)}))$, where t_h is the final number of iterations.

Selection step: From H candidate initial values $\{\dot{\mathbf{Q}}^{(t_h,h)}\}_{h=1,\dots,H}$, select an initial value such that

$$\mathbf{Q}^{(0)} = \arg\min\{\sigma(\mathbf{X}(\dot{\mathbf{Q}}^{(t_h,h)}), \mathbf{M}(\dot{\boldsymbol{\theta}}^{(t_h,h)}))\}_{h=1,\dots,H}.$$

The Short running step using the $v\varepsilon$-ALS algorithm obtains candidate initial values $\{\dot{\mathbf{Q}}^{(t_h,h)}\}_{h=1,\dots,H}$ after the smaller total number of iterations. By using $\mathbf{Q}^{(0)}$, the $v\varepsilon$-ALS algorithm estimates $\hat{\mathbf{Q}}$ and $\hat{\boldsymbol{\theta}}$ minimizing $\sigma(\mathbf{X}(\mathbf{Q}), \mathbf{M}(\boldsymbol{\theta}))$. Then, in both the initial value selection procedure and least squares estimation of $\mathbf{Q}$ and $\boldsymbol{\theta}$ for NPCA, the $v\varepsilon$-ALS algorithm reduces their number of iterations and computation time.

5 Numerical Experiments

Numerical experiments examine the performance of the initial value selection procedures using the ALS and $v\varepsilon$-ALS algorithms for NPCA. The first purpose of the experiments is to investigate whether these procedures can find an initial value leading to the global minimum of $\sigma(\mathbf{X}(\mathbf{Q}), \mathbf{M}(\boldsymbol{\theta}))$. The second is to evaluate how much faster the $v\varepsilon$-ALS algorithm computes than the ALS algorithm used in the initial value selection procedures. All computations are performed with the statistical package R (R Development Core Team 2015) executing on Intel Core i5 3.20 GHz with 4 GB of memory. CPU times (in seconds) are measured by the function `proc.time`.

We describe the setup of the initial value selection procedures. The Random initialization step generates 100 random initial values $\{\mathbf{Q}^{(0,h)}\}_{h=1,\dots,100}$. The ALS and $v\varepsilon$-ALS algorithms are used in the Short running step. We set $\delta_{ini} = 10^{-3}, 10^{-4}, 10^{-5}$ for Conditions 0 and 1 and $\delta_{ini} = 10^{-2}, 10^{-4}$ for Condition 2, and $T_{max} = 10^4$. Then, we utilize the C0.ALS, C1.$v\varepsilon$, and C2.$v\varepsilon$ procedures as the initial value selection procedure shown in Table 1. These procedures starting with $\{\mathbf{Q}^{(0,h)}\}_{h=1,\dots,100}$ obtain 100 candidate initial values of $\mathbf{Q}$ and select $\mathbf{Q}^{(0)}$ from them.

In the experiments, we consider a random data matrix $\mathbf{X}$ of 100 objects by 50 nominal variables with 5 levels each and NPCA for 10 principal components. $\mathbf{X}$ is generated 100 times, $\{\mathbf{X}^{(l)}\}_{l=1,\dots,100}$. For each $\mathbf{X}^{(l)}$, the random initialization ALS algorithm starting with $\{\mathbf{Q}^{(0,h)}\}_{h=1,\dots,100}$ finds the minimum value of $\sigma(\mathbf{X}(\mathbf{Q}), \mathbf{M}(\boldsymbol{\theta}))$ under the convergence criterion (2) of $\delta = 10^{-12}$. Then we regard

Table 1 The initial value selection procedures: the C0.ALS, C1.$v\varepsilon$, and C2.$v\varepsilon$ procedures

	Algorithm	Stop condition	δ_{ini}
C0.ALS	ALS	Condition 0	$10^{-3}, 10^{-4}, 10^{-5}$
C1.$v\varepsilon$	$v\varepsilon$-ALS	Condition 1	$10^{-3}, 10^{-4}, 10^{-5}$
C2.$v\varepsilon$	$v\varepsilon$-ALS	Condition 2	$10^{-2}, 10^{-4}$

Table 2 The number of findings of an initial value $\mathbf{Q}^{(0)}$ getting the global minimum of $\sigma(\mathbf{X}(\mathbf{Q}), \mathbf{M}(\boldsymbol{\theta}))$ for the C0.ALS, C1.vε, and C2.vε procedures

	$\delta_{ini} = 10^{-2}$	$\delta_{ini} = 10^{-3}$	$\delta_{ini} = 10^{-4}$	$\delta_{ini} = 10^{-5}$
C0.ALS	–	97	99	100
C1.vε	–	96	97	100
C2.vε	97	–	100	–

Table 3 Summary statistics of the total numbers of iterations and total CPU times of the random initialization ALS algorithm and the C0.ALS, C1.vε, and C2.vε procedures

	The total number of iterations				Total CPU time			
	ALS	C0.ALS	C1.vε	C2.vε	ALS	C0.ALS	C1.vε	C2.vε
Min.	18,160	5228	3929	4958	245.7	71.69	34.06	43.01
1st Qu.	28,010	7738	5616	7813	377.1	105.20	48.90	67.90
Median	38,820	9100	6488	9560	521.8	124.10	56.70	84.05
Mean	43,050	9208	6510	10,020	580.7	125.30	56.78	87.31
3rd Qu.	51,510	10,520	7314	11,670	691.4	142.00	63.56	101.90
Max.	214,200	14,350	10,060	18,300	2860.0	194.80	87.65	161.90

the minimum value as the global minimum of $\sigma(\mathbf{X}(\mathbf{Q}), \mathbf{M}(\boldsymbol{\theta}))$. Under the above setup, we apply the C0.ALS, C1.vε, and C2.vε procedures to each $\mathbf{X}^{(l)}$.

Table 2 is the number of findings of an initial value $\mathbf{Q}^{(0)}$ getting the global minimum of $\sigma(\mathbf{X}(\mathbf{Q}), \mathbf{M}(\boldsymbol{\theta}))$ for the C0.ALS, C1.vε, and C2.vε procedures. We can see from the table that the C0.ALS and C1.vε procedures require at least $\delta_{ini} = 10^{-5}$ and the C2.vε procedure does at least $\delta_{ini} = 10^{-4}$ for these experiments. In the below experiments for evaluating the computation speed of these procedures, we give the results of $\delta_{ini} = 10^{-5}$ for the C0.ALS and C1.vε procedures and $\delta_{ini} = 10^{-4}$ for the C2.vε procedure.

Table 3 presents the summary statistics of the total numbers of iterations and total CPU times of the random initialization ALS algorithm and the C0.ALS, C1.vε, and C2.vε procedures. We see the mean values of the total numbers of iterations and total CPU times of these algorithms. The random initialization ALS algorithm requires 43050 iterations, while the total number iterations of the C0.ALS and C2.vε procedures are about 10000, and that of the C1.vε procedure is 6510. The table also indicates that the CPU times of these procedures are much shorter than that of the random initialization ALS algorithm. Although the CPU time of the random initialization ALS algorithm is 580.7 s, the C1.vε procedure only takes 56.78 s. Figure 2 is the scatterplots of the C0.ALS, C1.vε, and C2.vε procedures by the random initialization ALS algorithm for the total numbers of iterations and total CPU times. The figure indicates that all the procedures enable efficiently reducing the total number of iterations and total CPU time.

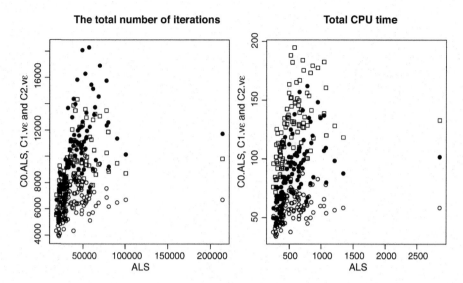

Fig. 2 Scatterplots of the C0.ALS (open square), C1.vε (open circle), and C2.vε (filled circle) procedures by the random initialization ALS algorithms for the total number of iterations and total CPU time

In order to compare the computation speeds of the C0.ALS, C1.vε, and C2.vε procedures, we calculate iteration and CPU time speed-ups. The iteration speed-up of the C0.ALS procedure is defined as

$$\frac{\text{The total number of iterations of the random initialization ALS algorithm}}{\text{The total number of iterations of the C0.ALS procedure}}.$$

The iteration speed-ups of the other procedures and the CPU time speed-ups of all procedures are calculated similarly to the iteration speed-up. Table 4 is the summary statistics of the iteration and CPU time speed-ups of the initial value selection procedures. The table indicates that the C1.vε procedure provides the best performance among them in both the total number of iterations and total CPU time. In particular, the CPU time of the C1.vε procedure is 10 times shorter than that of the random initialization ALS algorithm and also is about 2.2 and 1.5 times shorter than those of the C0.ALS and C2.vε procedures in these mean values. The boxplots of Fig. 3a also show that the C1.vε procedure clearly outperforms the other two procedures. Figure 3b is the scatterplot of the C0.ALS, C1.vε, and C2.vε procedures by the random initialization ALS algorithm for the CPU time speed-ups. The figure indicates that the computation speed of the C1.vε procedure is much faster than those of the other two procedures when the initialization ALS algorithm takes much computation time.

Table 4 Summary statistics of the iteration and CPU time speed-ups of the C0.ALS, C1.vε, and C2.vε procedures

	The total number of iterations			Total CPU time		
	C0.ALS	C1.vε	C2.vε	C0.ALS	C1.vε	C2.vε
Min.	2.39	3.15	2.34	2.39	4.90	3.64
1st Qu.	3.43	4.73	3.28	3.41	7.33	5.08
Median	4.00	5.56	3.87	3.97	8.61	6.01
Mean	4.66	6.58	4.29	4.62	10.18	6.64
3rd Qu.	5.15	7.49	4.69	5.13	11.57	7.25
Max.	21.79	31.86	18.24	21.53	49.06	28.16

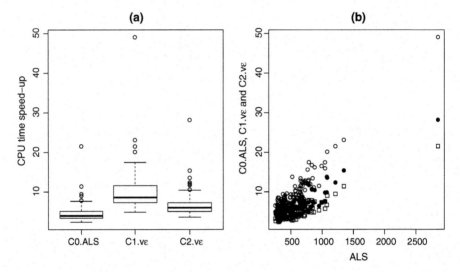

Fig. 3 (**a**) Boxplots of the CPU time speed-ups of the C0.ALS, C1.vε, and C2.vε procedures. (**b**) Scatterplots of the C0.ALS (open square), C1.vε (open circle), and C2.vε (filled circle) procedures by the random initialization ALS algorithms for the CPU time speed-ups

6 Discussion

In this paper, we proposed simple and efficient initial selection procedures of the ALS algorithm for NPCA. The procedures consist of the Random initialization, Short running, and Selection steps. The Random initialization and Short running steps iterate a prespecified number of times to obtain candidate initial values, and the Selection step selects an initial value from these values. Then we utilize the vε-ALS algorithm for speeding up the computation of the initial value selection procedure. Numerical experiments demonstrate that all the initial selection procedures greatly shorten the total number of iterations and total computation time. Especially, the C1.vε procedure using the vε-ALS algorithm improves the computational efficiency, and then its total CPU time is much shorter than that of the random initialization ALS algorithm. Once $\mathbf{Q}^{(0)}$ is selected by the initial value selection procedures, the $v\varepsilon$-ALS algorithm runs to find $\hat{\mathbf{Q}}$ and $\hat{\theta}$ that globally minimize $\sigma(\mathbf{X}(\mathbf{Q}), \mathbf{M}(\theta))$.

In the use of the initial value selection procedure, we require to determine the H number of initial values and the stop condition δ_{ini}. It seems that they depend on NMVA and the size of the data matrix. Then we may set large H and small δ_{ini} for safety, although it takes long computation time. The advantage of the $v\varepsilon$-ALS algorithm is that its computation speed is fast, and this property is very useful and important in real data analysis.

We discuss the initial value selection of the ALS algorithm for NPCA. The proposed procedures are also available to ALS algorithms for various NMVA given in the ALSOS algorithms of Young (1981).

Acknowledgments The authors would like to thank the editor and referees for their valuable comments and helpful suggestions. This work was supported by JSPS KAKENHI Grant Number JP16K00061.

References

Biernacki, C., Celeux, G., Govaert, G.: Choosing starting values for the EM algorithm for getting the highest likelihood in multivariate Gaussian mixture models. Comput. Stat. Data Anal. **41**, 561–575 (2003)

De Leeuw, J., Young, F.W., Takane, Y.: Additive structure in qualitative data: an alternating least squares method with optimal scaling features. Psychometrika **41**, 471–504 (1976)

Gifi, A.: Nonlinear Multivariate Analysis. Wiley, Chichester (1990)

Krijnen, W.P.: Convergence of the sequence of parameters generated by alternating least squares algorithms. Comput. Stat. Data Anal. **51**, 481–489 (2006)

Kruskal, J.B.: Nonmetric multidimensional scaling: a numerical method. Psychometrika **29**, 115–129 (1964)

Kuroda, M., Mori, Y., Iizuka, M., Sakakihara, M.: Acceleration of the alternating least squares algorithm for principal components analysis. Comput. Stat. Data Anal. **55**, 143–153 (2011)

R Core Team: R: A Language and Environment for Statistical Computing. R Foundation for Statistical Computing, Vienna (2015). http://www.R-project.org/

Young, F.W.: Quantitative analysis of qualitative data. Psychometrika **46**, 357–388 (1981)

Young, F.W., Takane, Y., de Leeuw, J.: Principal components of mixed measurement level multivariate data: an alternating least squares method with optimal scaling features. Psychometrika **43**, 279–281 (1978)

Wynn, P.: Acceleration techniques for iterated vector and matrix problems. Math. Comput. **16**, 301–322 (1962)

Inference for General MANOVA Based on ANOVA-Type Statistic

Łukasz Smaga

Abstract Inference methods for general multivariate analysis of variance (MANOVA) are studied. Homoscedasticity and any particular distribution are not assumed in general factorial designs under consideration. In that general framework, the existing methods based on the Wald-type statistic may behave poorly under finite samples, e.g., they are often too liberal under unbalanced designs and skewed distributions. In this paper, the testing procedures and confidence regions based on the ANOVA-type statistic and its standardized version are proposed, which usually perform very satisfactorily in cases where the known tests fail. Different approaches to approximate the null distribution of test statistics are developed. They are based on asymptotic distribution and bootstrap and permutation methods. The consistency of the asymptotic tests under fixed alternatives is proved. In simulation studies, it is shown that some of the new procedures possess good size and power characteristics, and they are competitive to existing procedures.

1 Introduction

Following the notation of Konietschke et al. (2015), we consider the multivariate linear model, which covers various factorial designs of interest in a unified way. Let $\mathbf{X}_{ij} = \boldsymbol{\mu}_i + \boldsymbol{\varepsilon}_{ij}$, $i = 1, \ldots, d$, $j = 1, \ldots, n_i$ be $N = n_1 + \cdots + n_d$ independent observations, where for fixed $i = 1, \ldots, d$, $\boldsymbol{\varepsilon}_{ij}$ are independent and identically distributed p-dimensional random vectors. Moreover, we assume that the error terms satisfy the following conditions:

$$E(\boldsymbol{\varepsilon}_{ij}) = 0, \qquad Cov(\boldsymbol{\varepsilon}_{ij}) = \Sigma_i > 0, \qquad E(\|\boldsymbol{\varepsilon}_{ij}\|^4) < \infty, \qquad (1)$$

Ł. Smaga (✉)
Faculty of Mathematics and Computer Science, Adam Mickiewicz University, Poznań, Poland
e-mail: ls@amu.edu.pl

© Springer Nature Singapore Pte Ltd. 2020
T. Imaizumi et al. (eds.), *Advanced Studies in Classification and Data Science*,
Studies in Classification, Data Analysis, and Knowledge Organization,
https://doi.org/10.1007/978-981-15-3311-2_19

$i = 1, \ldots, d, j = 1, \ldots, n_i$. The sample or treatment group (resp. the experimental unit) is represented by the index i (resp. j). The above model can be expressed as $\mathbf{X} = \mathrm{diag}(\mathbf{D}_1, \ldots, \mathbf{D}_d)\boldsymbol{\mu} + \boldsymbol{\varepsilon}$, where $\mathbf{X}^\top = (\mathbf{X}_{11}^\top, \ldots, \mathbf{X}_{dn_d}^\top)$, $\mathbf{D}_i = \mathbf{1}_{n_i} \otimes \mathbf{I}_p$, $\mathbf{1}_{n_i}$ is the $n_i \times 1$ vector of ones, $\mathbf{I}_p$ is the unit matrix of size p, $\otimes$ denotes the Kronecker product, $\boldsymbol{\mu}^\top = (\boldsymbol{\mu}_1^\top, \ldots, \boldsymbol{\mu}_d^\top)$, and $\boldsymbol{\varepsilon}^\top = (\boldsymbol{\varepsilon}_{11}^\top, \ldots, \boldsymbol{\varepsilon}_{dn_d}^\top)$. As we mentioned, the considered general design covers such special cases as multivariate Behrens–Fisher problem, heteroscedastic one- and two-way MANOVA or hierarchically nested designs (Konietschke et al. 2015, Sect. 4). The reason for this is that the factorial structure within the components of the vector $\boldsymbol{\mu}$ by splitting up the indices is allowed. Moreover, note that different sample sizes, covariance matrices, and distributions of error vectors are allowed.

In the above model, we can consider general hypotheses about the mean vectors. We formulate them by using appropriate contrast and projection matrices. Let $\mathbf{H}$ be a contrast matrix, i.e., $\mathbf{H1} = \mathbf{0}$, where $\mathbf{0}$ is the column vector of zeros. Then $\mathbf{T} = \mathbf{H}^\top(\mathbf{HH}^\top)^+\mathbf{H}$ is the unique projection matrix (symmetric and idempotent), where $\mathbf{A}^+$ is the Moore–Penrose inverse of $\mathbf{A}$. The null hypothesis of interest is $H_0 : \mathbf{H}\boldsymbol{\mu} = \mathbf{0} \Leftrightarrow \mathbf{T}\boldsymbol{\mu} = \mathbf{0}$. For example, in the one-way MANOVA, the null hypothesis $H_0 : \boldsymbol{\mu}_1 = \cdots = \boldsymbol{\mu}_d$ is equivalent to $H_0 : \mathbf{T}\boldsymbol{\mu} = \mathbf{0}$ with $\mathbf{T} = (\mathbf{I}_d - d^{-1}\mathbf{1}_d\mathbf{1}_d^\top) \otimes \mathbf{I}_p$.

For testing the null hypothesis $H_0 : \mathbf{T}\boldsymbol{\mu} = \mathbf{0}$, Konietschke et al. (2015) considered asymptotically exact tests based on the Wald-type test statistic (WTS) of the form

$$Q_N(\mathbf{T}) = N\bar{\mathbf{X}}^\top \mathbf{T}(\mathbf{T}\hat{\boldsymbol{\Sigma}}_N\mathbf{T})^+\mathbf{T}\bar{\mathbf{X}}, \qquad (2)$$

where $\bar{\mathbf{X}}^\top = (\bar{\mathbf{X}}_1^\top, \ldots, \bar{\mathbf{X}}_d^\top)$ is the vector of sample means $\bar{\mathbf{X}}_i = n_i^{-1}\sum_{j=1}^{n_i}\mathbf{X}_{ij}$, and $\hat{\boldsymbol{\Sigma}}_N = N\mathrm{diag}(n_1^{-1}\hat{\boldsymbol{\Sigma}}_1, \ldots, n_d^{-1}\hat{\boldsymbol{\Sigma}}_d)$ is the block diagonal matrix of the sample covariance matrices $\hat{\boldsymbol{\Sigma}}_i = (n_i - 1)^{-1}\sum_{j=1}^{n_i}(\mathbf{X}_{ij} - \bar{\mathbf{X}}_i)(\mathbf{X}_{ij} - \bar{\mathbf{X}}_i)^\top$, $i = 1, \ldots, d$. They proposed the asymptotic as well as nonparametric and parametric bootstrap testing procedures based on $Q_N(\mathbf{T})$. The asymptotic one, however, requires large sample sizes to maintain the preassigned type I error level. In general, the nonparametric bootstrap method also behaves unsatisfactorily under small sample sizes (It is too liberal or conservative in many settings.). The best small sample performance is presented by the parametric bootstrap test, but it may be unacceptably liberal under extremely skewed distributions.

In this paper, we study testing procedures using modification of the test statistic $Q_N(\mathbf{T})$. Namely, we consider the following statistic:

$$Q_N^A(\mathbf{T}) = N\bar{\mathbf{X}}^\top \mathbf{T}\bar{\mathbf{X}}, \qquad (3)$$

which can be seen as the Wald-type statistic with weight matrix being the identity matrix. It is called the ANOVA-type statistic or unscaled Wald-type statistic (Brunner et al. 1997; Chen and Qin 2010; Duchesne and Francq 2015; Smaga 2017b). Based on it and its standardized version, we construct different testing procedures, i.e., asymptotic, bootstrap, and permutation. Some of them behave quite well or even better than the tests of Konietschke et al. (2015) in terms of size control and power under small sample sizes.

The plan of the paper is as follows. In Sect. 2, we derive the asymptotic tests based on the ANOVA-type statistic (3) and its standardized version. Their theoretical properties are also stated there. The permutation and bootstrap versions of these testing procedures are presented in Sect. 3. Section 4 contains the confidence regions for contrasts based on the results of previous sections. In Sect. 5, the proposed tests and confidence regions are compared with certain competing ones via simulation studies. Finally, Sect. 6 concludes the paper.

2 Asymptotic Tests

In this section, the asymptotic null distributions of the test statistic $Q_N^A(\mathbf{T})$ and its standardized version are investigated. Based on them and appropriate approximations, the asymptotic testing procedures for testing $H_0 : \mathbf{T}\boldsymbol{\mu} = \mathbf{0}$ are proposed. Their properties are also established.

The asymptotic null distribution of $Q_N^A(\mathbf{T})$ is given in the following theorem.

Theorem 1 *Under assumptions (1), $n_i/N \rightarrow \kappa_i > 0$, as $N \rightarrow \infty$, $Q_N^A(\mathbf{T})$ has under the null hypothesis $H_0 : \mathbf{T}\boldsymbol{\mu} = \mathbf{0}$, asymptotically, as $N \rightarrow \infty$, the same distribution as $\sum_{i=1}^r \lambda_i N_i^2$, where $r = \mathrm{rank}(\mathbf{T})$, $\lambda_1, \ldots, \lambda_r$ are the nonzero eigenvalues of $\mathbf{T}\boldsymbol{\Sigma}\mathbf{T}$, $\boldsymbol{\Sigma} = \mathrm{diag}(\kappa_1^{-1}\boldsymbol{\Sigma}_1, \ldots, \kappa_d^{-1}\boldsymbol{\Sigma}_d)$, and $N_1, \ldots, N_r$ are the independent standard normal variables.*

Proof By the multivariate Central Limit Theorem, $n_i^{1/2}(\bar{\mathbf{X}}_i - \boldsymbol{\mu}_i) \xrightarrow{D} N_p(\mathbf{0}_p, \boldsymbol{\Sigma}_i)$, as $n_i \rightarrow \infty$, for $i = 1, \ldots, d$ ($\xrightarrow{D}$ denotes the converge in distribution.). Thus $N^{1/2}(\bar{\mathbf{X}} - \boldsymbol{\mu}) \xrightarrow{D} N_{dp}(\mathbf{0}_{dp}, \boldsymbol{\Sigma})$, as $N \rightarrow \infty$, since all the observations are independent. Hence, under $H_0 : \mathbf{T}\boldsymbol{\mu} = \mathbf{0}$, we obtain $N^{1/2}\mathbf{T}\bar{\mathbf{X}} \xrightarrow{D} \mathbf{N} \sim N_{dp}(\mathbf{0}_{dp}, \mathbf{T}\boldsymbol{\Sigma}\mathbf{T})$, as $N \rightarrow \infty$. The continuous mapping theorem implies $Q_N^A(\mathbf{T}) \xrightarrow{D} \mathbf{N}^{\top}\mathbf{N}$, as $N \rightarrow \infty$. Finally, we conclude that $Q_N^A(\mathbf{T}) \xrightarrow{D} \sum_{i=1}^r \lambda_i N_i^2$, as $N \rightarrow \infty$, by the representation theorem of the quadratic forms in normal variables.   □

In Theorem 1, it is shown that the asymptotic null distribution of $Q_N^A(\mathbf{T})$ is that of a central χ^2-type mixture (Zhang 2005), which can be approximated by the Box-type approximation (Box 1954) also called two-cumulant approximation (Zhang 2013, Chapter 4). The idea of this method is to approximate the distribution of $\sum_{i=1}^r \lambda_i N_i^2$ by that of $\beta\chi_\gamma^2$ (scaled χ^2-distribution). The parameters β and γ are determined by matching the first two moments of these random variables as follows: From the proof of Theorem 1, it follows that $\sum_{i=1}^r \lambda_i N_i^2 = \mathbf{N}^{\top}\mathbf{N}$, where $\mathbf{N} \sim N_{dp}(\mathbf{0}_{dp}, \mathbf{T}\boldsymbol{\Sigma}\mathbf{T})$, and hence theorem on the moments of quadratic forms (Mathai and Provost 1992, p. 55) implies $\mathrm{E}\left(\sum_{i=1}^r \lambda_i N_i^2\right) = \mathrm{trace}(\mathbf{T}\boldsymbol{\Sigma}\mathbf{T})$ and

$\text{Var}\left(\sum_{i=1}^{r}\lambda_i N_i^2\right) = 2\text{trace}((\mathbf{T}\Sigma\mathbf{T})^2)$. Thus, we obtain

$$\beta = \text{trace}((\mathbf{T}\Sigma\mathbf{T})^2)/\text{trace}(\mathbf{T}\Sigma\mathbf{T}), \qquad \gamma = (\text{trace}(\mathbf{T}\Sigma\mathbf{T}))^2/\text{trace}((\mathbf{T}\Sigma\mathbf{T})^2),$$

and $\beta\chi_{\gamma,1-\alpha}^2$ is $(1-\alpha)$-quantile of the distribution of $\beta\chi_\gamma^2$. (When the degrees of freedom γ is a real number, the distribution χ_γ^2 is the gamma distribution with scale parameter 2 and shape parameter $\gamma/2$.) These quantities have to be estimated in practice. The simple estimators as well as the asymptotic $Q_N^A(\mathbf{T})$ test are presented in the following result, where their consistency is shown.

Theorem 2 *Under assumptions (1), $n_i/N \to \kappa_i > 0$, as $N \to \infty$,*

$$\hat{\beta} = \text{trace}((\mathbf{T}\hat{\Sigma}_N\mathbf{T})^2)/\text{trace}(\mathbf{T}\hat{\Sigma}_N\mathbf{T}), \quad \hat{\gamma} = (\text{trace}(\mathbf{T}\hat{\Sigma}_N\mathbf{T}))^2/\text{trace}((\mathbf{T}\hat{\Sigma}_N\mathbf{T})^2) \tag{4}$$

and $\hat{\beta}\chi_{\hat{\gamma},1-\alpha}^2$ are consistent estimators of β, γ, and $\beta\chi_{\gamma,1-\alpha}^2$, respectively. Moreover, the test $\varphi_N^A = I\{Q_N^A(\mathbf{T}) > \hat{\beta}\chi_{\hat{\gamma},1-\alpha}^2\}$ for the null hypothesis $H_0 : \mathbf{T}\boldsymbol{\mu} = \mathbf{0}$ is consistent under fixed alternatives.

Proof The consistency of $\hat{\Sigma}_N$ for Σ immediately implies the consistency the estimators given by (4). Under the alternative hypothesis $H_1 : \mathbf{T}\boldsymbol{\mu} = \mathbf{h}$, where $\mathbf{h}$ is fixed nonzero vector, we have $\bar{\mathbf{X}}^\top\mathbf{T}\bar{\mathbf{X}} \xrightarrow{P} \mathbf{h}^\top\mathbf{h} > 0$ as $N \to \infty$ ($\xrightarrow{P}$ denotes the converge in probability.). Thus, we obtain $Q_N^A(\mathbf{T}) \xrightarrow{P} \infty$ as $N \to \infty$. Therefore, $P(Q_N^A(\mathbf{T}) > \hat{\beta}\chi_{\hat{\gamma},1-\alpha}^2|H_1) \to 1$, as $N \to \infty$. □

As we will see in Sect. 5, the asymptotic test based on the ANOVA-type statistic (3) behaves very well under finite samples, when the number of groups d is small. Unfortunately, this testing procedure may tend to result in conservative decisions, when d is greater. Therefore, we consider tests based on standardization of $Q_N^A(\mathbf{T})$, which demonstrate more accurate performance. Pauly et al. (2015) used similar ideas for repeated measures designs. To simplify the analysis, we derive the expected value and variance of the ANOVA-type statistic under assumption of multivariate normality. More precisely, we assume that $\boldsymbol{\varepsilon}_{i1} \sim N_p(\mathbf{0}_p, \Sigma_i)$, $i = 1, \ldots, d$. Then, under the null hypothesis, $N^{1/2}\mathbf{T}\bar{\mathbf{X}} \sim N_{pd}(\mathbf{0}_{pd}, \mathbf{T}\Sigma_N\mathbf{T})$, where $\Sigma_N = N\text{diag}(n_1^{-1}\Sigma_1, \ldots, n_d^{-1}\Sigma_d)$. By theorem on the moments of quadratic forms, we conclude that $\text{E}(Q_N^A(\mathbf{T})) = \text{trace}(\mathbf{T}\Sigma_N\mathbf{T})$ and $\text{Var}(Q_N^A(\mathbf{T})) = 2\text{trace}((\mathbf{T}\Sigma_N\mathbf{T})^2)$. Therefore, we consider the following test statistic:

$$Q_N^{A,s}(\mathbf{T}) = \left(Q_N^A(\mathbf{T}) - \text{trace}(\mathbf{T}\hat{\Sigma}_N\mathbf{T})\right)/\sqrt{2\text{trace}((\mathbf{T}\hat{\Sigma}_N\mathbf{T})^2)}. \tag{5}$$

The asymptotic null distribution of $Q_N^{A,s}(\mathbf{T})$ is derived in the following theorem. It follows from Theorem 1 and the consistency of the estimator $\hat{\Sigma}_N$.

Theorem 3 *Under assumptions and notation of Theorem 1, and under the null hypothesis $H_0 : \mathbf{T}\boldsymbol{\mu} = \mathbf{0}$, we have*

$$Q_N^{A,s}(\mathbf{T}) \xrightarrow{D} \left(\sum_{i=1}^{r} \lambda_i N_i^2 - \operatorname{trace}(\mathbf{T}\Sigma\mathbf{T}) \right) / \sqrt{2\operatorname{trace}((\mathbf{T}\Sigma\mathbf{T})^2)}$$

as $N \to \infty$.

By the results of Zhang (2005), the asymptotic null distribution of $Q_N^{A,s}(\mathbf{T})$ given in Theorem 3 can be approximated by a standardized χ^2-distribution of the form $(\chi_\delta^2 - \delta)/(2\delta)^{1/2}$, where $\delta = (\operatorname{trace}((\mathbf{T}\Sigma\mathbf{T})^2))^3/(\operatorname{trace}((\mathbf{T}\Sigma\mathbf{T})^3))^2$, and $(\chi_{\delta,1-\alpha}^2 - \delta)/(2\delta)^{1/2}$ is $(1 - \alpha)$-quantile of this distribution. The estimators of δ and this quantile, the asymptotic $Q_N^{A,s}(\mathbf{T})$ test, and their consistency are established in the following theorem.

Theorem 4 *Under assumptions of Theorem 3,*

$$\hat{\delta} = (\operatorname{trace}((\mathbf{T}\hat{\Sigma}_N\mathbf{T})^2))^3/(\operatorname{trace}((\mathbf{T}\hat{\Sigma}_N\mathbf{T})^3))^2, \quad \left(\chi_{\hat{\delta},1-\alpha}^2 - \hat{\delta} \right)/\sqrt{2\hat{\delta}} \qquad (6)$$

are consistent estimators of δ and $(\chi_{\delta,1-\alpha}^2 - \delta)/(2\delta)^{1/2}$, respectively. Moreover, the test $\varphi_N^{A,s} = I\{Q_N^{A,s}(\mathbf{T}) > (\chi_{\hat{\delta},1-\alpha}^2 - \hat{\delta})/(2\hat{\delta})^{1/2}\}$ for the null hypothesis $H_0 : \mathbf{T}\boldsymbol{\mu} = \mathbf{0}$ is consistent under fixed alternatives.

The proof of Theorem 4 is similar to that of Theorem 2, and hence it is omitted. The asymptotic $Q_N^A(\mathbf{T})$ and $Q_N^{A,s}(\mathbf{T})$ tests perform very similarly under finite samples. However, the permutation and bootstrap testing procedures based on $Q_N^{A,s}(\mathbf{T})$ proposed in the next section improve the finite sample behavior of the asymptotic one, when the number of samples is greater (in contrast to those procedures based on $Q_N^A(\mathbf{T})$, see Sect. 5, for more details).

3 Permutation and Bootstrap Tests

In this section, we approximate the null distribution the ANOVA-type statistics by using the permutation and bootstrap procedures. Some of these methods will result in better finite sample performance of the tests based on these statistics.

Permutation approach often gives good results and it is even finitely exact under exchangeability (Chung and Romano 2013, Janssen 1997). Recently, it has successfully been used in univariate general factorial designs by Pauly et al. (2015) and Smaga (2015), Smaga (2017b). Below, we extend this approach to multivariate case. Let S_N be a test statistic for $H_0 : \mathbf{T}\boldsymbol{\mu} = \mathbf{0}$, e.g., $Q_N(\mathbf{T})$, $Q_N^A(\mathbf{T})$ or $Q_N^{A,s}(\mathbf{T})$. The permutation testing procedure based on test statistic S_N is as follows:

1. Compute S_N for original data $\mathbf{X}_{ij}$, $i = 1, \ldots, d$, $j = 1, \ldots, n_i$.

2. Create a permutation sample from the given data in the following way: From all observations $\mathbf{X}_{ij}$, $i = 1, \ldots, d$, $j = 1, \ldots, n_i$, select randomly without replacement n_1 observations for the first new sample, then from the remainder of the observations choose randomly without replacement n_2 observations for the second new sample, and so on.

3. Repeat step 2 a large number of times, e.g., $B^\pi = 10,000$, and obtain B^π independent permutation samples $\mathbf{X}_{ij}^{\pi,l}$, $i = 1, \ldots, d$, $j = 1, \ldots, n_i$, $l = 1, \ldots, B^\pi$.

4. For each permutation sample, compute the value of the test statistic S_N. Denote them by $S_N^{\pi,l}$, $l = 1, \ldots, B^\pi$.

5. The final p-value of the permutation test is defined by $(B^\pi)^{-1} \sum_{l=1}^{B^\pi} I(S_N^{\pi,l} > S_N)$.

The permutation tests based on statistics $Q_N(\mathbf{T})$, $Q_N^A(\mathbf{T})$, and $Q_N^{A,s}(\mathbf{T})$ will be referred to as the permutation $Q_N^\pi(\mathbf{T})$, $Q_N^{A,\pi}(\mathbf{T})$, and $Q_N^{A,s,\pi}(\mathbf{T})$ tests, respectively.

Bootstrap is another approach, which may improve finite sample behavior of test statistics considered in Sect. 1 and Sect. 2. Konietschke et al. (2015) and Smaga (2017a) investigated different bootstrap methods for hypothesis testing in the case of multivariate data. The nonparametric bootstrap procedure of Konietschke et al. (2015) in general framework under consideration can be described as follows:

1. Compute S_N for given data $\mathbf{X}_{ij}$, $i = 1, \ldots, d$, $j = 1, \ldots, n_i$.

2. Create a bootstrap sample from the original data in the following way: From all observations $\mathbf{X}_{ij}$, $i = 1, \ldots, d$, $j = 1, \ldots, n_i$, select randomly with replacement n_1 observations for the first new sample, then from all observations choose randomly with replacement n_2 observations for the second new sample, and so on.

3. Repeat step 2 a large number of times, e.g., $B^{nb} = 5000$, and obtain B^{nb} independent bootstrap samples $\mathbf{X}_{ij}^{nb,l}$, $i = 1, \ldots, d$, $j = 1, \ldots, n_i$, $l = 1, \ldots, B^{nb}$.

4. For each bootstrap sample, compute the value of the test statistic S_N. Denote them by $S_N^{nb,l}$, $l = 1, \ldots, B^{nb}$.

5. The final p-value of the bootstrap test is defined by $(B^{nb})^{-1} \sum_{l=1}^{B^{nb}} I(S_N^{nb,l} > S_N)$.

The nonparametric bootstrap testing procedures based on statistics $Q_N(\mathbf{T})$, $Q_N^A(\mathbf{T})$, and $Q_N^{A,s}(\mathbf{T})$ will be referred to as the nonparametric bootstrap $Q_N^{nb}(\mathbf{T})$, $Q_N^{A,nb}(\mathbf{T})$, and $Q_N^{A,s,nb}(\mathbf{T})$ tests, respectively. Taking into account the denominator of $Q_N^{A,s}(\mathbf{T})$, we propose to use only those bootstrap samples, for which all diagonal elements of the matrix $\mathbf{T}\hat{\Sigma}_N^{nb}\mathbf{T}$ are nonzero.

Konietschke et al. (2015) also considered another resampling approach, i.e., the parametric bootstrap. In this method, the parametric bootstrap samples are generated from multivariate normal distributions with zero expectation and covariance matrices $\hat{\Sigma}_i$, $i = 1, \ldots, d$. The parametric bootstrap $Q_N^{pb}(\mathbf{T})$ test is the best one proposed by them. However, by simulations not included in the article, the parametric bootstrap approach applied to the ANOVA-type statistics resulted in even more conservative tests than the asymptotic ones. So we do not consider them in this paper.

The finite sample performance of the testing procedures based on asymptotic, bootstrap, and permutation approaches is investigated in Sect. 5.

4 Confidence Regions

In this section, the confidence regions for contrasts based on the results of previous sections are derived. In particular for a single contrast, the confidence intervals are presented.

In statistical analysis, it is usually helpful to complete the information from p-value of statistical test by this received from the confidence regions and vice versa. Statistical tests provide mechanisms for making quantitative decisions and give the information about significance or insignificance of effects of factors under consideration. On the other hand, confidence regions for the unknown parameters and their functions (e.g., contrasts) describe the magnitude and variability of effects.

In the general framework of Sect. 1, we consider contrasts of the form $\mathbf{H}\boldsymbol{\mu}$, where $\mathbf{H} = (\mathbf{h}_1, \ldots, \mathbf{h}_h)^\top$ is a $h \times pd$ contrast matrix of full row rank. The ANOVA-type statistic (3) for $H_0 : \mathbf{H}\boldsymbol{\mu} = \mathbf{0}$ takes the form $Q_N^A(\mathbf{H}) = N(\mathbf{H}\bar{\mathbf{X}})^\top(\mathbf{H}\mathbf{H}^\top)^+(\mathbf{H}\bar{\mathbf{X}})$. Let $q_{1-\alpha}^{A,a}$, $q_{1-\alpha}^{A,\pi}$, $q_{1-\alpha}^{A,nb}$ denote $(1 - \alpha)$-quantiles of the asymptotic, permutation, and bootstrap distributions of $Q_N^A(\mathbf{H})$, respectively. As described in Sects. 2 and 3, we reject $H_0 : \mathbf{H}\boldsymbol{\mu} = \mathbf{0}$ when $Q_N^A(\mathbf{H})$ is greater than the appropriate quantile (Depending on the method, we choose one of the quantiles $q_{1-\alpha}^{A,a}$, $q_{1-\alpha}^{A,\pi}$, $q_{1-\alpha}^{A,nb}$.). Then, the confidence regions for contrasts $\mathbf{H}\boldsymbol{\mu}$ based on $Q_N^A(\mathbf{H})$ are as follows:

$$\left\{ \mathbf{H}\boldsymbol{\mu} : N(\mathbf{H}\bar{\mathbf{X}} - \mathbf{H}\boldsymbol{\mu})^\top(\mathbf{H}\mathbf{H}^\top)^+(\mathbf{H}\bar{\mathbf{X}} - \mathbf{H}\boldsymbol{\mu}) \leq q_{1-\alpha}^{A,M} \right\}, \tag{7}$$

where $M \in \{a, \pi, nb\}$ denotes the method chosen (see, Anderson 2003, for similar results in multivariate normal distribution). The axes of the confidence ellipsoid (7) are $\mathbf{H}\bar{\mathbf{X}} \pm (\lambda_i q_{1-\alpha}^{A,M} N^{-1})^{1/2} \mathbf{v}_i$, where λ_i and $\mathbf{v}_i$ are the eigenvalues and eigenvectors of the matrix $\mathbf{H}\mathbf{H}^\top$. For a single contrast $\mathbf{h}_k^\top \boldsymbol{\mu}$, the confidence interval is of the form

$$\left(\mathbf{h}_k^\top \bar{\mathbf{X}} - \sqrt{\mathbf{h}_k^\top \mathbf{h}_k q_{1-\alpha}^{A,M}/N}, \mathbf{h}_k^\top \bar{\mathbf{X}} + \sqrt{\mathbf{h}_k^\top \mathbf{h}_k q_{1-\alpha}^{A,M}/N} \right).$$

In much the same way as for $Q_N^A(\mathbf{H})$, we derive the confidence regions for $\mathbf{H}\boldsymbol{\mu}$ based on the standardized ANOVA-type statistic (5). Let $q_{1-\alpha}^{A,s,a}$, $q_{1-\alpha}^{A,s,\pi}$, $q_{1-\alpha}^{A,s,nb}$

denote $(1 - \alpha)$-quantiles of the asymptotic, permutation, and bootstrap distributions of

$$Q_N^{A,s}(\mathbf{H}) = \frac{N(\mathbf{H}\bar{\mathbf{X}})^\top (\mathbf{H}\mathbf{H}^\top)^+ (\mathbf{H}\bar{\mathbf{X}}) - \text{trace}((\mathbf{H}\mathbf{H}^\top)^+ \mathbf{H}\hat{\Sigma}_N \mathbf{H}^\top)}{\sqrt{2\text{trace}(((\mathbf{H}\mathbf{H}^\top)^+ \mathbf{H}\hat{\Sigma}_N \mathbf{H}^\top)^2)}},$$

respectively. For $M \in \{a, \pi, nb\}$, the confidence regions for contrasts $\mathbf{H}\boldsymbol{\mu}$ based on $Q_N^{A,s}(\mathbf{H})$ are as follows:

$$\left\{ \mathbf{H}\boldsymbol{\mu} : \frac{N(\mathbf{H}\bar{\mathbf{X}} - \mathbf{H}\boldsymbol{\mu})^\top (\mathbf{H}\mathbf{H}^\top)^+ (\mathbf{H}\bar{\mathbf{X}} - \mathbf{H}\boldsymbol{\mu}) - \text{trace}((\mathbf{H}\mathbf{H}^\top)^+ \mathbf{H}\hat{\Sigma}_N \mathbf{H}^\top)}{\sqrt{2\text{trace}(((\mathbf{H}\mathbf{H}^\top)^+ \mathbf{H}\hat{\Sigma}_N \mathbf{H}^\top)^2)}} \leq q_{1-\alpha}^{A,s,M} \right\}.$$

The axes of this confidence ellipsoid are

$$\mathbf{H}\bar{\mathbf{X}} \pm \sqrt{\lambda_i [q_{1-\alpha}^{A,s,M} \sqrt{2\text{trace}(((\mathbf{H}\mathbf{H}^\top)^+ \mathbf{H}\hat{\Sigma}_N \mathbf{H}^\top)^2)} + \text{trace}((\mathbf{H}\mathbf{H}^\top)^+ \mathbf{H}\hat{\Sigma}_N \mathbf{H}^\top)]/N} \mathbf{v}_i.$$

For a single contrast $\mathbf{h}_k^\top \boldsymbol{\mu}$, the confidence interval based on $Q_N^{A,s}(\mathbf{H})$ is of the form

$$\left(\mathbf{h}_k^\top \bar{\mathbf{X}} - \sqrt{(q_{1-\alpha}^{A,s,M} \sqrt{2} + 1)(\mathbf{h}_k^\top \hat{\Sigma}_N \mathbf{h}_k)/N}, \ \mathbf{h}_k^\top \bar{\mathbf{X}} + \sqrt{(q_{1-\alpha}^{A,s,M} \sqrt{2} + 1)(\mathbf{h}_k^\top \hat{\Sigma}_N \mathbf{h}_k)/N} \right).$$

5 Simulation Studies

Monte Carlo simulation studies based on the Wald-type statistic (2) and the ANOVA-type statistics (3) and (5) were performed to investigate their performance in finite samples in regard to controlling the type I error under the null hypothesis $H_0 : \mathbf{T}\boldsymbol{\mu} = \mathbf{0}$, the power of the statistics under the alternatives and maintaining the preassigned coverage probability. We conducted all computations in the R program (R Core Team 2017).

5.1 Simulation Setup

To be consistent with the numerical results of Konietschke et al. (2015), we conducted simulations in the multivariate one-way layouts with $d = 2, 4$ groups, similar to those in that paper. We generated the four-dimensional ($p = 4$) independent observations according to the model $\mathbf{X}_{ij} = \boldsymbol{\mu}_i + \Sigma_i^{1/2} \boldsymbol{\varepsilon}_{ij}, i = 1, \ldots, d$, $j = 1, \ldots, n_i$. The random error vectors $\boldsymbol{\varepsilon}_{ij}^\top = (\varepsilon_{ij1}, \ldots, \varepsilon_{ijp})$ were independent and identically distributed, where $\varepsilon_{ijk} = (Z_{ijk} - \mathrm{E}(Z_{ijk}))/(\mathrm{Var}(Z_{ijk}))^{1/2}$ and Z_{ijk}

were taken from normal, Laplace, t_7, χ_{20}^2, and χ_3^2 distributions. For $d = 2$ (resp. $d = 4$), the following sample size vectors were used: $\mathbf{n}_1 = (10, 10)$, $\mathbf{n}_2 = (15, 15)$, $\mathbf{n}_3 = (10, 20)$, $\mathbf{n}_4 = (20, 10)$ (resp. $\mathbf{n}_5 = (10, 10, 10, 10)$, $\mathbf{n}_6 = (20, 20, 20, 20)$, $\mathbf{n}_7 = (7, 10, 13, 16)$, $\mathbf{n}_8 = (16, 13, 10, 7)$). The covariance structures described in the following Settings 1–4 (resp. 5–8) were investigated in case of $d = 2$ (resp. $d = 4$):

Setting 1 $\Sigma_1 = \Sigma_2 = \mathbf{I}_4 + 0.5(\mathbf{1}_4\mathbf{1}_4^\top - \mathbf{I}_4)$
Setting 2 $\Sigma_1 = \Sigma_2 = (0.6)^{|k-l|}$
Setting 3 $\Sigma_1 = \mathbf{I}_4 + 0.5(\mathbf{1}_4\mathbf{1}_4^\top - \mathbf{I}_4)$, $\Sigma_2 = 3\mathbf{I}_4 + 0.5(\mathbf{1}_4\mathbf{1}_4^\top - \mathbf{I}_4)$
Setting 4 $\Sigma_1 = (0.6)^{|k-l|}$, $\Sigma_2 = (0.6)^{|k-l|} + 2\mathbf{I}_4$
Setting 5 $\Sigma_i = \mathbf{I}_4 + 0.5(\mathbf{1}_4\mathbf{1}_4^\top - \mathbf{I}_4)$
Setting 6 $\Sigma_i = (0.6)^{|k-l|}$
Setting 7 $\Sigma_i = i\mathbf{I}_4 + 0.5(\mathbf{1}_4\mathbf{1}_4^\top - \mathbf{I}_4)$
Setting 8 $\Sigma_i = (0.6)^{|k-l|} + i\mathbf{I}_4$

for $i = 1, \ldots, 4$. Therefore, in our simulation designs, different symmetric and skewed distributions as well as balanced, unbalanced, homoscedastic, and heteroscedastic settings were considered. Moreover, the positive (increasing sample sizes combined with increasing variances) and negative (increasing sample sizes combined with decreasing variances) pairings were also included.

The type I error rate and power were obtained by running 1000 simulations under each scenario. The p-values of the bootstrap (resp. permutation) tests were estimated from 5000 (resp. 10,000) replications. The resulting empirical sizes and powers under the normal and χ_3^2 distributions are displayed in Tables 1, 2, 3, and 4. The results obtained in other settings give similar conclusions, and therefore are omitted for space saving, but available from the author.

Let us remind that the binomial proportion confidence interval is of the form $(\hat{p} - z_{1-\beta/2}(\hat{p}(1-\hat{p})/n)^{1/2}, \hat{p} + z_{1-\beta/2}(\hat{p}(1-\hat{p})/n)^{1/2})$, where $\hat{p}$ is the proportion of successes in a Bernoulli trial process (measured with n trials), and $z_{1-\beta/2}$ is $(1 - \beta/2)$-quantile of the standard normal distribution ($\beta \in (0, 1)$). Hence, by Duchesne and Francq (2015, Sect. 5.2), the empirical sizes of a test for a given nominal level α should belong to $(\alpha - z_{1-\beta/2}(\alpha(1 - \alpha)/n)^{1/2}, \alpha + z_{1-\beta/2}(\alpha(1-\alpha)/n)^{1/2})$ with probability $1 - \beta$, where n denotes the number of replications. Therefore, for $\alpha = 5\%$, the empirical size over the 1000 independent replications should belong to [3.6%, 6.4%] (resp. [3.2%, 6.8%]) with probability 95% (resp. 99%). Thus in Tables 1 and 2, when the rejection proportions are outside the 95% (resp. 99%) significance limits, they are displayed in bold (resp. underlined).

5.2   Discussion of the Simulation Results

In simulations, we considered twelve tests based on statistics (2), (3), and (5) and asymptotic, nonparametric, and parametric bootstrap, and permutation methods. However, we only present results for five testing procedures, i.e., the $Q_N^{pb}(\mathbf{T})$ test,

Table 1 Empirical sizes (as percentages) of the $Q_N^{pb}(\mathbf{T})$, $Q_N^{A}(\mathbf{T})$, $Q_N^{A,s}(\mathbf{T})$, $Q_N^{A,s,nb}(\mathbf{T})$, and $Q_N^{A,s,\pi}(\mathbf{T})$ tests obtained under the normal and χ_3^2 distributions ($d = 2$, Set.: Setting)

Set.	n	Normal					χ_3^2				
		Q_N^{pb}	Q_N^{A}	$Q_N^{A,s}$	$Q_N^{A,s,nb}$	$Q_N^{A,s,\pi}$	Q_N^{pb}	Q_N^{A}	$Q_N^{A,s}$	$Q_N^{A,s,nb}$	$Q_N^{A,s,\pi}$
1	n_1	4.7	6.4	6.4	5.2	5.3	4.3	5.3	5.2	4.6	4.8
	n_2	4.2	5.1	5.1	4.9	5.0	**3.4**	4.8	4.7	4.8	4.7
	n_3	4.8	5.2	5.2	4.6	5.0	5.8	4.8	4.8	4.1	4.1
	n_4	4.9	6.1	6.0	5.2	5.4	6.4	4.6	4.6	4.3	4.5
2	n_1	4.5	5.7	5.5	4.4	4.6	4.1	4.9	4.8	4.3	4.5
	n_2	4.5	4.9	4.9	4.7	4.6	**3.3**	4.8	4.8	4.5	4.6
	n_3	4.8	5.1	5.1	4.6	4.9	5.8	4.5	4.5	4.0	4.2
	n_4	5.0	5.5	5.4	4.9	5.0	6.2	4.7	4.5	4.0	4.4
3	n_1	5.4	5.4	5.2	5.0	5.6	6.0	5.5	5.4	4.8	5.7
	n_2	5.1	4.4	4.1	3.9	4.5	**6.5**	4.1	4.1	4.2	4.6
	n_3	4.7	5.3	5.2	5.1	5.6	4.8	4.8	4.6	4.5	4.9
	n_4	5.9	4.9	4.7	4.2	5.0	**<u>10.8</u>**	4.9	4.8	4.8	5.2
4	n_1	5.7	5.8	5.7	5.4	5.8	6.0	5.2	5.1	4.7	5.4
	n_2	4.1	4.4	4.2	4.4	4.6	**6.7**	4.3	4.2	4.5	4.7
	n_3	4.4	6.1	6.0	5.4	5.9	4.8	4.2	4.1	3.9	4.5
	n_4	6.2	5.2	5.2	5.0	5.0	**<u>11.0</u>**	5.0	4.9	4.9	5.3

Table 2 Empirical sizes (as percentages) of the $Q_N^{pb}(\mathbf{T})$, $Q_N^{A}(\mathbf{T})$, $Q_N^{A,s}(\mathbf{T})$, $Q_N^{A,s,nb}(\mathbf{T})$, and $Q_N^{A,s,\pi}(\mathbf{T})$ tests obtained under the normal and χ_3^2 distributions ($d = 4$, Set.: Setting)

Set.	n	Normal					χ_3^2				
		Q_N^{pb}	Q_N^{A}	$Q_N^{A,s}$	$Q_N^{A,s,nb}$	$Q_N^{A,s,\pi}$	Q_N^{pb}	Q_N^{A}	$Q_N^{A,s}$	$Q_N^{A,s,nb}$	$Q_N^{A,s,\pi}$
5	n_5	3.7	5.9	5.7	5.3	5.5	5.5	4.8	4.4	5.2	5.2
	n_6	5.5	5.8	5.7	5.7	5.6	5.6	4.5	4.3	4.7	4.8
	n_7	6.0	5.1	4.8	4.3	4.5	**6.5**	4.6	4.3	4.8	4.8
	n_8	5.1	5.3	5.1	4.4	4.6	**<u>7.3</u>**	5.1	5.0	4.8	5.0
6	n_5	4.0	5.7	5.6	5.5	5.3	5.7	4.9	4.6	5.5	5.5
	n_6	5.4	6.2	6.1	6.1	6.1	5.6	5.1	4.7	5.4	5.8
	n_7	6.4	5.4	5.0	4.6	4.8	**6.6**	4.8	4.5	4.7	4.9
	n_8	5.2	5.0	4.8	4.7	4.6	**<u>7.1</u>**	4.6	4.5	4.6	4.8
7	n_5	5.4	4.8	4.6	4.9	5.3	**<u>7.3</u>**	3.8	3.7	5.2	5.8
	n_6	4.7	5.3	5.2	6.3	**6.7**	**6.6**	3.9	3.8	5.9	6.2
	n_7	5.0	4.3	4.2	4.7	4.8	5.7	**3.4**	**3.3**	4.8	5.1
	n_8	**6.7**	3.7	**3.3**	4.6	5.1	**<u>10.8</u>**	3.2	**<u>2.9</u>**	4.2	4.5
8	n_5	5.1	5.2	4.9	6.0	**6.5**	**<u>6.9</u>**	3.2	**<u>3.0</u>**	5.0	6.0
	n_6	**6.5**	5.4	5.3	5.9	6.1	**6.6**	4.8	4.2	6.3	6.4
	n_7	5.8	4.6	4.1	5.0	5.4	5.9	**<u>3.1</u>**	**<u>3.1</u>**	3.8	4.7
	n_8	**<u>6.9</u>**	3.5	**<u>3.1</u>**	4.3	4.8	**<u>9.5</u>**	**<u>3.1</u>**	**<u>2.7</u>**	4.6	4.8

Table 3 Empirical powers (as percentages) of the $Q_N^{pb}(\mathbf{T})$, $Q_N^A(\mathbf{T})$, $Q_N^{A,s}(\mathbf{T})$, $Q_N^{A,s,nb}(\mathbf{T})$, and $Q_N^{A,s,\pi}(\mathbf{T})$ tests obtained under the normal and χ_3^2 distributions ($d = 2$, $\mathbf{n} = (15, 15)$, $\boldsymbol{\mu}_1 = \mathbf{0}_4$, $\mathbf{a}_1 = \mathbf{1}_4$, $\mathbf{a}_2 = (\mathbf{1}_3^\top, 0)^\top$, $\mathbf{a}_3 = (\mathbf{1}_2^\top, \mathbf{0}_2^\top)^\top$, Set.: Setting)

Set.	$\boldsymbol{\mu}_2$	Normal					χ_3^2				
		Q_N^{pb}	Q_N^A	$Q_N^{A,s}$	$Q_N^{A,s,nb}$	$Q_N^{A,s,\pi}$	Q_N^{pb}	Q_N^A	$Q_N^{A,s}$	$Q_N^{A,s,nb}$	$Q_N^{A,s,\pi}$
1	$\mathbf{a}_1$	68.4	91.7	91.6	91.0	90.7	75.9	91.9	91.9	91.6	91.4
	$\mathbf{a}_2$	87.5	85.6	85.6	84.8	84.6	90.8	86.6	86.4	85.7	86.2
	$\mathbf{a}_3$	88.0	73.8	73.7	71.1	71.2	89.9	75.2	74.9	73.7	74.0
2	$\mathbf{a}_1$	73.9	92.2	92.1	91.7	91.9	79.2	93.1	93.1	92.5	93.0
	$\mathbf{a}_2$	82.3	86.0	85.5	84.0	84.3	87.1	85.9	85.9	85.2	85.6
	$\mathbf{a}_3$	78.2	72.7	72.3	70.8	71.2	80.2	74.7	74.4	72.8	73.4
3	$\mathbf{a}_1$	52.9	74.3	74.1	73.7	74.2	48.6	78.5	78.5	78.7	79.9
	$\mathbf{a}_2$	50.5	63.2	62.8	62.9	63.8	51.1	66.8	66.2	67.0	68.7
	$\mathbf{a}_3$	45.0	48.4	48.0	48.2	48.8	46.9	49.4	49.1	51.0	51.7
4	$\mathbf{a}_1$	55.9	75.1	74.7	74.4	75.2	50.8	79.4	79.1	79.8	80.9
	$\mathbf{a}_2$	48.7	65.6	64.8	64.4	65.4	47.9	66.5	66.0	66.7	69.1
	$\mathbf{a}_3$	41.4	48.1	47.8	47.7	48.3	40.1	49.0	48.7	49.7	51.1

Table 4 Empirical powers (as percentages) of the $Q_N^{pb}(\mathbf{T})$, $Q_N^A(\mathbf{T})$, $Q_N^{A,s}(\mathbf{T})$, $Q_N^{A,s,nb}(\mathbf{T})$, and $Q_N^{A,s,\pi}(\mathbf{T})$ tests obtained under the normal and χ_3^2 distributions ($d = 4$, $\boldsymbol{\mu}_i = \mathbf{0}_4$, $i = 1, 2, 3$, $\mathbf{a}_1 = 2\mathbf{1}_4$, $\mathbf{a}_2 = 2(\mathbf{1}_3^\top, 0)^\top$, $\mathbf{a}_3 = 2(\mathbf{1}_2^\top, \mathbf{0}_2^\top)^\top$, Set.: Setting)

Set.	$\mathbf{n}$	$\boldsymbol{\mu}_4$	Normal					χ_3^2				
			Q_N^{pb}	Q_N^A	$Q_N^{A,s}$	$Q_N^{A,s,nb}$	$Q_N^{A,s,\pi}$	Q_N^{pb}	Q_N^A	$Q_N^{A,s}$	$Q_N^{A,s,nb}$	$Q_N^{A,s,\pi}$
7	$\mathbf{n}_5$	$\mathbf{a}_1$	69.4	97.9	97.9	98.4	98.8	77.3	98.4	98.2	98.7	98.8
		$\mathbf{a}_2$	64.3	93.7	93.4	94.6	95.0	73.7	93.8	93.4	95.0	95.4
		$\mathbf{a}_3$	50.3	77.9	77.2	80.7	81.1	57.9	78.9	77.9	83.0	84.5
	$\mathbf{n}_7$	$\mathbf{a}_1$	86.3	99.4	99.4	99.4	99.4	94.5	99.2	99.2	99.3	99.3
		$\mathbf{a}_2$	82.3	97.2	97.0	97.2	97.2	91.2	96.6	96.4	97.3	97.5
		$\mathbf{a}_3$	70.3	89.6	89.2	90.2	90.5	80.1	88.3	87.6	90.6	91.4
	$\mathbf{n}_8$	$\mathbf{a}_1$	58.1	92.8	92.3	93.1	93.8	57.7	96.2	96.0	96.8	97.2
		$\mathbf{a}_2$	52.8	83.0	82.4	85.0	86.0	56.9	89.6	89.2	92.2	93.1
		$\mathbf{a}_3$	42.7	65.9	64.3	68.3	69.4	51.0	69.9	68.5	74.5	75.9
8	$\mathbf{n}_5$	$\mathbf{a}_1$	63.2	93.6	93.3	94.1	94.6	69.2	93.0	92.6	95.0	95.4
		$\mathbf{a}_2$	53.7	86.0	85.9	87.4	88.0	61.2	84.1	83.5	87.8	88.5
		$\mathbf{a}_3$	39.6	68.3	66.8	71.4	72.4	42.8	63.3	62.8	69.3	70.8
	$\mathbf{n}_7$	$\mathbf{a}_1$	81.3	97.9	97.8	98.3	98.3	88.6	96.3	96.0	96.7	96.9
		$\mathbf{a}_2$	70.4	93.0	92.7	93.4	94.1	78.2	89.2	88.7	91.2	91.7
		$\mathbf{a}_3$	54.7	77.9	77.4	80.0	80.7	61.6	72.9	71.6	77.5	78.8
	$\mathbf{n}_8$	$\mathbf{a}_1$	50.6	85.8	84.8	87.6	88.2	51.8	90.3	89.7	92.7	93.6
		$\mathbf{a}_2$	45.4	74.7	73.5	76.6	77.6	47.3	79.3	78.0	82.7	84.3
		$\mathbf{a}_3$	35.2	54.2	52.7	57.0	58.3	40.6	54.5	52.8	61.0	63.6

which performed best in Konietschke et al. (2015), and the $Q_N^A(\mathbf{T})$, $Q_N^{A,s}(\mathbf{T})$, $Q_N^{A,s,nb}(\mathbf{T})$, and $Q_N^{A,s,\pi}(\mathbf{T})$ tests. The other tests are too liberal or conservative in many cases, and therefore they are not recommended.

In the generalized multivariate Behrens–Fisher problem ($d = 2$), the $Q_N^{pb}(\mathbf{T})$, $Q_N^A(\mathbf{T})$, $Q_N^{A,s}(\mathbf{T})$, $Q_N^{A,s,nb}(\mathbf{T})$, and $Q_N^{A,s,\pi}(\mathbf{T})$ tests demonstrate accurate control of the nominal type I error level under symmetric (see Table 1) and moderately skewed distributions. However, the $Q_N^{pb}(\mathbf{T})$ test may tend to result in slightly liberal decisions in negative pairing. Unfortunately, this testing procedure performs much more liberal under extremely skewed distributions as χ_3^2-distribution considered (see Table 1). The tests based on the ANOVA-type statistics still control the type I error quite well under such distributions.

When the number of groups increases ($d = 4$), the finite sample behavior of the tests under the null may change a little (see Table 2). The parametric bootstrap Wald-type test tends to over-reject the null hypothesis in more situations than for $d = 2$, e.g., under extremely skewed distributions and homoscedastic settings or balanced designs. The asymptotic ANOVA-type testing procedures control the nominal level in most cases, but they show a tendency to conservativity in positive and negative pairings, especially under extremely skewed distributions. Fortunately, the $Q_N^{A,s,nb}(\mathbf{T})$ and $Q_N^{A,s,\pi}(\mathbf{T})$ tests keep the preassigned type I error in all scenarios.

The power results for all tests are quite satisfactory. In case of two samples, the empirical power of the tests was computed for balanced designs with fifteen observations in each group (see Table 3), since all competing testing procedures control the type I error rate in such scenarios. The empirical powers of the tests based on the ANOVA-type statistic are very similar in all cases. Under Settings 3 and 4 as well as the case of $\mu_2 = \mathbf{a}_1$ in Settings 1–2, these tests have considerably larger power than the $Q_N^{pb}(\mathbf{T})$ procedure. In other cases, it is the other way around or the empirical power of all tests is comparable.

In contrast to power analysis in case of $d = 2$, we also consider unbalanced designs for $d = 4$ to emphasize some interesting observations (see Table 4). First of all, the empirical powers of the parametric bootstrap Wald-type test are (usually much) smaller than these of the tests based on the ANOVA-type statistics. Interestingly, it is very evident in negative pairing, where although this test is too liberal, it has very low power in comparison to the other ones. The empirical power of the ANOVA-type tests is not so similar as for $d = 2$. The empirical powers of the $Q_N^{A,s,nb}(\mathbf{T})$ and $Q_N^{A,s,\pi}(\mathbf{T})$ tests are even a few percent greater than these of the $Q_N^A(\mathbf{T})$ and $Q_N^{A,s}(\mathbf{T})$ procedures, which is particularly noticeable under χ_3^2-distribution. This can be explained by slightly conservative character of the asymptotic ANOVA-type tests for greater number of groups.

We also conducted simulation studies to check the performance of confidence regions of Sect. 4 in terms of maintaining the preassigned coverage probability. The results of these simulation studies were consistent with the relationship between confidence regions and hypothesis testing (Lehmann and Romano 2005, Sect. 6.11). (For this reason and to save space, the particular simulation results are omitted.)

More precisely, the confidence regions based on $Q_N^{A,s,nb}(\mathbf{H})$ and $Q_N^{A,s,\pi}(\mathbf{H})$ usually maintain the preassigned coverage probability quite accurately, since the $Q_N^{A,s,nb}(\mathbf{T})$ and $Q_N^{A,s,\pi}(\mathbf{T})$ tests keep the preassigned type I error level. On the other hand, the empirical coverage probabilities of the confidence regions based on $Q_N^{pb}(\mathbf{H})$ (resp. $Q_N^{A}(\mathbf{H})$ and $Q_N^{A,s}(\mathbf{H})$) may be smaller or even much smaller (resp. greater) than the preassigned coverage probability in the cases, when the $Q_N^{pb}(\mathbf{T})$ test is too liberal (resp. the $Q_N^{A}(\mathbf{T})$ and $Q_N^{A,s}(\mathbf{T})$ tests have conservative character).

6 Concluding Remarks

We have studied testing procedures and confidence regions in general MANOVA designs without assuming homoscedasticity or a particular multivariate distribution. Inference methods were based on the ANOVA-type statistic and its standardized version, which are convenient to apply in general framework under consideration. By approximating asymptotic distributions of these statistics by scaled and standardized χ^2-distributions, the asymptotic tests were proposed and proved to be consistent under fixed alternatives. For small number of samples, these testing procedures controlled the type I error very well, in general even better than (and less time-consuming than) the parametric bootstrap Wald-type test proposed by Konietschke et al. (2015). However, for greater number of treatment groups, the asymptotic ANOVA-type tests may be slightly conservative. On the other hand, the bootstrap and permutation tests based on the standardized ANOVA-type statistic maintained the preassigned type I error rate even for greater number of samples. Neither test dominates the other in terms of its ability to find true rejections. However, in most cases of our simulation experiments, the ANOVA-type tests offered larger power than the parametric bootstrap Wald-type test. Of course, the performance of the proposed methods needs to be further evaluated on additional real and artificial data sets and other particular designs of general MANOVA framework.

References

Anderson, T.W.: An Introduction to Multivariate Statistical Analysis, 3rd edn. Wiley, Hoboken (2003)

Box, G.E.P.: Some theorems on quadratic forms applied in the study of analysis of variance problems, I. Effect of inequality of variance in the one-way classification. Ann. Math. Stat. **25**, 290–302 (1954)

Brunner, E., Dette, H., Munk, A.: Box-type approximations in nonparametric factorial designs. J. Am. Stat. Assoc. **92**, 1494–1502 (1997)

Chen, S.X., Qin, Y.L.: A two-sample test for high-dimensional data with applications to gene-set testing. Ann. Stat. **38**, 808–835 (2010)

Chung, E.Y., Romano, J.P.: Exact and asymptotically robust permutation tests. Ann. Stat. **41**, 484–507 (2013)

Duchesne, P., Francq, C.: Multivariate hypothesis testing using generalized and {2}-inverses—with applications. Statistics **49**, 475–496 (2015)

Janssen, A.: Studentized permutation tests for non-i.i.d. hypotheses and the generalized Behrens–Fisher problem. Stat. Probab. Lett. **36**, 9–21 (1997)

Konietschke, F., Bathke, A.C., Harrar, S.W., Pauly, M.: Parametric and nonparametric bootstrap methods for general MANOVA. J. Multivar. Anal. **140**, 291–301 (2015)

Lehmann, E.L., Romano, J.P.: Testing Statistical Hypotheses, 3rd edn. Springer Texts in Statistics. Springer, Berlin (2005)

Mathai, A.M., Provost, S.B.: Quadratic Forms in Random Variables. Marcel Dekker, New York (1992)

Pauly, M., Brunner, E., Konietschke, F.: Asymptotic permutation tests in general factorial designs. J. R. Stat. Soc. Ser. B Stat Methodol. **77**, 461–473 (2015)

Pauly, M., Ellenberger, D., Brunner, E.: Analysis of high-dimensional one group repeated measures designs. Statistics **49**, 1243–1261 (2015)

R Core Team.: R: A language and environment for statistical computing. In: R foundation for statistical computing. Vienna, Austria (2017). http://www.R-project.org/

Smaga, Ł.: Wald-type statistics using {2}-inverses for hypothesis testing in general factorial designs. Stat. Probab. Lett. **107**, 215–220 (2015)

Smaga, Ł.: Bootstrap methods for multivariate hypothesis testing. Commun. Stat. Simul. Comput. **46**, 7654–7667 (2017a)

Smaga, Ł.: Diagonal and unscaled Wald-type tests in general factorial designs. Electron. J. Stat. **11**, 2613–2646 (2017b)

Zhang, J. T.: Approximate and asymptotic distributions of chi-squared-type mixtures with applications. J. Am. Stat. Assoc. **100**, 273–285 (2005)

Zhang, J.T.: Analysis of Variance for Functional Data. Chapman & Hall, Boca Raton (2013)

How To Cross the River? New "Distance" Measures

Andrzej Sokolowski, Malgorzata Markowska, Sabina Denkowska, and Dominik Rozkrut

Abstract In this paper, we propose three new "distance" measures which are the adjustments of Euclidean distance. First two are given for one-dimensional space, and the third one—called tube distance—provides the generalization for multidimensional case. New measures take into account points lying between those for which we calculate the distance as well as points lying in the close neighbourhood of the ones considered. The aim of the paper is to present the idea together with some basic simulation studies under uniform and standard normal generating distributions.

1 Introduction

If you are on one side of the river and have to arrive at some point on the other side, the distance you have to cover can be calculated in different ways. You can go directly to the other side which is equivalent to Euclidean distance, or cross the river perpendicularly and follow your way on the other side to a given point, along Manhattan distance. If there are stones in the river, you can jump from one to another. First you should reposition the stones by moving them to the line connecting two points, but you can move only these points which are close to this line. If you want to jump from one stone to another, the main problem is created by the biggest distance between consecutive stones. This story lead us to the new

A. Sokolowski (✉) · S. Denkowska
Cracow University of Economics, Kraków, Poland
e-mail: sokolows@uek.krakow.pl; sabina.denkowska@uek.krakow.pl

M. Markowska
Wroclaw University of Economics, Wrocław, Poland
e-mail: Malgorzata.Markowska@ue.wroc.pl

D. Rozkrut
Central Statistical Office, Warsaw, Poland
e-mail: d.rozkrut@stat.gov.pl

© Springer Nature Singapore Pte Ltd. 2020
T. Imaizumi et al. (eds.), *Advanced Studies in Classification and Data Science*,
Studies in Classification, Data Analysis, and Knowledge Organization,
https://doi.org/10.1007/978-981-15-3311-2_20

propositions of "distance" measures. They take into account points in-between and point lying in the close outside neighbourhood. The formulas for R^1 are given and generalization for multidimensional space is proposed. The basic behaviour of proposed measures is studied through simulation analysis. A story about crossing the river is maybe not an extravaganza. Fan and Raichel (2017) explain Frechet gap distance by considering a man walking the dog along two different curves or two military units moving on two roads. Deza and Deza (2006, 2009) published two excellent reference books on distance measures. In the Preface to *Dictionary of Distances*, they wrote "The concept of distance is basic to human experience. In everyday life it usually means some degree of closeness of two physical objects or ideas, i.e., length, time interval, gap, rank difference, coolness or remoteness (...)". Our propositions rely on intuition and not on formal definition of a distance with its non-negativity, symmetry, reflexivity and triangular inequality.

The basic idea for proposing new distance measures between two points is to take into account what is going on in the space between these points and in their neighbourhood. If we consider points on the line, then two points seem to be more distant if there are no other points between them.

2   Gap Adjusted Euclidean Distance in R^1

Let us consider two points A and B, on the line. The other points lying in-between and distributed uniformly create some kind of a chain connecting A and B, and insist the impression that A and B belong to the same group, because they are somehow "connected". The biggest distance between two consecutive "inside" points creates an impression of separability.

In Fig. 1, we have $A = 1$ and $B = 7$. The Euclidean distance is 6. We propose to adjust this distance by the biggest gap between two consecutive points lying between A and B. The adjustment is in the form of geometric average:

$$AS1(A, B) = \sqrt{d(A, B) \cdot \max_{x_{(i)}, x_{(i+1)} \in \overline{AB}} d(x_{(i)}, x_{(i+1)})} \tag{1}$$

where $x_{(i)}, x_{(i+1)}$ are consecutive points.

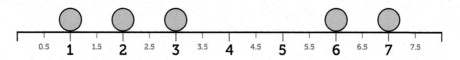

Fig. 1 Example 1

In Fig. 1, the biggest gap is between 3 and 6, so the adjusted Euclidean distance between A and B is:

$$AS1(A, B) = \sqrt{3 \cdot 6} = \sqrt{18} \cong 4.24$$

Of course, the smaller the gap, the smaller the $AS1$, because A and B are somehow "more" connected. I order to study the distribution of $AS1$ under null (homogenous distribution, no clusters) a simulation study was done, according to the following scheme:

1. generate n points from the null distribution,
2. select randomly A and B from this sample,
3. calculate $AS1$.

Each simulation went through 10,000 runs, and four sample sizes have been considered: 10 (definitely small sample), 16 (number of provinces in Poland), 28 (number of EU countries) and 1000 (definitely the large sample). Figure 2 presents sample distribution of $AS1$ under $U[0, 1]$ uniform distribution, and Fig. 3—under $N(0, 1)$ standard normal distribution.

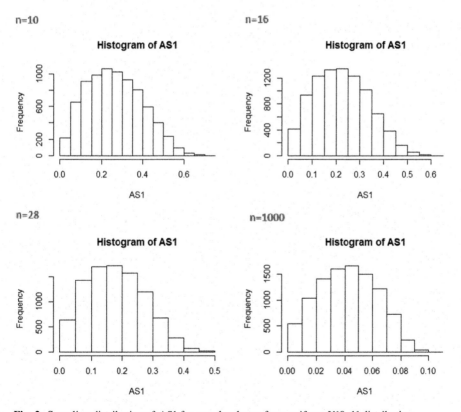

Fig. 2 Sampling distribution of $AS1$ for samples drawn from uniform $U[0, 1]$ distribution

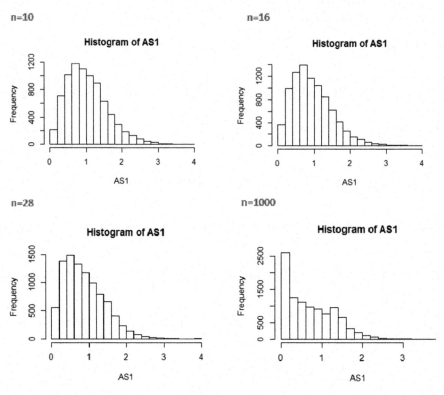

Fig. 3 Sampling distribution of $AS1$ for samples drawn from standard normal $N(0, 1)$ distribution

In Fig. 3 for samples smaller than 30, distributions look like log-normal, but the actual fit was not satisfactory.

3 Euclidean Distance in R^1 Adjusted for Gap and Outside Neighbourhood

This proposition takes into account points lying outside line segment between A and B, but only points relatively close to A and B. These outside neighbour points should be no further than r (radius) respectively from A or B.

At the moment, we considered three ways of choosing the radius:

- as a fixed value,
- as a percentage of $d(A, B)$,
- as a maximum value of minimal values from the rows of Euclidean distance matrix (of course not taking into account the main diagonal).

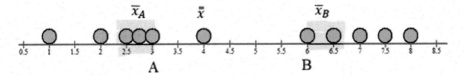

Fig. 4 Example 2

Example 2 in Fig. 4 explains the idea, and the formula for adjustment for gap and close outside neighbourhood is given as:

$$AS2(A, B)$$

$$= \sqrt{d(A, B) \cdot \max_{x_{(i)}, x_{(i+1)} \in \overline{AB}} d(x_{(i)}, x_{(i+1)}) \cdot \text{median}\{d(\bar{x}_A, \bar{x}_B), d(\bar{x}_A, \bar{\bar{x}}), d(\bar{x}_A, \bar{\bar{x}})\}}$$

(2)

$AS2$ measure is calculated as a geometric average of Euclidean distance, the biggest gap and median distance between averages of neighbourhoods of A and B and the joint average (the same idea as in median agglomerative clustering strategy).

In Example 2 (Fig. 4), we have $d(A, B) = 3, r = 0.2 \cdot d(A, B) = 0.6, gap = 2, AS1 = 2.449, \bar{x}_A = 2.625, \bar{x}_B = 6.5, \bar{\bar{x}} = 3.917, median = 2.583, AS2 = 2.493$.

Simulation studies have been performed under the same scheme as for $AS1$ measure. For uniform distribution, small samples produce distributions quite symmetric (Fig. 5). For normal distribution, we have $AS2$ distributions skewed to the right (Fig. 6).

4 Tube Distance

Measures $AS1$ and $AS2$ have been defined for one-dimensional space. How we can generalize them onto multidimensional space? The idea is to form a tube (cylinder) with r-radius and symmetry line connecting A and B points. Points lying in the cylinder are projected on this line and thus, the problem is transformed to one-dimensional space. Outside points are defined as those lying within r distance to (respectively) A and B. So $AS3$ is a measure obtained by the formula similar to (2), after projecting points in the tube on AB line.

Simulation study for tube distance has sense for at least two-dimensional classification space. First, we considered $2D$ uniform distribution with independent marginals. Since we considered only small samples, it was possible to take $r = \max_i \min_j D$, where D is distance matrix (without the main diagonal).

Distributions of $AS3$ under uniform generating scheme look symmetric (Fig. 7), like it was for one-dimensional case illustrated in Fig. 5. The same similarity—

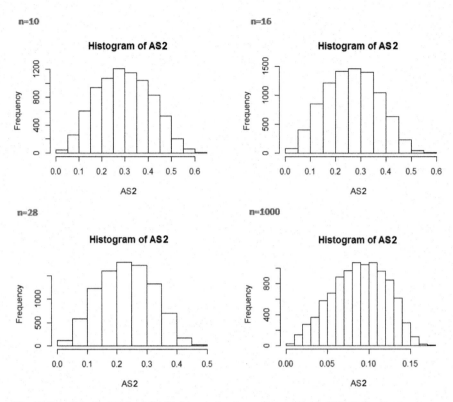

Fig. 5 Sampling distribution of $AS2$ for samples drawn from uniform $U[0, 1]$ distribution

this time with skewness—is observed for normal distribution (compare Fig. 6 with Fig. 8).

The last simulation experiment presented here considers a non-homogeneous population. A model consists of four standard normal distributions "moving" along axes in $2D$ co-ordinate system: $N1(a, 0, 1, 1, 0)$, $N2(0, a, 1, 1, 0)$, $N3(-a, 0, 1, 1, 0)$ and $N4(0, -a, 1, 1, 0)$. First, we assume that 16 points are generated, 4 from each distribution. The results are shown in Fig. 9.

If four distributions are relatively close to each other—partially overlapping ($a \leq 3$)—distributions remain unimodal. When normal distributions move further along the axes, the distributions of $A3$ become two-modal. Tube distance for points lying within one group is small, and for points from different groups—large. Of course this is also true for classical Euclidean distance, but $A3$ seems to facilitate the effect of separability and "emptiness" between points. Generally, the same is observed for 28 objects (Fig. 10).

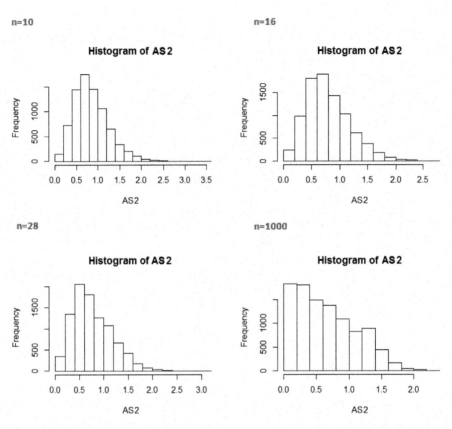

Fig. 6 Sampling distribution of $AS2$ for samples drawn from standard normal $N(0, 1)$ distribution

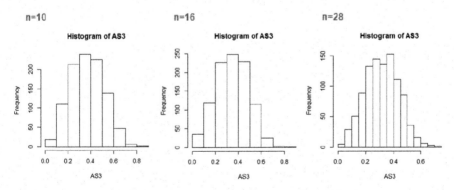

Fig. 7 Sampling distribution of $AS3$ under $2D$ uniform distribution

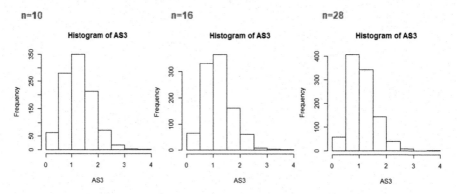

Fig. 8 Sampling distribution of $AS3$ under $2D$ normal distribution

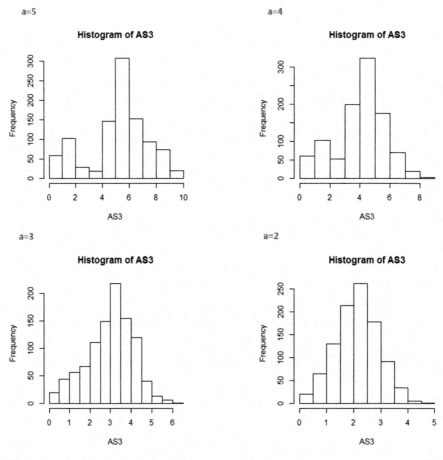

Fig. 9 Sampling distributions of tube distance for four-group experiment, $n_1 = n_2 = n_3 = n_4 = 4$

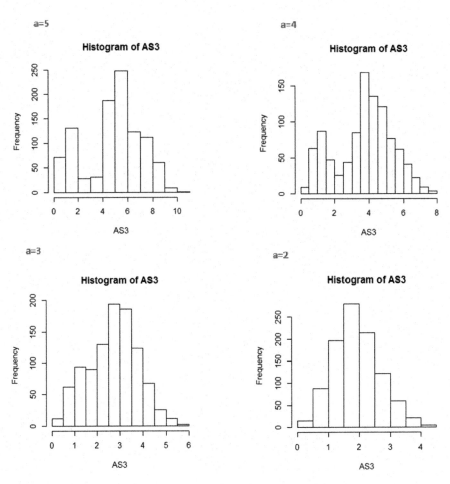

Fig. 10 Sampling distributions of tube distance for four-group experiment, $n_1 = n_2 = n_3 = n_4 = 7$

5 Closing Remarks

Two adjustments of Euclidean distance have been proposed. They take into account points lying between considered objects and in their close neighbourhood. Tube distance is the generalization of an idea onto multidimensional space. This is just the first presentation and the first simulation results. Much more has to be done in future research. The most important is to study the behaviour of proposed measures in classification and clustering tasks, as confronted to the classical distances. First partial results are promising. Further, we should compare three ways of establishing radius for neighbourhood with different simulation schemes, as well as with classical benchmark sets.

Acknowledgments The authors acknowledge support from the National Science Centre in Poland (grant no. 2015/17/B/HS4/01021) and research funds granted to the Faculty of Management at Cracow University of Economics, within the framework of the subsidy for the maintenance of research potential.

References

Deza, E., Deza, M.M.: Dictionary of Distances. Amsterdam, Elsevier (2006)

Deza, M.M., Deza, E.: Encyclopedia of Distances. Berlin, Springer (2009)

Fan, C., Raichel, B.: Computing the Frechet gap distance. In: 33rd International Symposium on Computational Geometry (SoGG 2017). Leibniz International Proceedings in Informatics (LIPIcs), vol. 77. Schloss Dagstuhl-Leibniz-Zentrum fuer Informatik, Daghtuhlpp (2017), pp. 42:1–42:16

New Statistical Matching Method Using Multinomial Logistic Regression Model

Isao Takabe and Satoshi Yamashita

Abstract Statistical matching techniques aim to build a dataset by combining different data sources. In recent years, matching techniques have been employed in various fields. However, because of some difficulties, there are only a few applications to company data. In this study, we proposed a new statistical matching methodology for company datasets by employing multinomial logistic regression model. The weighted distance was used to compute the probability of true match pairs through the model. The probability helps classify the record pairs as truly matched or not. We applied these techniques to a commercial company dataset and the official economic census microdata. The results showed that our method performs better than the nearest neighbor method used in the previous study in terms of true match rate.

1 Introduction

1.1 Background

Statistical matching techniques aim to build a dataset by combining different data sources (D'Orazio et al. 2006; Rässler 2002; Christen 2012). The most important objective of these techniques is to create useful and informative synthetic microdata without conducting any survey or collecting additional data. An accurate and efficiently linked database offers many benefits. In recent years, matching techniques have been employed in various fields including marketing, econometrics,

I. Takabe (✉)
Survey Planning Division, Statistics Bureau, Ministry of Internal Affairs and Communications, Shinjuku-ku, Tokyo, Japan
e-mail: i.takabe@soumu.go.jp

S. Yamashita
The Institute of Statistical Mathematics, Tachikawa, Tokyo, Japan
e-mail: yamasita@ism.ac.jp

© Springer Nature Singapore Pte Ltd. 2020
T. Imaizumi et al. (eds.), *Advanced Studies in Classification and Data Science*,
Studies in Classification, Data Analysis, and Knowledge Organization,
https://doi.org/10.1007/978-981-15-3311-2_21

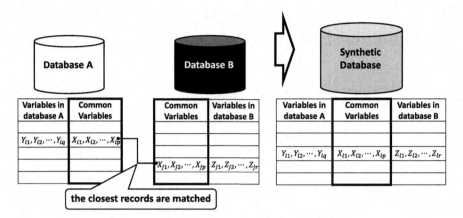

Fig. 1 Diagram of the statistical matching of two databases

epidemiology, and social sciences (Araki and Yoshizoe 2007; Herzog et al. 2007; Méray et al. 2007).

Currently, large databases of companies are needed in various fields. For example, a database is useful for a financial sector company to analyze the default probability of borrowers. A database may help a public-sector organization to discuss financial and tax regulations. In practice, however, individual databases are developed for each organization or company because of costs. In particular, a financial database generally has a large amount of financial index information, but such information is limited to borrower companies and biased. In Japan, economic census is conducted by the statistics bureau (Ministry of Internal Affairs and Communications), which has rich employment information and offers unbiased data of all companies, but does not have sufficient financial information (Fig. 1).

By combining these data types, a useful unbiased informative database can be obtained without additional costs for data collection. However, owing to difficulties in dealing with big datasets and skewed variables, there are few applications that utilize company data (Lie 2001; Kurihara 2015). Sometimes we cannot use detailed information of companies (name, location, etc.) or an organization's database because of confidentiality.

1.2 Statistical Matching Methods

There are many statistical matching techniques. One of these methods uses U-statistics (the probability of matching two different records by chance) and M-statistics (the probability that equal pairs match), and judges propriety of matching based on a threshold (Newcombe et al. 1959; Fellegi and Sunter 1969; Harron et al. 2015). The method usually uses many variables including the company's name and location information for matching. However, in many cases, it is considered that confidential information cannot be used and only few pieces of information can be used; therefore, these methods are not considered to be appropriate.

Another method assumes a multivariate normal distribution in the structure of enterprise data and estimating variables, which is not common in two databases as there are missing values in the framework of multiple regression models and Bayesian modeling (D'Orazio et al. 2006; Rässler 2002). However, company data usually has very skewed variables (employment, turnover, etc.) and categorical variables, which makes it inappropriate to assume a multivariate normal distribution.

The distance hot deck method has been used in early research and is still widely used in practice. The method often uses the following weighted distance D_{ij}, and the smallest distance records are matched:

$$D_{ij} = \sum_{k=1}^{K} \beta_k |X_{ik} - X_{jk}|$$

where D_{ij} is the distance between record i of one dataset A and record j of another dataset B. X_{ik} is the field k in record i and j. β_k is the weight of the common field k in both datasets.

Regarding information necessary for discriminating companies like capital amount, it can be considered that the matching precision is improved by simply using its difference (distance). The inverse of the standard deviation or range of variables are often used as the weights of distance (D'Orazio et al. 2006). In the presence of a set of mixed modes (including both continuous and categorical variables), the matching procedure has difficulty in determining the weight of each variable.

Next, we will discuss the statistical matching method based on the weighted distance function. The method works well without any distribution assumption, and it also works well in the situation that there are only a few common variables in two datasets.

1.3 Proposed Method

In this paper, a new statistical matching method using multinomial logistic model (McCullagh and Nelder 1989; Hosmer et al. 2013) combined with the weighted distance function is introduced. The weighted distance between records in two datasets was used to build a multinomial logistic model. This method has the following merits.

- The assumption of distributions is not needed.
- The methodology makes it possible to determine weights of distances statistically, and to evaluate matching probabilities.

- The model can deal with continuous and categorical variables equally in a multinomial logistic model framework.
- The model can compute the matching probabilities such that the record pairs can be classified as truly matched or not.
- The t-value makes it possible to analyze the estimation accuracy of the weights.
- The pseudo-R-squared value obtained by the model can be used to compare different model fitting techniques.

2 Model

In this section, details of the statistical matching method are explained using multinomial logistic regression model framework. Suppose that there are two datasets, say datasets A and B, having M and N records, respectively. The probability P_{ij}, which is defined as the likelihood that company i in dataset A matches company j in dataset B, is expressed as the following formula using weighted distance D_{ij} where the weights of the distance $\beta = (\beta_1, \beta_2, \ldots, \beta_p)$ are treated as the parameters in the multinomial logistic regression model.

$$P_{ij} = \frac{exp(-D_{ij})}{\sum_{j=1}^{N} exp(-D_{ij})}$$

The maximum likelihood method was used for estimating these parameters (weights). The following $L(\beta)$ is the likelihood function of the model:

$$L(\beta) = \prod_{i=1}^{M} \prod_{j=1}^{N} P_{ij}^{\delta_{ij}}$$

where $\delta_{ij} = \begin{cases} 1 & \text{when record } i \text{ (in dataset } A) = \text{record } j \text{ (in dataset } B) \\ 0 & \text{otherwise} \end{cases}$

As explained in the Sect. 3 later, training data and test data were selected from exactly matched data in which each δ_{ij} was known. All the weights in the model were estimated based on the training data, and model performances were compared and verified by using the test data.

Hence, the log-likelihood function $l(\beta)$ is as follows:

$$l(\beta) = \sum_{i=1}^{M} \sum_{j=1}^{N} \delta_{ij} \ln \left(\frac{\exp\left(-\sum_{k=1}^{K} \beta_k |X_{ik} - X_{jk}|\right)}{\sum_{j=1}^{N} \exp\left(-\sum_{k=1}^{K} \beta_k |X_{ik} - X_{jk}|\right)} \right)$$

The maximum likelihood estimators of the weights $\hat{\beta} = (\hat{\beta}_1, \hat{\beta}_2, \ldots, \hat{\beta}_p)$ were obtained by maximizing the above log-likelihood function. We used the software R and its optim function (quasi-Newton method by BFGS algorithm) to calculate the numerical optimization for maximum likelihood estimation.

The t-value t_k of the coefficient β_k is as follows (Hosmer et al. 2013):

$$t_k = \frac{\hat{\beta}_k}{\sqrt{[\nabla^2 l(\hat{\boldsymbol{\beta}})]_{kk}^{-1}}}$$

where, $[\nabla^2 l(\hat{\boldsymbol{\beta}})]_{kk} = \partial^2 l(\hat{\boldsymbol{\beta}})/\partial \beta_k{}^2$ is the diagonal elements of the Hessian matrix which is the output of optim function (argument hessian = TRUE).

After determining the model parameters (weights) $\hat{\boldsymbol{\beta}}$, the matching probability of each record pair P_{ij} was estimated. Using these probabilities, for each record in dataset A, the record with the highest estimated matching probability in dataset B was matched. Figure 2 illustrates this procedure.

3 Applications

3.1 Data

We applied the proposed method to a commercial firm dataset and the official economic census microdata. The dataset A is the Teikoku Databank data (TDB data) of the Japanese commercial database company. It is maintained by the interview survey. The dataset B is the Economic Census microdata. The economic census is the official survey conducted by the statistics bureau. In this study, we used the data in an area. Approximately 5000 records in dataset A and approximately 13,000 records dataset B were used.

These datasets were exactly matched by using name and location information in advance. Next, these key information were eliminated. Then, the training and test datasets were selected from the datasets. We used two-thirds of the samples as training dataset. The remaining data were used for the evaluation of the model performance as the test dataset.

We used both continuous and categorical variables. The continuous variables are as follows:

- number of employees;
- capital amount; and
- turnover.

The number of employees of the TDB data sometimes includes part-timers. The economic census has two types of number of employees (with and without part-timers); we calculated two patterns of distances for the number of employees (including part-timers or not) and used its minimum value as the distance of the number of employees.

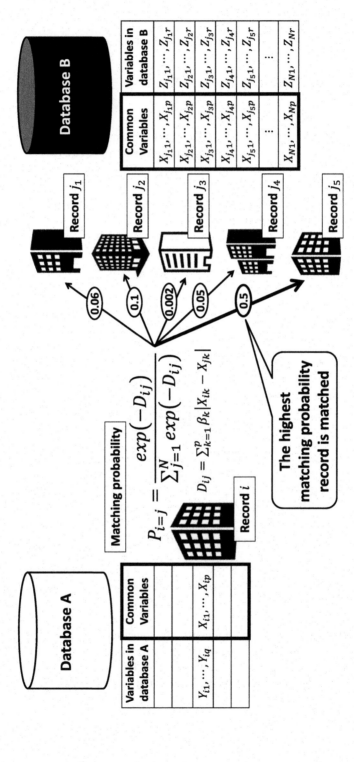

Fig. 2 Diagram of the statistical matching procedure with matching probabilities P_{ij} estimated by the multinomial regression model framework

We used the following categorical variables. If two companies in a record pair belong to the same category, then the distance $= 0$, otherwise the distance $= 1$.

- Year of establishments: four categories (1984, 1985–1994, 1995–2004, 2005)
- Industry: Japan Standard Industrial Classification, Major groups
- Region code (County level).

There is a difference in the industry definition in both databases; therefore, we reorganized the Industry of the TDB data to adapt to the major group of Japan Standard Industry Classification adopted by the Economic Census.

Each dataset contains missing values; therefore, we imputed these values using imputation by chained equations (ICE) (Van Buuren 2012) before the estimation of the models. The ICE imputation procedure was implemented using R package `mice`. The continuous and categorical values were imputed by using the multinomial logistic model and the predictive mean matching method, respectively.

3.2 Comparison Methods

We compared the proposed models (with the following distances) with the nearest neighbor matching as the benchmark method.

1. Weighted absolute distance: $D_{ij} = \sum_{k=1}^{K} \beta_k |X_{ik} - X_{jk}|$
2. Weighted absolute distance (log): $D_{ij} = \sum_{k=1}^{K} \beta_k \log(|X_{ik} - X_{jk}| + 1)$
3. Weighted Euclid distance (squared): $D_{ij} = \sum_{k=1}^{K} \beta_k (X_{ik} - X_{jk})^2$
4. Weighted Canberra distance: $D_{ij} = \sum_{k=1}^{K} \beta_k |X_{ik} - X_{jk}|/(X_{ik} + X_{jk} + 1)$
5. Nearest neighbor method

Nearest neighbor method searches the smallest distance record for matching. Nearest neighbor method, which was often used in previous studies, used fixed weights defined as the inverse of the standard deviation of each variable and did not estimate these weights by data. On the other hand, our proposed methods estimated optimal weights of weighted distances by data using maximum likelihood method. We measured performances of proposed methods by using the nearest neighbor method as a benchmark.

Sometimes the non-linear optimization calculation failed in the weighted Euclid distance; therefore, each continuous variable was divided by 100 to stabilize the estimation procedure.

3.3 Matching Performance Evaluation

Table 1 presents the estimation results. The coefficients of the model were indicated with their standard errors. To see the model fitting, the pseudo-R-square ρ^2 was

Table 1 Estimate results of each model: coefficient, t-value, and pseudo-R-square

	WAD		WAD(log)	
Variables	Coeff	t-value	Coeff	t-value
Employees	0.1868	-25.95^a	1.0596	-34.68^a
Capital amounts	0.0062	-46.79^a	0.8598	-58.54^a
Turnover	0.0089	-36.09^a	0.9384	-66.19^a
Industry	3.5368	-66.37^a	3.4545	-62.87^a
Establishment year	1.5943	-40.89^a	1.5758	-37.53^a
Region code	11.2817	-15.04^a	8.9814	-23.62^a
Initial log-likelihood	-37934			-37934
log-likelihood	-13238		-10041	
Pseudo R-sq	0.6510		0.7353	
	WED		WCD	
Variables	Coeff	t-value	Coeff	t-value
Employees	4.9156	-10.55^a	3.6371	-33.13^a
Capital amounts	0.0126	-32.07^a	14.605	-45.01^a
Turnover	0.0037	-9.32^a	5.5528	-55.31^a
Industry	1.6580	-46.15^a	3.4407	-64.91^a
Establishment year	1.5943	-40.89^a	1.5434	-39.11^a
Region code	9.2648	-22.41^a	8.9672	-23.69^a
Initial log-likelihood	-37934		-37934	
log-likelihood	-18729		-11277	
Pseudo R-sq	0.5063		0.7027	

Significance level [a]0.1%
WAD weighted absolute distance, *WAD(log)* weighted absolute distance (log-transformed), *WED* weighted Euclid distance, *WCD* weighted Canberra distance

also indicated. The initial log-likelihood was defined as the log-likelihood with all parameters set to 0.

$$\rho^2 = 1 - \frac{l(\hat{\boldsymbol{\beta}})}{l(\mathbf{0})}$$

According to Table. 1, all coefficients were significant at 0.1% significant level. In particular, the coefficients of Industry and region code had higher values than that of other variables since these categorical variables have a strong influence on the matching probabilities estimated by the model. According to the pseudo-R-squared value, the log-transformed weighted absolute distance achieved the best fitting for the data.

We evaluated the model performance from the viewpoint of the true match rate. Based on the method explained in Yoshikawa et al. (2015), we evaluated the model performances using the probabilities that the true record is included in the top R candidates for matching pairs. The details of the calculation are as follows. First, let $t(i)$ be the index of the true record in dataset B corresponding to the record i in dataset A. Next, for each company i in dataset A, let $C(i, R)$ be a set of records of the top R cases in the order of the highest matching probability among the companies of dataset B. The precision rate $P(R)$ when the correct record is included in the top R candidates is expressed as follows:

$$P(R) = \frac{1}{M_{test}} \sum_{i=1}^{M_{test}} I(t(i) \in C(i, R))$$

where M_{test} is the size of the test data in datasets A. $I(\cdot)$ is the indicator function, which takes the value 1 if the inside of $(\cdot)$ is true, otherwise it takes the value 0. Figure 3 shows the relationship between $P(R)$ and R. The proposed method, with any distance function, showed a higher precision value $P(R)$ than that of the nearest neighbor method. The log-transformed absolute distance model is the best model.

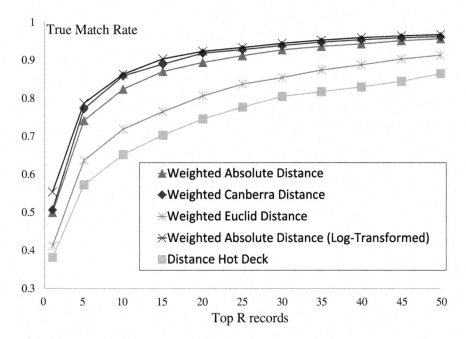

Fig. 3 True match rate of each statistical matching method

4 Conclusion

We proposed a new statistical matching methodology for company datasets using the multinomial logistic regression framework. The methodology makes it possible to determine the distance weights statistically, and to calculate the matching probabilities. The proposed method does not assume any distribution for the data, can evaluate the coefficient by the t-value, and can compare the models by the pseudo-R-square value. The model can be applied to data having few common variables. The model performed better than the nearest neighbor method in terms of true match precision rates. Once the model parameters have been estimated, they can be applied to the two other similar datasets to calculate the distance and matching probabilities.

Working with large datasets entails a considerable amount of time to calculate the distances of all possible pairs. To address the problem, we plan to utilize the principal component analysis method, which divides the datasets and creates strata to shrink the searching space of the record pairs. The method is considered to be useful for searching for a true match pair more efficiently.

Acknowledgement This work was supported by JSPS KAKENHI Grant Numbers JP16H02013 and 15H03390.

References

Araki, M., Yoshizoe, Y.: Exact matching and statistical matching using household survey data. Aoyama J. Bus. **42**(1), 175–209 (2007)

Christen, P.: Data Matching: Concepts and Techniques for Record Linkage, Entity Resolution, and Duplicate Detection. Springer, Berlin (2012)

Fellegi, I.P., Sunter, A.B.: A theory for record linkage. J. Am. Stat. Assoc. **64**(328), 1183–1210 (1969)

Harron, K., Goldstein, H., Dibben, C.: Methodological Developments in Data Linkage. Wiley, London (2015)

Herzog, T.N., Scheuren, F.J., Winkler, W.E.: Data Quality and Record Linkage Techniques. Springer, Berlin (2007)

Hosmer, Jr, D.W., Lemeshow, S., Sturdivant, R.X.: Applied Logistic Regression, 3rd edn. Wiley, London (2013)

Kurihara, Y.: Estimation of durability of profit of small and medium enterprises by statistical matching. J. Math. Syst. Sci. **5**, 173–182 (2015)

Lie, E.: Detecting abnormal operating performance: Revisited. Financ. Manag. **30**(2), 77–91 (2001)

Méray, N., Reitsma, J.B., Ravelli, A.C., Bonsel, G.J.: Probabilistic record linkage is a valid and transparent tool to combine databases without a patient identification number. J. Clin. Epidemiol. **60**(9), 883–891 (2007)

McCullagh, P., Nelder, J.A.: Generalized Linear Models, 2nd edn. CRC Press, Boca Raton (1989)

Newcombe, H.B., Kennedy, J.M., Axford, S.J., James, A.P.: Automatic Linkage of vital records. Science **130**, 954–959 (1959)

D'Orazio, M., Di Zio, M., Scanu, M.: Statistical Matching: Theory and Practice. Wiley, London (2006)

Rässler, S.: Statistical Matching. Springer, Berlin (2002)

Van Buuren, S.: Flexible Imputation of Missing Data. CRC Press, Boca Raton (2012)

Yoshikawa, Y., Iwata, T., Sawada, H., Yamada, T.E.: Cross-domain matching for bag-of-words data via kernel embeddings of latent distributions. In: Advances in Neural Information Processing Systems, pp. 1405–1413 (2015)

Constructing Graphical Models for Multi-Source Data: Sparse Network and Component Analysis

Pia Tio, Lourens Waldorp, and Katrijn Van Deun

Abstract Gaussian graphical models (GGMs) are a popular method for analysing complex data by modelling the unique relationships between variables. Recently, a shift in interest has taken place from investigating relationships within a (sub)discipline (e.g. genetics) to estimating relationships between variables from various subdisciplines (e.g. how gene expression relates to cognitive performance). It is thus not surprising that there is an increasing need for analysing large, so-called *multi-source* datasets, each containing detailed information from many data sources on the same individuals. GGMs are a straightforward statistical candidate for estimating *unique cross-source relationships* from such network-oriented data. However, the multi-source nature of the data poses two challenges: First, different sources may inherently differ from one another, biasing the estimation of the relations. Second, GGMs are not cut out for separating cross-source relationships from all other, source-specific relationships. In this paper we propose the addition of a simultaneous-component-model pre-processing step to the Gaussian graphical model, the combination of which is suitable for estimating cross-source relationships from multi-source data. Compared to the graphical lasso (a commonly used GGM technique), this Sparse Network and Component (SNAC) model more accurately estimates the unique cross-source relationships from multi-source data. This holds in particular when the multi-source data contains more variables than observations ($p > n$). Neither differences in

---

P. Tio (✉)
Department of Psychological Methods, University of Amsterdam, Amsterdam, The Netherlands

Department of Methodology, Tilburg University, Tilburg, The Netherlands
e-mail: p.tio@uva.nl

L. Waldorp
Department of Psychological Methods, University of Amsterdam, Amsterdam, The Netherlands
e-mail: L.J.Waldorp@uva.nl

K. Van Deun
Department of Methodology, Tilburg University, Tilburg, The Netherlands
e-mail: k.vandeun@uvt.nl

© Springer Nature Singapore Pte Ltd. 2020
T. Imaizumi et al. (eds.), *Advanced Studies in Classification and Data Science*,
Studies in Classification, Data Analysis, and Knowledge Organization,
https://doi.org/10.1007/978-981-15-3311-2_22

sparseness of the underlying component structure of the data nor in the relative dominance of the cross-source compared to source-specific relationships strongly affect the relationship estimates. Sparse Network and Component analysis, a hybrid component-graphical model, is a promising tool for modelling unique relationships between different data sources, thus providing insight into how various disciplines are connected to one another.

1 Background

Gaussian graphical models (GGMs) are a popular method for analysing complex data by modelling the unique relationships between variables (Koller and Friedman 2009). Both the estimates and their visualisation as a network provide valuable insights in the underlying structure of the data. However, understanding the pathways from genotype or physiology to phenotype or behaviour requires more than estimating GGMs of each individual discipline; one also needs to know how these various isolated fields are connected to one another via *cross-source relationships* between variables from different data sources. These findings are intrinsically valuable on their own. However, they do not paint the full picture of genetic-cognitive interaction. Being able to identify how information from additional sources such as (functional) brain data fits together with gene expression and cognitive functioning broadens our understanding of human functioning, opens new treatment venues, and increases prediction accuracy for those at risk for developing pathologies.

It is thus not surprising that emerging fields such as systems biology and network science emphasise the need for collecting and analysing large, so-called *multi-source* datasets, each containing detailed information from many data sources on the same individuals (Silverman and Loscalzo 2012; Bartel et al. 2013). Luckily, with increasingly more sophisticated instruments and growing interdisciplinary cooperation, multi-source datasets become more common. However, availability of multi-source data alone is not enough to answer questions about cross-source relationships; appropriate statistical tools are needed too.

GGMs are a straightforward statistical candidate for analysing such network-oriented data. However, the multi-source nature of our data poses a challenge. In non-multi-source data, unique linear relationships can usually be estimated straightforwardly using partial correlations. However, if the data come from multiple sources, it is possible that groups of variables have different characteristics. For example, one group of variables may contain more noise than another group because of different measuring techniques or lower granularity; or variables within a group may be highly correlated to one another (e.g., positive correlations amongst cognitive variables) compared to other groups of variables (e.g., genetic information). Disregarding such inherent differences between sources can lead to incorrect relationship estimates. Furthermore, it could also be the case that the cross-source relationships that we are interested in are weaker than the source-specific relationships. This problem is exacerbated by the fact that only few variables are relevant for the cross-source mechanism. GGMs, in particular those adapted to deal with high-dimensional data ($p > n$), are well suited for estimating the strongest relationships amongst variables and thus less cut out for separating weaker cross-source relationships from all other, stronger source-specific relationships. Given that we already know that we want to estimate cross-source relationships of data from multiple sources, it makes sense to use statistical analyses that can disentangle relevant cross-source information from irrelevant source-specific information while incorporating the multi-source data structure.

In this paper we propose determining the set of variables from different sources that are most likely to be connected; the GGM focusing on this subset will more accurately reflect unique cross-source relationships than without such a variable selection. We use a variant of sparse simultaneous component analysis to assess the subset of cross-source variables that are connected. In Sect. 2 we will describe GGM and the multi-source data structure in more detail, followed by introducing sparse simultaneous component analysis as a pre-processing step that enables GGMs to estimate cross-source relationships from multi-source data. Section 3 reports a simulation study investigating whether this Sparse Network And Component (SNAC) model outperforms regular GGMs. Lastly, in Sect. 4 we discuss current restrictions and possible improvements of SNAC.

2 Methods

First we present the graphical model as it appears in the literature, this is for data coming from a single source; second, we introduce the assumed data-generating model for multi-source data and show how to isolate the cross-source relations; and third, we show how to apply the graphical model (for single source data) to these cross-source relations in order to obtain the desired network.

Gaussian Graphical Models Graphical models (GMs) estimate conditional dependency relationships between variables using probability theory. Conditional dependence indicates that none of the remaining variables can explain away the relation between two variables. As such conditional dependence relations can be said to be *unique* to the pair of variables. A GM can be visualised as a graph whose nodes and edges represent variables and conditional dependency relationships, respectively. Many types of graphical models have been formulated; here we focus on Gaussian Graphical models (GGMs), which model undirected unique relationships in a multivariate Gaussian setting (Koller and Friedman 2009). Let $\mu = 0$ be a p-dimensional zero mean vector and Σ be a $p \times p$ positive definite covariance matrix. For a p-dimensional vector x, the multivariate Gaussian density is defined as $f(x) = (2\pi)^{-p/2}|\Sigma|^{-1/2}\exp\left[-\frac{1}{2}x^T\Sigma^{-1}x\right]$, with $|\Sigma|$ the determinant of Σ. Note that this equation is often expressed as $x \sim \mathcal{N}_p(\mu, \Sigma)$.

Under the assumption of normality a zero off-diagonal element is equivalent to the two corresponding variables being conditionally independent given all remaining variables, and a non-zero entry means conditional dependence (Lauritzen 1996; Koller and Friedman 2009). Gaussian graphical models thus provide a mathematical and visual representation of unique dependency relationships between variables that is straightforward to interpret. These dependency relations can be estimated using maximum likelihood. In multi-source data, it is very likely that there will be fewer observations than variables ($n < p$). A popular solution to deal with this situation is applying sparse modelling through an ℓ_1 penalty. This Least Absolute Shrinkage and Selection Operator (lasso) results in shrinkage of the parameters to zero with small values set exactly to zero (Tibshirani 1996). The underlying assumption is that only a small number of all possible parameters is non-zero, i.e., there is a sparse solution including only the *relevant* relations and variables. One of the algorithms used to obtain such a lasso estimate of $\boldsymbol{\Sigma}^{-1}$ is the graphical lasso (Friedman et al. 2008). Let $\hat{\boldsymbol{\Sigma}} = \frac{1}{n} \sum_{i=1}^{n} \boldsymbol{x}_i \boldsymbol{x}_i^T$ be the empirical covariance matrix; the graphical lasso optimises the function $\log |\boldsymbol{\Sigma}^{-1}| - \mathrm{tr}\boldsymbol{\Sigma}^{-1}\hat{\boldsymbol{\Sigma}} - \lambda \sum_{k \neq j}^{p} |(\boldsymbol{\Sigma}^{-1})_{jk}|$ where $|(\boldsymbol{\Sigma}^{-1})_{jk}|$ is the absolute value of the jkth ($j, k = 1, \ldots, p$) entry of $\boldsymbol{\Sigma}^{-1}$ and $\lambda \geq 0$ is a tuning parameter for the lasso penalty. Note that we do not use the diagonal elements in the penalty as this implies shrinking the variance (Bühlmann and Van De Geer 2011).

Multi-source Data Of interest to this paper are multi-source data and, in particular, the cross-source relationships existing between the variables of different sources. We assume that there are several sources of structural variation that give rise to the interconnections within and possibly between different sources. Together, these sources form a low-rank representation of the correlation matrix as used in factor and principal component models where the correlations can be reproduced on the basis of the loadings of the variables on the factors (or, latent variables). Let $\boldsymbol{\Lambda}$ be the matrix of loadings of the p variables on the R components; then, the factor analytic model assumes the following data-generating mechanism for the observed data $\boldsymbol{x} \sim \mathcal{N}(0, \boldsymbol{\Lambda}\boldsymbol{\Lambda}^T + \mathrm{diag}([\sigma_1^2, \ldots, \sigma_j^2, \ldots, \sigma_p^2])$ (1), with σ_j^2 the residual variance of the jth variable and $\boldsymbol{\Lambda}\boldsymbol{\Lambda}^T$ the covariance (correlation) matrix as reproduced by the factor analytic model. Here, to account for the multi-source structure $p = \sum_k p_k$ for $k = 1, \ldots, K$ sources with p_k the number of variables in source k; and also $\boldsymbol{x} = [\boldsymbol{x}_1^T \ldots \boldsymbol{x}_K^T]^T$ so $\boldsymbol{\Lambda} = [\boldsymbol{\Lambda}_1^T \ldots \boldsymbol{\Lambda}_K^T]^T$. Under this model the covariance matrix $\boldsymbol{\Sigma} = \boldsymbol{\Lambda}\boldsymbol{\Lambda}^T + \mathrm{diag}([\sigma_1^2, \ldots, \sigma_j^2, \ldots, \sigma_p^2]$ and, making use of the Woodbury formula, its inverse is $\boldsymbol{\Sigma}^{-1} = \boldsymbol{D} - \boldsymbol{D}\boldsymbol{\Lambda} \left(\boldsymbol{I} - \boldsymbol{\Lambda}^T \boldsymbol{D}\boldsymbol{\Lambda} \right)^{-1} \boldsymbol{\Lambda}^T \boldsymbol{D}$ (2) with $\boldsymbol{D} = \mathrm{diag}([\sigma_1^{-2}, \ldots, \sigma_j^{-2}, \ldots, \sigma_p^{-2}]$ and $\boldsymbol{I}$ the $R \times R$ identity matrix.

What we aim for is to determine the connection strengths of the bridge nodes (variables) that connect between different sources. Application of the GGM to the observed data will not reach this aim. The reason for this is twofold. First, connection strengths between variables of different sources are low because of the different nature of the data between different sources. Under sparseness restrictions, such as a lasso penalty, such weak correlations will be set to zero. Second, several sources of structural variation may underlie a variable whereby it may be involved both in cross-source and within-source relations. Let Λ_C represent the factors associated with the cross-source structural variation and Λ_S the source-specific variation, then $\Lambda = [\Lambda_C|\Lambda_S]$ and $\Lambda\Lambda^T = [\Lambda_C|\Lambda_S][\Lambda_C|\Lambda_S]^T = \Lambda_C\Lambda_C^T + \Lambda_S\Lambda_S^T$ (3), showing that the covariances consist of both cross-source and source-specific contributions.

To illustrate this, think of a hypothetical dataset containing data from three sources: four cognitive, four genetic, and three cardiovascular variables (Fig. 1a). Here we not only need to a) differentiate irrelevant (orange nodes) from relevant variables and b) identify which of the relevant variables form cross-source relationships (nodes within at least two circles), but additionally c) separate source-specific (S1 and S2) from shared (or common; C) variation. In the next paragraph we show how the cross-source relations can be singled out.

Isolating the Cross-Source Relations To single out the cross-source relations, we need to disentangle the common sources of variation shared between the different data blocks (with each block containing the data or variables of one source) from the sources of variation that are specific for a single or a few data blocks only. To do this, we perform a so-called sparse DIStinctive and COmmon Simultaneous Component Analysis decomposition of the data (sparse DISCO SCA; see Gu and Van Deun 2016, 2019), a method that was developed for the integrated analysis of multi-source data with the specific aim of separating block-specific sources of variation from common sources of variation. Sparse DISCO SCA models common and specific components by using specific constraints on the loadings of each of the components in Λ. Next, we explain how this is done.

First, note that the loading of a variable on the component reflects the correlation of that variable with the component and this property is used to define common and specific components: A **common component** is associated with all data blocks and thus *each of the data blocks should have some variables with non-zero loadings* on this component; on the other hand, a **source-specific component** has no association at all with one (some) of the blocks and, consequently, for that (these) blocks *all variables of this block (these blocks)* should have a zero loading on this component. Note that sparse components are assumed, meaning that only a few relevant variables make up the component and thus also for the common component zero loadings do show up as well as in the non-zero part of the specific components.

To obtain such constrained structures, sparse DISCO SCA makes use of a combination of penalties and/or hard constraints. Sparseness of the common component and of the non-zero part of the specific components is imposed by a lasso penalty on the loadings. The blocks of zero loadings of the specific components can be obtained in two ways: either by using a constrained approach or by using a group lasso penalty (Van Deun et al. 2011). The former approach requires prior knowledge of the structure of the components. In absence of such prior knowledge, when the number of blocks and components is small—which is often the case in empirical applications—an exhaustive strategy can be used that compares all possible combinations of common and specific components. We refer to Schouteden

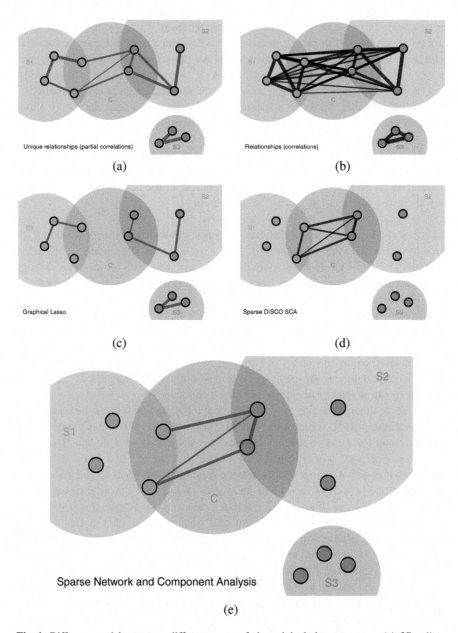

Fig. 1 Different models capture different parts of the original data structure. (**a**) Visualises the unique relationships between 4 cognitive, 4 genetic, and 3 cardiovascular variables. Pink circle contains variables involved in cross-source relationships. (**b**) The corresponding correlation structure. (**c**) Graphical lasso recovers a sparse solution of the unique relationships structure which does not contain the unique cross-source relationships. (**d**) Sparse DISCO SCA disentangles source-specific (S) from common (C) sources of variation, retaining only cross-source correlations. (**e**) SNAC combines the strengths of graphical lasso and sparse DISCO SCA and can therefore estimate the unique cross-source relationships

et al. (2014) and Gu and Van Deun (2019) for further discussion of this model selection issue.

Once the data have been modelled with common and specific components through a sparse DISCO SCA analysis, the cross-source relationships can be obtained from the common components as follows: Let $\hat{\boldsymbol{\Lambda}}_C$ denote the estimated loadings associated with the common components, then $\hat{\boldsymbol{\Sigma}}_C = \hat{\boldsymbol{\Lambda}}_C \hat{\boldsymbol{\Lambda}}_C^T$ reflects the cross-source correlation. For computational efficiency, only the variables containing at least one non-zero loading on a common component should be included.

SNAC: Sparse Network and Component Model When estimating unique cross-source relationships from multi-source data, a statistical procedure is needed that (a) combines data from different sources where each source may have different characteristics, (b) selects variables to determine which variables are involved in cross-source relationships, (c) estimates these unique cross-source relationships, and (d) presents results that can be interpreted in a meaningful, substantive way. Sparse DISCO SCA covers both (a) and (b) and graphical models cover (c) and (d). Therefore, when taken together both models address all the points necessary to provide effective estimation.

We therefore introduce Sparse Network And Component model (SNAC; Fig. 1e), a two-step component-graphical model for estimating cross-source relationships from multi-source, multivariate Gaussian distributed data. First, sparse DISCO SCA is used to reveal the underlying common and source-specific sources of variation; as discussed in the previous section this allows us to single out the common source of variation by calculating $\hat{\boldsymbol{\Lambda}}_C \hat{\boldsymbol{\Lambda}}_C^T$ which is the matrix containing the non-unique cross-source relations (see Fig. 1d). However, our interest is in the unique cross-source relations. A straightforward way to obtain these may seem to calculate the inverse of $\hat{\boldsymbol{\Lambda}}_C \hat{\boldsymbol{\Lambda}}_C^T$. Yet, $\hat{\boldsymbol{\Lambda}}_C$ is a low-rank matrix of rank R_C, the number of common components. Furthermore, generalised inverses such as the Moore–Penrose inverse and the regularised estimation of inverse covariance matrices (shrinkage estimator; Schäfer et al. 2005)—although able to deal with the singularity incurred by the low rank—also do not give the desired result. This is because the pre-processing of the data in the sparse DISCO SCA removes the information in the data on the residual variances σ_j^2; see the data-generating model for multi-source data given by expression (1). Sparse DISCO SCA models the covariance matrix as $\boldsymbol{\Lambda}\boldsymbol{\Lambda}^T$ with inverse $(\boldsymbol{\Lambda}\boldsymbol{\Lambda}^T)^{-1}$ while the inverse of the population covariance matrix is given by Eq. (2). Considering the decomposition into common and source-specific components $\boldsymbol{\Lambda} = [\boldsymbol{\Lambda}_C | \boldsymbol{\Lambda}_S]$, we find for the inverse $(\boldsymbol{\Lambda}\boldsymbol{\Lambda}^T)^{-1} = \boldsymbol{B}^+ - \boldsymbol{B}^+ \boldsymbol{\Lambda}_C (\boldsymbol{I} - \boldsymbol{\Lambda}_C^T (\boldsymbol{\Lambda}_S \boldsymbol{\Lambda}_S^T)^{-1} \boldsymbol{\Lambda}_C)^{-1} \boldsymbol{\Lambda}_C^T \boldsymbol{B}^+$ where $\boldsymbol{B} = \boldsymbol{\Lambda}_S \boldsymbol{\Lambda}_S^T$ and $^+$ denote the Moore–Penrose inverse. We can see that the source-specific components incorrectly scale the common components.

Let us now inspect the generalised inverses of this covariance matrix a bit closer. First, to study the Moore–Penrose inverse, we consider the eigenvalue decomposition: $\boldsymbol{\Lambda}\boldsymbol{\Lambda}^T = \boldsymbol{V} \boldsymbol{S}^2 \boldsymbol{V}^T$ with $\boldsymbol{V}$ containing the R eigenvectors and $\boldsymbol{S}^2$ a diagonal matrix containing the eigenvalues. The Moore–Penrose inverse of $\boldsymbol{\Lambda}\boldsymbol{\Lambda}^T$

is then equal to $VS^{-2}V^T$ while for the population covariance matrix, replacing Λ by VS in Eq. (2), is given by $\Sigma^{-1} = D - DVS(I - SV^T DVS)^{-1} SV^T D$. This implies that the Moore–Penrose inverse is not suitable to estimate the direct cross-source relations. The regularised estimation of inverse covariance matrices as presented by Schäfer et al. (2005) is not suitable either as it estimates the variances on the basis of the given covariance matrix and these are biased downwards as a result of the sparse DISCO SCA step. The graphical lasso, on the other hand, inflates the given variances by fixing the variances as follows in the first step of the iterative estimation procedure used to estimate the off-diagonal elements of the inverse covariance matrix $\hat{\Sigma} = \hat{\Lambda}\hat{\Lambda}^T + \rho I$, with $\rho \geq 0$, see page 3 in Friedman et al. (2008). Note that this expression is very similar to the expression of the population covariance matrix (1). The inverse of this covariance matrix closely resembles the inverse population covariance matrix $\hat{\Sigma}^{-1} = (\rho I) - (\rho I)VS(I - SV^T(\rho I)VS)^{-1} SV^T(\rho I)$, showing that the GGM will correctly estimate the direct relations when $\sigma_j^2 = \rho$ for all j. Hence, the way to go in constructing networks for cross-source relations is to combine a sparse DISCO SCA analysis with a Gaussian Graphical model.

3 Simulation Study

In this section, we describe the simulation study designed to compare the performance of graphical lasso, a commonly used GGM technique, and SNAC in estimating the unique cross-source relationships amongst many variables from two multivariate normal sources. Additionally we demonstrate that a different covariance-inversion technique such as Moore–Penrose or shrinkage estimator inaccurately estimates unique cross-source relationships, and thus are inappropriate statistical analyses for such research questions.

Design Three factors were systematically manipulated in a factorial design:

- *Sparsity* of the component structure. Either 20 or 50% of all non-zero component loadings were set to 0.
- *Common component importance ratio*. It is not unlikely that, in comparison to the common source of variation, the source-specific sources of variation dominate the relationships between variables. To investigate whether cross-source relationships from a relative weaker common source of variation can be detected amongst stronger source-specific relationships, we manipulate the common component such that it is either equally strong (S1) or weaker compared to the source-specific components (S2).
- *n/p ratio*. The ratio of number of individuals (n) to number of variables (p): 2/3 ($n = 200$, $p = 300$; $n < p$), 1 ($n = 300$, $p = 300$; $n = p$), and 3/2 ($n = 300$, $p = 200$; $n > p$).

Data Generation For this simulation study we generated multivariate normal data with variance-covariance matrix Σ, which reflects a sparse three-component, two-source structure (see Eq. 3). On two of the three components, only the variables from one of two groups load; these two components reflect source-specific sources of variation (*source-specific or S-components*). The third component contains loadings for variables of both groups and reflects multi-source or common sources of variation (*common or C-component*).

The sparse three-component structure is generated as follows. First, a singular value decomposition is applied to a randomly generated, standard-normal distributed data matrix $L = UWV^T$ with dimensions $n \times p$, where the first half of the variables were set to belong to the first data source and the rest to the second source. From the resulting decomposition we derive orthogonal component loadings Λ by setting them equal to the three right singular vectors associated with the three largest singular values. Component loadings in Λ are set to zero such that its first component reflects a common source of variation and the other two components reflect source-specific sources of variation. Lastly, sparseness was introduced on the remaining non-zero loadings.

A 3×3 diagonal matrix served as singular values matrix S, and indicates how relative important each component is. When all components are equally important, S is an identity matrix. To create a data structure that is dominated by source-specific sources of variation, their singular values will be set to 2 (resulting in the diagonal 1 2 2). In empirical data it is unlikely that two groups of variables have source-specific correlation structures of equal strength. To incorporate this complexity in the data, the scaling vector g was manipulated to be $g_1 = 0.8$ for source 1 and $g_2 = 0.3$ for source 2:

$$
\lambda_{kr}^{true} = \sqrt{g_k / \sum_{r=1}^{R} \lambda_{kr}^2 s_{rr}^2 \lambda_{kr}}, \tag{4}
$$

where s_{rr} is the singular value of the rth component and $k = 1, 2$. Using Λ^{true} and S, the true variance-covariance matrix Σ^{true} was calculated, $\Sigma^{true} = \Lambda^{true} S^2 (\Lambda^{true})^T$, where the diagonal of Σ^{true} was manually set to 1 (see also Eq. 1), resulting in lower correlations. The true off-diagonal values in the covariance matrix representing the source-specific sources can be calculated in similar fashion $\Sigma_S^{true} = V_S^{true} S_S^2 (V_S^{true})^T$ noting that the diagonal values in Σ_S^{true} are not the variances of the variables. The true covariance matrix of the common source of variability between variables (the common component) is then the difference between the previous two equations: $\Sigma_C^{true} = \Sigma^{true} - \Sigma_S^{true}$. Finally, data was generated from a multivariate normal distribution with $\mu = 0$ and variance-covariance matrix Σ^{true}.

Analyses The recovery performance of graphical lasso, SNAC and the combination of sparse DISCO SCA with either Moore–Penrose or shrinkage estimator was assessed on the inverse covariance matrix of the unique cross-source relationships Σ_C^{-1}. Two fit indices were used: (1) the percentage of correctly estimated zeros and non-zeros $SSR = $ (number of 0 and non-0 edges)$/(p(p-1)/2)$ where p is the number of variables, and (2) Tucker's $\phi = U^T Z / \sqrt{(U^T U)(Z^T Z)}$ (Tucker 1951), where U is the upper triangle of the population (inverse) covariance matrix, and Z is the upper triangle of the estimated (inverse) covariance matrix. Values between 0.85 and 0.95 indicate "a fair similarity" and values above 0.95 indicate that the two covariance matrices can be considered equal (Abdi 2007). With the exception of the Moore–Penrose inverse, all statistical methods used in this study require input for a lasso-tuning parameter. Additionally, sparse DISCO SCA requires the status of the components (common or distinctive), and their sparseness. To avoid confounding influence of potentially mis-specifying the lasso-tuning parameter, we set it such that the estimated (inverse) covariance matrix recovers the true amount of zeros as much as possible. In addition, when running sparse DISCO SCA we assumed the component structure was known.

Graphical lasso was performed using the function *glasso* (*R*-package *glasso*, Friedman et al. 2014). Sparse DISCO SCA was performed using the *R*-package *Regularised SCA* (Gu and Van Deun 2019). Moore–Penrose inverse was performed using the function *ginv* (*R*-package MASS). Shrinkage estimation was performed using the function *pcor.shrink* (*R*-package *corpcor*, Schäfer et al. 2005).

4 Results

Graphical Lasso The graphical lasso has been developed to estimate sparse (inverse) covariance matrices from datasets. However, because it has no way of identifying which information is part of the cross-source relationships (captured in the common component) and which is not, it does not estimate the inverse covariance matrix of the common source of variance Σ_C^{-1} very accurately (SSR 0.64–0.91; ϕ 0.59–0.64), especially when $p > n$ (SSR 0.50–0.65; ϕ 0.17–0.28; see Table 1). However, one may wonder what would happen if information on which variables are part of cross-source relationships is available. Adding such a priori information to the graphical lasso results in a decrease of the percentage correctly estimated zeros (SSR 0.55–0.67) but the accuracy of the estimates increases (ϕ 0.56–0.84). Again, for $p > n$ estimates are less accurate (SSR 0.24–0.54; ϕ 0.21–0.42). For both analyses SSR and ϕ are higher when the data-generating component structure was sparser. Common component importance ratio does not have much impact on graphical lasso's performance.

SNAC The Sparse Network And Component analysis adequately estimates the inverse covariance matrix of the common source of variation Σ_C^{-1} (SSR 0.62–0.90; ϕ 0.73–0.89; see Table 1). It outperforms graphical lasso in all conditions when considering ϕ, and in $n > p$ and $p > n$ when considering SSR. SSR is higher when the data-generating component structure was sparser, while common component importance ratio does not have much impact on SNAC's performance.

Table 1 Performance of the various methods on estimating unique cross-source relationships (Σ_C^{-1})

n/p		Selection status recovery				Tucker's ϕ			
		Sparse 20		Sparse 50		Sparse 20		Sparse 50	
		S1	S2	S1	S2	S1	S2	S1	S2
2/3	Graphical lasso	0.50	0.50	0.65	0.63	0.17	0.17	0.28	0.26
	Graphical lasso a priori	0.54	0.54	0.24	0.29	0.21	0.21	0.42	0.41
	SNAC	**0.75**	**0.69**	**0.90**	**0.89**	**0.86**	**0.89**	**0.87**	**0.83**
	Moore–Penrose	0.99	0.91	1.00	1.00	−0.86	−0.85	−0.86	−0.86
	Shrink	0.99	0.91	1.00	0.99	−0.57	−0.52	−0.60	−0.56
1	Graphical lasso	0.80	0.79	0.91	0.90	0.59	0.59	0.62	0.62
	Graphical lasso a priori	0.60	0.55	0.59	0.56	0.71	0.56	0.82	0.81
	SNAC	**0.68**	**0.62**	**0.86**	**0.84**	**0.73**	**0.75**	**0.75**	**0.76**
	Moore–Penrose	0.99	0.91	1.00	1.00	−0.86	−0.85	−0.86	−0.86
	Shrink	0.99	0.91	1.00	0.99	−0.56	−0.52	0.60	−0.56
3/2	Graphical lasso	0.69	0.64	0.83	0.81	0.64	0.64	0.63	0.63
	Graphical lasso a priori	0.67	0.60	0.65	0.62	0.76	0.74	0.84	0.84
	SNAC	**0.75**	**0.69**	**0.90**	**0.89**	**0.86**	**0.89**	**0.87**	**0.83**
	Moore–Penrose	0.99	0.90	1.00	0.99	−0.86	−0.84	−0.86	−0.86
	Shrink	0.99	0.90	0.99	0.99	−0.57	−0.51	−0.60	−0.56

A priori indicates that information on which variables are part of cross-source relationships is known. n/p-*ratio* observation-variables ratio. Results from Sparse Network and Component (SNAC) analysis are shown in bold

Sparse DISCO SCA Sparse DISCO Simultaneous Component Analysis has been developed to find source-specific and common components that together maximise the amount of variance explained. Because we assume a sparse- component data-generating mechanism, Σ and Σ_C are singular and thus cannot be inverted. In such cases, however, it is possible to calculate a pseudoinverse such as the Moore–Penrose pseudoinverse (MP) and shrunken partial correlations (Shrink). While both methods almost perfectly identify the zeroes (SSR 0.90–1), the actual estimates are misspecified (ϕ −0.51 to −0.86; see Table 1). One may think that an explanation for these negative ϕ might be a change in signs; however, this seems unlikely given that the component model is sign invariant (the change of sign in loadings is compensated by a change of sign in the component scores).

5   Conclusion and Discussion

In this paper, we propose a component-network hybrid method that is suited for estimating the unique cross-source relationships amongst multivariate normally distributed variables found in multi-source datasets. This Sparse Network And Component (SNAC) model combines the strengths of two existing methods: the variable selecting and information dis-entangling properties of sparse DISCO Simultaneous Component Analysis (SCA) with the conditional dependence focused theoretical framework of Gaussian graphical models (GGMs). We have shown that SNAC outperforms a regular graphical model technique in accurately estimating unique cross-source relationships, especially when the data consists of more variables than observations. These results hold even when a priori knowledge about which variables are part of cross-source relations is available. Neither variations in sparseness of the estimated structure nor in the relative dominance of source-specific sources of variation influence SNACs performance.

Using the GGM framework, we can model unique relationships by estimating the inverse covariance matrix. However, adding sparse DISCO SCA to GGM has two consequences. First, it is not possible to use GGM to estimate an inverse covariance matrix based on a limited number of SCA-estimated components. This means that a pseudoinverse is required to gain insight into the relationships between variables. Second, adding the component pre-processing step changes the assumed underlying data-generating structure of our model. As demonstrated, inverse techniques that assume correctly estimated variances, such as Moore–Penrose and shrinkage estimator, are inaccurate as they do not take the assumed underlying factor analytic structure, this is including residual variances in the data-generating model, into account.

Applying the SNAC model to data requires input for several hyper-parameters: the number of common and source-specific components, and tuning parameters for several lassos. Because the purpose of this paper is to provide a proof of concept, we set these parameters as close to their optimal value as possible. Selecting the proper parameters in a non-simulation study, especially when analysing high-dimensional data, is challenging. As demonstrated by Gu and Van Deun (2016), the tuning parameter of $l1$-Lasso can be successfully selected using (Meinshausen and Bühlmann 2010)'s resample-based stability selection method, although the multi-component structure of both (Gu and Van Deun 2016)'s and our work complicates matters further.

Finally, in this paper we have only considered Gaussian distributed data, which while common are not the only data-type in multi-source data. Both sparse DISCO Simultaneous Component Analysis and graphical lasso are able to handle non-Gaussian variables, and as such we expect that this characteristic will be transferred to Sparse Network And Component analysis. Further research will have to investigate the translation from non-Gaussian-based component scores to non-Gaussian-based partial correlations.

Acknowledgments Support for this study was provided by European Research Council Consolidator Grant 647209 (PT, LW) and Netherlands Organisation for Scientific Research Aspasia Grant 015.011.034 (PT, KVD).

References

Abdi, H.: RV coefficient and congruence coefficient. Encyclopedia Meas. Stat. **849**, 853 (2007)

Bartel, J., Krumsiek, J., Theis, F.J.: Statistical methods for the analysis of high-throughput metabolomics data. Comput. Struct. Biotechnol. J. **4**(5), 1–9 (2013)

Bühlmann, P., Van De Geer, S.: Statistics for High-Dimensional Data: Methods, Theory and Applications. Springer, Berlin (2011)

Friedman, J., Hastie, T., Tibshirani, R.: Sparse inverse covariance estimation with the graphical lasso. Biostatistics **9**(3), 432–441 (2008)

Friedman, J., Hastie, T., Tibshirani, R.: Glasso: Graphical Lasso-estimation of Gaussian graphical models (2014). https://CRAN.R-project.org/package=glasso. R package version 1.8

Gu, Z., Van Deun, K.: A variable selection method for simultaneous component based data integration. Chemom. Intell. Lab. Syst. **158**, 187–199 (2016)

Gu, Z., Van Deun, K.: RSCA: Regularized simultaneous component analysis of multiblock data in R. Behav. Res. **51**, 2268–2289 (2019)

Johnson, M.R., Shkura, K., Langley, S.R., Delahaye-Duriez, A., Srivastava, P., Hill, W.D., Rackham, O.J., Davies, G., Harris, S.E., Moreno-Moral, A., Rotival, M.: Systems genetics identifies a convergent gene network for cognition and neurodevelopmental disease. Nat. Neurosci. **19**(2), 223 (2015)

Koller, D., Nir Friedman N.: Probabilistic Graphical Models: Principles And Techniques. MIT Press, Cambridge (2009)

Lauritzen, S.L.: Graphical Models. Clarendon Press, Oxford (1996)

Meinshausen, N., Bühlmann, P.: Stability selection. J. R. Stat. Soc. Ser. B Stat Methodol. **72**(4), 417–473 (2010)

Schäfer, J., Strimmer K.: A shrinkage approach to large-scale covariance matrix estimation and implications for functional genomics. Stat. Appl. Genet. Mol. Biol. **4**(1), 32 (2005)

Schouteden, M., Van Deun, K., Wilderjans, T.F., Van Mechelen, I.: Performing disco-SCA to search for distinctive and common information in linked data. Behav. Res. Methods **46**(2), 576–587 (2014)

Silverman, E.K., Loscalzo, J.: Network medicine approaches to the genetics of complex diseases. Discov. Med. **14**(75), 143 (2012)

Tibshirani, R.: Regression shrinkage and selection via the lasso. J. R. Stat. Soc. Ser. B Methodol. **58**(1), 267–288 (1996)

Tucker, L.R.: A method for synthesis of factor analysis studies. Technical report, DTIC Document (1951)

Van Deun, K., Wilderjans, T.F., Van den Berg, R.A., Antoniadis, A., Van Mechelen, I.: A flexible framework for sparse simultaneous component based data integration. BMC Bioinform. **12**(1), 448 (2011)

Understanding Malvestuto's Normalized Mutual Information

Hanneke van der Hoef and Matthijs J. Warrens

Abstract Malvestuto's version of the normalized mutual information is a well-known information theoretic index for quantifying agreement between two partitions. To further our understanding of what information on agreement between the clusters the index may reflect, we study components of the index that contain information on individual clusters, using mathematical analysis and numerical examples. The indices for individual clusters provide useful information on what is going on with specific clusters.

1 Introduction

Cluster analysis is the collection of techniques that can be used to divide unlabeled data objects into meaningful groups (Hennig et al. 2015; Rezaei and Fränti 2016). Cluster analysis is applied in many scientific disciplines, e.g. biology, information retrieval, psychology and medicine (Kumar 2005). All these domains make use of different data types and seek for different types of clusters, and hence require different clustering methods (Kaufman and Rousseeuw 1990). This has led to the development of numerous clustering techniques and algorithms. So far, there is no 'best' clustering algorithm that dominates over all other algorithms across all application domains (Fisher and Van Ness 1971; Jain 2010).

An important topic in cluster analysis research is which partition best fits the data set. A large number of both internal and external validity indices have been proposed to address the important, but challenging task of cluster validation (Rendón et al. 2011). Internal indices assess the clustering itself by measuring characteristics as

H. van der Hoef (✉)
Faculty of Behavioral and Social Sciences, University of Groningen, Groningen, TS, Netherlands
e-mail: h.van.der.hoef@student.rug.nl

M. J. Warrens
GION, University of Groningen, Groningen, TG, Netherlands
e-mail: m.j.warrens@rug.nl

© Springer Nature Singapore Pte Ltd. 2020
T. Imaizumi et al. (eds.), *Advanced Studies in Classification and Data Science*,
Studies in Classification, Data Analysis, and Knowledge Organization,
https://doi.org/10.1007/978-981-15-3311-2_23

cohesion, likelihood and distortion of the data objects in the clusters (Pfitzner et al. 2009). External validity indices on the other hand are used to compare different clusterings of the same set of objects, and to assess the similarity between the different partitions (Pfitzner et al. 2009; Rezaei and Fränti 2016).

External validity indices can be categorized into three approaches, namely (1) counting object pairs, (2) indices based on information theory and (3) set-matching measures (Rezaei and Fränti 2016). Most external validity indices are of the pair-counting approach, which is based on counting pairs of objects placed in identical and different clusters. Commonly used indices based on the pair-counting approach are the Rand index (Rand 1971) and the adjusted Rand index (Hubert and Arabie 1985; Steinley 2004; Steinley et al. 2016; Warrens 2008a).

Information theoretic indices are based on the concepts of mutual information and Shannon entropy (Kvalseth 1987; Shannon 1948). In recent years, information theoretic indices have received increasing attention due to their strong mathematical foundation, ability to detect non-linear similarities and applicability to soft clustering (Lei et al. 2016; Vinh et al. 2010). Commonly used information theoretic indices are the variation of information and different normalizations of the mutual information (Meilă 2007; Pfitzner et al. 2009).

Because there are many validity indices that one may use to assess similarity between two partitions, various authors have studied properties of the indices. Indices based on the pair-counting approach have been studied quite extensively (Albatineh et al. 2006; Albatineh and Niewiadomska-Bugaj 2011; Baulieu 1989; Milligan et al. 1996; Milligan and Cooper 1986; Steinley 2004; Warrens 2008,b). Only a few authors have studied indices from information theory (Pfitzner et al. 2009; Vinh et al. 2010).

Many validity indices quantify agreement between two partitions for all clusters simultaneously. Since these overall measures give a general notion of what is going on, their value (usually between 0 and 1) is often hard to interpret (except, perhaps, for values close to 0 or 1). In this paper we study a version of normalized mutual information already considered in Malvestuto (1986). To further our understanding of what information on agreement between the clusters the index may reflect, we study components of the index that contain information on individual clusters. These indices for individual clusters provide useful information on what is going on with specific clusters.

The paper is organized as follows. In Sect. 2 we introduce the notation and define Malvestuto's version of the normalized mutual information (Malvestuto 1986). In Sect. 3 we present two different decompositions of the index, one based on indices for the clusters of the first partition, and one based on indices for the clusters of the second partition. The decompositions indicate that Malvestuto's index may be interpreted as an overall measure that summarizes the information of both the cluster indices of the first partition, or the cluster indices of the second partition.

In Sect. 4 we study with numerical examples the various relationships between the indices and weights presented in Sect. 3. In the presented examples clusters that are completely mixed up between partitions tend to be more important in the calculation of the overall Malvestuto's index than clusters that are a perfect match between the partitions. In Sect. 5 we use additional numerical examples to illustrate that the maximum value of Malvestuto's index of unity is not easily attained. Section 6 contains a discussion. It would be good practice to report the measures for the individual clusters, since they provide more information than a single overall number.

2 Normalized Mutual Information

Suppose the data are scores of N objects. Let $U = \{U_1, U_2, \ldots, U_I\}$ and $V = \{V_1, V_2, \ldots, V_J\}$ be two partitions of the N objects in, respectively, I and J clusters. One partition could, for example, be a reference partition that purports to represent the true cluster structure of the objects, while the second partition may have been obtained with a clustering method that is being evaluated.

Let $P = \{p_{ij}\}$ be a matching table of size $I \times J$ where p_{ij} indicates the proportion of objects (with respect to N) placed in cluster U_i of the first partition and in cluster V_j of the second partition. The relative cluster sizes of the partitions are reflected in the row and column totals of P, denoted by p_{i+} and p_{+j}, respectively.

The joint Shannon entropy of partitions U and V (Shannon 1948) is given by

$$H(U, V) = - \sum_{i=1}^{I} \sum_{j=1}^{J} p_{ij} \log p_{ij}, \tag{1}$$

in which log denotes the base two logarithm as is common use in information theory, and $p_{ij} \log p_{ij} = 0$ if $p_{ij} = 0$. The entropy of a partition is a measure of the amount of randomness of a partition. The joint entropy quantifies the amount of joint randomness of the two partitions.

The joint entropy (1) is always non-negative. Furthermore, we have $H(U, V) = 0$ if all objects are in one cluster of the first partition and one cluster of the second partition, i.e. $|U_i| = N$ and $|V_j| = N$ for some i and j. Moreover, since $\log(1/2) = -1$, we have $-(1/2)\log(1/2) = 1/2$, and it follows that we have $H(U, V) = 1$ if $|U_i| = |U_j| = N/2$ and $|V_k| = |V_\ell| = N/2$ for some i, j, k and ℓ.

Next, the mutual information of partitions U and V is given by

$$I(U, V) = \sum_{i=1}^{I} \sum_{j=1}^{J} p_{ij} \log \frac{p_{ij}}{p_{i+}p_{+j}}. \tag{2}$$

The mutual information quantifies how much information the two partitions have in common (Pfitzner et al. 2009). Mutual information is occasionally referred to as the 'correlation measure' in information theory (Malvestuto 1986). It is always non-negative and has a value of 0 if and only if the partitions are statistically independent,

Table 1 Two example
matching tables of size 3×3
with corresponding statistics

		V_1	V_2	V_3	Total	Statistics
(a)	U_1	0.50			0.50	$H = 1.49$
	U_2			0.20	0.20	$I = 1.49$
	U_3		0.30		0.30	$M = 1.00$
	Total	0.50	0.30	0.20	1.00	
(b)	U_1	0.02	0.04	0.04	0.10	$H = 2.82$
	U_2	0.06	0.12	0.12	0.30	$I = 0.00$
	U_3	0.12	0.24	0.24	0.60	$M = 0.00$
	Total	0.20	0.40	0.40	1.00	

i.e. $p_{ij} = p_{i+}p_{+j}$ for all i and j. Higher values of mutual information indicate more shared information.

The mutual information (2) is bounded below by zero, but does not have an upper bound. Various authors have proposed normalizations of (2) such that the maximum value of the normalized index is equal to unity (Pfitzner et al. 2009). Malvestuto (1986) considers, among other things, the normalization of (2) given by

$$M(U, V) = \frac{I(U, V)}{H(U, V)}. \tag{3}$$

Index M can be used to assess how well the clusters of the two partitions match. The index takes on values in the unit interval. We have $M = 1$ if each cluster of U only contains objects from a single cluster of V and, vice versa, if each cluster of V only contains objects from a single cluster of U. Furthermore, we have $M = 0$ if the partitions are statistically independent, i.e. $p_{ij} = p_{i+}p_{+j}$ for all i and j. In general, higher values of index M imply higher similarity between U and V. Index M is a normalization of the mutual information that is frequently used in cluster analysis research (Kvalseth 1987; Malvestuto 1986; Quinlan 1986).

To illustrate the extreme values of index M, consider the two matching tables in Table 1. Both tables have size 3×3. We have $H = I = 1.49$ and $M = 1$ for panel (a) of Table 1 because there is a perfect match between the clusters of U and V. Furthermore, we have $H = 2.82$, $I = 0$ and $M = 0$ for panel (b) of Table 1 since the two partitions are statistically independent.

3 Decompositions

In this section we present two decompositions of index M into indices that contain information on individual clusters. Define for $U_i \in U$ the weight

$$u_i := -\sum_{j=1}^{J} p_{ij} \log p_{ij}, \tag{4}$$

and the index

$$R_i := \frac{\sum_{j=1}^{J} p_{ij} \log \frac{p_{ij}}{p_{i+} p_{+j}}}{-\sum_{j=1}^{J} p_{ij} \log p_{ij}}. \tag{5}$$

The numerator of (5) consists of the part of the mutual information between U and V that is associated with cluster U_i only. Furthermore, the denominator of (5) is the part of the joint entropy of U and V that is associated with cluster U_i only. The latter quantity is also identical to the weight in (4).

Index R_i can be used to assess how well cluster U_i matches to the clusters of partition V. The index takes on values in the unit interval. We have $R_i = 1$ if there is a perfect match between U_i and some cluster in V, i.e. all objects from U_i are in precisely one cluster of V and the latter cluster contains no other objects from partition U. Furthermore, we have $R_i = 0$ if $p_{ij} = p_{i+} p_{+j}$ for all j. This is the case if the objects of cluster U_i are randomly assigned (in accordance with the p_{+j}'s) to the clusters of partition V.

We have the following decomposition for index M. Index M is a weighted average of the indices in (5) using the u_i's in (4) as weights:

$$M = \frac{\sum_{i=1}^{I} u_i R_i}{\sum_{i=1}^{I} u_i}. \tag{6}$$

Since M is a weighted average of the R_i-values the M-value lies somewhere between the minimum and maximum of the R_i-values. Equation (6) shows that the M-value is largely determined by the R_i-values of clusters with high u_i-values. The M-value will be high if R_i-values corresponding to high u_i-values are themselves high. Vice versa, the M-value will be low if R_i-values corresponding to high u_i-values are low.

Next, define for $V_j \in V$ the weight

$$v_j := -\sum_{i=1}^{I} p_{ij} \log p_{ij}, \tag{7}$$

and the index

$$C_j := \frac{\sum_{i=1}^{I} p_{ij} \log \frac{p_{ij}}{p_{i+} p_{+j}}}{-\sum_{i=1}^{I} p_{ij} \log p_{ij}}. \tag{8}$$

The numerator of (8) consists of the part of the mutual information between U and V that is associated with cluster V_j only. Furthermore, the denominator of (8) is the part of the joint entropy of U and V that is associated with cluster V_j only. The latter quantity is also identical to the weight in (7).

We have the following second decomposition for index (3). Index (3) is also a weighted average of the indices in (8) using the v_j's in (7) as weights:

$$M = \frac{\sum\limits_{j=1}^{J} v_j C_j}{\sum\limits_{j=1}^{J} v_j}. \tag{9}$$

Since M is a weighted average of the C_j-values the M-value lies somewhere between the minimum and maximum of the C_j-values. Formula (9) shows that the M-value is largely determined by the C_j-values of the clusters with high v_j-values. The M-value will be high if C_j-values corresponding to high v_j-values are themselves high. Vice versa, the M-value will be low if C_j-values corresponding to high v_j-values are low.

4 Numerical Examples

In this section we explore with numerical examples the relationships between overall index (3), the indices with cluster information in (5) and their corresponding weights in (4). Fourteen numerical examples are presented in Tables 2, 3, and 4, four examples in the first two tables, and six in the third. To simplify the discussion for Tables 2 and 3 a little bit all example matching tables are symmetric so that we have $R_i = C_i$ and $u_i = v_i$ for all i. Thus, the discussion on Tables 2 and 3 can be limited to the weights in (4) and the cluster indices in (5). The overall indices, cluster indices and weights associated with a particular example are presented behind each example table in the same panel.

Table 2 presents four example matching tables of size 3×3. In each panel of Table 2 there is a perfect match on one cluster ($U_1 = V_1$), while the other two clusters of the partitions are completely mixed up. The size of the perfect match cluster decreases from panel (a) to (d). The cluster is the largest cluster in panels (a) and (b) and the smallest cluster in (c) and (d).

In all tables of Table 2 the cluster index associated with the first cluster shows that there is perfect agreement on the perfect match cluster ($R_1 = 1.00$). Furthermore, the R_i-values show that there is less than perfect agreement or rather poor agreement on the other two clusters ($R_2 = R_3$). The R_2- and R_3-value strictly decrease from panel (a) to panel (d). If we combine this with the fact that the mixed up clusters become larger from panel (a) to panel (d), it follows that for mixed up clusters the corresponding cluster index decreases with the relative cluster size.

Table 2 Four additional example matching tables of size 3×3 with corresponding statistics

		V_1	V_2	V_3	Total	Overall	Cluster	Weights
(a)	U_1	0.92			0.92	$H = 0.56$	$R_1 = 1.00$	$u_1 = 0.11$
	U_2		0.02	0.02	0.04	$I = 0.40$	$R_2 = 0.65$	$u_2 = 0.23$
	U_3		0.02	0.02	0.04	$M = 0.72$	$R_3 = 0.65$	$u_3 = 0.23$
	Total	0.92	0.04	0.04	1.00			
(b)	U_1	0.40			0.40	$H = 2.17$	$R_1 = 1.00$	$u_1 = 0.53$
	U_2		0.15	0.15	0.30	$I = 0.97$	$R_2 = 0.27$	$u_2 = 0.82$
	U_3		0.15	0.15	0.30	$M = 0.45$	$R_3 = 0.27$	$u_3 = 0.82$
	Total	0.40	0.30	0.30	1.00			
(c)	U_1	0.20			0.20	$H = 2.32$	$R_1 = 1.00$	$u_1 = 0.46$
	U_2		0.20	0.20	0.40	$I = 0.72$	$R_2 = 0.14$	$u_2 = 0.93$
	U_3		0.20	0.20	0.40	$M = 0.31$	$R_3 = 0.14$	$u_3 = 0.93$
	Total	0.20	0.40	0.40	1.00			
(d)	U_1	0.08			0.08	$H = 2.24$	$R_1 = 1.00$	$u_1 = 0.29$
	U_2		0.23	0.23	0.46	$I = 0.40$	$R_2 = 0.06$	$u_2 = 0.98$
	U_3		0.23	0.23	0.46	$M = 0.18$	$R_3 = 0.06$	$u_3 = 0.98$
	Total	0.08	0.46	0.46	1.00			

Table 3 Four example agreement tables of size 4×4 with corresponding statistics

		V_1	V_2	V_3	V_4	Total	Overall	Cluster	Weights
(a)	U_1	0.46				0.46	$H = 1.48$	$R_1 = 1.00$	$u_1 = 0.52$
	U_2		0.46			0.46	$I = 1.32$	$R_2 = 1.00$	$u_2 = 0.52$
	U_3			0.02	0.02	0.04	$M = 0.89$	$R_3 = 0.65$	$u_3 = 0.23$
	U_4			0.02	0.02	0.04		$R_4 = 0.65$	$u_4 = 0.23$
	Total	0.46	0.46	0.04	0.04	1.00			
(b)	U_1	0.40				0.40	$H = 1.92$	$R_1 = 1.00$	$u_1 = 0.53$
	U_2		0.40			0.40	$I = 1.52$	$R_2 = 1.00$	$u_2 = 0.53$
	U_3			0.05	0.05	0.10	$M = 0.79$	$R_3 = 0.54$	$u_3 = 0.43$
	U_4			0.05	0.05	0.10		$R_4 = 0.54$	$u_4 = 0.43$
	Total	0.40	0.40	0.10	0.10	1.00			
(c)	U_1	0.30				0.30	$H = 2.37$	$R_1 = 1.00$	$u_1 = 0.52$
	U_2		0.30			0.30	$I = 1.57$	$R_2 = 1.00$	$u_2 = 0.52$
	U_3			0.10	0.10	0.20	$M = 0.66$	$R_3 = 0.40$	$u_3 = 0.66$
	U_4			0.10	0.10	0.20		$R_4 = 0.40$	$u_4 = 0.66$
	Total	0.30	0.30	0.20	0.20	1.00			
(d)	U_1	0.20				0.20	$H = 2.57$	$R_1 = 1.00$	$u_1 = 0.46$
	U_2		0.20			0.20	$I = 1.37$	$R_2 = 1.00$	$u_2 = 0.46$
	U_3			0.15	0.15	0.30	$M = 0.53$	$R_3 = 0.27$	$u_3 = 0.82$
	U_4			0.15	0.15	0.30		$R_4 = 0.27$	$u_4 = 0.82$
	Total	0.20	0.20	0.30	0.30	1.00			

Table 4 Four additional example matching tables of size 3×3 with corresponding statistics

		V_1	V_2	V_3	Total	Overall	Rows	Columns
(a)	U_1		0.33		0.33	$H = 1.65$	$R_1 = 1.00$	$C_1 = 0.97$
	U_2	0.32		0.01	0.33	$I = 1.52$	$R_2 = 0.80$	$C_2 = 1.00$
	U_3			0.34	0.34	$M = 0.92$	$R_3 = 0.97$	$C_3 = 0.81$
	Total	0.32	0.33	0.35	1.00			
(b)	U_1	0.01	0.33		0.34	$H = 1.71$	$R_1 = 0.81$	$C_1 = 0.78$
	U_2	0.32		0.01	0.33	$I = 1.46$	$R_2 = 0.78$	$C_2 = 0.97$
	U_3			0.33	0.33	$M = 0.85$	$R_3 = 0.97$	$C_3 = 0.81$
	Total	0.33	0.33	0.34	1.00			
(c)	U_1	0.01	0.32		0.33	$H = 1.78$	$R_1 = 0.78$	$C_1 = 0.76$
	U_2	0.32	0.01	0.01	0.34	$I = 1.39$	$R_2 = 0.63$	$C_2 = 0.78$
	U_3			0.33	0.33	$M = 0.78$	$R_3 = 0.97$	$C_3 = 0.80$
	Total	0.33	0.33	0.34	1.00			
(d)	U_1	0.01	0.32		0.33	$H = 1.84$	$R_1 = 0.78$	$C_1 = 0.61$
	U_2	0.32	0.01	0.01	0.34	$I = 1.33$	$R_2 = 0.61$	$C_2 = 0.78$
	U_3	0.01		0.32	0.33	$M = 0.72$	$R_3 = 0.78$	$C_3 = 0.78$
	Total	0.34	0.33	0.33	1.00			
(e)	U_1	0.01	0.32		0.33	$H = 1.91$	$R_1 = 0.76$	$C_1 = 0.61$
	U_2	0.32	0.01	0.01	0.34	$I = 1.26$	$R_2 = 0.61$	$C_2 = 0.63$
	U_3	0.01	0.01	0.31	0.33	$M = 0.66$	$R_3 = 0.63$	$C_3 = 0.76$
	Total	0.34	0.34	0.32	1.00			
(f)	U_1	0.01	0.31	0.01	0.33	$H = 1.97$	$R_1 = 0.61$	$C_1 = 0.61$
	U_2	0.32	0.01	0.01	0.34	$I = 1.20$	$R_2 = 0.61$	$C_2 = 0.61$
	U_3	0.01	0.01	0.31	0.33	$M = 0.61$	$R_3 = 0.61$	$C_3 = 0.61$
	Total	0.34	0.33	0.33	1.00			

The weight u_i reflects the importance of each R_i-value in the calculation of index (3) in (6). In all four panels the weight associated with the perfect match cluster is substantially lower than the weights associated with the mixed up clusters. In panels (a) and (c) the latter weights are twice as large, while in panel (d) they are more than three times as large. It appears that weights are larger if the joint entropy associated with the clusters is larger. Furthermore, the weights corresponding to the mixed up clusters increase from panel (a) to panel (d). If we combine this with the fact that the mixed up clusters become larger from panel (a) to panel (d), it appears that for mixed up clusters the cluster weights increase with the relative cluster size.

In Table 2, the M-value strictly decreases from panel (a) to panel (d). The decline may be explained as follows. In panel (a) the perfect match cluster is the largest cluster by far. The relative size of the perfect match cluster decreases from panel (a) to panel (d), and the M-value decreases accordingly. The decline may also be explained in terms of the information related to the individual clusters. The M-value is a weighted average of the cluster indices R_i in (5) using the weights u_i in (4). Each M-value lies somewhere between the three R_i-values. Since for the two mixed up clusters the corresponding cluster index decreases with the relative cluster size, the M-value, as a weighted average, decreases with the R_2- and R_3-values.

To better understand the M-value one should consider the values of the cluster weights. Since the weight associated with the perfect match cluster is substantially lower than the weights associated with the mixed up clusters, the M-value is pulled 'more' towards the cluster indices of the mixed up cluster than the cluster index of the perfect match cluster. For example, in panel (a) of Table 2 the larger clusters of the two partitions are perfectly matched, yet the M-value is only 0.72. The explanation is that the M-value is closer to the values $R_2 = R_3 = 0.65$ than the value $R_1 = 1.00$ due to the weighting.

Table 3 presents four example matching tables of size 4×4. In each panel of Table 3 there are two perfect matches between the clusters ($U_1 = V_1$ and $U_2 = V_2$) of the same size, while the other two clusters of the partitions are completely mixed up. The size of the perfect matched clusters decreases from panel (a) to (d). The clusters are the largest cluster in panels (a), (b), and (c), and the smallest cluster in (d).

In all tables of Table 3 the cluster indices associated with the first two clusters show that there is perfect agreement on the first two clusters ($R_1 = R_2 = 1.00$). Furthermore, the R_i-values show that there is less than perfect agreement or rather poor agreement on the other two clusters ($R_3 = R_4$). The R_3- and R_4-value strictly decrease from panel (a) to panel (d). Thus, for the mixed up clusters the corresponding cluster index decreases with the relative cluster size.

In the top two panels of Table 3 the weights associated with the two perfect matched clusters are higher than the weights associated with the mixed up clusters. For the bottom two panels the roles are interchanged. The weights corresponding to the mixed up clusters increase from panel (a) to panel (d). The weights corresponding to the perfect matched clusters show less variability. Thus, for mixed up clusters the cluster weights increase with the relative cluster size. Furthermore, panels (a) and (b) show that the perfect matched clusters can have the highest weights. Apparently, it is required that the M-value is relatively high.

In Table 3, the M-value strictly decreases from panel (a) to panel (d). An explanation is that the relative size of the perfect matched clusters decreases from panel (a) to panel (d), and the M-value decreases accordingly. The decline of the M-value may also be explained in terms of the information reflected in the individual clusters. Each M-value lies somewhere between the four R_i-values. Since for the two mixed up clusters the corresponding cluster index decreases with the relative cluster size and the corresponding weight increases with the relative cluster size as well, the M-value, as a weighted average, decreases with the R_2- and R_3-values.

5 More Numerical Examples

The examples in Tables 2 and 3, especially those in panel (a), suggest that an M-value of unity is not easily attained. This is indeed what we found in many examples. Table 4 presents six example matching tables of size 3×3. In each panel of Table 4 there is a very good match for all clusters from the first partition with one cluster from the second partition ($U_1 \approx V_2$ (but $U_1 = V_2$ for panel (a)), $U_2 \approx V_1$ and $U_3 \approx V_3$). Panel (a) is closest to a perfect match between the partitions. There is disagreement on only 1% of the objects. From panel (a) to panel (f) the disagreement between the partitions increases with small steps of 1%.

As the disagreement increases from panel (a) to (f) of Table 4 the value of the joint entropy H goes up and the value of the mutual information I. Consequently, the M-value decreases as well, from 0.92 to 0.85 to 0.78 to 0.72 to 0.66 to 0.61. What is striking about these M-values is that the decline is rather steep, while the disagreement step is rather small. Apparently, index M can attain high values, say 0.90 or higher, only if the matching table contains a lot of zero cells.

6 Discussion

For assessing similarity between two partitions researchers usually use and report overall measures that quantify similarity for all clusters simultaneously. Since, overall indices only give a general notion of what is going on their value is generally difficult to interpret. In this paper we took a closer look at a version of normalized mutual information already considered in Malvestuto (1986). Mutual information is a concept from information theory (Kvalseth 1987; Shannon 1948).

In Sect. 3 we presented two decompositions of Malvestuto's index from information theory, one into indices for the clusters of the first partition, and one into indices for the clusters of the second partition. The decompositions show that Malvestuto's index may be interpreted as an overall measure that summarizes the information of both the cluster indices of the first partition, or the cluster indices of the second partition.

We presented numerical examples that showed that clusters that are completely mixed up between partitions tend to be more important in the calculation of the overall index than clusters that are a perfect match between the partitions. It would probably be good practice to report the measures for the individual clusters, since they provide more (detailed) information than a single overall number. Furthermore, other numerical examples illustrated that the maximum value of Malvestuto's index of unity is not easily attained.

References

Albatineh, A.N., Niewiadomska-Bugaj, M.: Correcting Jaccard and other similarity indices for chance agreement in cluster analysis. Adv. Data Anal. Classif. **5**, 179–200 (2011)
Albatineh, A.N., Niewiadomska-Bugaj, M., Mihalko, D.: On similarity indices and correction for chance agreement. J. Classif. **23**, 301–313 (2006)
Baulieu, F.B.: A classification of presence/absence based dissimilarity coefficients. J. Classif. **6**, 233–246 (1989)

Fisher, L., Van Ness, J.W.: Admissible clustering procedures. Biometrika **58**, 91–104 (1971)

Hennig, C., Meilă, M., Murtagh, F., Rocci, R.: Handbook of Cluster Analysis. Chapman and Hall/CRC, New York (2015)

Hubert, L.J., Arabie, P.: Comparing partitions. J. Classif. **2**, 193–218 (1985)

Jain, A.K.: Data clustering: 50 years beyond K-means. Pattern Recogn. Lett. **31**, 651–666 (2010)

Kaufman, L., Rousseeuw, P.: Finding groups in data: an introduction to cluster analysis. Wiley, New York (1990)

Kumar, V.: Cluster analysis: basic concepts and algorithms. In: Tan, P., Steinbach, M., Kumar, V. (eds.) Introduction to Data Mining, pp. 487–568. Pearson Education, New York (2005)

Kvalseth, T.O.: Entropy and correlation: some comments. IEEE Trans. Syst. Man Cybern. **17**, 517–519 (1987)

Lei, Y., Bezdek, J.C., Chan, J., Vinh, N., Romano, S., Bailey, J.: Extending information-theoretic validity indices for fuzzy clustering. IEEE Trans. Fuzzy Syst. **99**, 1–1 (2016)

Malvestuto, F.M.: Statistical treatment of the information content of a database. Inf. Syst. **11**, 211–233 (1986)

Meilă, M.: Comparing clusterings. An information based distance. J. Multivar. Anal. **98**, 873–895 (2007)

Milligan, G.W.: Clustering validation: results and implications for applied analyses. In: Arabie, P., Hubert, L.J., De Soete, G. (eds.) Clustering and Classification, pp. 341–375. World Scientific, River Edge (1996)

Milligan, G.W., Cooper, M.C.: A study of the comparability of external criteria for hierarchical cluster analysis. Multivar. Behav. Res. **21**, 441–458 (1986)

Pfitzner, D., Leibbrandt, R., Powers, D.: Characterization and evaluation of similarity measures for pairs of clusterings. Knowl. Inf. Syst. **19**, 361–394 (2009)

Quinlan, J.R.: Induction of decision trees. Mach. Learn. **1**, 81–106 (1986)

Rand, W.M.: Objective criteria for the evaluation of clustering methods. J. Am. Stat. Assoc. **66**, 846–850 (1971)

Rendón, E., Abundez, I., Arizmendi, A., Quiroz, E.: Internal versus external cluster validation indexes. Int. J. Comput. Commun. **5**, 27–34 (2011)

Rezaei, M., Fränti, P.: Set matching measures for external cluster validity. IEEE Trans. Knowl. Data Eng. **28**, 2173–2186 (2016)

Shannon, C.E.: A mathematical theory of communication. Bell Syst. Tech. J. **27**, 623–656 (1948)

Steinley, D.: Properties of the Hubert-Arabie adjusted Rand index. Psychol. Methods **9**, 386–396 (2004)

Steinley, D., Brusco, M.J., Hubert, L.J.: The variance of the adjusted Rand index. Psychol. Methods **21**, 261–272 (2016)

Vinh, N.X., Epps, J., Bailey, J.: Information theoretic measures for clustering comparison: variants, properties, normalization and correction for chance. J. Mach. Learn. Res. **11**, 2837–2854 (2010)

Warrens, M.J.: On similarity coefficients for 2×2 tables and correction for chance. Psychometrika **73**, 487–502 (2008)

Warrens, M.J.: On the equivalence of Cohen's kappa and the Hubert-Arabie adjusted Rand index. J. Classif. **25**, 177–183 (2008a)

Warrens, M.J.: On association coefficients for 2×2 tables and properties that do not depend on the marginal distributions. Psychometrika **73**, 777–789 (2008b)

Understanding the Rand Index

Matthijs J. Warrens and Hanneke van der Hoef

Abstract The Rand index continues to be one of the most popular indices for assessing agreement between two partitions. The Rand index combines two sources of information, object pairs put together, and object pairs assigned to different clusters, in both partitions. Via a decomposition of the Rand index into four asymmetric indices, we show that in many situations object pairs that were assigned to different clusters have considerable impact on the value of the overall Rand index.

1 Introduction

In research domains like medical classification, image segmentation, taxonomy, and web data analysis, it is frequently of interest to find meaningful groupings or partitions of a set of objects (Dubey et al. 2016; Huo et al. 2016; Katiyar et al. 2016; Luo et al. 2014; Zeng et al. 2014). Partitions are commonly found with the so-called clustering methods (Hennig et al. 2015; Jain 2010; Kaufman and Rousseeuw 1990; Kumar 2005). These methods can, for example, be used to find subtypes of cancer in tissue samples, and to group consumers based on attitudes, knowledge or uses concerning a product.

To empirically assess the performance of clustering methods researchers typically assess the agreement between a reference standard partition that purports to represent the true cluster structure of the objects, and a trial partition produced by the method that is being evaluated. High agreement between the two partitions indicates good recovery of the reference cluster structure. Agreement between the partitions

M. J. Warrens (✉)
GION, University of Groningen, Groningen, TG, Netherlands
e-mail: m.j.warrens@rug.nl

H. van der Hoef
Faculty of Behavioral and Social Sciences, University of Groningen, Groningen, TS, Netherlands
e-mail: h.van.der.hoef@student.rug.nl

© Springer Nature Singapore Pte Ltd. 2020
T. Imaizumi et al. (eds.), *Advanced Studies in Classification and Data Science*,
Studies in Classification, Data Analysis, and Knowledge Organization,
https://doi.org/10.1007/978-981-15-3311-2_24

can be assessed with the so-called external validity indices (Albatineh et al. 2006; Albatineh and Niewiadomska-Bugaj 2011; Brun et al. 2007; Pfitzner et al. 2009).

Indices for assessing agreement between partitions can be categorized into three approaches, namely (1) counting object pairs, (2) indices based on information theory, and (3) set-matching measures (Pfitzner et al. 2009; Rezaei and Fränti 2016; Vinh et al. 2010). Most indices are of the pair-counting approach, which is based on counting pairs of objects placed in identical and different clusters. Commonly used examples are the Rand index (Rand 1971) and the Hubert-Arabie adjusted Rand index (Hubert and Arabie 1985; Steinley et al. 2016; Warrens 2008d). The latter corrects the Rand index for agreement due to chance (Albatineh et al. 2006; Warrens 2008c). Several authors proposed to use the adjusted Rand index as a standard tool in cluster validation research (Milligan 1996; Milligan and Cooper 1986; Steinley 2004).

The Rand index continues to be one of the most popular indices for assessing agreement between partitions, probably because it has a simple interpretation (Anderson et al. 2010). The Rand index may be interpreted as the ratio of the number of object pairs placed together in a cluster in each of the two partitions and the number of object pairs assigned to different clusters in both partitions, relative to the total number of object pairs. Thus, the Rand index combines two sources of information, object pairs put together, and object pairs assigned to different clusters, in both partitions.

To understand what the value of the Rand index may actually mean requires knowledge of how the two sources of information on object pairs contribute to the overall value of the Rand index. The above interpretation suggests that both sources may contribute equally. It turns out that this is generally not the case. Via a decomposition of the Rand index into four asymmetric indices, we show in this paper that in many situations object pairs that were assigned to different clusters in both partitions have (a lot) more impact on the value of the overall Rand index than object pairs that were combined in both partitions.

A problem with the Rand index is that it does not range over the entire $[0, 1]$ interval, where one indicates perfect agreement between the partitions. Fowlkes and Mallows (1983) illustrated with simulated data that the Rand index concentrates in a small interval near one (Meilă 2007). The decomposition of the Rand index presented in this paper will also be used to provide some insight into this feature of the index.

The paper is organized as follows. In Sect. 2 we introduce the notation. In Sect. 3 we define the Rand index and present a decomposition of the index into four asymmetric indices. In Sect. 4 we show with numerical examples that the value of the Rand index is determined to a large extent by the number of object pairs that are not joined in either of the partitions. In Sect. 5 we present several results that suggest the importance of these object pairs increases as the number of clusters increase. Finally, in Sect. 6 we show what values of the Rand index one may expect under statistical independence of the partitions.

2 Notation

In this section we introduce the notation. Suppose we have n objects. Let $A = \{A_1, A_2, \ldots, A_I\}$ and $B = \{B_1, B_2, \ldots, B_J\}$ denote two partitions of the objects, where $I \geq 2$ and $J \geq 2$ are the number of clusters. The two partitions may, for example, be a reference standard partition and a trial partition that was obtained with a clustering method that is being evaluated. Let $\mathbf{M} = \{m_{ij}\}$ be a matching table of size $I \times J$ where m_{ij} indicates the number of objects placed in cluster A_i of the first partition and in cluster B_j of the second partition. The cluster sizes of the respective partitions are the row and column totals of $\mathbf{M}$:

$$|A_i| = m_{i+} = \sum_{j=1}^{J} m_{ij} \quad \text{and} \quad |B_j| = m_{+j} = \sum_{i=1}^{I} m_{ij}.$$

Following Fowlkes and Mallows (1983) the information in matching table $\mathbf{M}$ can be summarized in a fourfold contingency table by counting several different types of pairs of objects:

$$N := \frac{n(n-1)}{2}$$

is the total number of pairs of objects,

$$T := \sum_{i=1}^{I} \sum_{j=1}^{J} \binom{m_{ij}}{2}$$

is the number of object pairs that were placed in the same cluster in both partitions,

$$P := \sum_{i=1}^{I} \binom{m_{i+}}{2}$$

is the number of object pairs that were placed in the same cluster in the first partition A, and

$$Q := \sum_{j=1}^{J} \binom{m_{+j}}{2}$$

is the number of object pairs that were placed in the same cluster in the second partition B. Furthermore, define $a := T$, $b := P - T$, $c := Q - T$ and $d := N + T - P - Q$. Quantity b (c) is the number of object pairs that were placed in the same cluster in the first (second) partition but in different clusters in the second

Table 1 Two 2 × 2 contingency table representations of matching table **M**

First partition	Second partition		
	Pair in the same cluster	Pair in different cluster	Totals
Representation 1			
Pair in the same cluster	T	$P - T$	P
Pair in different clusters	$Q - T$	$N + T - P - Q$	$N - P$
Totals	Q	$N - Q$	N
Representation 2			
Pair in the same cluster	a	b	$a + b$
Pair in different clusters	c	d	$c + d$
Totals	$a + c$	$b + d$	N

(first) partition. The quantity d is the number of object pairs that are not joined in either of the partitions.

Table 1 presents two representations of the fourfold contingency table that summarizes matching table **M**. The upper panel of Table 1 gives a representation in terms of the counts N, T, P, and Q, whereas the lower panel of Table 1 gives a representation using the counts a, b, c, and d. The latter notational system is commonly used for expressing similarity measures for 2 × 2 tables (Albatineh et al. 2006; Baulieu 1989; Gower and Warrens 2017; Pfitzner et al. 2009; Warrens 2008a,b,c,d,e, 2019).

3 Rand Index and a Decomposition

The Rand index (Rand 1971) is defined as

$$R = \frac{N + 2T - P - Q}{N} = \frac{a + d}{a + b + c + d}.$$

In the context of 2 × 2 tables (Albatineh et al. 2006; Baulieu 1989; Heiser and Warrens 2010; Pfitzner et al. 2009; Warrens 2008a,b,c,d,e) the formula on the right-hand side can be found in Sokal and Michener (1958). We have $R = 1$ and perfect agreement if the partitions are identical. Furthermore, we have $R = 0$ when one partition consists of a single cluster and the other partition only consists of clusters containing a single object (Rand 1971).

Wallace (1983) considers the following two asymmetric indices. The first index

$$U = \frac{T}{P} = \frac{a}{a + b}$$

is the proportion of object pairs in the first partition that are also joined in the second partition (Severiano et al. 2011). The second index

$$V = \frac{T}{Q} = \frac{a}{a+c}$$

is the proportion of object pairs in the second partition that are also joined in the first partition. In addition to Wallace indices U and V, we may consider the following two asymmetric indices. The third index

$$W = \frac{N+T-P-Q}{N-P} = \frac{d}{c+d}$$

is the proportion of object pairs not placed together in the first partition that are also not joined in the second partition. The fourth index

$$Z = \frac{N+T-P-Q}{N-Q} = \frac{d}{b+d}$$

is the proportion of object pairs not placed together in the second partition that are also not joined in the first partition. The quantity $N + T - P - Q$ in the numerator of W and Z is the number of pairs that are not joined in either of the partitions. As an indication of agreement between the partitions, this quantity is rather neutral, counting pairs that are perhaps not clearly indicative of agreement (Wallace 1983).

The Rand index is a weighted average of the four fractions U, V, W, and Z using the denominators of the indices, respectively, P, Q, $N - P$, and $N - Q$, as weights:

$$R = \frac{UP + VQ + W(N-P) + Z(N-Q)}{P + Q + (N-P) + (N-Q)}.$$

An equivalent expression with relative weights is given by

$$R = \frac{1}{2}\left[U\left(\frac{P}{N}\right) + V\left(\frac{Q}{N}\right) + W\left(1 - \frac{P}{N}\right) + Z\left(1 - \frac{Q}{N}\right) \right].$$

Many agreement indices are functions of the two Wallace indices U and V (Albatineh et al. 2006; Baulieu 1989; Warrens 2008a,b). The Wallace indices contain information on object pairs that were combined in both partitions. The above decomposition illustrates that the Rand index combines the information in U and V with the information in W and Z. The latter asymmetric indices contain information on pairs of objects that were not put together in the same cluster in both partitions. To obtain more insight into the Rand index and its behavior we study the four indices U, V, W, and Z together with the relative weights P/N, Q/N, $1 - P/N$, and $1 - Q/N$.

4 Numerical Examples

In this section we use numerical examples to illustrate the relationships between the Rand index and the four asymmetric indices and their corresponding weights. Table 2 presents four numerical example tables of size 4×4. The five agreement indices are presented behind each example table in the same panel. Behind each asymmetric index we also present the value of the corresponding relative weight, P/N, Q/N, $1 - P/N$, or $1 - Q/N$.

All partitions in the example tables of Table 2 consist of four clusters of about the same size ($n = 24, 25$, or 26). Because the four example tables are almost symmetric we have $U = V$ and $W = Z$ in each case. In panel (a) the clusters can be well separated. The separateness decreases from panel (a) to panel (d). This is reflected in all five indices to some extent, since the Rand index and the four asymmetric indices decrease accordingly. However, the Wallace indices U and V decrease much faster than the indices W and Z. Apparently, the number of objects that are not paired in both partitions remains relatively high. Even for panel (d), where the joint distribution is close to uniform, and we may say that the two partitions are close to statistically independent, we have $W = Z = 0.75$. To obtain further insight into how this affects the overall Rand index we incorporate the relative weights into the discussion.

Table 2 Four example matching tables of size 4×4 with corresponding statistics

		B_1	B_2	B_3	B_4	Total	Indices		Relative weights
(a)	A_1	20	2	0	2	24	$R = 0.83$	$U = 0.64$	0.24
	A_2	2	20	2	2	26		$V = 0.64$	0.24
	A_3	2	2	20	2	26		$W = 0.88$	0.76
	A_4	0	2	2	20	24		$Z = 0.88$	0.76
	Total	24	26	24	26	100			
(b)	A_1	15	4	3	3	25	$R = 0.70$	$U = 0.39$	0.24
	A_2	3	15	3	3	24		$V = 0.39$	0.24
	A_3	3	4	15	3	25		$W = 0.80$	0.76
	A_4	4	3	4	15	26		$Z = 0.80$	0.76
	Total	25	26	25	24	100			
(c)	A_1	10	5	5	5	25	$R = 0.64$	$U = 0.25$	0.24
	A_2	5	10	5	5	25		$V = 0.25$	0.24
	A_3	6	5	10	5	26		$W = 0.76$	0.76
	A_4	4	5	5	10	24		$Z = 0.76$	0.76
	Total	25	25	25	25	100			
(d)	A_1	7	6	6	6	25	$R = 0.62$	$U = 0.22$	0.24
	A_2	6	7	5	6	24		$V = 0.22$	0.24
	A_3	6	7	7	6	26		$W = 0.75$	0.76
	A_4	6	6	6	7	25		$Z = 0.75$	0.76
	Total	25	26	24	25	100			

Table 3 Four additional example matching tables of size 4 × 4 with corresponding statistics

		B_1	B_2	B_3	B_4	Total	Indices		Relative weights
(a)	A_1	64	0	0	0	64	$R = 0.94$	$U = 0.93$	0.45
	A_2	0	4	4	4	12		$V = 0.93$	0.45
	A_3	0	4	4	4	12		$W = 0.95$	0.55
	A_4	0	4	4	4	12		$Z = 0.95$	0.55
	Total	64	12	12	12	100			
(b)	A_1	37	0	0	0	37	$R = 0.82$	$U = 0.66$	0.26
	A_2	0	7	7	7	21		$V = 0.66$	0.26
	A_3	0	7	7	7	21		$W = 0.88$	0.74
	A_4	0	7	7	7	21		$Z = 0.88$	0.74
	Total	37	21	21	21	100			
(c)	A_1	19	0	0	0	19	$R = 0.71$	$U = 0.40$	0.25
	A_2	0	9	9	9	27		$V = 0.40$	0.25
	A_3	0	9	9	9	27		$W = 0.80$	0.75
	A_4	0	9	9	9	27		$Z = 0.80$	0.75
	Total	19	27	27	27	100			
(d)	A_1	10	0	0	0	10	$R = 0.64$	$U = 0.33$	0.27
	A_2	0	10	10	10	30		$V = 0.33$	0.27
	A_3	0	10	10	10	30		$W = 0.75$	0.73
	A_4	0	10	10	10	30		$Z = 0.75$	0.73
	Total	10	30	30	30	100			

The Rand index R is a weighted average of the four indices U, V, W, and Z. For all example tables the relative weights are identical. For U and V the relative weight is 0.24, whereas for W and Z the relative weight is identical to 0.76. Thus, in all example tables the weight of W and Z is three times as high as the weight of U and V. The R-value is pulled more towards the values of W and Z than to the values of U and V. In other words, in the calculation of the Rand index in these examples, pairs of objects that were not put together in the same cluster in both partitions are three times more important than pairs of objects that were paired in both partitions.

Table 3 presents four additional numerical example tables of size 4 × 4. Because all three tables are symmetric we have $U = V$ and $W = Z$ in each case. All partitions in the example tables of Table 3 consist of three clusters that have the same size and one cluster that has a different size. In all examples there is a perfect match on the one single cluster, while the other three clusters are completely mixed up. The size of the perfect match cluster decreases from panel (a) to panel (d). At the same time the size of the three mixed up clusters increases. The values of the five indices decrease with the size of the single cluster. Again, the Wallace indices U and V decrease much faster than the indices W and Z. It seems that the number of objects that are not paired in both partitions remains relatively high. Even for panel (d) we have $W = Z = 0.75$.

Because all example tables of Table 3 are symmetric the weights for U and V, and the weights for W and Z, are identical. In panel (a) the relative weights are not that far apart. In panel (a) all four asymmetric indices contribute more or less the same to the overall Rand index. For panels (b), (c), and (d) the weights are very similar to the weights in Table 2. The weight of W and Z is about three times higher than the weight of U and V. In the calculation of the Rand index in panels (b), (c), and (d), pairs of objects that were not put together in the same cluster in both partitions are three times more important than pairs of objects that were paired in both partitions.

5 Relative Weights

The numerical examples in Sect. 4 show that the value of the Rand index is in many situations primarily determined by the asymmetric indices W and Z that focus on object pairs not placed together in clusters, since these two indices have higher relative weights than Wallace indices U and V that focus on object pairs put together in the same clusters. However, the numerical examples in Sect. 4 were limited to tables of size 4×4. In other simulated examples that are not reported here we found that the values of indices W and Z tend to become more important in the calculation of the Rand index as the number of clusters increase. In this section we present several theorems that formalize this observation to some extent.

For three distributions we show that, for a fixed number of objects, the quantities P and Q strictly decrease in, respectively, I and J. If the quantities P and Q decrease, and if n, and thus N, is kept fixed, the relative weights

$$1 - \frac{P}{N} \quad \text{and} \quad 1 - \frac{Q}{N}$$

increase with the number of clusters, and the indices W and Z become more important for the overall R-value.

In the theorems below we use that the number of objects in a cluster can be a real number. Technically this is not a problem, since binomial coefficients can also be applied to real numbers. With real-life data the number of objects is of course a positive integer.

In Theorem 1, we consider the case in which the objects are uniformly distributed over the clusters.

Theorem 1 *Suppose n is fixed. If we set $m_{i+} = n/I$ and $m_{+j} = n/J$ for all i and j, then P and Q are strictly decreasing in, respectively, I and J.*

Proof Under the conditions of the theorem we have

$$P = \sum_{i=1}^{I} \binom{n/I}{2} = \frac{n^2}{2I} - \frac{n}{2} \quad \text{and} \quad Q = \sum_{j=1}^{J} \binom{n/J}{2} = \frac{n^2}{2J} - \frac{n}{2}.$$

Thus, for fixed n, P is strictly decreasing in I, and Q is strictly decreasing in J.

$\square$

Understanding the Rand Index

In Theorem 2, we consider a case in which the cluster sizes form an increasing triangular distribution.

Theorem 2 *Suppose n is fixed. If we set*

$$m_{i+} = \frac{2ni}{I(I+1)} \quad \text{and} \quad m_{+j} = \frac{2nj}{J(J+1)}$$

for, respectively, $i \in \{1, 2, \ldots, I\}$ and $j \in \{1, 2, \ldots, J\}$, then P and Q are strictly decreasing in, respectively, I and J.

Proof Under the conditions of the theorem we have

$$P = \sum_{i=1}^{I} \frac{ni}{I(I+1)} \left(\frac{2ni}{I(I+1)} - 1 \right) = \frac{2n^2}{I^2(I+1)^2} \sum_{i=1}^{I} i^2 - \frac{n}{I(I+1)} \sum_{i=1}^{I} i.$$

Using the identities

$$\sum_{i=1}^{I} i^2 = \frac{I(I+1)(2I+1)}{6} \quad \text{and} \quad \sum_{i=1}^{I} i = \frac{I(I+1)}{2},$$

we obtain

$$P = \frac{n^2(2I+1)}{3I(I+1)} - \frac{n}{2}.$$

The first fraction has an I term in the numerator and an I^2 term in the denominator. Thus, for fixed n, P is strictly decreasing in I. Using similar arguments for Q yields

$$Q = \frac{n^2(2J+1)}{3J(J+1)} - \frac{n}{2},$$

which is strictly decreasing in J. □

In Theorem 3, we consider a case in which the cluster sizes form a symmetric triangular distribution with its peak in the middle.

Theorem 3 *Suppose n is fixed, and that I and J are even. If we set*

$$m_{i+} = m_{(I-i)+} = \frac{2ni}{I(I/2+1)} \quad \text{and} \quad m_{+j} = m_{+(J-j)} = \frac{2nj}{J(J/2+1)}$$

for, respectively, $i \in \{1, 2, \ldots, I/2\}$ and $j \in \{1, 2, \ldots, J/2\}$, then P and Q are strictly decreasing in, respectively, I and J.

Proof Under the conditions of the theorem we have

$$P = \sum_{i=1}^{I/2} \frac{2ni}{I(I/2+1)} \left(\frac{2ni}{I(I+1)} - 1 \right) = \frac{4n^2}{I^2(I/2+1)^2} \sum_{i=1}^{I/2} i^2 - \frac{2n}{I(I/2+1)} \sum_{i=1}^{I/2} i.$$

Using the identities

$$\sum_{i=1}^{I/2} i^2 = \frac{I}{12} \left(\frac{I}{2} + 1 \right) (I+1) \qquad \text{and} \qquad \sum_{i=1}^{I/2} i = \frac{I}{4} \left(\frac{I}{2} + 1 \right),$$

we obtain

$$P = \frac{n^2(I+1)}{3I(I/2+1)} - \frac{n}{2}.$$

The first fraction has an I term in the numerator and an I^2 term in the denominator. Thus, for fixed n, P is strictly decreasing in I. Using similar arguments for Q yields

$$Q = \frac{n^2(J+1)}{3J(J/2+1)} - \frac{n}{2},$$

which is strictly decreasing in J.   □

Table 4 presents the values of relative weights P/N and $1 - P/N$ for $n = 100$, the three distributions considered in Theorems 1, 2, and 3, and different values of I. The numbers in Table 4 show that in most cases $1 - P/N$ is higher than P/N, and that in many cases $1 - P/N$ is even much higher than P/N.

Table 4 Relative weights for $n = 100$, three distributions and different numbers of clusters

# clusters $I =$	2	4	6	8	10	12	14	16	20	30	40	50
Theorem 1: uniform distribution												
P/N	0.49	0.24	0.16	0.12	0.09	0.07	0.06	0.05	0.04	0.02	0.02	0.01
$1 - P/N$	0.51	0.76	0.84	0.88	0.91	0.93	0.94	0.95	0.96	0.98	0.98	0.99
Theorem 2: skewed triangular distribution												
P/N	0.55	0.29	0.20	0.15	0.12	0.10	0.08	0.07	0.06	0.03	0.02	0.02
$1 - P/N$	0.45	0.71	0.80	0.85	0.88	0.90	0.92	0.93	0.94	0.97	0.98	0.98
Theorem 3: symmetric triangular distribution												
P/N	0.49	0.27	0.19	0.14	0.11	0.09	0.08	0.07	0.05	0.03	0.02	0.02
$1 - P/N$	0.51	0.73	0.81	0.86	0.89	0.91	0.92	0.93	0.95	0.97	0.98	0.98

6 Statistical Independence

It has been noted in the literature that the value of the Rand index concentrates in a small interval near one (Fowlkes and Mallows 1983; Meilă 2007). The results from the previous sections can be used to explain this phenomenon to some extent. In this final section we show that the Rand index in many situations already has a high value under statistical independence of the partitions.

If the partitions are statistically independent, we have the expectation $E(T) = PQ/N$ (Albatineh et al. 2006; Warrens 2008c,d). In this case the Wallace indices are identical to

$$U = \frac{Q}{N} \quad \text{and} \quad V = \frac{P}{N},$$

which are the relative weights of, respectively, V and U in the decomposition of the Rand index presented in Sect. 3. Furthermore, the other two asymmetric indices become

$$W = 1 - \frac{Q}{N} \quad \text{and} \quad Z = 1 - \frac{P}{N},$$

which are the relative weights of, respectively, Z and W in the decomposition of the Rand index. Thus, under statistical independence the value of the Rand index is given by

$$R = \frac{PQ}{N^2} + \left(1 - \frac{P}{N}\right)\left(1 - \frac{Q}{N}\right).$$

In Sect. 5 we showed for three distributions that the values of P/N and Q/N decrease when the numbers of clusters I and J increase. As both I and J become larger the first term on the right-hand side becomes smaller and goes to zero, while the second term on the right-hand side becomes higher and goes to unity.

Table 5 presents the values of the Rand index under statistical independence for $n = 100$, the uniform distribution considered in Theorem 1, and for various values of I and J. If either $I = 2$ or $J = 2$, we have approximately $R = 0.50$.

Table 5 shows that the value of R increases with the number of clusters. The R-value may already be quite high (say ≥ 0.60) for a moderate number of clusters, despite the fact that Table 5 gives R-values that hold under statistical independence. We obtain a very similar table if we replace the uniform distribution in Table 5 by the distributions considered in Theorems 2 and 3, or any pairing of the three distributions. Thus, even in the case of statistical independence, the value of the Rand index concentrates in the upper half of the unit interval $[0, 1]$.

312 M. J. Warrens and H. van der Hoef

Table 5 Values of the Rand index under statistical independence for uniform distributed objects and different numbers of clusters I and J

I	J	3	4	5	6	8	10	20	30	50
3		0.56	0.59	0.61	0.62	0.63	0.64	0.66	0.67	0.67
4		0.59	0.63	0.66	0.68	0.70	0.71	0.74	0.75	0.75
5		0.61	0.66	0.69	0.71	0.74	0.75	0.78	0.79	0.80
6		0.62	0.68	0.71	0.73	0.76	0.78	0.81	0.83	0.83
8		0.63	0.70	0.74	0.76	0.79	0.81	0.85	0.87	0.88
10		0.64	0.71	0.75	0.78	0.81	0.83	0.88	0.89	0.90
20		0.66	0.74	0.78	0.81	0.85	0.88	0.92	0.94	0.95
30		0.67	0.75	0.79	0.83	0.87	0.89	0.94	0.95	0.97
50		0.67	0.75	0.80	0.83	0.88	0.90	0.95	0.97	0.98

References

Albatineh, A.N., Niewiadomska-Bugaj, M., Mihalko, D.: On similarity indices and correction for chance agreement. J. Classif. **23**, 301–313 (2006)

Albatineh, A.N., Niewiadomska-Bugaj, M.: Correcting Jaccard and other similarity indices for chance agreement in cluster analysis. Adv. Data Anal. Classif. **5**, 179–200 (2011)

Anderson, D.T., Bezdek, J.C., Popescu, M., Keller, J.M.: Comparing fuzzy, probabilistic, and possibilistic partitions. IEEE Trans. Fuzzy Syst. **18**, 906–917 (2010)

Baulieu, F.B.: A classification of presence/absence based dissimilarity coefficients. J. Classif. **6**, 233–246 (1989)

Brun, M., Sima, C., Hua, J., Lowey, J., Carroll, B., Suh, E., Dougherty, E.R.: Model-based evaluation of clustering validation measures. Pattern Recogn. **40**, 807–824 (2007)

Dubey, A.K., Gupta, U., Jain, S.: Analysis of k-means clustering approach on the breast cancer Wisconsin dataset. Int. J. Comput. Assist. Radiol. Surg. **11**, 2033–2047 (2016)

Fowlkes, E.B., Mallows, C.L.: A method for comparing two hierarchical clusterings. J. Am. Stat. Assoc. **78**, 553–569 (1983)

Gower, J.C., Warrens, M.J.: Similarity, dissimilarity, and distance, measures of. Wiley StatsRef: Statistics Reference Online (2017)

Heiser, W.J., Warrens, M.J.: Families of relational statistics for 2×2 tables. In: Kaul, H., Mulder, H.M. (eds.) Advances in Interdisciplinary Applied Discrete Mathematics, pp. 25–52. World Scientific, Singapore (2010)

Hennig, C., Meilă, M., Murtagh, F., Rocci, R.: Handbook of Cluster Analysis. Chapman and Hall/CRC, New York (2015)

Hubert, L.J., Arabie, P.: Comparing partitions. J. Classif. **2**, 193–218 (1985)

Huo, Z., Ding, Y., Liu, S., Oesterreich, S., Tseng, G.: Meta-analytic framework for sparse K-means to identify disease subtypes in multiple transcriptomic studies. J. Am. Stat. Assoc. **111**, 27–52 (2016)

Jain, A.K.: Data clustering: 50 years beyond K-means. Pattern Recogn. Lett. **31**, 651–666 (2010)

Katiyar, P., Divine, M.R., Kohlhofer, U., Quintanilla-Martinez, L., Schölkopf, B., Pichler, B.J., Disselhorst, J.A.: Spectral clustering predicts tumor tissue heterogeneity using dynamic 18F-FDG PET: a complement to the standard compartmental modeling approach. J. Nucl. Med. **57**, 651–657 (2016)

Kaufman, L., Rousseeuw, P.: Finding groups in data: an introduction to cluster analysis. Wiley, New York (1990)

Kumar, V.: Cluster analysis: basic concepts and algorithms. In: Tan, P., Steinbach, M., Kumar, V. (eds.) Introduction to Data Mining, pp. 487–568. Pearson Education, New York (2005)

Luo, C., Pang, W., Wang, Z.: Semi-supervised clustering on heterogeneous information networks. In: Tseng, V.S., Ho, T.B., Zhou, Z., Chen, A.L.P., Kao, H. (eds.) Advances in Knowledge Discovery and Data Mining, pp. 548–559. Springer, Berlin (2014)

Meilă, M.: Comparing clusterings. An information based distance. J. Multivar. Anal. **98**, 873–895 (2007)

Milligan, G.W.: Clustering validation: results and implications for applied analyses. In: Arabie, P., Hubert, L.J., De Soete, G. (eds.) Clustering and Classification, pp. 341–375. World Scientific, River Edge (1996)

Milligan, G.W., Cooper, M.C.: A study of the comparability of external criteria for hierarchical cluster analysis. Multivar. Behav. Res. **21**, 441–458 (1986)

Pfitzner, D., Leibbrandt, R., Powers, D.: Characterization and evaluation of similarity measures for pairs of clusterings. Knowl. Inf. Syst. **19**, 361–394 (2009)

Rand, W.M.: Objective criteria for the evaluation of clustering methods. J. Am. Stat. Assoc. **66**, 846–850 (1971)

Rezaei, M., Fränti, P.: Set matching measures for external cluster validity. IEEE Trans. Knowl. Data Eng. **28**, 2173–2186 (2016)

Severiano, A., Pinto, F.R., Ramirez, M., Carrio, J.A.: Adjusted Wallace coefficient as a measure of congruence between typing methods. J. Clin. Microbiol. **49**, 3997–4000 (2011)

Sokal, R.R., Michener, C.D.: A statistical method for evaluating systematic relationships. Univ. Kansas Sci. Bull. **38**, 1409–1438 (1958)

Steinley, D.: Properties of the Hubert-Arabie adjusted Rand index. Psychol. Method **9**, 386–396 (2004)

Steinley, D., Brusco, M.J., Hubert, L.J.: The variance of the adjusted Rand index. Psychol. Methods **21**, 261–272 (2016)

Vinh, N.X., Epps, J., Bailey, J.: Information theoretic measures for clustering comparison: variants, properties, normalization and correction for chance. J. Mach. Learn. Res. **11**, 2837–2854 (2010)

Wallace, D.L.: Comment on a method for comparing two hierarchical clusterings. J. Am. Stat. Assoc. **78**, 569–576 (1983)

Warrens, M.J.: On the indeterminacy of resemblance measures for binary (presence/absence) data. J. Classif. **25**, 125–136 (2008)

Warrens, M.J.: Bounds of resemblance measures for binary (presence/absence) variables. J. Classif. **25**, 195–208 (2008)

Warrens, M.J.: On similarity coefficients for 2×2 tables and correction for chance. Psychometrika **73**, 487–502 (2008)

Warrens, M.J.: On the equivalence of Cohen's kappa and the Hubert-Arabie adjusted Rand index. J. Classif. **25**, 177–183 (2008)

Warrens, M.J.: On association coefficients for 2×2 tables and properties that do not depend on the marginal distributions. Psychometrika **73**, 777–789 (2008)

Warrens, M.J.: Similarity measures for 2×2 tables. J. Intell. Fuzzy Syst. **36**, 3005–3018 (2019)

Zeng, S., Huang, R., Kang, Z., Sang, N.: Image segmentation using spectral clustering of Gaussian mixture models. Neurocomputing **144**, 346–356 (2014)

Layered Multivariate Regression with Its Applications

Naoto Yamashita and Kohei Adachi

Abstract Multivariate regression is known as a multivariate extension of multiple regression, which explain/predict the variations in multiple dependent variables by multiple independent variables. Recently, various procedures for Sparse Multivariate Regression (SMR) have been proposed, in which a sparse regression coefficient matrix (having a number of zero elements) is obtained aiming to facilitate its interpretation. The procedures for SMR can be classified into the following two types; penalized and cardinality-constrained procedures. In them, the resulting number of zeros in the regression coefficient matrix is controlled/constrained by a prespecified penalty parameter or cardinality value. In this research, we propose another approach for SMR, referred to as Layered Multivariate Regression (LMR). In LMR, the regression coefficient matrix is assumed to be a sum of several sparse matrices, which is called layer. Therefore, the sparseness of the resulting coefficient matrix is controlled by how many layers are used. In LMR, k-th layer can be viewed as the coefficient matrix in the regression of a partial residual (i.e., the residual for all but k-th layer) on independent variables, and thus the variance explained by LMR gets closer to that for the unconstrained regression as the number of layers increases. We present an alternating least squares algorithm for LMR and a procedure for determining how many layers should be used. LMR is assessed in a simulation study and illustrated with a real data example. As an application of LMR, procedures for sparse estimation in some multivariate analysis techniques (e.g., principal component analysis) are also presented.

N. Yamashita (✉) · K. Adachi
Graduate School of Human Sciences, Osaka University, Suita, Japan
e-mail: nyamashita@hus.osaka-u.ac.jp; adachi@hus.osaka-u.ac.jp

© Springer Nature Singapore Pte Ltd. 2020
T. Imaizumi et al. (eds.), *Advanced Studies in Classification and Data Science*,
Studies in Classification, Data Analysis, and Knowledge Organization,
https://doi.org/10.1007/978-981-15-3311-2_25

1 Introduction

Multivariate regression is commonly used for extracting the relationship between multiple independent and dependent variables. Let $\mathbf{X}$ be an N (observations) $\times$ P (variables) matrix of independent variables and $\mathbf{Y}$ be a N (observations) $\times$ Q (variables) matrix of dependent variables. The multivariate regression of P dependent variables on Q dependent variables is formulated as the minimization of the least squares loss function $f(\mathbf{W})$ defined by

$$f(\mathbf{W}) = ||\mathbf{Y} - \mathbf{XW}||^2 \tag{1}$$

where $\mathbf{W}$ denotes a $P \times Q$ matrix of regression coefficients (Izenman 2008). The loss function can be rewritten as $f(\mathbf{W}) = [\mathbf{w}_1, \ldots, \mathbf{w}_Q,] \sum_q (\mathbf{y}_q - \mathbf{Xw}_q)$ with $\mathbf{y}_q$ and $\mathbf{w}_q$ being the q-th column vector of $\mathbf{Y}$ and $\mathbf{W}$, respectively. The solution of the minimization of (1) is simply given by $\mathbf{W} = (\mathbf{X'X})^{-1}\mathbf{X'Y}$ and we call the solution as unconstrained solution in this article. The pth element of $\mathbf{w}_q$ denoted as w_{pq} represents the relationship between p-th independent and q-th dependent variable, and therefore the matrix $\mathbf{W}$ shows the correspondences between the set of variables. As a small example of multivariate regression, multivariate regression is applied to Tobacco dataset. The total of six variables of chemical compounds, Nitrogen (N), Chlorine (Cl), Potassium (K), Phosphorus (P), Calcium (Cl), and Magnesium (Mg), are regressed on the three characteristics of tobacco leaves. Table 1 shows the estimated regression coefficient matrix.

To interpret a coefficient matrix, it is required to find the correspondence between a single dependent variable and a reduced number of independent variables. For example, one can find such correspondence between the dependent variable "Brate" and the independent variables "Cl," "K," and "Ca," which are considered to be important for explaining the variation of the dependent variable. In order to find easily such correspondences, simple structure (Thurstone 1947; Ullman 2006) is a useful property which coefficient matrices should possess, because it highlights the between/within-column contrast in the coefficient matrix and facilitate to match a reduced number of independent variables to each of the dependent variables. However, the estimated coefficient matrix does not always possess such helpful

Table 1 Estimated coefficient matrix by unconstrained multivariate regression		Brate	Sug	Nic
	N	0.10	−0.58	0.29
	Cl	−0.58	0.39	−0.32
	K	0.45	0.20	0.11
	P	−0.13	0.22	−0.05
	Ca	0.41	0.11	0.24
	Mg	−0.32	−0.22	0.48
	Vexp.	0.744		

structure and thus it is difficult to find the correspondence between the variable sets, because no constraint is imposed on **W** with respect to its structure.

In this research, we newly propose a procedure of multivariate regression in order to facilitate the interpretation of the solution. In this procedure, the coefficient matrix is reparametrized as a sum of matrices, which are called *layers*, and each layer is constrained to have a certain simple structure. This is called *Layered Multivariate Regression* (LMR) formulated as

$$\min_{\mathbf{W}} f_{LMR}(\mathbf{W}_1, \ldots, \mathbf{W}_L) = ||\mathbf{Y} - \mathbf{X} \sum_l \mathbf{W}_l||^2 \tag{2}$$

where $\mathbf{W}_l$ denotes the l-th layer for $l = 1, \ldots, L$. As a structure of layers, we consider a perfect cluster structure where each independent variable is associated with a single dependent variable; i.e., only one nonzero element in each row of **W**. As a result, **W** contains several elements equaling to zero. The resulting coefficient matrix $\mathbf{W} = \sum_l \mathbf{W}_l$ is thus easy to interpret because the between/within-column contrasts are highlighted.

In order to facilitate the interpretation of **W**, several approaches have been proposed in which the structure of **W** is constrained in some ways. Table 2 shows the estimated coefficient matrices obtained by the two existing procedures which we consider here. The first approach is penalized optimization of (1) which is defined as the minimization of

$$f_{PR}(\mathbf{W}) = ||\mathbf{Y} - \mathbf{XW}||^2 + P(\mathbf{W}) \tag{3}$$

where $P(\mathbf{W})$ is a certain penalty function (Hastie et al. 2015; Zou et al. 2006). For example, one can consider $P(\mathbf{W}) = \lambda \sum_q ||\mathbf{w}_q||_1$ as a lasso penalty with $\mathbf{w}_q$ being the q-th column vector of **W** and λ is the penalty parameter used for controlling the amount of shrinkage of the entries of **W**. The minimization of $f_{PR}(\mathbf{W})$ aims to shrink several elements in **W** toward zero and therefore element-wise contrast is highlighted. The penalty approach does not constrain matrix-wise structure of **W** and therefore the resulting solution is often difficult to interpret. The first column

Table 2 Estimated coefficient matrices and the rate of variance explained (Vexp.) obtained by penalty (left) and perfect cluster approach (right); a blank cell shows zero element

	Penalty approach			Perfect cluster approach		
	Brate	Sug	Nic	Brate	Sug	Nic
N	0.09	−0.51	0.37		−0.69	
Cl	−0.54	0.39	−0.23	−0.61		
K	0.40	−0.12		0.62		
P	−0.13	0.20			0.17	
Ca	0.22		0.15	0.33		
Mg	−0.17	−0.21	0.36			0.73
Vexp.	0.611			0.569		

in the right side of Table 3 gives an example of such difficulty, in that the within-column contrast is lower compared with the other columns. The second approach is "perfect cluster approach" in which $\mathbf{W}$ is constrained to have a perfect cluster structure. Because of this structure, each independent variable corresponds to the single dependent variable, and therefore the independent variables are classified into the disjoint sets. As seen in left side of Table 2, the procedure proposed by Yamashita (2012) produces a coefficient matrix with a perfect cluster structure. The estimated solution fits poorly to the data matrix, however, as seen in the rate of variance explained (Vexp.) in the table. This is because the constraint imposed on the structure of $\mathbf{W}$ is too restrictive to explain the given data sufficiently. The proposed procedure is considered to remedy the drawbacks of these procedures; LMR achieves a simple structure by restricting layer-level structure with better fit to the data matrix by using multiple layers.

2 Proposed Method

2.1 Algorithm

Our proposed LMR is formulated as the minimization of $f_{LMR}(\mathbf{W}_1, \ldots, \mathbf{W}_L)$ in (2) subject to the constraint that each layer has a perfect cluster structure. The loss function is minimized by alternately updating each layer with keeping the other layers fixed until convergence is reached. Namely, for updating the l-th layer, the problem to be solved is the minimization of

$$f_{LMR}(\mathbf{W}_l | \mathbf{W}_{l' \neq l}) = \left\| \left(\mathbf{Y} - \mathbf{X} \sum_{l' \neq l} \mathbf{W}_l' \right) - \mathbf{X}\mathbf{W}_l \right\|^2 = ||\mathbf{Y}_l^{\#} - \mathbf{X}\mathbf{W}_l||^2 \qquad (4)$$

with $\mathbf{Y}_l^{\#} = \mathbf{Y} - \mathbf{X} \sum_{l' \neq l} \mathbf{W}_l'$, the partial residual of the other $L - 1$ layers. Because $\mathbf{W}_l$ is constrained to have a perfect cluster structure, $\mathbf{W}_l$ is reparametrized as $\mathbf{W}_l = \mathbf{D}_l \mathbf{R}_l$ where $\mathbf{D}_l$ is a $P \times P$ diagonal matrix and $\mathbf{R}$ is a $P \times Q$ binary and row stochastic matrix (Vichi 2016; Vichi and Saporta 2009). The diagonal elements in $\mathbf{D}_l$ indicate the values of nonzero elements and $\mathbf{R}_l$ indicates these position in each of P rows. Therefore, $\mathbf{D}_l$ and $\mathbf{R}_l$ are jointly updated in order to update the l-th layer, at the D- and R-step which are described in detailed as follows.

In the D-step for the l-th layer, $\mathbf{D}_l$ is updated with fixed $\mathbf{R}_l$. For the minimization, with the partial derivative of the function $f_{LMR}(\mathbf{W}_l | \mathbf{W}_{k \neq l}) = f_{LMR}(\mathbf{D}_l | \mathbf{D}_{k \neq l}, \mathbf{R}_1, \ldots, \mathbf{R}_L)$ with respect to $\mathbf{D}_l$ being $P \times P$ matrix of zeros denoted as $\mathbf{O}$, we get

$$2\text{diag}(\mathbf{R}_l \mathbf{Y}_l^{\#'} \mathbf{X}) - 2\mathbf{D}_l \text{diag}(\mathbf{X}'\mathbf{X}) = \mathbf{O} \iff \mathbf{D}_l = \text{diag}(\mathbf{R}_l \mathbf{Y}_l^{\#'} \mathbf{X}) \text{diag}(\mathbf{X}'\mathbf{X})^{-1} \quad (5)$$

Table 3 Estimated coefficient matrices by LMR with ($L = 1, \ldots, 4$); element equaling to zero shows a blank cell

	L = 1			L = 2			L = 3			L = 4		
	Brate	Sug	Nic	Brate	Sug	Nic	Brate	Sug	Nic	Brate	Sug	Nic
N		-0.69			-0.65	0.51		-0.52	0.39		-0.58	0.30
Cl	-0.61			-0.60	0.38		-0.6	0.41	-0.31	-0.60	0.39	-0.32
K	0.62			0.52	0.26		0.52	0.14		0.52	0.20	0.11
P		0.17		-0.14	0.20		-0.14	0.22		-0.14	0.22	
Ca	0.33			0.47			0.47		0.18	0.47	0.11	0.25
Mg			0.73	-0.26		0.43	-0.26	-0.21	0.40	-0.26	-0.21	0.47
Vexp.	0.569			0.701			0.740			0.742		

where the notation diag($\mathbf{a}$) specifies a diagonal matrix with diagonal equal to the vector $\mathbf{a}$. Further, $\mathbf{R}_l$ is updated in the R-step with other parameter matrices kept fixed. Let, α_p^l be the column number of the nonzero element in p-th row of $\mathbf{W}_l$ and it takes an integer within the range $[1, Q]$. The optimal α_p^l can be obtained by computing the attained function values for all possible values for α_p^l, and update α_p^l as the one which attains the minimum of the function values. Namely, the update formula of α_p^l is simply given by

$$\alpha_p^l = \underset{\alpha_p^l = 1,\ldots,Q}{\arg\min} \; f_{LMR}(\alpha_p^l | \alpha_{p' \neq p}^l, \mathbf{R}_{l' \neq l}, \mathbf{D}_1, \ldots, \mathbf{D}_L). \tag{6}$$

The above two steps are applied for all ps in order to update the k-th layer, and layer-wise update is repeated for all layers until convergence is reached. The whole algorithm is therefore described as follows.

Step 1.   Set $t = 0$.
Step 2.   Set initial values for $\mathbf{D}_1, \ldots, \mathbf{D}_L$ and $\mathbf{R}_1, \ldots, \mathbf{R}_L$ and construct $\mathbf{W}_1, \ldots, \mathbf{W}_L$.
Step 3.   Repeat Step. 3.1. to 3.3. for $l = 1, \cdots L$.

  Step 3.1.   (D-step) Update $\mathbf{D}_l$ by (5).
  Step 3.2.   (R-step) Update α_p^l by (6) for all ps and construct $\mathbf{R}_l$.
  Step 3.3.   Update $\mathbf{W}_l$.

Step 4.   Increase t by one and go to Step 5, if the decrease in the $f_{LMR}(\mathbf{W}_1, \ldots, \mathbf{W}_L)$ value from the previous round is less than 1.0×10^{-7}; otherwise, return to Step 3.
Step 5.   Set $\hat{\mathbf{W}}_l$ by the current $\mathbf{W}_l$ for all ls if $f_{LMR}(\hat{\mathbf{W}}_1, \ldots, \hat{\mathbf{W}}_L) < f_{LMR}(\mathbf{W}_1, \ldots, \mathbf{W}_L)$ or $t = 1$.
Step 6.   If $t = t_{\max}$ accept $\hat{\mathbf{W}}_1, \ldots, \hat{\mathbf{W}}_P$ as a final solution and $\mathbf{W} = \sum_l \mathbf{W}_l$; otherwise go to Step 2.

Because the above algorithm involves the optimization with respect to the discrete values α_p^l, the algorithm is considered to be sensitive to local minimizers. In order to avoid accepting a local minimum as a final solution, the algorithm is started from $t_{\max}$ sets of initial values. In the following simulation study and the real data example, we set $t_{\max} = 300$, which presents no significant problem from our experience.

2.2   Layer Number Selection

In the LMR algorithm, the number of layers has to be specified beforehand as a positive integer L. As an example, Fig. 1 shows the plot of the goodness of fit against the number of layers. As shown in the figure variance explained gets closer to the one of unconstrained solution as the number of layers increases. Therefore,

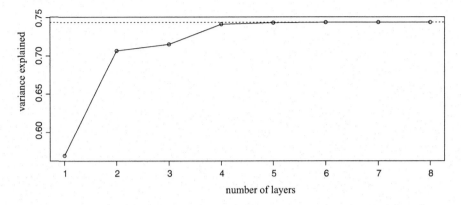

Fig. 1 Scree plot of variance explained against the number of layers; variance explained of unconstrained solution is shown as dotted line

we propose to use a scree plot of variance explained for how many layers should be employed. In the figure, the scree point can be found at the number of layer is equal to two and this implies that two layers should be employed in the LMR. In addition, if the one requires a solution which explains at least 72% of total variance, which is slightly lower than the one of unconstrained solution, the number of layers can be set at three.

3 Numerical Simulation

In order to assess the behaviors of the LMR algorithm, we performed a numerical simulation study. The simulation was designed to assess how well the LMR recovers the true parameters from which the artificial datasets were generated. The number of observations and independent/dependent variables are set at 100, 20, and 3. The 20×3 true coefficient matrix $\mathbf{W}^*$ was synthesized by $\mathbf{W}^* = \sum_l \mathbf{W}_l^* = \sum_l \mathbf{D}_l^* \mathbf{R}_l^*$. The elements of $\mathbf{D}_l^*$ and $\mathbf{R}_l^*$, which denote the true value of $\mathbf{D}_l$ and $\mathbf{R}_l$, were randomly drawn from the standard normal distribution $N(0, 1)$ and the range of $[1, Q]$, respectively. The sign of the elements in $\mathbf{D}$ was randomly determined. The elements of 100×20 matrix of independent variables $\mathbf{X}$ and error matrix $\mathbf{E}$ were also drawn from $N(0, 1)$. The matrix of dependent variable $\mathbf{Y}$ was synthesized with

$$\mathbf{Y} = \mathbf{X}\mathbf{W}^* + \mu(\rho)\mathbf{E} \tag{7}$$

where ρ is the ratio of the variance explained by the model part $\mathbf{XW}^*$ against the error variance. The function $\mu(\rho)$ is defined as

$$\mu(\rho) = \sqrt{\frac{1-\rho}{\rho} \times \frac{||\mathbf{XW}^*||^2}{||\mathbf{E}||^2}} \qquad (8)$$

and it is used to control the level of errors (adachi 2011). In this simulation study, we considered low error ($\rho = 0.8$) and high error condition ($\rho = 0.6$) indicating that 80% and 60% of variance of $\mathbf{Y}$ is explained by the one of $\mathbf{XW}$, respectively. In each condition, a hundred of $\mathbf{X}$ were synthesized and the LMR with the layer number $L = 1, 2, \ldots, 5$ were independently applied. The estimated coefficient matrix $\hat{\mathbf{W}}$ was compared with $\mathbf{W}_*$ with respect to their proximity. As a measure of accuracy of parameter recovery, we used Averaged Absolute Error (AAE) defined as $AAE = (PQ)^{-1}|\text{abs}(\hat{\mathbf{W}}) - \text{abs}(\mathbf{W}_{true})|$ where $\text{abs}(\cdot)$ denotes a matrix with elements replaced by these absolute values. Also, as a measure of goodness of fit, variance accounted for the model part $\mathbf{X}\hat{\mathbf{W}}$ which is defined as $VE = ||\mathbf{X}\hat{\mathbf{W}}||^2/||\mathbf{Y}||^2$ is used.

Figure 2 shows the resulting AAE values and it is clear from the figure that AAE value is decreased at a reasonable level (i.e., less than 0.1) as the number of layers increases. However, too much layers slightly degenerate the quality of recovery, as AAE increases in the cases of four or more layers were employed. This is because the zero elements in a layer are overwritten by nonzero element in the successive layers, although the corresponding element in $\mathbf{W}^*_{true}$ is equal to zero. In addition, Fig. 3 shows the boxplots of VE in two error conditions. As indicated in the figure, VE is improved as the number of layers increases in both conditions. It is worthy of note that VE of $\hat{\mathbf{W}}$ gets closer to one by the unconstrained multivariate regression when three or more layers employed. The above results indicate that the

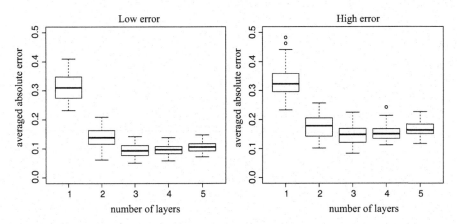

Fig. 2 Boxplots of AAE values against layer numbers in low error condition (left) and high error condition (right)

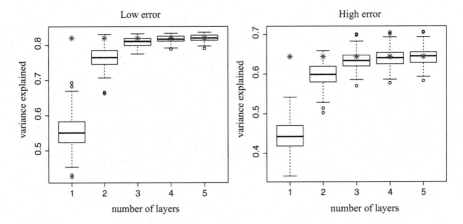

Fig. 3 Boxplots of VE values against layer numbers in low error (left) and high error (right) condition; * indicates the VE for the unconstrained solution

proposed method fairly recovers true parameter, when a sufficient number of layers are employed.

4 Real Data Example: Tobacco Data

As a demonstration of the proposed method, LMR is applied Tobacco dataset refereed in Sect. 1. Table 3 shows the results of LMR with $L = 1, \ldots, 4$ and the resulting variance explained. It can be seen that the number of the exact zero elements increases as L increases, and clear correspondences between the independent variables (rows) and the dependent variables (columns) are observed in the cases of fewer layers, although the goodness of fit is worse than the unconstrained solution in Table 1. In other words, the result suggests "trade-off" of the better fit and the simplicity, because variance explained gets higher but fewer exact zero elements as the number of layers increases. It is reasonable to assume that $L = 3$ is appropriate, since the variance explained is sufficient compared with unconstrained solution and also inter-/within-column contrast is highlighted.

5 Extension of LMR: Layered Principal Component Analysis

The concept of LMR can be applied to various procedures for multivariate analysis in order to improve the interpretability of their solutions. In this article, we present an application to principal component analysis, in which loading matrix is expressed as a sum of layers. It is called Layered PCA (LPCA) and formulated as the minimization of

$$f_{LPCA}(\mathbf{F}, \mathbf{A}_1, \ldots, \mathbf{A}_L) = ||\mathbf{X} - \mathbf{FA}'||^2 = ||\mathbf{X} - \mathbf{F}\sum_l \mathbf{A}_l'||^2 \qquad (9)$$

where the N (observations) $\times$ P (variables) data matrix $\mathbf{X}$ is approximated by the product of the $N \times M$ (components) matrix of component scores $\mathbf{F}$ and the $P \times M$ loading matrix $\mathbf{A}$.

The loss function $f_{LPCA}(\mathbf{F}, \mathbf{A}_1, \ldots, \mathbf{A}_L)$ is minimized under the constraint $n^{-1}\mathbf{F}'\mathbf{F} = \mathbf{I}_M$, as well as in the standard PCA. For the minimization, the loss function can be rewritten as

$$f(\mathbf{F}, \mathbf{A}_1, \ldots, \mathbf{A}_L) = -2\mathrm{tr}\mathbf{X}'\mathbf{FA}' + c \qquad (10)$$

where c denotes the constant irrelevant to $\mathbf{F}$. Here, let $N^{-1/2}\mathbf{XA} = \mathbf{UDV}'$ be the singular value decomposition of $N^{-1/2}\mathbf{XA}$, where $\mathbf{U}$ and $\mathbf{V}$ denote matrices of the left and right singular vector of $N^{-1/2}\mathbf{XA}$, and $\mathbf{D}$ is a diagonal matrix of the corresponding singular values. $\mathbf{F}$ which minimizes the loss function is given by

$$\mathbf{F} = N^{1/2}\mathbf{UV}'. \qquad (11)$$

In addition, we have

$$f(\mathbf{F}, \mathbf{A}_1, \ldots, \mathbf{A}_L) = \left\|\left(\mathbf{X} - \mathbf{F}\sum_{k \neq l}\mathbf{A}_k'\right) - \mathbf{FA}_l'\right\|^2 \qquad (12)$$

for $l = 1, \ldots, L$, and it implies the l-th layer is obtained by the D- and R-step in the optimization of LMR we proposed above. Therefore, $\mathbf{F}$ and $\mathbf{A}$ which minimize the loss function can be found by repeating the following until the convergence reached; (1) update $\mathbf{F}$ by (11) based on the SVD of $N^{-1/2}\mathbf{XA}$, (2) LMR for the matrices of dependent/independent variables $(\mathbf{X}, \mathbf{F})$ and update $\mathbf{A}$.

LPCA with three layers and four components is applied to Wine dataset (Lichman 2013), where the 173 wines are described by the 13 chemical features. Also, we applied Sparse PCA by Zou et al. (2006) to the dataset for comparison and the estimated loading matrix is shown in Table 4. We can see the clear correspondence of each component and reduced number of variables; the third component obtained by the LPCA is characterized by "MAc," "Ash," and "AAsh" and therefore the components are easy to be characterized. The values of variance explained indicate that the solution obtained by LPCA fits to the dataset better than the one of the existing sparse PCA procedure.

Table 4 Loading matrices obtained by LPCA with three layers (left) and Sparse PCA (right) by Zou et al. (2006); exact zero elements are shown as blank cells

	Layered PCA (L = 3)				Sparse PCA			
	PC1	PC2	PC3	PC4	PC1	PC2	PC3	PC4
Alc	0.16	0.85				0.53		
MAc	0.58		0.31					−0.74
Ash		0.34	0.85		−0.02	0.15	0.78	
AAsh		−0.39	0.80			−0.19	0.62	
Mg		0.39		0.80	−0.05	0.33		
TotP	−0.71	0.53			−0.44	0.07		
Flv	−0.82	0.46			−0.50			
NFlv	0.53			−0.55	0.31		0.11	
Pro	−0.58	0.38			−0.41			
Col	0.60	0.63			0.14	0.54		−0.01
Hue	−0.77			−0.10	−0.10	−0.02		0.67
OD	−0.86	0.19			−0.50	−0.04		
Pro	−0.21	0.84			−0.11	0.50		
Vexp.	0.712				0.658			

6 Concluding Remarks

In this article, we considered to improve interpretability of regression coefficient matrix and proposed Layered Multivariate Regression in which the coefficient matrix is expressed as a sum of layers having simple structure. The proposed procedure allows to capture the relationship between two variable sets more easier than the unconstrained regression, because a reduced number of independent variables corresponds to each dependent variables. The results of simulation study show that LMR correctly recovers true parameters when the sufficient number of layers are used. Also, it is empirically shown that the goodness of fit of LMR solution gets closer to the one of the standard multivariate regression as the number of layers increases. As an extension of the proposed method, layered principal component analysis is also proposed.

Throughout this article, we considered a perfect cluster structure as the layer-level structure, in both LMR and LPCA. Another simple structure for layers should be considered in order to improve interpretability of the resulting solution. Further, application of the proposed method to another multivariate analysis should be considered. For example, simple structure in canonical correlation analysis, multiple correspondence analysis, K-means clustering might be helpful for easier interpretation of those solutions.

References

Adachi, K.: Constrained principal component analysis of standardized data for biplots with unit-length variable vectors. Adv. Data Anal. and Classif. **5**, 23–36 (2011)

Hastie, T., Tibshirani, R., Wainwright, M.: Statistical learning with sparsity: the lasso and generalizations. CRC press, West Palm Beach (2015)

Izenman, A.J.: Modern Multivariate Statistical Techniques: Regression, Classification and Manifold Learning. Springer, New York (2008)

Lichman, M.: UCI Machine Learning Repository. https://archive.ics.uci.edu/ml/index.php (2013)

Thurstone, L.L.: Multiple-Factor Analysis. University of Chicago Press, Chicago (1947)

Ullman, J.B.: Structural equation modeling: reviewing the basics and moving forward. J. Personality Assess. **87**, 33–50 (2006)

Vichi, M.: Disjoint factor analysis with cross-loadings. Adv. Data Anal. and Classif. **11**, 563–591 (2016)

Vichi, M., Saporta, G.: Clustering and disjoint principal component analysis. Comp. Stat. and Data Anal. **53**, 3194–3208 (2009)

Yamashita, N.: Sparse multivariate multiple regression analysis and its application. In: The 40th Conference of the Behaviormetric Society of Japan (2012)

Zou, H., Hastie, T., Tibshirani, R.: Sparse principal component analysis. J. Comp. Graph. Stat. **15**, 265–286 (2006)

Part IV
Statistical Data Analysis

An Exploratory Study on the Clumpiness Measure of Intertransaction Times: How Is It Useful for Customer Relationship Management?

Yuji Nakayama and Nagateru Araki

Abstract In the field of marketing science, customer relationship management (CRM) is an important area of research. One of the essential goals of CRM is to measure customers' behavior based on their purchase history to predict their future behavior and find valuable customers. The recency/frequency/monetary value (RFM) framework that has been used since the 1960s is well known in the industry, and is still valuable for summarizing customers' purchase history (Blattberg, Kim, and Neslin 2008). Recently, Zhang, Bradlow, and Small (2015) proposed the clumpiness (C) measure of inter-event times, defined as the degree of nonconformity of equal spacing, and showed that adding C to an RFM framework enhanced the predictive power of their empirical applications in relation to customers' future behavior. In this study, we examine the robustness of their results, and explore whether it may prove useful in finding profitable customers in the future. For this purpose, we use the ISMS Durable Goods Dataset 1, a panel dataset of households' purchase history at a major US consumer electronics retailer, provided by Ni, Neslin, and Sun (2012), and conduct Logit and Tobit regression analyses for prediction. We find that clumpiness does not necessarily have predictive power in relation to customers' future visits and their spending. However, we also find that a low C value, which means the regularity of shopping intervals, predicts customers' future spending. Moreover, we also find that the interaction effect between clumpiness and frequency reduces the prediction error in customers' future spending (i.e., a customer with clumpy shopping behavior who frequently visits the retailer will spend more than other customers in the future). Thus, Zhang, Bradlow, and Small's (2015) findings do not hold exactly as they are, but we find the C measure has helpful aspects that they do not mention.

Y. Nakayama (✉) · N. Araki
Osaka Prefecture University, Sakai, Osaka, Japan
e-mail: nakayama@eco.osakafu-u.ac.jp; araki@eco.osakafu-u.ac.jp

© Springer Nature Singapore Pte Ltd. 2020
T. Imaizumi et al. (eds.), *Advanced Studies in Classification and Data Science*,
Studies in Classification, Data Analysis, and Knowledge Organization,
https://doi.org/10.1007/978-981-15-3311-2_26

1 Introduction

Customer relationship management (CRM) is an established area in the field of marketing science (Neslin 2014; Winer and Neslin 2014). Kumar and Reinartz (2012, p. 4) define CRM from a customer value perspective as "the practice of analyzing and using marketing databases and leveraging communication technologies to determine corporate practices and methods that maximize the lifetime value of each customer to the firm."

Nowadays, large costs are necessary for acquiring new customers, so it is indispensable for a firm to retain loyal customers who repeatedly purchase products from the firm, bringing large profits. These customers have high lifetime value. Managing the relationship with excellent customers is crucial for a firm to maximize its corporate value. For that purpose, a firm needs to measure customers' behavior based on their purchase history to predict their future behavior and to find valuable customers.

In this study, we focus on a new concept in customer relationship management proposed by Zhang, Bradlow, and Small (2015) —clumpiness— which is a measure of aspects of customers' purchasing behavior that uses their intertransaction times. They showed that adding the new measure to the existing framework for summarizing customers' purchase histories in their empirical applications enhanced the power to predict future behavior. However, more validation is needed to confirm the robustness of their results, which is the objective of our study.

The remainder of this chapter is as follows. In Sect. 2, we briefly review the recency/frequency/monetary value (RFM) framework to summarize customer behavior. In Sect. 3, we explain the clumpiness measure, which captures an important aspect of customers' behavior that the RFM framework cannot. In Sect. 4, we explain the data and method used in our empirical analysis. The results are provided in Sect. 5. Concluding remarks are given in Sect. 6.

2 RFM Measure

In this section, we review the the RFM value framework and point out that it overlooks an important aspect of customers' purchasing behavior.

RFM is a summary of the purchasing behavior of a customer from a store in a specific time period. Recency (R) is the duration between the most recent day of purchase and the current day. It measures time since a customer's last purchase. Frequency (F) is the purchase frequency in the time period. It measures how often a customer makes purchases at the store. Monetary value (M) is total expenditure in the period.[1] If a customer has low R, high F, and/or high M values, these indicate that she is an excellent customer.

---

[1] Alternatively, average expenditure at one time may be used.

Fig. 1 Example of customers' purchasing histories

Figure 1 shows purchasing behaviors by four hypothetical people who become customers of a given store at $t = 0$. There are 30 potential opportunities (e.g., days or weeks) for shopping ($t = 1, 2, \ldots, 30$). A black circle in the figure indicates a purchase incidence. It is assumed that they spend 30 dollars on each purchase incidence.

Table 1 summarizes their RFM values. Customer A is the best customer of the four because this customer's R is the lowest and F and M are the highest. Customer B is the second best. On the other hand, based on RFM measures, we cannot distinguish between the purchasing behaviors of Customers C and D. However, in the figure, their intertransaction times are very different. Customer C's intertransaction times increase over time, while Customer D does make repeat purchases in a short period of time; then, after a long period of no purchases, she recommences repeat shopping. That is, Customer D's purchase event looks like a clump, collected in a short period of time. Such phenomena motivated Zhang, Bradlow, and Small (2015) to construct a new concept of purchasing behavior.

Table 1 RFM measures of the four customers in Fig. 1

	Recency	Frequency	Monetary value
Customer A	2	14	420
Customer B	10	10	300
Customer C	10	5	150
Customer D	10	5	150

3   Clumpiness Measure

Zhang, Bradlow, and Small (2015), hereafter denoted as ZBS, defined clumpiness as
the degree of nonconformity with equal spacing of visit/purchase incidences (ZBS,
p. 196).[2] ZBS provide the following desirable properties of a clumpiness measure
(ZBS, p. 197):

- Minimum. The measure should be the minimum if the events are equally spaced.
- Maximum. The measure should be the maximum if all of the events are gathered
  together.
- Continuity. Shifting event times by a very small amount should only change the
  measure by a small amount.
- Convergence. As events move closer, the measure should increase.

ZBS provided a metric for clumpiness to satisfy the above properties:

$$C = 1 + \frac{\sum_{i=1}^{n+1} \log(x_i) x_i}{\log(n+1)}. \tag{1}$$

N and n are the number of purchase opportunities and the number of actual
purchases in a given period, respectively, while x_i is intertransaction time, which
is standardized as the sum of x_i becomes equal to one in a given period defined as:

$$x_i = \begin{cases} \frac{t_1}{N+1} & \text{if } i = 1, \\ \frac{t_i - t_{i-1}}{N+1} & \text{if } i = 2, \ldots, n, \\ \frac{N+1-t_n}{N+1} & \text{if } i = n+1, \end{cases} \tag{2}$$

where t_i is the ith occurrence of the event time from the start of the period.

Based on formula (1), the clumpiness measures of Customers A, B, C, and D are
calculated as 0.002, 0.109, 0.081, and 0.315, respectively. As these numbers show,
Customer D's C value is the largest, which means her purchasing behavior is very
clumpy, while Customer A's C value is the smallest, which means her purchasing
behavior has a regular pattern. On the other hand, Customer B's C value is larger
than that of Customer C, which seems to be inconsistent with their pattern of
purchasing behaviors. However, strictly speaking, a large C value indicates clumpy
purchasing behavior given N and n. Under different pairs of N and n, clumpiness is
judged using a statistical test. Under the null hypothesis that intertransaction times
are random given N and n, we can make an empirical distribution of the C measure
using a Monte Carlo simulation,[3] then, find the critical value of $\alpha\%$ on the upper side
of the distribution. If the C value is larger than the critical value, then the customer's
purchasing behavior is judged to be clumpy. The 5% percentiles on the upper side

---

[2]Zhang, Bradlow, and Small (2013) describe the details of this measure.
[3]See Appendix B of Zhang, Bradlow, and Small (2015) for the test in detail.

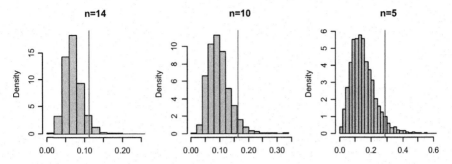

Fig. 2 Empirical distributions of the C measure given $N = 30$

of the distribution under the null hypotheses of $n = 14$, 10, and 5 given $N = 30$ are calculated as 0.112, 0.163, and 0.289, respectively.[4] Thus, only Customer D's purchasing behavior is considered to be clumpy because her C value, 0.315, is larger than the critical value of 0.289. Figure 2 shows three empirical distributions of the C measure given $N = 30$ under the above null hypothesis. The vertical lines represent the critical value of 5% on the upper side of distributions. We found that the shape of distributions and the corresponding critical values are stable to repeated simulations.[5]

In their empirical applications, ZBS show that clumpiness is useful in many but not all cases for predicting customers' future purchasing behavior in terms of visits and spending.[6] However, ZBS also suggest that the context-specific nature of clumpiness must be examined in a more systematic manner in future. In the following sections, we conduct an empirical analysis for such a direction.

4   Data and Method

In this section, we explain the data and method used in our empirical analysis. We also describe the modification of original metric (1) to adapt to the features of the data.

4.1   Data

We use the ISMS Durable Goods Dataset 1 provided by Ni, Neslin, and Sun (2012) for academic research. This contains customers' purchase history at a major

---

[4]These figures are approximately obtained by conducting a Monte Carlo simulation of 20 thousand repetitions with R Statistical Software (R Core Team 2017).

[5]Based on our 100 simulations, the standard deviations of the critical values for the cases of $n = 14$, 10, and 5 were about 1.0×10^{-4}, 1.4×10^{-3}, and 1.7×10^{-3}, respectively.

[6]They measure C values not only by purchase intervals (Purchase-C) but also by visit intervals (Visit-C) in their empirical applications.

(anonymous) US consumer electronic retailer for 6 years from December 1998 to November 2004. The retailer operates multiple stores. Not only transactions of electronic durables but also those of products such as music CDs, bottled water, and snack foods are included in this dataset.

To examine the usefulness of the C measure with RFM values, hereafter denoted as RFMC measure, we divide the data into two periods: the period from the start of December 1998 to the end of October 2002 is used for calculating RFMC measures, while the period from the start of November 2002 to the end of November 2004 is used for computing the number of purchase incidences and expenditure. We call the former the in-sample period, and the latter the out-of-sample period.

To choose customers shopping at the retailer during relatively long periods in similar situations, we set the following criteria: the first purchase was before the end of January 2001, and shopping was done at only one store at least four times after the first purchase in the estimation period. We find that 653 customers satisfy the above criteria.

4.2   Method

We made some changes to the calculation of the C measure. In our dataset, the purchase frequency is usually not high. Thus, the original formula by ZBS tends to create a large C value if we use fixed start and end days for all customers.[7] Then, we decided to use customer-specific start and end days, and to calculate intertransaction times between the first and the last purchase days for measuring C defined in the following modified formula.

$$C = 1 + \frac{\sum_{i=1}^{n} \log(x_i) \, x_i}{\log(n)}, \tag{3}$$

$$x_i = \begin{cases} \frac{t_1}{N} & \text{if } i = 1, \\ \frac{t_i - t_{i-1}}{N} & \text{if } i = 2, \dots, n, \end{cases} \tag{4}$$

where the first purchase incidence is excluded when calculating x_i. That is, t_1 is the interval between the first and the second purchase incidences, while t_n is the interval between the first and the last purchase occurrences. N is the number of potential purchase opportunities between the first and the last purchase incidences.

Importantly, we judge not only clumpiness but also regularity of purchases by each customer using Eq. (3). In particular, the critical value of $\alpha = 2.5\%$ at both sides of an empirical distribution is computed under a null hypothesis of random intertransaction times. For each customer, if her measure (3) is higher (lower) than

---

[7]Rao (2015) pointed out a related problem in his commentary on ZBS's original paper.

the 2.5% upper (lower) side of a critical value, then she is judged to have a clumpy (regular) purchasing pattern.

We calculate the chosen 653 customers' RFM values and dummy variables, identifying clumpiness and regularity by using in-sample data, and predict their purchasing behavior; in particular, whether they purchase at least once and how much they spend in the out-of-sample period.

For the prediction of purchase incidence, we conduct a Logit regression in which the dependent variable is binary: one if customer i purchases at least once ($Buy_i = 1$), or zero if she never purchases ($Buy_i = 0$) in the out-of-sample period. In Logit regression model, the probability that customer i purchases at least once ($Buy_i = 1$) during the out-of-sample period is given as

$$Pr\,(Buy_i = 1) = \frac{\exp\left(\overline{\mu}_i^*\right)}{1 + \exp\left(\overline{\mu}_i^*\right)}, \tag{5}$$

$$\overline{\mu}_i^* = \beta_1 + \beta_2 R_i + \beta_3 F_i + \beta_4 M_i + \beta_5 Clum_i + \beta_6 Reg_i, \tag{6}$$

where R_i, F_i, and M_i are her recency, frequency, and monetary values, respectively during the in-sample period, while $Clum_i$ or Reg_i is a dummy variable identifying whether she has a clumpy or regular purchasing pattern during that period. That is, if customer i's measure (3) is higher (lower) than the 2.5% upper (lower) side of a critical value, then $Clum_i = 1$ ($Reg_i = 1$). Otherwise, $Clum_i = 0$ ($Reg_i = 0$).

For the prediction of money amount, we conduct a Tobit regression whose dependent variable is the money spending in the out-of-sample period, which is the mix of continuous and discrete values: it is positive and varies continuously ($Y_i > 0$) if customer i purchases at least once during the out-of-sample period, while it is zero ($Y_i = 0$) if she never purchases in that period. Tobit regression model is formulated as

$$Y_i = \begin{cases} Y_i^* & \text{if } Y_i^* > 0, \\ 0 & \text{if } Y_i^* \le 0, \end{cases} \tag{7}$$

$$Y_i^* = \beta_1 + \beta_2 R_i + \beta_3 F_i + \beta_4 M_i + \beta_5 Clum_i + \beta_6 Reg_i + \varepsilon_i, \quad \varepsilon_i \sim N\left(0, \sigma^2\right), \tag{8}$$

where Y_i^* is the unobserved latent variable expressing the tendency of money spending and ε_i is the error term following a normal distribution. The same explanatory variables as in (6) are also used in the Tobit regression. The parameters $\{\beta_i, \ i = 1, 2, \ldots, 6\}$ in the two regression models are estimated by the maximum likelihood method.

5 Result

Table 2 contains the summary statistics of explanatory variables during the in-sample period and dependent variables during the out-of-sample period. On average, a customer spends $ 1800 and makes purchases about 6 times in the in-sample period. The mean duration between the most recent day of customers' purchase and the end of October 2002 is about 286 days. About 16% of customers have a clumpy purchasing pattern, while about 2% of customers have a regular purchasing pattern. About 64% of customers return to the store during the out-of-sample period. Note that the average monetary value of $590 during that period is based on the purchase history of all customers including those not returning to the store. Based on the return customers' purchase history, the average monetary value is about $927.

The first column in Table 3 shows the estimated parameters of Logit regression (5)–(6). We find that recency and frequency are statistically significant. That is, they have the predictive power about the probability of return purchase during the out-of-sample period. If a customer has a low recency (high frequency) value, then the probability increases. On the other hand, other variables including clumpiness do not have that power. Table 4 shows the actual and predicted outcomes of purchase incidence during the out-of-sample period based on the Logit model. The result is reasonable since the rates of precision and recall are 69% ($376/(166 + 376)$) and 90% ($376/(40 + 376)$), respectively.

The second column in Table 3 shows the estimated parameters of Tobit regression (7)–(8). We find that recency, frequency, and monetary value are statistically significant. That is, they have the predictive power about the future money spending. In addition, we find that regularity is significant, which means that a customer who has a regular shopping pattern spends more in the future. On the other hand, clumpiness is not significant as in the Logit model, and has no predictive power for the future money spending.

Table 2 Summary statistics

	Mean	S.D.	Max	Min
In-sample period				
Recency (days)	285.80	243.50	1115	0
Frequency (the number of purchases)	5.87	3.22	33	4
Monetary value ($)	1846.95	1618.13	11449.84	66
Clumpiness	0.16	0.36	1	0
Regularity	0.02	0.13	1	0
Out-of-sample period				
Buy	0.64	0.48	1	0
Monetary value	590.50	1039.20	7980.90	0

Table 3 Results of estimation by Logit and Tobit regressions

	Logit	Tobit 1	Tobit 2
Intercept	0.78***	−156.69	−144.53
	(0.26)	(142.81)	(140.80)
Recency	-2.51×10^{-3}***	−1.46***	−1.45***
	(3.72×10^{-4})	(0.25)	(0.25)
Frequency	0.10**	93.00***	91.00***
	(0.04)	(19.68)	(19.63)
Monetary value	-1.88×10^{-5}	0.09**	0.09**
	(6.56×10^{-5})	(0.04)	(0.04)
Clumpiness	−0.18	217.77	
	(0.23)	(151.29)	
Regularity	0.52	1388.77***	1388.36***
	(0.70)	(389.64)	(388.93)
Frequency × Clumpiness			42.40*
			(24.81)

Note: standard errors are reported in parentheses
*** $p < 0.01$, ** $p < 0.05$, * $p < 0.1$

Table 4 Actual and predicted purchase incidence based on the Logit model

		Predicted outcome		
		$Buy_i = 0$	$Buy_i = 1$	Total
Actual outcome	$Buy_i = 0$	71	166	237
	$Buy_i = 1$	40	376	416
	Total	111	542	653

However, if we remove clumpiness and add the interaction term between frequency and clumpiness as the explanatory variable as follows:

$$Y_i^* = \beta_1 + \beta_2 R_i + \beta_3 F_i + \beta_4 M_i \qquad (9)$$

$$+ \beta_6 Reg_i + \beta_7 (F_i \times Clum_i) + \varepsilon_i, \qquad (10)$$

then, as shown in the third column in Table 3, we find the interaction effect is weakly significant, which means that a customer with clumpy shopping behavior who frequently visits the retailer during the in-sample period spends more than other customers in the out-of-sample period. Moreover, this modification reduces the mean absolute error of the money spending during the out-of-sample period from 1527.27 to 544.52. This result seems noteworthy.

6 Conclusion

In this study, we focus on a new concept in customer relationship management proposed by Zhang, Bradlow, and Small (2015) —clumpiness— which is a measure of aspects of customers' purchasing behavior that uses their intertransaction times.

Contrary to Zhang, Bradlow, and Small's (2015) findings, we do not observe that customers with high C values (i.e., those whose purchasing behavior is clumpy) purchase more in the prediction period than do other customers. Our empirical application used transaction data for durable products (mainly consumer electronics) at a physical retailer. The same products might have been available from competitors. Thus, customers did not necessarily need to return to the retailer for new transaction. However, we also find that the interaction effect between clumpiness and frequency reduces the prediction error in customers' future spending. That is, a customer with clumpy shopping behavior who frequently visits the retailer will spend more than other customers in the future. It suggests that clumpiness when combined with frequency may become useful.

In addition, we find that customers with low C values (i.e., their purchasing behavior has regular intervals) purchase more in the prediction period than do other customers. Zhang, Bradlow, and Small (2015) did not focus on regularity in transaction intervals. They may have considered that it is not surprising that customers with regular transaction intervals return. However, the measure that they developed (1), or that we modified (3), can be used for identifying regularity in intervals. RFM measures could not identify such regularity in customers' purchasing behavior. Therefore, we consider that measures (1) or (3) are valuable tools for summarizing an aspect of customers' purchasing behavior.

To clarify the potential of the C measure, more empirical applications are needed. Moreover, we should explore the cause of heterogeneous intertransaction times among customers of products in the same category to develop a new model of purchasing behavior of customers with heterogeneous intertransaction times. Platzer and Reutterer (2016) proceeded in this direction. However, there remains much research to be conducted to develop a new model of customers' purchasing behavior.

Acknowledgments The authors would like to thank those who made comments and suggestions at IFCS-2017 and the 2017 Annual Spring Conference of the Japan Institute of Marketing Science held at Keio University. In particular, comments and suggestions by Professors Takahiro Hoshino, Takeshi Moriguchi, and Nozomi Nakajima are gratefully acknowledged. The authors would like to thank the INFORMS Society of Marketing Science (ISMS) (https://connect.informs.org/isms/) for providing the data used in this research. They also would like to thank the anonymous reviewer whose comments improved the content of this chapter. All remaining errors are the authors' own responsibility. This work was supported by the Japan Society for the Promotion of Science (JSPS) through a Grant-in-Aid for Scientific Research (C) (Grant Number: JP15K03729), which is gratefully acknowledged.

References

Blattberg, R.C., Kim, B.D., Neslin, S.A.: Database Marketing: Analyzing and Managing Customers (International Series In Quantitative Marketing). Springer, Berlin (2008)

Kumar, V., Reinartz, W.: Customer Relationship Management: Concept, Strategy, and Tools, 2nd edn. (Springer Texts in Business and Economics). Springer, Berlin (2012)

Neslin, S.A.: Customer relationship management (CRM). In: Winer, R.S., Neslin, S.A. (eds.) The History of Marketing Science, Chapter 11. World Scientific Publishing, Singapore (2014)

Ni, J., Neslin, S.A., Sun, B.: The ISMS durable goods data sets. Mark. Sci. **31**(6), 1008–101 (2012)

Platzer, M., Reutterer, T.: Ticking away the moments: timing regularity helps to better predict customer activity. Mark. Sci. **35**(5), 779–799 (2016)

Rao, V.R.: Comments on "Predicting customer value using clumpiness: from RFM to RFMC". Mark. Sci. **34**(2), 213–215 (2015)

R Core Team.: R: A language and environment for statistical computing. R Foundation for Statistical Computing, Vienna. https://www.R-project.org/ (2017)

Winer, R.S., Neslin, S.A.: The History of Marketing Science (World Scientific-Now Publishers Series in Business), vol. 3. World Scientific Publishing, Singapore (2014)

Zhang, Y., Bradlow, E.T., Small, D.S.: New measures of clumpiness for incidence data. J. Appl. Stat. **40**(11), 2533–2548 (2013)

Zhang, Y., Bradlow, E.T., Small, D.S.: Predicting customer value using clumpiness: From RFM to RFMC. Mark. Sci. **34**(2), 195–208 (2015)

A Data Quality Management of Chain Stores based on Outlier Detection

Linh Nguyen and Tsukasa Ishigaki

Abstract For successfully analyzing data in the business of chain stores, the quality of data recorded in their shops or factories is a key factor. Data quality management is an important practical issue because data qualities widely vary depending on the managers or workers of many stores in the chain. In this paper, we present a data quality evaluation method for shops in chain businesses based on outlier detection and then, we apply this method to a dataset observed in real chain stores, which provide tire maintenance for vehicles. To evaluate the data quality of each shop, we use data about truckŠs tire information such as tread depth, tread pattern, and distance which was recorded by the shops at maintenance time to calculate low-quality data by using outlier detection methods with reliable experimental data and practical knowledge. Some outlier detection methods such as Isolation Forest and one-class Support Vector Machine are applied to detect anomalous tire information, which is used to calculate dataŠs abnormal rate in each shop. Our result showed that with this kind of data, Isolation Forest is outstanding than other methods because Isolation Forest is designed to detect Šfew and differentŤ outliers. The proposed method can support better maintenance services for customers as well as be able to get more correct data from these shops, which will be useful for the next research.

1 Introduction

Customer satisfaction is interested topic for not only industry, but also academics (Morgan et al. 2005; Anderson and Sullivan 1993; Anderson et al. 1997; Guo et al. 2009). A numerous researches and practical experiments are presented that service quality, service value, and customer satisfaction may all be directly related to customer behavior (Jr et al. 2000; Choi et al. 2004; Lai et al. 2009; Grönroos

L. Nguyen (✉) · T. Ishigaki
Tohoku University, Sendai, Japan
e-mail: igsk@tohoku.ac.jp

© Springer Nature Singapore Pte Ltd. 2020
T. Imaizumi et al. (eds.), *Advanced Studies in Classification and Data Science*,
Studies in Classification, Data Analysis, and Knowledge Organization,
https://doi.org/10.1007/978-981-15-3311-2_27

1984). Therefore, remaining and improving customer satisfaction is an important goal of many companies.

In chain stores, as human and device resources are in distributed area, retaining service quality uniform is even harder (Beckman and Nolen 1938). If a customer comes to a store and experiences a service, next time, when he comes to another store in this chain, he hopes to receive at least same service quality. If the service quality he received at second time does not match with it in first time, his disconfirmation belief will be negative and affects directly to satisfaction, following expectation and confirmation theory (Oliver 1977). An effective way to avoid this situation is standardizing whole chain (Polo-Redondo and Cambra-Fierro 2008; Smith et al. 2015).

Standardization takes the form of manuals, operating procedures, and other blue prints to regulate individual behaviors so as to control, predict, and minimize mistakes and deviation among employees (Wang et al. 2010).

Besides, data have become more and more essential in business process (chin Chen et al. 2012; Chengalur-Smith et al. 1999). They contribute an objective view in assessment besides human factor. In data quality management, we use outlier detection algorithms to observe anomaly data. Anomalies can come from many sources in collecting process such as failure sensor, imprecise measurement tool, or inexact reader. This process allows us to track oddity rate of each store and to highlight unreliable stores. Then, managers in headquarter cooperate with employees in these unreliable stores to find the cause and give solutions. For example, if the problem is human factor, they may re-train for employees, or if device has trouble, they can supply a new one. This step also helps us clean data for other research purpose as well as improve raw data quality in future.

In this paper, to standardization chain stores, we present a data quality evaluation method for shops in chain businesses based on outlier detection and then, we apply this method to a dataset observed in real chain stores, which provide tire maintenance for vehicles. To evaluate the data quality of each shop, we use data about truck's tire information such as tread depth and distance which was recorded by shop at maintenance time to calculate obviously low quality data by using outlier detection methods with reliable experimental data and practical knowledge. Some of outlier detection methods such as Isolation Forest and one-class Support Vector Machine are applied to detect anomalous tire information, which make use for calculate data's anomaly rate in each shop.

2 Problem Setting

In this section, we give a high-level overview of our research and data problems, which lead to our goals and solutions. Although the context is within our company, these insights may be same with other chain stores.

2.1 Research Problem

As we wrote in Sect. 1, if shops in a chain have not same service quality, it will affect customer satisfaction of the whole system. To solve this problem, our research goal is detecting unreliable shops which have low data quality, through dataset collected from chain stores, so that managers and employees of these shops can propose solution to improve it.

2.2 Data Description

We use a dataset observed in tire maintenance service chain. It contains 2.5M+ rows with more than 200 stores in Japan. Because stores are in distributed areas, data they collected have many problems such as missing values, anomalies, and uncorrected label. We found that high proportion data (40%) which have enough information is untrustworthy. Therefore, we need to review our data quality carefully.

2.3 Research Method

There are many data problems, but we focus on two characteristics which are not easy to solve, and usually happen in real data, they are:

- High rated untrustworthy data
- Train data only has trustworthy data which can be generated from experiments

With these characteristics, in this work, we try to identify anomalies in high rated noisy data to detect unreliable shops through semi-supervisor outlier detection algorithms. To do this, we propose a Data Quality Management Method.

We provide the method in two steps:

- **Step 1:** Outlier Detection with the following algorithms:

  - Isolation Forest
  - One-class SVM
  - Local outlier factor

- **Step 2:** Assessment

  - Assess algorithms by F-score
  - Assess shops

3 Outlier Detection Algorithms

There are many outlier detection algorithms divided mainly into four categories: statistical outlier detection methods, distance-based approaches, tree-based approaches, outlier detection using kernel functions and Fuzzy approach with the use of kernel functions (Petrovskiy 2003).

In statistical outlier detection, there are some famous methods such as tests and regression analysis. All these methods require to construct a probabilistic data model with supervisor data, so they are not suitable with noisy practical semi-supervisor data.

Distance-based approaches are different. They do not require any probabilistic model, but based on distance between data. Two well-known branches of this approach are k-nearest neighbors (kNN) and local outlier factor (LOF). kNN predicts class of a point by voting class of k-nearest points in training set. Therefore, with one-class training set, kNN will consider every point in test set be normal and cannot work. Different with kNN, LOF is only based on distance of a point with the nearest points in same set. However, test set has a high outlier rate, and outliers are near each other which leads LOF predicts them as normal.

Outlier detection using kernel function and Fuzzy approach with the use of kernel functions, which Support Vector Machine (SVM) is the most notable algorithm, focus on building the hyperplane bounds normal data. There are some studies about SVM for outlier detection and we use Bernhard Scholkopf's method (Scholkopf et al. 1999). This method concentrates to create hyperplane, which has maximum distance to the origin, to separate and bound the normal data. Because our data has high untrustworthy ratio, and train data only has trustworthy data which generated from experiments, it seems not appropriate methods.

There are not much tree-based approaches for outlier detection. However, recently, Isolation Forest (Liu et al. 2008) which builds a number of binary trees achieves the-state-of-the-art result. It gives two assumptions about anomalies characteristics: minority and different. Despite the fact that our data do not have these characteristics, characteristics of Isolation Tree allow us to beat them. In Isolation Tree, we do not use the whole data to construct, but only a sub-sample. Through sub-sample, we can control data size as well as anomalies set in each tree. Not only novelty but also experiment shows that Isolation Forest be suitable with real data which has characteristics described in Sect. 2.

In our research, we mainly use Isolation Forest, which belongs to tree-based approaches. We also add one-class SVM with RBF kernel and poly kernel as well as LOF to compare.

Table 1 Assessment method

	Predict positive	Predict negative
Condition positive	True positive (TP)	False negative (FN)
Condition negative	False positive (FP)	True negative (TN)

4 Assessment Method

4.1 Algorithm Assessment

To highlight the accuracy to detect outlier, we use F-score and accuracy which are calculated following:

$$precision = \frac{tp}{tp + fp}; \quad recall = \frac{tp}{tp + fn}; \quad accuracy = \frac{tp + tn}{tp + tn + fp + fn}$$

(1)

where tp, tn, fp, tn are defined in Table 1. From that, we calculate F-score:

$$F - score = 2 * \frac{precision * recall}{precision + recall} = \frac{2 * tp}{2tp + fn + fp}$$

(2)

4.2 Shop Assessment

After applying outlier algorithms, we can classify data into two groups: anomalies and normal data. We will assess shops based on outlier percentage of each shop.

$$outlier\_pc_i = \frac{n\_anomalies_i}{n\_data_i} * 100\%$$

(3)

where $\begin{cases} outlier\_pc_i : \text{outlier percentage of shop } i \\ n\_anomalies_i : \text{number of anomalies of shop } i \\ n\_data_i : \text{number of data in shop } i \end{cases}$

Good shops have low outlier percentage and vice versa, unreliable shops have high outlier percentage.

5 Experiment

In this section, we present our detailed data as well as our result.

Fig. 1 Tread and tread depth of tires

5.1 Data

We have a real dataset about tire maintenance information with 2.5M+ rows. One of main purpose of tire maintenance service is checking tire and tread status as well as tread depth as Fig. 1. Staffs in tire service will compare tread depth of checking day and of the day started to use this tread to decide whether that tire is needed a new tread or not as well as predict when it need a new tread. We use this dataset as test set. And after removing missing values, we have:

- 256,788 rows
- Belong to 198 stores

Besides, we also have a train dataset which has 838 rows and contains only normal data.
We use two main data features:

- Distance which is number of kilometers the truck run since a new tread was used
- Erosion (which is equal to difference between tread depth at new and current average tread depth)

5.2 Implement

We use Isolation Forest,[1] OneClassSVM[2], and LOF[3] models in sklearn package. We implement the following four steps.

---

[1] http://scikit-learn.org/stable/modules/generated/sklearn.ensemble.IsolationForest.html.

[2] http://scikit-learn.org/stable/modules/svm.html#svm-outlier-detection.

[3] https://scikit-learn.org/stable/modules/generated/sklearn.neighbors.LocalOutlierFactor.html#sklearn.neighbors.LocalOutlierFactor.

Table 2 Hyperparameter settings for our model

Isolation forest	One-class SVM(1)	One-classSVM(2)
$n\_estimator = 100$	$v = 0.1$	$v = 0.1$
$\max\_samples = 500$	$kernel = RBF$	$kernel = poly$
$random\_state = 42$	$\gamma = 1/n\_features$	$\gamma = 1/n\_features$

5.2.1 Step 1: Train Models

We use the dataset which contains only normal data to train. We construct hyperparameter settings as Table 2.

5.2.2 Step 2: Classify Data

We divide test dataset following shops and apply the models obtains in step 1 into dataset of each shop. The predict result will belong to $\{1, -1\}$.

- -1 means this datum is anomaly
- 1 mean this datum is normal datum.

5.2.3 Step 3: Classify Shops into Good and Unreliable Groups

We applied Eq. (3) then acquire the percentage of outlier in each shop. We define that if 0%–23% data of a shop are anomalies, it will be a good shop and if this percentage is in range of 75%–100%, it will be an unreliable shop.

Figure 2 shows an example of good shop and two examples of unreliable shop. We can see that unreliable shops have high proportion of data which have short distance but extremely high erosion (Fig. 2b) or low erosion but extraordinarily long distance (Fig. 2c). These data are reasonable to be considered as inexact data.

5.2.4 Step 4: Calculate Algorithm Accuracy

We create an obvious outlier area which based on technical information supplied by company. The lines separated outlier area and normal area are defined as:

$$d = 2151 \times e \text{ and} \tag{4}$$

$$d = 27727 \times e \quad \text{which} \begin{cases} \text{2151 and 27727 are provided by company} \\ d : \text{distance (km)} \\ e : \text{erosion (mm)} \end{cases}$$

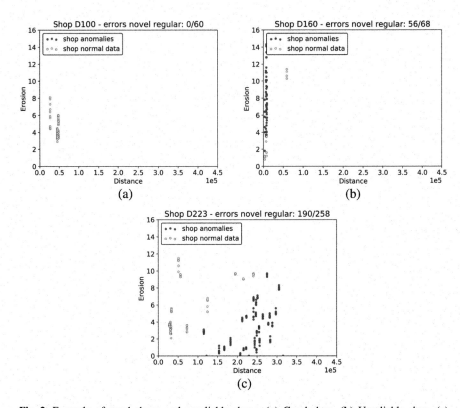

Fig. 2 Example of good shops and unreliable shops. (**a**) Good shop. (**b**) Unreliable shop. (**c**) Unreliable shop

Then normal area is the area inside two lines which distance d and erosion e satisfy:

$$d \geq 2151 \times e \text{ and } d \leq 27727 \times e \tag{5}$$

and obvious outlier area is the area outside two lines which distance d and erosion e satisfy:

$$d < 2151 \times e \text{ or } d > 27727 \times e \tag{6}$$

Then we calculate tp, fp, fn, tn following Table 1 with:

- Condition Positive: data which are not belong to obvious outlier area in 6
- Condition Negative: data which are belong to obvious outlier area in 6
- Predict Positive: data which are predicted as normal data by algorithms (Isolation Forest, one-class SVM with RBF kernel, or one-class SVM with poly kernel)
- Predict Negative: data which are predicted as anomalies by algorithms.

Therefore,

- *tp*: number of data which are not belong to obvious outlier area and are predicted as normal data by algorithms
- *fp*: number of data which are belong to obvious outlier area but are predicted as normal data by algorithms
- *fn*: number of data which are not belong to obvious outlier area but are predicted as anomalies by algorithms
- *tn*: number of data which are belong to obvious outlier area and are predicted as anomalies by algorithms

In addition, we calculate precision, recall, accuracy, and F-score for each shop with each algorithm as in Eqs. (1) and (2).

Finally, we take the average of precision, recall, accuracy, and F-score of all shops then compare results among algorithms.

5.3 Result

5.3.1 Result about Algorithm

We make Table 3 to compare the precision, recall, accuracy, and F-score of Isolation Forest, SVM with RBF kernel, SVM with poly kernel, and LOF. Besides, we also compare the results when using distance and taking logarithm of distance. In the result, we can see that sometimes SVM looks like better than Isolation Forest. However, the reason is that SVM with RBF kernel considers all data as outliers and opposite, SVM with poly kernel considers all data as normal data. As we show above, in this case, one-class SVM determined a bad original point which leads all data space to become outlier or normal. With LOF, as we analyzed in Sect. 3, LOF considers outlier data as normal because they are much near together. It leads its results are much lower than other algorithms. We show an example in Fig. 3 to make clear.

Table 3 Algorithm accuracy

		Recall	Precision	Accuracy	F-score
log(distance)	Isolation forest	0.91	0.76	0.89	0.82
	SVM RBF	0.98	0.70	0.88	0.82
	SVM poly	0.23	0.48	0.62	0.25
	LOF	0.07	0.36	0.31	0.39
Distance	Isolation forest	0.56	0.72	0.77	0.57
	SVM RBF	1	0.38	0.38	0.56
	SVM poly	0.38	0.62	0.69	0.38
	LOF	0.16	0.36	0.59	0.18

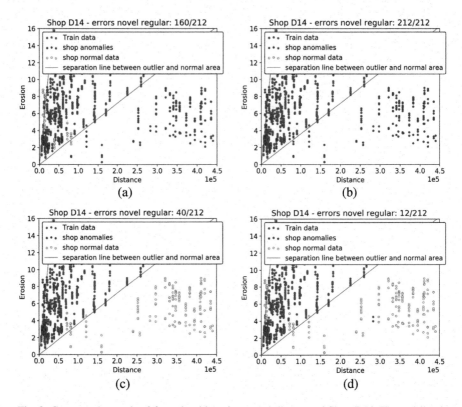

Fig. 3 Compare the result of four algorithms in normal distance of Shop D14. The red line is defined in Eq. (4). (**a**) Isolation forest. (**b**) SVM with RBF kernel. (**c**) SVM with poly kernel. (**d**) LOF

Because of Isolation Forest outstanding algorithm, from now, we only use its results.

5.3.2 Result of Unreliable Shops

We made a rank of shops and plot to see the relationship between outlier percentage and distance in Fig. 4

6 Discussion

After discussion with managers of company we divide shops into groups like Fig. 4.

Best group contains shops have large data amount ($\geq$250), but low outlier percentage (<30%). And safe group contains shops have few data amount (<250) and low outlier percentage (<23%). Company's managers need to focus on highest

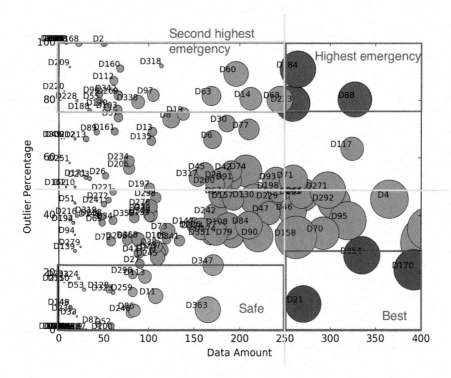

Fig. 4 Group of shops based on emergency rate

emergency group and second highest emergency group. The reason we divided unreliable shops into two groups is that company has many unreliable stores (~41), so it needs much finance, human resource, and effort to solve problems of all at the same time. Corresponding with two groups, we divided our solutions into two stages as Fig. 5.

First is the highest emergency group. It includes shops which have large data amount (>250) and high outlier percentage (>75%). Company's managers need to concern it immediately. We proposed these solutions for company's managers:

- Go to meet the managers of these shops and find the reason.
- Hold training for managers as well as employees of these shops.
- Let managers of these shops go to best shops to ask about experience along with good processes.

Second, after solving problems of the highest emergency group, company's managers need to continue focusing on shops in second highest emergency group. At that time, they need to run the model again to check the current situation of this group. Shops in this group may move following two directions as Fig. 5b:

- Case 1: shops move to safe area or even best area. Shops in second highest emergency group are new shops. Maybe, at first, they are not familiar with

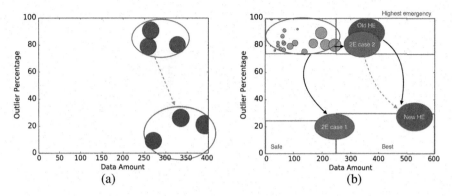

Fig. 5 Solution for company's managers. (**a**) Deal with shops in Highest Emergency Group. (**b**) Deal with shops in Second Highest Emergency Group

process, but after, they can modify by themselves and become good shops. With this case, we do not need to do anything.

- Case 2: shops move to highest emergency area. At this case, we do as same as Stage 1, but instead of asking experience of best shops, they will ask the experience of shops which were in highest emergency group first, but now are in best area. These shops have much experience about modifying themselves from unreliable shops to good ones.

Our method not only uses once, but also has ability to apply in lifetime to control and manage shop quality. Company's managers can run the model quarterly, monthly, or even weekly to assess quality improvement of shops in their chain.

7 Conclusion

This paper proposes a method to manage data quality of chain stores based on outlier algorithms. In addition, we apply this method to a tire maintenance dataset.

Next, we present main groups of outlier algorithms as well as suggest two algorithms: Isolation Forest and one-class Support Vector Machine. We also show the performance of two algorithms and see that Isolation Forest is more suitable with noisy data.

Moreover, we classify shops into groups and propose strategies for improvement chain quality.

The proposed method can support better maintenance services for customer as well as be able to get more correct data from these shops, which will be useful for next research.

References

Anderson, E.W., Sullivan, M.W.: The antecedents and consequences of customer satisfaction for firms. Mark. Sci. **12**, 125–143 (1993)

Anderson, E.W., Fornell, C., Rust, R.T.: Customer satisfaction, productivity, and profitability: difference between goods and services. Mark. Sci. **16**, 129–145 (1997)

Beckman, T.N., Nolen, H.C.: Chain store disadvantages and limitations. In: The Chain Store Problem: A Critical Analysis, chap. 5, pp. 62–67. McGraw-Hill, New York (1938)

Chengalur-Smith, I., Ballou, D., Pazer, H.: The impact of data quality information on decision making: an exploratory analysis. IEEE Trans. Knowl. Data Eng. **11**, 853–864 (1999)

chin Chen, H., Chiang, R.H.L., Storey, V.C.: Business intelligence and analytics: from big data to big impact. MIS Q. **36**, 1165–1188 (2012)

Choi, K.S., Cho, W.H., Lee, S., Lee, H., Kim, C.: The relationships among quality, value, satisfaction and behavioral intention in health care provider choice: a South Korean study. J. Bus. Res. **57**, 913–921 (2004)

Cronin, J.C. Jr., Brady, M.K., Hult, G.M.: Assessing the effects of quality, value, and customer satisfaction on consumer behavioral intentions in service environments. J. Retail. **76**, 193–218 (2000)

Grönroos, C.: A service quality model and its marketing implications. Eur. J. Mark. **18**, 36–44 (1984)

Guo, L., Xiao, J.J., Tang, C.: Understanding the psychological process underlying customer satisfaction and retention in a relational service. J. Bus. Res. **62**, 1152–1159 (2009)

Lai, F., Griffin, M., Babin, B.J.: How quality, value, image and satisfaction create loyalty at a Chinese telecom. J. Bus. Res. **62**, 980–986 (2009)

Liu, F.T., Ting, K.M., Zhou, Z.H.: Isolation forest. In: Proceedings of the 8th IEEE International Conference on Data Mining, ICDM '08, pp. 413–422. ICDM, New York (2008)

Morgan, N.A., Anderson, E.W., Mittal, V.: Understanding firms' customer satisfaction information usage. J. Mark. **69**, 131–151 (2005)

Oliver, R.L.: Effect of expectation and disconfirmation on postexposure product evaluations: an alternative interpretation. J. Appl. Psychol. **62**, 480–486 (1977)

Petrovskiy, M.I.: Outlier detection algorithms in data mining systems. Program. Comput. Softw. **29**, 228–237 (2003)

Polo-Redondo, Y., Cambra-Fierro, J.: Influence of the standardization of a firm's productive process on the long-term orientation of its supply relationships: an empirical study. Ind. Mark. Manag. **37**, 407–420 (2008)

Scholkopf, B., Williamson, R.C., Smola, A.J., John Shawe-Taylor, J.C.P.: Support vector method for novelty detection. NIPS **12**, 582–588 (1999)

Smith, B.K., Nachtmann, H., Pohl, E.A.: Improving healthcare supply chain processes via data standardization. Eng. Manag. J. **24**, 3–10 (2015)

Wang, G., Wang, J., Ma, X., Qiu, R.G.: The effect of standardization and customization on service satisfaction. J. Serv. Sci. **2**, 1–23 (2010)

Analysis of Expenditure Patterns of Virtual Marriage Households Consisting of Working Couples Synthesized by Statistical Matching Method

Mikio Suga and Yasuo Nakatani

Abstract This paper explores the impact of male and female living together, namely "marriage." We applied the statistical matching method and estimate the scale merit of living together. In this method, different single households of the same statistical survey are matched referring to information of members in the existing household and create "virtual" households.

1 Introduction

Although benefits of marriage are generally recognized, the marriage rate has been declining in Japan. The reasons could attribute to unclear advantages of settling down. However, there is little research which examines the benefits of married households by comparing with unmarried households with a relevant statistical method. The unit of consumption expenditure is "household" in the ordinary economic analysis. "Household" is a group of individuals (household members) who live together with their residences. These household members are expected to enjoy the benefits of saving consumption expenditure, that is, the scale merit, by living together and share living costs. We applied the statistical matching method and estimate the scale merit of living together. In this method, different single households of the same statistical survey are matched by referring to information of members in the existing households and create "virtual" households. In this way, virtual households are synthesized, and compared with the existing households. The feature of this method is that it is possible to efficiently analyze the impact of differences in household structure on consumption expenditure for detailed items. This paper explores the impact of male and female living together, namely

M. Suga (✉) · Y. Nakatani
Faculty of Economics, Hosei University, Tokyo, Japan
e-mail: msuga@hosei.ac.jp

© Springer Nature Singapore Pte Ltd. 2020
T. Imaizumi et al. (eds.), *Advanced Studies in Classification and Data Science*,
Studies in Classification, Data Analysis, and Knowledge Organization,
https://doi.org/10.1007/978-981-15-3311-2_28

"marriage", on income and expenditure by using the micro-data of *National Survey of Family Income and Expenditure* for the year 1989, 1994, 1999 and 2004.

2 Methodology

Following explanations of the statistical matching method and the nearest neighbor matching is based on D'Orazio et al. (2006). First, the statistical matching method of micro data is roughly divided into parametric approaches and non-parametric approaches. Next there are three ways of non-parametric approaches: random hot deck, rank hot deck and distance hot deck. Of these, the distance hot deck is also called the "nearest neighbor matching". Regarding the distance hot deck, there are studies such as Okner (1982); Ruggles and Ruggle (1982); Rodgers (1982), etc. In the nearest neighbor matching, each record of the recipient file is imputed by the record of donor file that is nearest to the recipient. In this paper, we adopt the nearest neighbor matching. First of all, y is work and salaries income and a is age class. The superscripts of variables c denotes dual-income couple household, s denotes single households, m denotes male household member, and f denotes female household member. The lower subscript is the household identification number. For example, y_j^{cm} is the wage and salaries income of the male member of the j-th dual-income couple household. In the anonymous sample data of the NSFIE, the wage and salaries income is continuous data and the age class is discrete data.

Here, the distance of the male member of the j-th dual-income couple household and the k-th single male household is defined as follows:

$$d_{jk}^m = w_y \cdot \left| y_j^{cm} - y_k^{sm} \right| + w_a \cdot \left| a_j^{cm} - a_k^{sm} \right| \quad j = 1, \cdots, n^c, k = 1, \cdots, n^{sm}$$
$$w_y \sim N\left(\mu_y, \sigma_y^2\right)$$
$$w_a \sim N\left(\mu_a, \sigma_a^2\right)$$

Here, w is weight, and n is the number of households. Note that, w is a normal random variable that follows the normal distribution with mean μ and variance σ^2. Therefore, the distance changes stochastically and the calculation result changes with each simulation. For this reason, we calculate 100 times for each year. The minimum distance is expressed by the following equation.

$$\hat{d}_{j\hat{k}}^m = \min\left\{d_{j1}^m, \cdots, d_{jn^{sm}}^m\right\} \quad j = 1, \cdots, n^c.$$

However, the hat ($\wedge$) indicates that it is related to the minimum distance, $\hat{d}$ is the minimum distance, and $\hat{k}$ shows the household number of a single household with the minimum distance.

Next, the distance of the female member of the j-th dual-income couple household and the l-th single female household is defined as follows:

$$d_{jl}^f = w_y \cdot \left| y_j^{cf} - y_l^{sf} \right| + w_a \cdot \left| a_j^{cf} - a_l^{sf} \right| \quad j = 1, \cdots, n^c, l = 1, \cdots, n^{sf}$$
$$w_y \sim N \left(\mu_y, \sigma_y^2 \right)$$
$$w_a \sim N \left(\mu_a, \sigma_a^2 \right)$$

The minimum distance is expressed by the following equation.

$$\hat{d}_{j\hat{l}}^f = \min \left\{ d_{j1}^f, \cdots, d_{jn^{sf}}^f \right\} \quad j = 1, \cdots, n^c.$$

The expenditure of the i-th item of the virtual household is defined as follows.

$$E_{ij}^{Virtual} = E_{i\hat{k}}^{sm} + E_{i\hat{l}}^{sf} \quad i = 1, \cdots, n^E, j = 1, \cdots, n^c$$

Here, E is the expenditure. By calculating the difference between the consumption expenditure of this virtual household and the consumption expenditure of an actual household, it is possible to estimate the influence of the difference in household structure on item-specific consumption expenditure.

$$D_{ij} = E_{ij}^{Actial} - E_{ij}^{Virtual} \quad i = 1, \cdots, n^E, j = 1, \cdots, n^c$$

For example, suppose there is a dual-income couple household consisting of male whose age is 35 years old, and wage and salaries income is 250,000 yen, and women whose age is 30 years old, and wage and salaries income is 200,000 yen. Male member with age of 35 years old and wage and salaries income of 250,000 yen of the dual-income couple household is matched with male of a single household whose age is 35 years, and wage and salaries income is 250,000 yen. Females are matched similarly. As a result, virtual dual-income couple households are synthesized, and their expenditure pattern can be compared with households with actual households.

3 Data

In order to promote the use of statistical data, according to Article 36 of the Statistical Law, which was fully implemented from April 2009, anonymous data produced by administrative institutions who conduct statistical surveys is expected to contribute to the development of academic research and higher education. It was decided that the anonymous data could be supplied to user when the Director-General for Policy Planning (Statistical Standards) admit it. Anonymous data is processed in the way that the respondents cannot be identified by the specific individual, corporation or other organization (including identification by matching to other information) In this paper, anonymous data of 1989, 1994, 1999, 2004 of the *National Survey of Family Income and Expenditure* (NSFIE) were used.

The households to be analyzed are "dual-income couple households". Its definition is a household consisted with couple only, and both male and female members earn wages and salaries income. And the definition of single male household is a single household of male who earn work and salaries income. Likewise, the definition of a single female household is a single household of a female who earn work and salaries income.

The purpose of the analysis is to measure the scale merit by living together. If a single household who has non-living together family members, that single household may bear the cost of living more than one person. And thus the scale merit cannot be measured accurately. Likewise, if a couple household who has non-living together family members, they may bear the living expenses of people other than household members (other than that couple). The scale merit cannot be measured accurately. Therefore, it is necessary to exclude households who have non-living together family members from the analysis.

In the NSFIE, the non-living together family members and their breakdown(absence of main income earner, absence of non-main income earner, in hospital, in school, and other cases) are being surveyed. There are few single households who have non-living together family members, but about 10 to 20% of dual-income households have non-living together family members, by the reason of "non-main income earner in school". This seems to be a case where a child separates from parents while going to college or university far from house. In such a case parents may pay tuition fees directly to college or university.

Also, it is well known that consumption expenditure differs greatly even at the same income level depending on owning houses. Therefore, in order to accurately measure the scale merit, it is necessary to choose whether household with owned house or household without it for the target of analysis. According to the NSFIE survey, small number of single households have their houses. Therefore, households who owe their houses need to be excluded from targets of analysis.

As already mentioned, the NSFIE surveys non-living together families, and thus households with non-living together families have been excluded from the target of analysis. Moreover, there are some households who answered that they are "paying remittance" while saying "non-living together family member is zero" at the same time. Such households bear the living expenses of people other than household members and should be excluded.

In the NSFIE, payment of remittances is surveyed, and 10 to 20% of dual-income couple households paid remittances. Small number of single households paid remittances. Even "households without remittance payment" do not necessarily have "non-living together family." There are cases where parents are paying rent and education expenses directly.

As described above, as a result of excluding households with non-living together families, households with owned households, or households with remittances payment, the number of households to be analyzed has been determined (Table 1).

Table 1 The number of households to be analyzed

Year	Dual income couple HH	Single HH Male	Single HH Female
1989	501	997	701
1994	667	1174	710
1999	760	1084	633
2004	608	782	571

Note: HH is the abbreviation of "Household"

Table 2 Descriptive statistics of earned income by gender (2004)

	Male		Female	
	Double income couple HH	Single HH	Double income couple HH	Single HH
Number of HHs	608	782	608	571
Mean	301,837	297,207	148,929	220,494
Standard deviation	118,793	114,261	96,630	104,199
Skewness	0.781	1.388	0.803	0.844
Kurtosis	2.329	4.833	0.018	1.438

Note: "HH" is the abbreviation of "Household"

Table 2 is the descriptive statistics of earned income by gender for the year 2004. The distribution of male's earned income of single household is more skewed than that of double income couple household. The mean of female's earned income of single household is larger than that of double income household.

4 Results of Analysis

4.1 Estimation of Weights by Regression Analysis

In performing the nearest neighbor matching, it is necessary to determine the weight of the distance. In the case of the missing value imputation, at first we can experimentally create a missing value in the complete data, and on the second impute the missing value by changing the weight variously. Finally, we adopt the weight that can impute the missing value to be the nearest to the complete data. But in this case, this approach is impossible, since this is not missing value imputation. Therefore, weight estimation by regression analysis was performed. That is, for a dual-income couple household, a regression analysis with no intercept such as the following formula was conducted.

$$x_j^c = w_y \cdot y_j^c + w_a \cdot a_j^c + u_j \quad j = 1, \cdots, n^c$$

Table 3 Estimation results of regression analysis (1989–2004)

		1989	1994	1999	2004
R square		0.807	0.844	0.823	0.738
Adjusted R square		0.804	0.842	0.821	0.736
Number of observations		501	667	677	608
Total income of couple	Coeff.	0.554	0.510	0.488	0.415
	Std. error	0.033	0.026	0.026	0.031
	t-value	16.62	19.67	18.96	13.37
Average age of couple	Coeff.	5953	7321	6782	11834
	Std. error	1749	1638	1599	1797
	t-value	3.40	4.47	4.24	6.59

Here, y_j^c denotes the total earned income of the j-th dual-income couple household, a_j^c is the average age class of the j-th couple, and x_j^c is the consumption expenditure of the j-th dual-income couple household, and u_j is the error term. Table 3 is the estimation results of regression analysis. All estimated parameters are 1% statistically significant and adjusted R squares are over 0.8 for 3 years and over 0.7 for all years.

5 Matching

Weight is set based on the regression coefficient estimated in the previous section and its standard error. First, uniform random numbers from 0 to 1 are generated. Next, the value of the inverse function of the cumulative distribution function of the normal distribution with respect to the average (regression coefficient) and the standard deviation (standard error) specified as the probability is obtained and set as each weight. This was done 100 times for each of 4 years. Based on the weights set in this manner, we matched male and female single households with the closest distance to each of the male and female in the dual-income couple households.

5.1 Comparison of Income and Expenditure of Virtual Households and Actual Households

Table 4 shows the income and expenditures of virtual households and actual households in 1989. Since the dual-income couple households and single households of male and female are matched by wage and salaries income, the wages and salaries income of the actual households and that of the virtual households are almost same. The consumption expenditure of the actual household is 9% smaller than that of the virtual household, indicating that consumption expenditure can be saved

Table 4 The difference between actual and virtual households (1989), unit: JPY per month

	Actual HH	Virtual HH Mean	Std. Dev	Difference	Difference rate
Earned income	401,355	401,582	334	−227	0%
Consumption expenditure	267,633	294,767	665	−27,133	−9%
Food	60,709	85,034	249	−24,324	−29%
Raw fish and shellfish	4051	1903	28	2148	113%
Raw meat	4291	2213	33	2078	94%
Fresh vegetables	5156	2731	25	2425	89%
Cooked food	3887	5870	90	−1983	−34%
Beverage	2368	4421	12	−2053	−46%
Eating out	18,170	43,909	254	−25,739	−59%
Housing expenditures	37,883	41,452	178	−3569	−9%
Fuel, light and water charges	9827	8885	55	942	11%
Furniture and household utensils	7794	5853	71	1941	33%
Clothing and footwear	20,506	29,170	96	−8664	−30%
Health care	6067	4322	108	1745	40%
Transportation & communication	37,538	39,706	203	−2168	−5%
Education	144	456	51	−312	–
Culture and recreation	26,736	40,053	357	−13,317	−33%
Other consumption expenditures	60,429	39,836	213	20,593	52%
Tobacco	2787	5218	63	−2430	−47%
Pocket money	21,650	132	23	21,518	–

Note: "HH" is the abbreviation of "Household"

if single household male and female lives together. Looking at the breakdown of consumption expenditure, food can be largely saved. The actual households' food expenses are 29% smaller than the virtual households. Looking further at breakdown of food, expenditure on perishable food expenditure is larger. On the other hand, cooked food expenses, beverage expenses, eating out expenses are smaller. The expenditure that can be saved next to food cost is the cultural and recreation expenses. The third biggest expenditure is the clothing and footwear cost. On the contrary, the other consumption expenditure of the actual household is larger than that of the virtual household. Among the breakdown of them, the tobacco expenses is smaller and pocket money is larger. Such a tendency can be observed similarly for the other 3 years (Tables 5, 6, and 7).

Table 5 The difference between actual and virtual households (1994), unit: JPY per month

	Actual HH	Virtual HH Mean	Std. Dev	Difference	Difference rate
Earned income	478,290	478,782	86	−492	0%
Consumption expenditure	302,194	342,662	586	−40,468	−12%
Food	63,662	91,010	234	−27,348	−30%
Raw fish and shellfish	3721	1841	11	1880	102%
Raw meat	4043	2333	9	1710	73%
Fresh vegetables	5232	3125	16	2108	67%
Cooked food	5391	9175	20	−3784	−41%
Beverage	2778	5598	15	−2820	−50%
Eating out	19,106	45,173	199	−26,067	−58%
Housing expenditures	51,918	62,996	127	−11,079	−18%
Fuel, light and water charges	11,977	12,009	21	−32	0%
Furniture and household utensils	7833	7811	84	22	0%
Clothing and footwear	17,655	21,865	78	−4210	−19%
Health care	6179	4787	16	1392	29%
Transportation & communication	42,246	41,957	162	289	1%
Education	6	212	0	−206	−
Culture and recreation	33,858	53,362	76	−19,505	−37%
Other consumption expenditures	66,860	46,653	133	20,207	43%
Tobacco	1669	4914	47	−3245	−66%
Pocket money	28,390	559	2	27,831	−

Note: "HH" is the abbreviation of "Household"

Table 6 The difference between actual and virtual households (1999), unit: JPY per month

	Actual HH	Virtual HH Mean	Std. Dev	Difference	Difference rate
Earned income	483,208	485,237	345	−2030	0%
Consumption expenditure	295,145	336,483	957	−41,337	−12%
Food	59,505	85,619	269	−26,114	−31%
Raw fish and shellfish	3166	1420	16	1747	123%
Raw meat	3785	1877	11	1908	102%
Fresh vegetables	4497	2705	35	1792	66%
Cooked food	5525	10,532	64	−5007	−48%
Beverage	2908	6714	25	−3806	−57%
Eating out	18,258	37,729	329	−19,471	−52%

(continued)

Table 6 (continued)

	Actual HH	Virtual HH			Difference rate
		Mean	Std. Dev	Difference	
Housing expenditures	58,726	63,988	320	−5262	−8%
Fuel, light and water charges	13,226	14,402	65	−1176	−8%
Furniture and household utensils	6875	4933	49	1943	39%
Clothing and footwear	14,103	19,863	316	−5760	−29%
Health care	7060	6672	53	388	6%
Transportation & communication	43,977	51,998	229	−8021	−15%
Education	156	6	0	150	–
Culture and recreation	29,528	49,067	622	−19,540	−40%
Other consumption expenditures	61,989	39,934	268	22,055	55%
Tobacco	1865	3786	54	−1921	−51%
Pocket money	25,135	594	33	24,541	–

Note: "HH" is the abbreviation of "Household"

Table 7 The difference between actual and virtual households (2004), unit: JPY per month

	Actual HH	Virtual HH			Difference rate
		Mean	Std. Dev	Difference	
Earned income	462,809	464,058	67	−1249	0%
Consumption expenditure	295,199	324,829	975	−29,629	−9%
Food	57,457	77,221	130	−19,765	−26%
Raw fish and shellfish	2521	1240	3	1281	103%
Raw meat	3026	1354	5	1672	123%
Fresh vegetables	3795	2278	2	1517	67%
Cooked food	6547	12,113	18	−5565	−46%
Beverage	3466	7943	8	−4477	−56%
Eating out	17,989	28,767	78	−10,779	−37%
Housing expenditures	51,796	61,720	994	−9924	−16%
Fuel, light and water charges	12,349	14,842	14	−2493	−17%
Furniture and household utensils	6393	5335	7	1058	20%
Clothing and footwear	13,750	16,745	18	−2996	−18%
Health care	8434	8582	16	−148	−2%
Transportation & communication	52,139	52,932	73	−793	−1%
Education	2	0	0	2	–

(continued)

Table 7 (continued)

	Actual HH	Virtual HH		Difference	Difference rate
		Mean	Std. Dev		
Culture and recreation	30,779	43,769	108	−12,990	−30%
Other consumption expenditures	62,101	43,683	27	18,418	42%
Tobacco	1478	4442	28	−2964	−67%
Pocket money	19,093	818	14	18,274	–

Note: "HH" is the abbreviation of "Household"

6 Conclusion

Considering that food expenses of actual households are smaller than virtual households. Considering that the perishable foodstuffs expenses are smaller, and the cooked food expenses, the beverage expenses, and the eating out expenses are larger, if male and female live together. Presumably, instead of eating out, they purchase fresh food items and cook at home to have a meal. Since eating out is generally more expensive than eating at home, it is reasonable that you can save if you eat at home. On the other hand, there are two ways to interpret that the actual households' culture and recreation expenses, clothing and footwear expenses, tobacco expenses are smaller than the virtual households. One possibility is that each person's behavior changes as male and female live together. For instance, it can be interpreted that clothing footwear expenses will decrease because partner has already been found. It may be possible to say that tobacco expenses will decrease because the partner dislikes the tobacco. However, such interpretation cannot explain that the culture and recreation expenses are smaller. It is difficult to think that male and female living together becomes to reduce recreational activities. Another interpretation is that even if both male and female live together, social activities do not change, and only the expenditure categories are changed. In other words, culture and recreation expenses, clothing and footwear expenses, and tobacco expenses could be paid via "pocket money."

Acknowledgments We would like to thank National Statistical Center (NSTAC) for providing us the anonymous micro data of National Survey of Family Income and Expenditure. We also thank Yasumasa Baba, Mark Wooden, Rebecca Valenzuela and Nhan La for instructive comments.

References

D'Orazio, M., Di Zio, M., Scanu, M.: Statistical Matching: Theory and Practice. Wiley, Hoboken (2006)

Okner, B.: Constructing a new data base from existing microdata sets: the 1966 merge file. Ann. Econ. Soc. Meas. 1(3), 325–342 (1982)

Rodgers, W.L.: An evaluation of statistical matching. J. Bus. Econ. Stat. 2, 91–102 (1982)

Ruggles, N., Ruggle, R.: A strategy for merging and matching microdata sets. Ann. Econ. Soc. Meas. 32, 353–371 (1982)

The Effects of Natural Disasters on Household Income and Poverty in Rural Vietnam: An Analysis Using the Vietnam Household Living Standards Survey

Rui Takahashi

Abstract This study analyzes the relation among natural disasters, household income, and poverty using microeconometric methods. The effects of natural disasters on household income and poverty have received international attention from the United Nations, the World Bank, and similar organizations. In rural Vietnam, many people live in poverty, and Vietnam is one of the regions suffering from considerably great damage because of natural disasters. Further, we analyzed the effects of natural disasters on household income and poverty in rural Vietnam using Vietnam Household Living Standards Survey data. We conducted quantile regression to clarify the relation between natural disasters and household income. Therefore, we can confirm that droughts seriously affect all income classes, including those living in poverty. Therefore, we should consider countermeasures against droughts to alleviate poverty.

1 Introduction

Recently, the impact of natural disasters on economic development in many industrializing and globalizing countries has become a severe problem that has received significant interest from the international community. The 2010 *World Development Report*, published by the World Bank, indicated that the ability to cope with risk is poor in developing countries even though the risk of climate change is higher in these countries when compared with that in developed countries (World Bank 2010). Also, the Institute for Environment and Human Security, United Nations University has been actively investigating the effects of climate change on economic development and poverty in developing countries. The research results are published annually as the *World Risk Report*. According to the latest version *World Risk Report*

R. Takahashi (✉)
Department of Economics, School and Political Science and Economics, Tokai University, Hiratsuka, Kanagawa, Japan
e-mail: ruita@keyaki.cc.u-tokai.ac.jp

© Springer Nature Singapore Pte Ltd. 2020
T. Imaizumi et al. (eds.), *Advanced Studies in Classification and Data Science*,
Studies in Classification, Data Analysis, and Knowledge Organization,
https://doi.org/10.1007/978-981-15-3311-2_29

2016, Vietnam ranks 18th among 171 countries in terms of natural disasters risk as measured by the World Risk Index (WRI).

Although Vietnam has developed economically and has become a middle-income country in Southeast Asia, natural disasters still cause immense damage. For instance, the drought in the Mekong River Delta of southern Vietnam in early 2016 and the floods and storms in recent years in the North Central Coast and the Northern Midlands and Mountains are still fresh in our memory. Furthermore, almost all the shrimp farmers who have been surveyed belonging to the rural areas of the *Quang Ngai* province located in the South Central Coast of Vietnam reported losses in a natural disaster.[1] The large population and high poverty rate in rural Vietnam make the analysis of the effects of natural disasters on rural income an urgent issue. Therefore, this research examines the effects of natural disasters on the household income and poverty in rural Vietnam using microeconometrics based on the aforementioned background.

In Sect. 2, we briefly explain the local administrative units in Vietnam for defining a rural area. Section 3 gives information about the Vietnam Household Living Standards Survey (VHLSS) data that we used for our analysis. In Sect. 4, we examined the existing research and provide an overview of the climate patterns, household income, and poverty in various parts of Vietnam. In Sect. 5, we conducted empirical analysis using quantile regression. The final section provides our conclusions.

2   The Administrative Areas of Vietnam

Vietnam is divided into three administrative tiers by the official administrative system. The first tier is the province-level division; as of November 2017, this division consists of 58 provinces and 5 centrally governed cities. Rural districts, district-level towns, provincial cities, and urban districts make up the second tier. Rural districts and district-level towns are under the authority of both the provinces and the centrally governed cities. However, urban districts are only under the authority of centrally governed cities. Provincial cities, on the other hand, are only under provincial authority. The commune-level division is the third tier, and it includes communes, commune-level towns, and wards. Communes are under the authority of the second-tier rural districts, district-level towns, and provincial cities but not under urban districts. Commune-level towns are only under rural districts. Wards are under district-level towns, provincial cities, and urban districts. Urban districts include only wards as the commune-level divisions.[2] Since the term "rural

---

[1]We conducted a survey on shrimp aquaculture in rural areas of *Quang Ngai* province (*Duc Pho* district) in March 2015.

[2]The commune is traditionally formed by several natural villages and hamlets.

area" in the statistics of Vietnam is generally defined as "commune, " we also consider commune to have the same meaning.[3]

In addition to the local administrative units established by the Vietnamese government, there are six commonly known regions. In northern Vietnam, there are the Northern Midlands and Mountains and the Red River Delta. In central Vietnam, there are the North and South Central Coasts and the Central Highlands. The Southeast and the Mekong River Delta are in southern Vietnam.[4]

3 Data

The VHLSS provides most of the household-level microdata, especially the data in the income and expenditure survey included in the 2010 version (VHLSS 2010).[5]

The VHLSS is the most important survey concerning Vietnamese household living standards and poverty. Therefore, the household-level microdata in VHLSS datasets have been widely used for various research projects throughout the world. The VHLSS has been conducted 10 times, including the Vietnam Living Standard Survey (VLSS), which predates the VHLSS. The first survey, VLSS 1992/1993, was conducted from 1992 to 1993, and the second survey, VLSS 1997/1998, was conducted from 1997 to 1998. Since the sample households in VLSS 1997/1998 included the households surveyed in VLSS 1992/1993, we were able to create panel data by merging both datasets.

In the 2000s, VLSS was renamed the VHLSS, consisting of panel surveys, and the households have been surveyed every 2 years. VHLSS is a new research series that began in 2002 (VHLSS 2002). It is possible to construct panel data because the sample contains the household data from the previous surveys until VHLSS 2008, which is the sixth survey conducted beginning with the first VLSS. Unfortunately, VHLSS 2010 cannot be combined with VHLSS 2008 data to construct panel data. The subsequent surveys VHLSS 2012, VHLSS 2014, and VHLSS 2016 are necessary in order to prepare panel data using VHLSS 2010. Hence, we handled

---

[3]For example, the definition of "rural area" is "commune" in the *Rural, Agricultural and Fishery Censuses*, which have been implemented by the General Statistics Office of Vietnam (GSO) every 5 years since 2001. However, along with economic development, there are many cases where the area that was originally a commune is no longer a rural area. The definition of rural areas in Vietnam is still an important issue today.

[4]However, Vietnam was divided into the following eight regions before September 2006: the Northwest, the Northeast, the Red River Delta, the North Central Coast, the South Central Coast, the Central Highlands, the Southeast, the Mekong River Delta. In this article, we also use this regional division as needed.

[5]Because the VHLSS datasets are managed by the Department of Social and Environmental Statistics of the GSO, it is necessary to apply for use to this Department. For VHLSS 2006 and 2010 (resampling data), it can be used in an on-site laboratory of the Institute of Statistical Mathematics (IMS) in Japan (https://ds.rois.ac.jp/center3_micro/asia.html accessed on 11 Aug. 2019).

the data of VHLSS 2010 as cross-section data because we have not yet obtained VHLSS 2012, 2014 and 2016.[6]

Also, the VHLSS datasets have included commune data obtained by interviewing the head of each commune in addition to household-level data since the first VLSS. In general, it is difficult to obtain the microdata, including information on household income and natural disasters. For example, in the case of Cambodia, which is next to Vietnam geographically and is also exposed to high natural disaster risk (8th among 171 countries ranked by natural disaster risk measured by the WRI), the data from the census of agriculture contain the information on natural disasters (Govt. of Cambodia 2019). However, we are not able to obtain information on household income or expenditure via the data. On the other hand, with VHLSS 2010 we can provide information on both household income and natural disasters. Although only commune datasets can provide the data on occurrence and damage of natural disasters, we can analyze the causal relation between household income and natural disasters by merging the household data including the information on household income and the commune data including the information on natural disasters based on the commune ID. In addition to this useful feature for the purpose of our analysis, the VHLSS is highly reliable because the data have been inspected through peer reviews of the many empirical studies on household income in Vietnam. These are the major reasons why we regard VHLSS 2010 as appropriate data for our research.

4   Household Income and Climate Change in Vietnam

4.1   Climate Change and Natural Disasters in Vietnam

4.1.1   Regional Patterns of Climate

As mentioned in Sect. 2, Vietnam is a country that extends lengthwise from north to south with high geographical diversity. Therefore, the climate is different in the northern area, the central area, and the southern area of Vietnam. Figure 1 depicts hythergraphs using data on rainfall and temperature obtained at meteorological stations in various parts of Vietnam.

According to Fig. 1, the northern area of Vietnam is characterized by the temperature that varies greatly throughout the year, but the fluctuation in precipitation is relatively small. Hence, there is no clear distinction between a rainy season and a dry season. However, it is noteworthy that precipitation fluctuates greatly in the Northern Midlands and Mountains (*Lai Chau*) where flood damage has been noticeable in recent years.

---

[6]It is desirable to use panel data for the control of endogeneity in the analysis as described below; so we are now negotiating with the GSO for use of the VHLSS 2012, 2014, and 2016 datasets.

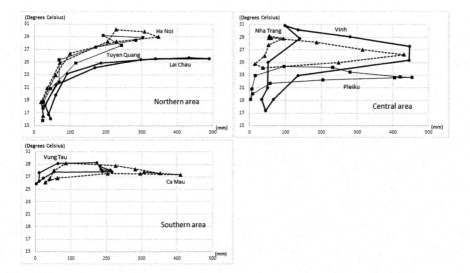

Fig. 1 Climate characteristics in Vietnam (Hythergraph). Source: https://www.gso.gov.vn accessed on 26 Nov. 2017. Note: The horizontal line indicates the precipitation averaged by month between 2002 and 2010. Further, the vertical line indicates the temperature averaged by month between the same period

Since the southern area of Vietnam belongs to the tropical zone, it is stable at around 30 °C throughout the year with a slight change in air temperature. On the other hand, the change in rainfall is large throughout the year, and it is possible to distinguish clearly between a rainy season and a dry season.

In central Vietnam both temperature and rainfall fluctuate greatly throughout the year. The hythergraph of *Vinh* on the North Central Coast demonstrates this feature clearly. This area is one of the regions in the path of typhoons in Vietnam, and flood damage has been considerable in recent years. It is a region where the risk of natural disasters is high.

4.1.2 Regional Risk of Natural Disasters

Le (2019) argued that climate change is a natural disaster risk. He measured the potential risks of natural disasters in various parts of Vietnam as the probability of the occurrence of storms, floods, droughts, and other natural disasters, and visualized them on a map.

According to Le's research, the potential risks of natural disasters are higher in the Northwest, the Northeast, the North Central Coast, and the Central Highlands than in other areas. These results were consistent with the observations in the hythergraphs where we confirmed that there was a large variation in the rainfall of *Lai Chau* which is located in the Northern Midlands and Mountains, and there were large temperature and rainfall fluctuations on the North Central Coast, in places such as *Vinh*.

4.2 Household Income and Poverty in Vietnam

Next, we observed the distribution of household income by region and by urban-rural areas in Vietnam and the differences. Table 1 confirms the change in the poverty rate by region and by urban-rural areas. It can be observed from this table that since 1998 the poverty rates in the Northern Midlands and Mountains, the North and South Central Coasts, and the Central Highlands have been consistently higher than in other areas. In other words, as confirmed in Sect. 4.1, these areas are regions with high disaster risks; therefore, we can confirm there is a correlation between risks of natural disasters and poverty rates.

In addition, by determining the poverty rate by urban-rural area in Table 1, we can conclude that the poverty rate of rural households is consistently higher than that of urban areas although the overall poverty rate has been decreasing.

Table 1 General poverty rate by region and residence (unit: %)

Regions and residence	1998	2002	2004	2006	2008	2010	2011	2012	2013	2014	2015
Whole country	37.4	28.9	18.1	15.5	13.4	14.2	12.6	11.1	9.8	8.4	7.0
Red River Delta	30.7	21.5	12.7	10.0	8.6	8.3	7.1	6.0	4.9	4.0	3.2
Northern Midlands and Mountains	64.5	47.9	29.4	27.5	25.1	29.4	26.7	23.8	21.9	18.4	16.0
North and South Central Coasts	42.5	35.7	25.3	22.2	19.2	20.4	18.5	16.1	14.0	11.8	9.8
Southeast	7.6	8.2	4.6	3.1	2.5	2.3	1.7	1.3	1.1	1.0	0.7
Central Highlands	52.4	51.8	29.2	24.0	21.0	22.2	20.3	17.8	16.2	13.8	11.3
Mekong River Delta	36.9	23.4	15.3	13.0	11.4	12.6	11.6	10.1	9.2	7.9	6.5
Rural	44.9	35.6	21.2	18.0	16.1	17.4	15.9	14.1	12.7	10.8	9.2
Urban	9.0	6.6	8.6	7.7	6.7	6.9	5.1	4.3	3.7	3.0	2.5

Source: https://www.gso.gov.vn accessed on 5 Aug. 2017
Note: Poverty rate is calculated by monthly average income per capita of household according to the government's poverty line for the 2011–2015 period, which is updated by the consumer price index (CPI) as follows:

- 2010: 400 thousand dongs for rural areas and 500 thousand dongs for urban areas
- 2012: 530 thousand dongs for rural areas and 660 thousand dongs for urban areas
- 2014: 605 thousand dongs for rural areas and 750 thousand dongs for urban areas
- 2015: 615 thousand dongs for rural areas and 760 thousand dongs for urban areas

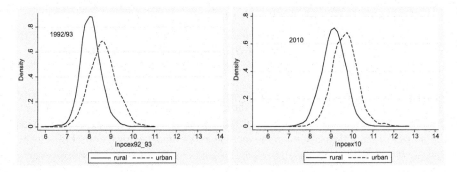

Fig. 2 Kernel densities of log real per capita expenditure in 1992/1993 and 2010. Source: VLSS 1992/1993 and VHLSS 2010, own calculations. Note: The data of real per capita expenditure are evaluated by Jan. 2010 price using CPI

Figure 2 shows the estimation of kernel density by urban-rural area on the per capita consumption expenditure (logarithmic value) of the whole of Vietnam based on VLSS 1992/1993 and VHLSS 2010.[7]

According to Fig. 2, comparing the distribution of 1992/1993 with 2006, it seems that the disparities between urban and rural areas did not change significantly since the difference in locations of both distributions was almost unchanged. However, the width of the distribution of per capita expenditure in rural areas has expanded from 1992/1993 to 2006. In other words, widening disparities within rural areas can be markedly observed.

In addition to this result, considering that nearly 70% of the population of Vietnam lives in rural areas (Takahashi 2019) and the poor have been concentrated in rural areas in recent years, it is extremely important to verify the impact of natural disasters on low-income classes in rural areas.

---

[7]Kernel density estimators (KDE) estimates the density $f(x)$ based upon observations on a continuous variable x. The kernel density estimate of $f(x)$ at $x = x_0$ is as follows:

$$\widehat{f}(x_0) = \frac{1}{Nh}\sum_{i=1}^{N}K\left(\frac{x_i - x_0}{h}\right)$$

We need to choose a kernel function $K(\cdot)$ and a bandwidth h to obtain kernel density plot. Hence, the following kernel function and bin width were used:

- The Epanechnikov kernel: $K(z) = \frac{(3/4)(1-z^2/5)}{\sqrt{5}}$ if $|z| < \sqrt{5}$ and $K(z) = 0$ otherwise.

- $h = \frac{0.9m}{n^{1/5}}$, where $m = (s_x, \ iqr_x/1.349)$. iqr_x is the interquartile range of x.

5   Effects of Natural Disasters on Household Income

5.1   Existing Research on Natural Disasters and Household Income

As mentioned above, it seems that there is a certain relation between natural disasters and rural poverty in Vietnam. Hence, many existing studies have verified the impact of natural disasters on the income of Vietnamese rural households using data such as VHLSS. Arouri et al. (2015) are among the most noteworthy research efforts in recent years. Their research used VHLSS 2004, 2006, 2008, and 2010 for the analysis. Its remarkable feature was the analysis of merged commune data and household data included in VHLSS. From VHLSS 2004 onward, only the commune data contains information on natural disasters that occurred over the past 3 years. Therefore, by merging commune data with household data, they could analyze the impacts of natural disasters on household income.[8]

Their empirical model used for analysis is as follows:

$$\ln\left(Y_{ijt}\right) = \beta_0 + X_{ijt}\beta_1 + C_{jt}\beta_2 + D_{jt}\beta_3 + X_{ijt}D_{jt}\beta_4 + C_{jt}D_{jt}\beta_5 + G_t\beta_6 + \varepsilon_{ijt}$$

$$(1)$$

Y_{ijt} is welfare indicator of household i at commune j in the year t; X_{ijt} is the vector of characteristics of households; C_{jt} is the vector of characteristics of communes; D_{jt} is the vector of natural disaster dummies (i.e., storms, floods, and droughts that happened in communes during the past 3 years); G_t indicates year dummies; and ε_{ijt} indicates unobserved variables.

The error term ε_{ijt} might include unobserved variables at the commune level. Thus, it can correlate with the natural disaster dummies, which are commune-level variables. Therefore, Arouri et al. (2015) dealt with these endogenous problems by conducting commune fixed-effect regression.

5.2   A Quantile Regression: Effects of Natural Disasters on Household Income and Poverty

Arouri et al. (2015) explained the impact of natural disasters on rural household incomes in Vietnam, coping with the endogeneity problem with a fixed-effect model. However, the fixed- effect model conducted by Arouri et al. (2015) only revealed the average effect of natural disasters on rural household income and poverty. Hence, we conducted quantile regression to verify the effect of three natural

---

[8]Therefore, as the households to be analyzed belong to commune, which is defined as a rural area in Vietnam statistics, the sample is limited to rural households.

disasters (storms, floods, and droughts) for each income class of rural households using VHLSS 2010. To be more precise, the following q th Koenker and Bassett estimator was estimated.

$$\widehat{\boldsymbol{\beta}}_q \overset{a}{\sim} N\left(\boldsymbol{\beta}_q \, , \, \mathbf{A}^{-1}\mathbf{B}\mathbf{A}^{-1}\right)$$
$$\mathbf{A} = \textstyle\sum_i f_{u_q}(0|\mathbf{x}_i)\mathbf{x}_i\mathbf{x}_i' \, , \quad \mathbf{B} = \textstyle\sum_i q(1-q)\mathbf{x}_i\mathbf{x}_i' \tag{2}$$

Although $f_{u_q}(0|\mathbf{x}_i)$ is the conditional density function of the error term $u_q = y - \mathbf{x}'\boldsymbol{\beta}_q$ at $u_q = 0$, estimating this density function is awkward. Thus, we use a bootstrap to estimate the variance-covariance matrix. The explanatory variables are the same as the model used by Arouri et al. (2015).

Table 2 shows the results of our quantile regression in which household income is used as a dependent variable. The household size, the proportion of adults aged 15–60 in households, the percentage of high school degree holders in the households, the proportion of upper-secondary degrees, the proportion of members with college or university degrees, the total crop land area, per capita living area, the number of communes with roads that are passable during all 12 months, the communes with a market, and communes with firms are explanatory variables that indicate significant positive effects on household income for any income class as expected. On the other hand, ethnic minority (of the household head) and age of the household head were significantly negative in any income class. The gender of the household head and communes with irrigation systems only make a difference depending on income class. From low-income class to middle-income class (Q10 to Q50), if the gender of household head is female, it has a significant negative effect on household income. However, a significant effect cannot be confirmed in the high-income classes (Q75 to Q90). This is consistent with the fact that many women are in poor and low-income categories, especially widows. In addition, communes with irrigation systems indicate a significant negative effect in all classes except for the low-income class (Q10). It seems that there are many households that depend on agriculture in these communes, and household income tends to be lower.

How do natural disasters impact household income? For storms, significant negative effects were seen in middle-income classes (Q50 and Q75). As for floods, significant negative effects are seen in the middle- and high-income classes (Q50 to Q90). Regarding droughts, significant negative effects were seen for all income classes.

Table 3 shows the results of quantile regression with the dependent variable as agricultural income (crop farming and animal husbandry income). We confirmed storms have significant negative effects in all classes except the lowest income class (Q10). However, low- and middle-income classes (Q10 to Q50) have significant positive effects for floods and droughts. Although these results might seem strange at a glance, it is possible to understand that farmers breed livestock with the expectation of its liquidity for consumption, so they are better equipped to cope with the shocks caused by floods and droughts.

Table 2 A quantile regression of household income in 2010

Explanatory variables	Household income (million VND)				
	Q10	Q25	Q50	Q75	Q90
Commune affected by storm	0.2103	−0.5515	−1.8275***	−2.2620**	−1.3461
	(0.5506)	(0.4964)	(0.5821)	(1.0113)	(2.2405)
Commune affected by flood	−0.1084	−0.7694	−2.3104***	−4.5950***	−8.8872***
	(0.5148)	(0.4901)	(0.5638)	(0.9842)	(2.1386)
Commune affected by drought	−1.3675**	−1.8190***	−2.2966***	−4.3552***	−7.9825***
	(0.5727)	(0.609)	(0.7047)	(1.076)	(2.1114)
Household size	4.3331***	5.7538***	7.8544***	10.6008***	14.4509***
	(0.1945)	(0.2048)	(0.2606)	(0.399)	(0.8569)
Proportion of adults aged 15–60 in households	5.0352***	8.8840***	12.7918***	18.1342***	18.6413***
	(0.7753)	(0.7815)	(1.0228)	(1.7961)	(3.6909)
Ethnic minorities (household head)	−8.3522***	−11.4451***	−14.7044***	−19.5583***	−24.2405***
	(0.5951)	(0.6113)	(0.6843)	(1.0795)	(2.2103)
Age of household head	−0.0419**	−0.0768***	−0.0733***	−0.0955***	−0.2137***
	(0.0181)	(0.0171)	(0.0198)	(0.0312)	(0.0723)
Gender of household head (female = 1)	−2.0897***	−2.1339***	−1.9458***	−1.9849	−0.7772
	(0.6012)	(0.5627)	(0.7082)	(1.2193)	(2.5082)
Proportion of members with upper-secondary degree	11.2756***	13.4486***	22.2334***	30.4729***	42.4958***
	(1.6936)	(1.2286)	(2.2378)	(3.3883)	(6.5776)
Proportion of members with college/university degree	28.7283***	59.0507***	75.7484***	103.8492***	97.4081***
	(7.5476)	(6.7543)	(5.8246)	(6.9992)	(18.156)
Total crop land area (1000 m^2)	0.3459***	0.4958***	0.6205***	0.9162***	1.6270**
	(0.0592)	(0.061)	(0.0694)	(0.001)	(0.7971)
Per capita living area (m^2)	0.1692***	0.2623***	0.4224***	0.7818***	1.4779***
	(0.0171)	(0.0268)	(0.0336)	(0.0696)	(0.1985)
Commune with road passable all 12 months	1.7506***	2.1797***	3.5945***	4.7396***	6.7747***
	(0.5106)	(0.5467)	(0.5902)	(1.1708)	(2.1816)
Commune with irrigation system	−0.7363	−1.0064**	−1.3290**	−3.6966***	−4.4383**
	(0.4701)	(0.4642)	(0.5909)	(1.0228)	(2.3098)
Commune with a market	0.3279	1.6295***	2.3443***	3.3519***	6.2833***
	(0.5548)	(0.515)	(0.6285)	(1.0484)	(2.2415)
Commune with firms	3.0298***	4.3828***	5.9719***	8.1810***	14.0125***
	(0.5333)	(0.5263)	(0.6501)	(1.1713)	(2.1017)
Constant	−4.0479**	−5.5955***	−9.7293***	−13.3335***	−17.6944***
	(1.7914)	(1.5115)	(2.2356)	(3.3885)	(6.526)
Observations	6594	6594	6594	6594	6594
Pseudo R^2	0.148	0.169	0.191	0.202	0.207
Number of communes	2198	2198	2198	2198	2198

Source: VHLSS 2010, own calculations

Note: Bootstrap standard errors with 1000 replications are in parenthesis

Table 3 A quantile regression of agricultural income per household in 2010

Explanatory variables	Agricultural income (million VND)				
	Q10	Q25	Q50	Q75	Q90
Commune affected by storm	−0.0426	−0.3503***	−0.8633***	−0.6243**	−0.4115
	(0.0358)	(0.1063)	(0.2008)	(0.2726)	(0.5375)
Commune affected by flood	0.0707*	0.8333***	0.6385***	0.4540	0.6802
	(0.0388)	(0.1259)	(0.2009)	(0.2966)	(0.5313)
Commune affected by drought	0.1352**	0.6052***	0.6400***	0.2014	−0.4262
	(0.0665)	(0.139)	(0.2346)	(0.3138)	(0.5476)
Household size	0.0276***	0.1857***	0.5715 ***	0.9798***	1.4658***
	(0.0097)	(0.0374)	(0.0794)	(0.1531)	(0.3462)
Proportion of adults aged 15–60 in households	0.2594***	1.3022***	2.9638***	3.9476***	4.0657***
	(0.0821)	(0.1899)	(0.3201)	(0.5124)	(1.0365)
Ethnic minorities (household head)	0.1830	0.4383**	−0.1912	−1.4406***	−3.0738***
	(0.1262)	(0.1924)	(0.0215)	(0.3406)	(0.6064)
Age of household head	−0.0012	0.0124***	0.0350***	0.0429***	−0.0114
	(0.0009)	(0.0035)	(0.0059)	(0.0096)	(0.018)
Gender of household head (female = 1)	−0.1000**	−0.5028***	−1.1200***	−1.8596***	−2.5068***
	(0.0437)	(0.0853)	(0.1735)	(0.3428)	(0.6496)
Proportion of members with upper-secondary degree	−0.0024	0.2070	1.0735***	2.2994**	8.0393***
	(0.0491)	(0.1847)	(0.4124)	(1.0065)	(2.2005)
Proportion of members with college/university degree	−0.1688	−0.8710***	−2.2543***	−1.8741*	−1.3790
	(0.146)	(0.2658)	(0.4584)	(1.0132)	(1.4479)
Total crop land area (1000 m^2)	0.6706***	1.1024***	1.8087***	2.8437***	4.2531***
	(0.0324)	(0.0447)	(0.0967)	(0.1782)	(0.6112)
Per capita living area (m^2)	−0.0022	−0.0014	0.0129**	0.0505***	0.1937***
	(0.0014)	(0.0026)	(0.0063)	(0.0171)	(0.034)
Commune with road passable all 12 months	−0.0085	0.0821	0.6018***	0.7932**	1.7445***
	(0.0232)	(0.1)	(0.2274)	(0.3336)	(0.4962)
Commune with irrigation system	0.0745**	0.3810***	0.7858***	1.0693***	0.4948
	(0.0345)	(0.0821)	(0.1669)	(0.2848)	(0.551)
Commune with a market	−0.0640**	−0.3682***	−0.8665***	−0.7236**	−0.7040
	(0.0309)	(0.0792)	(0.175)	(0.3007)	(0.5403)
Commune with firms	−0.0119	0.0583	0.2483	0.3075	1.5404***
	(0.0259)	(0.1136)	(0.1891)	(0.3227)	(0.5188)
Constant	−0.2461**	−1.9234***	−4.1566***	−5.0448***	−5.1723***
	(0.1085)	(0.3305)	(0.5729)	(0.9801)	(1.9192)
Observations	6594	6594	6594	6594	6594
Pseudo R^2	0.102	0.207	0.277	0.329	0.367
Number of communes	2198	2198	2198	2198	2198

Source: VHLSS 2010, own calculations

Note: Bootstrap standard errors with 1000 replications are in parenthesis. Agricultural income contains the income from crop production and husbandry

6 Conclusion

Based on the above analysis, it can be seen that storms had negative effects on the middle-income class, and floods had negative effects on the household income of the middle- and high-income classes and droughts had negative effects on the household income of all income classes. Therefore, we need to consider countermeasures against droughts for poverty alleviation. On the other hand, as far as agricultural income is concerned, the positive effects of floods and droughts could be confirmed in low- and middle-income classes. It is suggested that these low- and middle-income classes engaged in agriculture to smooth their consumption by holding livestock as liquidity against the shock of natural disasters.

Our research involves two tasks that have been left for the future. First, we could not deal with the endogeneity problem caused by commune-level unobservable variables, which Arouri et al. (2015) explained because the panel data could not be constructed. For this reason, we will need to use quantile regression with fixed-effect and instrumental variables using panel data in the future. Further data acquisition efforts to construct panel data are necessary. Second, natural disasters are regionally specific problems. Hence, we will need to analyze the relation between natural disasters and household income considering the spatial autocorrelation such as spatial econometric models.

Acknowledgments This study was carried out under the ISM Cooperative Research Program (2017 ISM CRP83: Study on the Promotion of Secondary Use of Household Micro Data in Asian Countries). Also, this study is supported in part by research and study Program of Tokai University Educational System General Research Organization. Furthermore, I would like to thank Enago (www.enago.jp) for the English language review.

References

Arouri, M., Nguyen, C., Youssef, A.B.: Natural disasters, household welfare, and resilience: evidence from rural Vietnam. World Dev. **70**, 59–77 (2015) https://doi.org/10.1016/j.worlddev.2014.12.017

Govt. of Cambodia: Census of Agriculture of the Kingdom of Cambodia 2013: National Report on Final Census Results, 2nd edn. Govt. of Cambodia, Phnom Penh (2015). https://www.nis.gov.kh/index.php/en/12-publications/15-agriculture-\census-in-cambodia-2013-final-result. Cited 11 Aug 2019

Le, D.T.: Identifying 6000 communes that are the most vulnerable to natural hazards for the Government CBDRM Programme. RTA (c2012). https://cms.rta.vn/identifying-6000-communes-that-are-the-most-\vulnerable-to-natural-hazards-for-the-government-cbdrm-programme. Cited 11 Aug 2019

Takahashi, R.: Features and problems of Vietnamese agriculture under industrialization. J. Fac. Political Sci. Econ. Tokai Univ. **48**, 177–196 (2016) (in Japanese). https://www.u-tokai.ac.jp/academics/undergraduate/political_science_and_eco/kiyou/2016.html. Cited 11 Aug 2019

World Bank: Development and Climate Change. World Bank, Washington (2010)

Generalizability of Relationship Between Number of Tweets About and Sales of New Beverage Products

Hiroyuki Tsurumi, Junya Masuda, and Atsuho Nakayama

1 Introduction

Today, many marketers analyze large online word-of-mouth (WOM) data sets of social networking services (SNSs), such as Twitter, to gather information about consumers' product evaluations. However, if online WOM were not related to product sales, which are performance indicators for many marketers, it could be considered to be merely "buzz". If, on the other hand, online WOM were related to sales, then it would provide valuable marketing information. Although we cannot directly control online WOM, whether there is a relationship between online WOM and product sales should be determined.

Therefore, we studied the relationship between online WOM on SNSs and product sales. Tsurumi et al. (2013) used path analysis to analyze the relationships among the sales of product A (a new low-malt beer), its gross rating points (GRPs), and the number of tweets about the product. The results revealed an indirect effect, whereby TV advertisements increased sales via tweets.

H. Tsurumi (✉)
Yokohama National University, Kanagawa, Japan
e-mail: tsurumi@ynu.ac.jp

J. Masuda
INTAGE Inc., Tokyo, Japan
e-mail: masuda-j@intage.co.jp

A. Nakayama
Tokyo Metropolitan University, Tokyo, Japan
e-mail: atsuho@tmu.ac.jp

© Springer Nature Singapore Pte Ltd. 2020   377
T. Imaizumi et al. (eds.), *Advanced Studies in Classification and Data Science*,
Studies in Classification, Data Analysis, and Knowledge Organization,
https://doi.org/10.1007/978-981-15-3311-2_30

In this study, we tested the generalizability of this indirect effect based on the results of a multigroup path analysis of data on new beverage products reported by Tsurumi et al. (2015). The results indicated that the indirect effect is generalizable to new beverage products, and we herein consider future directions for market research based on these findings.

2 literature Review

Many studies on the relationship between Internet communication and product sales rely on box-office records. Zufryden (2000) analyzed the association between the box-office records and online traffic to official movie websites. Additionally, Liu (2006) analyzed the association between negative/positive Yahoo! movie reviews and box-office records. Mishne and Glance (2006) analyzed the impact of negative blog posts on box-office records, and Gopinath et al. (2013) analyzed the relationship between box-office records and pre- and post-release advertisements and blog reviews.

Many previous studies focused on movies due to the ease of obtaining box-office data. Additionally, the movie experience includes not only watching, but also sharing reactions to movies both online and offline. Therefore, movie-related data are easy to obtain and can be used to determine causal relationships. However, movies are unique in their generation of widespread public comment.

Therefore, our previous study analyzed data on more generic products, i.e., consumer packaged goods (CPGs). Tsurumi et al.'s (2013) path analysis of data on a low-malt beer showed that television advertising had an indirect effect on increasing sales through Twitter posts. In this study, we analyzed data on four new beverage products using multigroup path analysis, and compared these analysis results, to determine the generalizability of this indirect effect.

3 Data

We analyzed data on the following products, which were all introduced in the fall of 2011, to rule out any impact of previous marketing activities: product A (low-malt beer), product B (milk tea), product C (green tea), and product D (non-alcoholic cocktail). All data were collected on a weekly basis from September 2011 to February 2012.

The two variables of interest, "sales performance" and "price," were calculated using the National POS Index (NPI). The NPI is national point of sales (POS) data on supermarkets in Japan provided by the Distribution Economics Institute of Japan. We extracted data on 234 stores in Tokyo, Nagoya, and Osaka from the NPI.

Sales performance was defined by the purchase index (PI), which is the number of a given product sold per 1000 visitors to a store. The GRPs of products in Tokyo,

Nagoya, and Osaka were provided by Video Research Co., Ltd. (Japan). The number of tweets (num. of tweets; defined as the number of tweets including the product name) was calculated by a web crawler. We included geographical area dummy variables. In a previous analysis, we showed that the fitness index was improved by natural logarithmic transformation of the data; therefore, we performed a natural logarithmic transformation of the variables.

4 Path Analysis

As the analysis was based on Tsurumi et al. (2013), we used a path analysis approach. To improve generalizability, we subjected the data for the four products to multigroup path analysis, where the grouping variable was the product. Tsurumi et al. (2013) treated sales results, which are indicators of market performance, as endogenous variables, and treated the number of tweets, GRP, price, and area dummy variable as exogenous variables. Our model included an additional path, from GRP to num. of tweets, to reflect the indirect effect of GRP on sales performance (via num. of tweets).

5 Results

Figures 1, 2, 3, and 4 show the results of applying multigroup path analysis modeling to the data for the four products. The goodness-of-fit (GFI) indices had relatively high values: root mean error of approximation (RMSEA), 0.047; GFI, 0.977; and adjusted GFI (AGFI), 0.840.

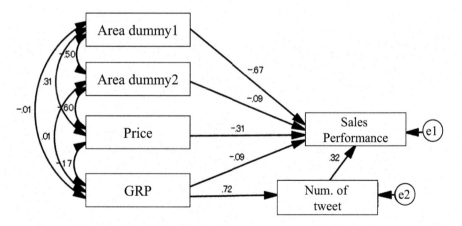

Fig. 1 Result of multigroup path analysis: product A (low-malt beer)

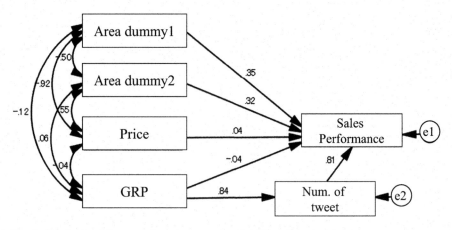

Fig. 2 Result of multigroup path analysis: product B (milk tea)

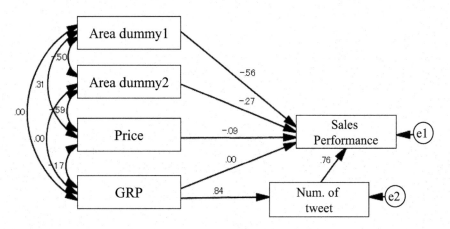

Fig. 3 Result of multigroup path analysis: product C (green tea)

5.1 Path From GRP to Num. of Tweets

The path from GRP to num. of tweets was positive and significant at the 0.05 level for all products. Moreover, in all cases, the standardized coefficients were close to 1, indicating that TV advertising had a strong influence on num. of tweets.

5.2 Path From Num. of Tweets to Sales Performance

The path from num. of tweets to sales performance was positive and significant at the 0.05 level for products A to C; only the path of product D was non-significant at

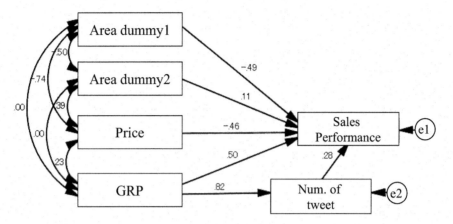

Fig. 4 Result of multigroup path analysis: product D (non-alcoholic cocktail)

the 0.05 level (p=0.11). The latter result is attributable to the manufacturer's decision to launch a consumer giveaway campaign when the product was released. This campaign involved automatically tweeting a predetermined phrase that included the product name when consumers draw lots. These non-spontaneous tweets generated an analytic noise.

5.3 Indirect Effect

The path from GRP to sales performance was non-significant at the 0.05 level for three of four products, being significant only for product D. Thus, there was no direct effect of GRP on sales performance for three of four products.

On the other hand, as noted above, the path from GRP to num. of tweets was positive and significant at the 0.05 level for all four products. Furthermore, the path from num. of tweets to sales performance was also positive and significant at the 0.05 level for all products except product D. Thus, GRP had an indirect effect on sales performance via num. of tweets for three of four products.

No indirect effect could be confirmed for product D, because the relevant data were influenced by the aforementioned Twitter campaign. Except for such special cases, it appears that while television advertising has no direct effect on improving sales performance, it does have an indirect effect via Twitter. Thus, the indirect effect suggested by Tsurumi et al. (2013) generalized to the new beverage products assessed in this study.

This phenomenon can be interpreted as follows. There are, at most, several thousand tweets regarding each of the new products in this study per month. Although it is difficult to understand how so few tweets could have sufficient influence to affect sales results, our data nevertheless showed that num. of tweets was significantly positively related to the sales performance of these products.

Based on these results, we hypothesized that Twitter communication can be considered a "proxy index of the topicality" of products. In other words, actual Twitter posts only represent the "tip of the iceberg," as they reflect both offline and online reviews by consumers. Indeed, it is reasonable to assume that products that generate numerous total reviews (i.e., reviews across all types of media) also generate numerous online reviews, and vice versa. If so, marketers should treat Twitter as not only a communication channel, but also as a marketing index that reflects customer opinion and is associated with product sales.

6 Conclusion

This research had several limitations. First, our results pertain to only a few new beverage products. Moreover, we did not consider the characteristics of Twitter users or the content of their tweets. However, as we observed the indirect effects of such Twitter traffic on the sales of several new beverage products, we hypothesize that such effects will be observed for other beverage products as well. Indeed, although additional verification is necessary, it is unlikely that this indirect effect is unique to only these four beverages. We expect that the same indirect impact will be in operation for many other types of consumer goods.

In the future, people will likely spend increasing amounts of time on SNSs, and massive amounts of information will be shared by consumers on such sites. As a result, the influence of online WOM on purchasing behavior will increase. Furthermore, various kinds of information, such as photographs, music, videos, and so on are shared on SNSs. In this context, marketers should place an even greater emphasis on collecting and analyzing diverse data from SNSs to understand what consumers are posting and reading on these sites. However, it will be difficult for marketers to analyze the increasingly massive amounts of information that will become available. Therefore, following the example of this study, marketers should first identify the factors with the strongest effects on sales performance. Second, they should focus on the SNS data with the strongest association with performance indicators. Understanding the relationship between purchasing behavior and SNSs will efficiently yield important marketing information.

Acknowledgments This work was supported by the Yoshida Hideo Memorial Foundation (45th research grant). We would like to thank Video Research Co., Ltd. and the Distribution Economics Institute of Japan, which provided the data.

References

Gopinath, S., Chintagunta, P.K., Venkataraman, S.: Blogs, advertising, and local-market movie box office performance. Manag. Sci. **59**(12), 2635–2654 (2013)

Liu, Y.: Word of mouth for movies: its dynamics and impact on box office revenue. J. Market. **70**(3), 74–89 (2006)

Mishne, G., Glance, N.: Predicting movie sales from blogger sentiment. In: AAAI 2006 Spring Symposium on Computational Approaches to Analyzing Weblogs (2006)

Tsurumi, H., Masuda, J., Nakayama, A.: Analysis on relationship between tweets and product sales. Oper. Res. **58**(8), 436–441 (2013)

Tsurumi, H., Masuda, J., Nakayama, A.: Possibilities and limitations of text data analysis on SNS in marketing (in Japanese). Japan Market. J. **35**(2), 38–54 (2015)

Zufryden, F.S.: New film website promotion and box-office performance. J. Advert. Res. **40**, 55–64 (2000)

Cluster Distance-Based Regression

J. Fernando Vera and Eva Boj del Val

Abstract In distance-based regression analysis, the vector of continuous responses is projected in a Euclidean space given by multidimensional scaling (MDS). One of the main problems in this methodology is that of determining the number of dimensions or latent predictor variables in this framework, particularly when a large dissimilarity data set is observed in the predictor space. It is well known that the combined use of cluster and MDS enables a better understanding of the data and reduces the number of parameters to be estimated. Moreover, the most appropriate approach is to integrate these procedures within a procedure that produces a cluster while simultaneously representing the cluster centres instead of the original responses, in reduced dimensionality. In this paper, we propose a combined methodology that uses cluster-MDS to determine a Euclidean configuration in a reduced latent predictor space. Taking into account the classification obtained in the response space, a distance-based regression model is fitted by projecting the weighted average vector within each cluster onto the continuous response variable in the clustered predictor space. The model is applied to a real-world data set obtained from the automobile insurance industry.

1 Introduction

In distance-based regression analysis (see Cuadras and Arenas 1990; Cuadras et al. 1996; Boj et al. 2010, 2016, 2017), the vector of continuous responses is projected in a Euclidean space given by multidimensional scaling (MDS). The

J. F. Vera (✉)
Department of Statistics and O.R., Faculty of Sciences, University of Granada, Granada, Spain
e-mail: jfvera@ugr.es

E. B. del Val
Department of Mathematics for Economics, Finance and Actuarial Sciences, Faculty of Economics and Business, University of Barcelona, Barcelona, Spain
e-mail: evaboj@ub.edu

© Springer Nature Singapore Pte Ltd. 2020
T. Imaizumi et al. (eds.), *Advanced Studies in Classification and Data Science*,
Studies in Classification, Data Analysis, and Knowledge Organization,
https://doi.org/10.1007/978-981-15-3311-2_31

MDS configuration is obtained by considering a dissimilarity matrix (typically Euclidean distances) between the observed elements in the predictor space. In a general procedure, a dissimilarity matrix between individuals is the only predictor information used, making the models applicable, for instance, to mixed (qualitative and quantitative) explanatory variables or to a functional-type regressor.

A problem that arise with this approach is that of determining the number of dimensions or latent predictor variables. This number rises as the sample size increases with respect to the original number of predictor variables (if any), or when the observed proximities are previously transformed by the well-known additive constant procedure. A possible solution is to choose the *effective rank* of the distance-based regression using ordinary or generalised cross-validation (OCV or GCV), or Akaike or Bayesian information criteria (AIC or BIC) (see Boj et al. 2016, 2017). Although these methods effectively obtain an optimal dimension, e.g., in order to account for a proportion of the total geometrical variability, they do not ensure an equally effective reduction in the number of latent variables.

It is well known that the combined use of cluster and MDS provides a better understanding of the data while reducing the number of parameters to be estimated (see Kruskal 1977). When the information in distance-based regression comes from a large dissimilarity matrix in the predictor space, the use of cluster analysis in conjunction with metric least squares MDS might, in situations such as those described above, produce a suitable configuration in low dimensionality. Although performing clustering and obtaining the MDS representation of cluster centres might seem to be an acceptable alternative in such a situation, it should be noted that the MDS reduced space is optimal for the embedded points, but it may not be optimal for the superimposed clusters, while the cluster structure is usually only optimal in the original space, but not in a reduced one.

To enhance the interpretation of the MDS solution and/or to obtain an adequate fit of the model when a large number of objects must be represented, cluster-MDS methods have proved to be useful, both in the classical and in the least squares frameworks, (see Bock 1986, 1987; Heiser 1993; Heiser and Groenen 1997; Vera et al. 2008), as well as in a probabilistic framework (Vera et al. 2009a,b). In this paper, we propose a combined methodology that uses cluster-MDS to determine a Euclidean configuration in a reduced latent predictor space. Then, assuming the same estimated classification in the response space, we fit a distance-based regression model by projecting the weighted average vector within each cluster onto the continuous response variable in the clustered predictor space.

The rest of this paper is structured as follows: the model is described in Sect. 2, after which we explain the cluster-MDS procedure in Sect. 3. Section 4 describes the distance-based regression models used for prediction, and Sect. 5 illustrates the performance of the proposed procedure by analysing a real-world data set. Some concluding remarks are then presented in Sect. 6.

2 The Model

Let us consider a sample of size N of objects $O = \{o_1, \ldots, o_N\}$, from a given population. Moreover, let us denote by $Y: Y_1, \ldots, Y_N$ the random response variable for which $y = (y_1, \ldots, y_N)^\top$ represents a realization, and $w = (w_1, \ldots, w_N)^\top$, a known positive weighting vector. On the other hand, Δ denotes a $N \times N$ symmetric matrix of dissimilarities δ_{ij} between the objects. This matrix is assumed to be the only information given in the predictor space, and can be obtained either directly or from a set Z of independent variables, possibly including quantitative and qualitative measurements or other nonstandard quantities, such as character strings or functions.

Let X be the $N \times M$ matrix of (usually Euclidean) coordinates in a space of dimension M, with $M \leq N - 1$, obtained by an MDS procedure, such that $d_{ij} \approx \delta_{ij}$, $\forall i, j = 1, \ldots, N$, where $d_{ij} = d(x_i, x_j)$, $X = (x_1^\top, \ldots, x_N^\top)^\top$, $x_i \in \mathbb{R}^M$. Distance-based linear models (DB-LM) have been developed by Cuadras and Arenas (1990), Cuadras et al. (1996) and Boj et al. (2010), and distance-based generalised linear models (DB-GLM), by Boj et al. (2016) and Boj et al. (2017). When M is large it is usually difficult to obtain a good fit of the model. Accordingly, our aim is to reduce the N original objects into a small number T, with $T \ll N$ while simultaneously representing the T cluster centres in a Euclidean space of dimension $\tilde{M}$, using a cluster-MDS procedure. By these means, the new MDS Euclidean configuration of the cluster centres, $\tilde{X}$, has the dimension $T \times \tilde{M}$, with $\tilde{M} \leq T - 1$.

To fit a distance-based regression model, T clusters of objects $\tilde{O}_t, t = 1, \ldots, T$, as presented in the above-described cluster-MDS procedure, are considered. To this end, the criteria proposed by Vera and Macías (2017) to select the number of clusters in a dissimilarity matrix are used. The positive weight vector is denoted by $\tilde{w} = (\tilde{w}_1, \ldots, \tilde{w}_T)^\top$, where $\tilde{w}_t, t = 1, \ldots, T$ is the sum of the original weights w for the n_t original individuals belonging to each cluster $\tilde{O}_t, t = 1, \ldots, T$. It is assumed that $\tilde{w} = 1$, where 1 is the $T \times 1$ vector of ones. The response variable in the clustered space is denoted by $\tilde{y} = (\tilde{y}_1, \ldots, \tilde{y}_T)^\top$, where $\tilde{y}_t, t = 1, \ldots, T$, are the weighted means of the original individuals belonging to each cluster $t, t = 1, \ldots, T$.

3 Reduction and Representation: Cluster Differences Scaling

Let us consider the set of sample units $O = \{o_1, \ldots, o_N\}$ and denote by $\Delta = (\delta_{ij})$, $i, j = 1 \ldots N$, a symmetric matrix of dissimilarities between them. Assume that, based on a partition $\mathscr{P}(\Delta)$ into $T(T + 1)/2$ classes Δ_{tl}, each individual is allocated to one and only one of T clusters, denoting by E an indicator matrix of order $N \times T$, whose elements e_{it} are equal to one if individual o_i, belongs to cluster t, or zero otherwise. Thus, if we denote by $\tilde{O}_t = \{o_i | e_{it} = 1\}$, for $t = 1, \ldots, T$, the hypothesis that the clusters form a partition is expressed as $\tilde{O}_t \cap \tilde{O}_l = \emptyset$, for $t \neq l$, and $\bigcup_t \tilde{O}_t = O$.

The aim in cluster difference scaling is to achieve a configuration, $\tilde{X}$, of T points, $\tilde{x}_t, t = 1, \ldots T$, in a Euclidean metric space of low dimension $\tilde{M} \leq T - 1$, which is optimal in the sense that the associated vector of distances, d, in $\mathbb{R}^{T(T-1)/2}$, approaches as closely as possible to the estimated latent dissimilarities between clusters. This model is based on the assumption that when the allocation leads to individual $o_i \in \tilde{O}_t$ and individual $o_j \in \tilde{O}_l$, their dissimilarity will be represented in the model as the Euclidean distance d_{tl} between cluster centres $\tilde{x}_t$ and $\tilde{x}_l$. For a partition, the dissimilarities are assumed to vary randomly within a cluster, while the corresponding distance is constant within the same cluster, whereas between clusters, differences in distance will reflect the tendency of the corresponding dissimilarities to vary systematically (see Heiser and Groenen 1997). Therefore, the objective here is to minimise the loss function, which is termed *stress* and defined as:

$$\sigma^2(E, \tilde{X}) = \sum_{t \leq l} \sum_{i \in O_t} \sum_{j \in O_l} w_{ij}(\delta_{ij} - d_{tl})^2, \tag{1}$$

where d_{tl} is the Euclidean distance between points $\tilde{x}_i$ and $\tilde{x}_j$. Considering the Sokal-Michener dissimilarities between blocks (see Sokal and Michener 1958) denoted by $\bar{\delta}_{tl}$, the stress can be written as follows:

$$\sigma^2(E, \tilde{X}) = \sum_{t \leq l}^{T} \sum_{i=1}^{N} \sum_{j=1}^{N} e_{ik} e_{jl} w_{ij}(\delta_{ij} - \bar{\delta}_{tl})^2 + \sum_{t \leq l}^{T} \tilde{w}_{tl}(\bar{\delta}_{tl} - \hat{\tilde{d}}_{tl})^2, \tag{2}$$

where

$$\bar{\delta}_{tl} = \frac{\displaystyle\sum_{i=1}^{N} \sum_{j=1}^{N} e_{it} e_{jl} w_{ij} \delta_{ij}}{\tilde{w}_{tl}}, \quad \text{and} \quad \tilde{w}_{tl} = \sum_{i=1}^{N} \sum_{j=1}^{N} e_{it} e_{jl} w_{ij}. \tag{3}$$

The parameters of the model are estimated in an alternating least squares procedure in which the following two steps can be distinguished:

Step 1 *Initialisation*: First a random classification $E^{(0)}$ is obtained for a value T previously set by the investigator. Then the Sokal-Michener dissimilarities are calculated and an initial configuration $\tilde{X}^0$ is obtained using classical MDS.

Step 2 *Allocation*: An indicator matrix $E = (e_{it})_{N \times T}$ is estimated, with $e_{it} = 1$ if $o_i \in O_t$ and zero otherwise for $i = 1, \ldots, N$, determining a partition $\mathscr{P}(\Delta)$ into $T(T + 1)/2$ classes Δ_{tl}. To this end, let us denote by $\kappa^2(E) = 2\sigma_<^2(E)$ with $w_{ij} = 1, \forall i, j$, except $w_{ii} = 0$, and let $\|\mathbf{a}_i - \mathbf{b}_t^{(i)}\|^2$ denote the squared Euclidean distance between the ith row of the $N \times c$ matrix $A = \{a_{ir}\}$ and the kth row of the $T \times c$ matrix $B = \{b_{kr}\}$, where $c = N - 1$. The elements of **A** and **B** are specified as $a_{ir} = \delta_{ir}^*$ and $b_{tr}^{(i)} = \overline{\delta}_{tr}^*$, where $\delta_{ir}^* = \delta_{is}$ and $\overline{\delta}_{tr}^* = \sum_l e_{sl} \overline{\delta}_{tl}$ for $r = 1, \ldots, N - 1$, with $s = r$ if $r < i$, and $s = r + 1$ if $r \geq i$. Then $\kappa^2(E)$ satisfies

$$\kappa^2(E) = \sum_i \sum_t e_{it} \|\mathbf{a}_i - \mathbf{b}_t^{(i)}\|^2,$$

that is, the classical k-means criterion for dissimilarities, and minimising (2) with respect to E is equivalent to minimising $\kappa^2(E)$.

Step 3 *Geometrical representation:* Given the estimated values for E, a configuration $\tilde{X} = (\tilde{x}_1^\top, \ldots, \tilde{x}_T^\top)^\top$ is estimated for the cluster centres such that

$$\delta_{ij} \approx d_{tl}, \forall \delta_{ij} \in \mathbf{\Delta}_{tl},$$

where $d_{tl} = d(\tilde{x}_t, \tilde{x}_l)$, $\forall t, l = 1 \ldots T$.

The stress function, (2), is then evaluated and the procedure concludes if the convergence criterion holds that the differences between two consecutive values of the stress should be below a previous fixed small quantity. If it does not, steps two and three are repeated until convergence is achieved. Finally, a configuration $\tilde{X}$, usually in low dimension $\tilde{M}$ is estimated.

4 Distance-Based Prediction

This section describes distance-based regression models, in terms of the responses $\tilde{Y}$, the weights $\tilde{w}$ and the coordinates $\tilde{X}$ of the cluster space.

4.1 Distance-Based Linear Model

The response $\tilde{Y}$, the weights $\tilde{w}$ and the distance matrix $\tilde{\mathbf{\Delta}}$ are said to follow a DB-LM when $\tilde{\mu} = E(\tilde{Y})$ $\tilde{w}$-centred (that is, $\tilde{J}\tilde{\mu}$, where $\tilde{J}$ is the $\tilde{w}$-centring matrix, defined as $\tilde{J} = I - 1\tilde{w}^\top$) belongs to the column space $\mathscr{G}$ of $\tilde{G}$ (see Boj et al. 2016).

The expression of DB-LM hat matrix is

$$\tilde{H} = \tilde{G} \left(D_{\tilde{w}}^{1/2} \tilde{F}^+ D_{\tilde{w}}^{1/2} \right),$$

where $D_{\tilde{w}} = \text{diag}(\tilde{w})$ is a diagonal matrix which diagonal entries the weights $\tilde{w}$,

$$\tilde{F} = D_{\tilde{w}}^{1/2} \tilde{G} D_{\tilde{w}}^{1/2},$$

and $\tilde{F}^+$ is the Moore–Penrose pseudo-inverse of $\tilde{F}$.

Giving an overparametrised model with unstable predictions, it is advisable to replace the pseudo-inverse $\tilde{F}^+$ with a lower rank approximation, implemented in Boj et al. (2017) by the singular value decomposition. Various criteria, such as OCV, GCV, AIC or BIC can be used to choose the dimension, termed *effective rank* (see Boj et al. 2017).

DB-LM contains weighted least squares (WLS) as a particular instance: in general, if we start from a w-centred $N \times M$ matrix X of M continuous predictors corresponding to N objects and we define Δ as the matrix of Euclidean distances between rows of X, then X is trivially a Euclidean configuration of Δ, hence the DB-LM hat matrix, response and predictions coincide with the corresponding WLS quantities of ordinary LM. The fitted values can be calculated as follows:

$$\hat{\tilde{y}} = \bar{\tilde{y}}\mathbf{1} + \tilde{H}\left(\tilde{y} - \bar{\tilde{y}}\mathbf{1}\right),$$

where $\bar{\tilde{y}} = \tilde{w}^\top \tilde{y}$ is the $\tilde{w}$ weighted mean of $\tilde{y}$.

Assume that a new original object o_{N+1} is available, and that it is given the $1 \times N$ vector δ_{N+1} of distances from o_{N+1} to the N previously known objects. In the cluster space, the new object is termed the new cluster object $\tilde{o}_{T+1}$. Now, the $1 \times T$ vector $\tilde{\delta}_{T+1}$ of distances from $\tilde{o}_{T+1}$ to the T cluster centres must be calculated. These can be obtained by using the distances δ_{N+1} to calculate the (weighted) means in each cluster t for $t = 1, \ldots, T$, that is, the weighted mean of the distances of δ_{N+1} for the objects belonging to each cluster. To this end, $\tilde{o}_{T+1}$ can be represented as an $\tilde{M}$-vector x_{T+1} in the row space of $\tilde{X}$ using the Gower interpolation or add-a-point formula (Gower 1968) as follows:

$$\hat{\tilde{y}}_{n+1} = \bar{\tilde{y}} + \frac{1}{2}(\tilde{g} - \tilde{\delta}_{T+1}^2)\left(D_{\tilde{w}}^{1/2}\tilde{F}^+ D_{\tilde{w}}^{1/2}\right)\left(\tilde{y} - \bar{\tilde{y}}\mathbf{1}\right),$$

where $\tilde{g}$ is the $1 \times T$ row vector with the diagonal entries of $\tilde{G}$.

4.2 Distance-Based Generalised Linear Model

A DB-GLM consists of random variables $\tilde{Y} = (\tilde{Y}_1, \ldots, \tilde{Y}_T)^\top$ whose expectation, $\tilde{\mu} = (\tilde{\mu}_1, \ldots, \tilde{\mu}_T)^\top$, transformed by the link function and $\tilde{w}$-centred, is a vector in the column space $\mathscr{G}$ of $\tilde{G}$, that coincides with the column space of any Euclidean configuration $\tilde{X}$ of $\tilde{\Delta}$ (see Boj et al. 2016).

Like GLM with respect to LM, DB-GLM differs from DB-LM in two aspects: The response distribution belongs to the exponential dispersion family, and the relation between the linear predictor $\tilde{\eta} = \tilde{X}\beta$ (where $\beta \in \mathbb{R}^{\tilde{M}}$ is an $\tilde{M} \times 1$ parameter vector of the latent Euclidean configuration $\tilde{X}$), and the response $\tilde{y}$ is given by a link function, $g(\tilde{\mu}_t) = \tilde{\eta}_t$. DB-GLM are fitted using an iterative weighted least squares (IWLS) algorithm where DB-LM substitutes LM.

As in DB-LM, a decision about the effective rank of the regression can be taken according to GCV, AIC or BIC criteria. Furthermore, DB-GLM contains GLM as a particular case.

A new original object o_{N+1} described by δ_{N+1}, the distances to the N previously known objects, can be predicted as follows: First, to calculate $\tilde{\delta}_{T+1}$ for $\tilde{o}_{T+1}$ in the cluster space. Second, to estimate $\hat{\tilde{\eta}}_{T+1}$ with the quantities of the last IWLS step by making use of the Gower interpolation add-a-point formula. Finally, the prediction is $\hat{\tilde{\mu}}_{T+1} = g^{-1}\left(\hat{\tilde{\eta}}_{T+1}\right)$.

DB-LM and DB-GLM can be fitted using functions dblm and dbglm of the *dbstats* package for R, see Boj et al. (2017).

5 Actuarial Application

The cluster distance-based procedure is illustrated with a real-world data set obtained from the automobile insurance industry. In this example, we study the random variable *claim severity* in the non-life insurance context. This variable is commonly used in the *a priori* rate-making (see, e.g., Haberman and Renshaw 1996 or Boj et al. 2004 among other references), where the objective is to determine:

Claim frequency = Number of claims/Number of policyholders or exposures,
Claim severity = Total claim amount/Number of claims,
Pure Premium = Claim frequency x Claim severity.

In the *a priori* ratemaking, the clusters correspond to the different risk groups in the final tariff, and therefore we wish to reduce the dimension. To standardise the data, a subset related to own-damage cover for passenger cars of the *sedan* category is selected. A period of 21 months from 01/06/2010 to 29/02/2012 is considered. This period includes 22,050 cases from which a subset of 1439 is chosen, which corresponds to insured vehicles without franchise and private use, excluding the missing data. The relevant information is

Response: Claim severity (or mean claim amount per claim).
Weights: Number of claims.
Mixed predictors: Vehicle age (continuous); Vehicle power (continuous); Vehicle weight (continuous); Vehicle price at new (continuous), Driver age (continuous); Vehicle brand (categorical nominal with 19 categories with the following levels: AMERICAN, CHEVROLET, CHINESE, CITROEN, FIAT, FORD, HONDA, HYUNDAI, KIA MOTORS, MAZDA, MITSUBISHI, NISSAN, PEUGEOT-RENAULT, SAMSUNG, SUBARU, SUZUKI, TOYOTA, VOLKSWAGEN, OTHER).

The original distance matrix is calculated from the predictor information using the Gower similarity index (see Gower (1971)). These data have 23 degrees of freedom in a classical GLM.

5.1   Fitting Distance-Based Regression

Two DB-GLM models are fitted using the function dbglm in the R package dbstats (Boj et al. (2017)). The first DB-GLM, dbglm1, is fitted using relative geometric variability equal to 1, i.e., taking into account all the dimensions, 162:

```
R> dbglm1 <- dbglm(D, CM_TOTAL, weights = NUM_TOTAL_SIN,
data = AUTOMOBILE, family = Gamma(link = "log"),
method = "rel.gvar", rel.gvar = 1); dbglm1

Call:  dbglm(D2 = D, y = CM_TOTAL, ... = pairlist(data =
 AUTOMOBILE), family = Gamma(link = "log"), method =
 "rel.gvar", weights = NUM_TOTAL_SIN, rel.gvar = 1)

family: Gamma

Degrees of Freedom: 1438 Total (i.e. Null);   1276 Residual
Null Deviance:        2305
Residual Deviance: 1963

AIC:        13383.48
BIC:        14242.77
GCV:        1.4315

R> dbglm1$eff.rank
[1] 162
```

The second DB-GLM, dbglmGCV, is fitted using the GCV criterion to choose the effective rank:

```
R> summary(dbglmGCV)

Call:  dbglm(D2 = D, y = CM_TOTAL, ... = pairlist(data =
 AUTOMOBILE), family = Gamma(link = "log"), method =
 "GCV", full.search = TRUE, weights = NUM_TOTAL_SIN,
 range.eff.rank = c(1, 162))

Deviance Residuals:
   Min. 1st Qu.  Median   Mean 3rd Qu.    Max.
 -5.120  -1.172  -0.494  -0.412   0.266   4.256

(Dispersion parameter for Gamma family taken to be 1.852228)

    Null deviance: 2304.6  on 1438 degrees of freedom
Residual deviance: 2265.63  on 1434 degrees of freedom

Number of Fisher Scoring iterations: 7
```

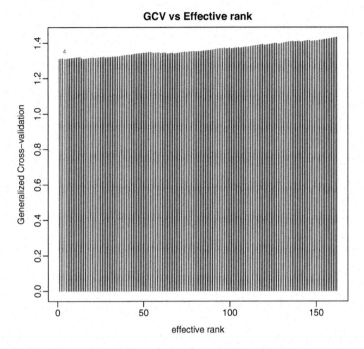

Fig. 1 DB-GLM results for the original data set using the Gower similarity index, for mixed predictors and Method = "GCV" dbglm. The values of the GCV statistic up to 162 dimensions are shown. The minimum is achieved for four dimensions (in red)

```
Convergence criterion: DevStat

AIC:        13362.95
BIC:        13389.31
GCV:        1.31
```

From the obtained results we see that the effective rank using the GCC criterion is equal to 4, with 57.82% explained geometric variability. Figure 1 had shown the values of the GCV statistic for different dimensions, from 1 to 162, with a minimum of 4 for a value of 1.31.

We then see that y and w are vectors of dimension $N = 1439$, and X is of dimension $N \times M$, where $M = 162$. When the GCV criterion is used, the effective rank is equal to 4.

5.2 Fitting Cluster Distance-Based Regression

This section illustrates the proposed cluster distance-based regression. First the cluster-MDS is applied to the data. Using the Hartigan statistic (see Vera and Macías

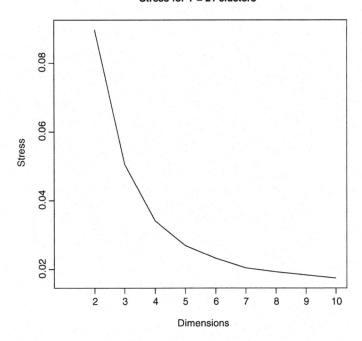

Fig. 2 Stress values for the cluster-MDS procedure, for $T = 21$ clusters and two to ten dimensions

2017), we obtain that for $T = 21$ clusters, the optimal value of the stress is 0.02042572 attained for $\tilde{M} = 7$, as shown in Fig. 2.

In the application, $T = 21$ clusters ($21 \ll 1439$) are obtained. The optimal dimension is $\tilde{M} = 7$, $1 \le \tilde{M} \le 20$, and the coordinates $\tilde{X}$ are of dimension $T \times \tilde{M} = 21 \times 7$. Thus, 21 clusters or risk groups are obtained. We then calculate the weighted means of the original response in each risk group to construct the new $\tilde{y}$. The original weights for the claims belonging to each risk group t, for $t = 1, \ldots, T$ are summed, to obtain $\tilde{w}$ of dimension 21×1. From the cluster-MDS we obtain $\tilde{X}$ of dimension 21×7. The associated inner products matrix $\tilde{G} = \tilde{X}\tilde{X}^{\top}$ is a positive semidefinite matrix.

A DB-GLM model is fitted assuming a Gamma distribution and the logarithmic link, termed dbglmAICclus. The AIC criterion is used in the dbglm function. The values of the AIC statistic are calculated for dimensions ranging from 1 to 7, reaching a minimum at 7, as expected, i.e., all the dimensions of the cluster-MDS are included, thus explaining the total geometric variability of $\tilde{G}$. Figure 3 shows the values of the AIC statistic for the different dimensions, reaching a minimum at 7 with a value of 8262.97. The command predict of the object dbglmAICclus

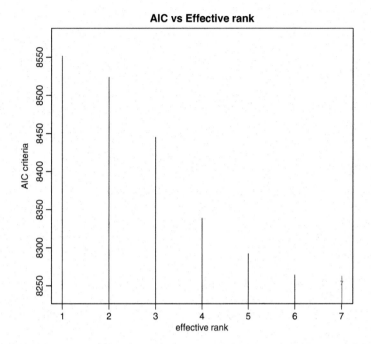

Fig. 3 DB-GLM fitted model for the cluster latent classes using Method = "AIC" dbglm Values of the AIC statistic for one to seven dimensions. The minimum value is reached at seven dimensions (in red)

allows us to make predictions for new cases. The resulting cluster distance-based model can be summarised as:

```
R> summary(dbglmAICclus)

Call:   dbglm(G = Gclus, y = yclus, family = Gamma(link = "log"),
        method = "AIC", full.search = TRUE, weights = wclus,
        range.eff.rank = c(1, 7))

Deviance Residuals:
    Min.   1st Qu.   Median     Mean   3rd Qu.      Max.
-3.01600  -1.21000  -0.29760  -0.09088  0.91100   2.61600

(Dispersion parameter for Gamma family taken to be 3.209092)

    Null deviance: 78.09  on 20 degrees of freedom
Residual deviance: 41.17  on 13 degrees of freedom

Number of Fisher Scoring iterations: 7
Convergence criterion: DevStat

AIC:         8262.97
BIC:         8271.33
GCV:         0.05
```

6   Concluding Remarks

Distance-based regression analysis is based on the projection of the vector of continuous responses in a Euclidean space given by multidimensional scaling (MDS). One of the main problems encountered in this approach is that of determining the number of dimensions or latent predictor variables. When this number becomes large, for example, when the sample size increases with respect to the original number of predictor variables (if any), or when the observed proximities are previously transformed by the additive constant procedure, it is usually difficult to obtain a good fit of the model. In this paper, a combined procedure of cluster-MDS and distance-based regression analysis is proposed for situations in which the information to be analysed is obtained from a large dissimilarity matrix.

It is well known that the combined use of cluster and multidimensional scaling provides a better understanding of the data and reduces the number of parameters to be estimated. Since the MDS reduced space is optimal for the embedded points, but not for the superimposed clusters, while the cluster structure is only optimal in the original, non-reduced space, integrating the two procedures to represent cluster centres instead of the original responses in reduced dimensionality is the most appropriate approach. Given the estimated classification in the response space and the cluster centres configuration, the fitting of a distance-based regression model is performed by projecting the weighted average vector within each cluster onto the continuous response variable within the clustered predictor space.

In our paper, the above procedure was applied to the *a priori* ratemaking process. Claim severity was been fitted in automobile insurance with respect to own-damage cover. First, the clusters and the MDS configuration were calculated, obtaining a total of 21 risk groups for the tariff in a low seven-dimensional space, and then a DB-GLM was fitted to the cluster data, assuming a Gamma distribution and the logarithmic link.

Although a two-step procedure in this framework works well, the formulation of a cluster distance-based regression model that simultaneously incorporates these two procedures within a structural equation model framework (Vera and Rivera 2014; Vera and Mair 2019) is a promising approach that is currently being addressed by the authors.

Supporting Agencies

Grants ECO2013-48413-R of the Ministerio de Economía y Competitividad of Spain, co-financed by FEDER, and RTI2018-099723-B-I00 of the Ministerio de Ciencia, Innovación y Universidades of Spain, co-financed by FEDER, (J. Fernando Vera), and MTM2014-56535-R of the Spanish Ministry of Economy and Competitiveness Spanish, co-financed by FEDER, and 2017SGR152 of the Generalitat de Catalunya, AGAUR (Eva Boj). Multivariate Statistics and Classification Group of the Spanish Society of Statistics and Operations Research, AMyC-SEIO.

References

Bock, H.H.: Multidimensional scaling in the framework of cluster analysis. In: Hermes, H.J., Optiz, O., Degens, P.O. (eds.) Studien zur Klassifikation: [Classification and its Environment], vol. 17, pp. 247–258. INDEKS-Verlag, Frankfurt (1986)

Bock, H.H.: On the interface between cluster analysis, principal components, and multidimensional scaling. In: Bozdogan, H., Gupta, A.J. (eds.) Multivariate Statistical Modelling and Data Analysis, pp. 17–34. Riedel, New York (1987)

Boj, E., Claramunt, M.M., Fortiana, J.: Análisis multivariante aplicado a la selección de factores de riesgo en la tarificación. Cuadernos de la Fundación MAPFRE, vol. 88. Fundación MAPFRE Estudios, Madrid (2004)

Boj, E., Delicado, P., Fortiana, J.: Distance-based local linear regression for functional predictors. Comput. Stat. Data Anal. 54(2), 429–437 (2010)

Boj, E., Delicado, P., Fortiana, J., Esteve A., Caballé, A.: Global and local distance-based generalized linear models. TEST 25(1), 170–195 (2016)

Boj, E., Caballé, A., Delicado, P., Fortiana, J.: dbstats: distance-based statistics (dbstats). R package version 1.0.5. https://CRAN.R-project.org/package=dbstats (2017)

Cuadras, C., Arenas, C.: A distance-based regression model for prediction with mixed data. Commun. Stat. Theory Methods 19, 2261–2279 (1990)

Cuadras, C.M., Arenas, C., Fortiana, J.: Some computational aspects of a distance-based model for prediction. Commun. Stat. Simul. Comput. 25(3), 593–609 (1996)

Gower, J.C.: Adding a point to vector diagrams in multivariate analysis. Biometrika 55, 582–585 (1968)

Gower, J.C.: A general coefficient of similarity and some of its properties. Biometrics 27, 857–874 (1971)

Haberman, S., Renshaw, A.E.: Generalized linear models and actuarial science. J. R. Stat. Soc. Ser. D 45(4), 407–436 (1996)

Heiser, W.J.: Clustering in low-dimensional space. In: Opitz, O., Lausen, B., Klar, R. (eds.) Information and classification, pp. 162–173. Springer, Berlin (1993)

Heiser, W.J., Groenen, P.J.F.: Cluster differences scaling with a within-clusters loss component and a fuzzy successive approximation strategy to avoid local minima. Psychometrika $62$1, 63–83 (1997)

Kruskal, J.B.: The relationship between multidimensional scaling and clustering. In: Van Ryzin, J. (ed.) Classification and clustering, pp. 17–44. Academic Press, New York (1977)

Sokal, R., Michener, C.: A statistical method for evaluating systematic relationships. Univ. Kansas Sci. Bull. 38, 1409–1438 (1958)

Vera, J.F., Macías, R.: Variance-based cluster selection criteria in a k-means framework for one-mode dissimilarity data. Psychometrika 82(2), 275–294 (2017)

Vera, J.F., Mair, P.: SEMDS: an R package for structural equation multidimensional scaling. Struct. Equ. Model. Multidiscip. J. 26, 803–818 (2019)

Vera, J.F., Rivera, C.D.: A structural equation multidimensional scaling model for one-mode asymmetric dissimilarity data. Struct. Equ. Model. Multidiscip. J. 21(1), 54–62 (2014)

Vera, J.F., Macías, R., Angulo, J.M.: Non-stationary spatial covariance structure estimation in oversampled domains by cluster differences scaling with spatial constraints. Stoch. Env. Res. Risk A. 22, 95–106 (2008)

Vera, J.F., Macías, R., Angulo, J.M.: A latent class MDS model with spatial constraints for non-stationary spatial covariance estimation. Stoch. Env. Res. Risk A. 23(6), 769–779 (2009)

Vera, J.F., Macías, R., Heiser, W.J.: A latent class multidimensional scaling model for two-way one-mode continuous rating dissimilarity data. Psychometrika 74(2), 297–315 (2009)

Bayesian Network Analysis of Fashion Behavior

Keiko Yamaguchi and Hiroshi Kumakura

Abstract Fashion behavior, i.e., consumer behavior with regard to choosing, purchasing, and wearing clothing, is diverse and intricate because it springs from consumers' complicated contexts. It is important for those in the fashion industry to understand the full picture of fashion behavior, find ways to sustain their customers' interest in fashion, and promote their products appropriately. However, as far as we know, there has been little comprehensive research on the diverse and intricate chain reactions involved in fashion behavior. In light of this gap in our understanding, the aim of this research is to describe consumer fashion behavior holistically using a Bayesian network approach with actual business data. We train four types of Bayesian network with different constraints, finding that different types of network are best suited to describing the behavior of men and women. Then, we validate the theoretical consistency of the selected networks by comparing their results with the findings of previous studies. In this paper, we mainly focus on fashion-related problems and consumer impulse buying behavior because of space limitations. In addition to these academic contributions, we demonstrate how businesses could apply Bayesian networks in describing fashion behavior, allowing marketers to use sensitivity analysis to devise potential marketing campaigns and change their customers' behavior.

K. Yamaguchi (✉)
Nagoya University, Nagoya, Japan
e-mail: keiko.yamaguchi@soec.nagoya-u.ac.jp

H. Kumakura
Chuo University, Tokyo, Japan
e-mail: kumakura@tamacc.chuo-u.ac.jp

© Springer Nature Singapore Pte Ltd. 2020
T. Imaizumi et al. (eds.), *Advanced Studies in Classification and Data Science*,
Studies in Classification, Data Analysis, and Knowledge Organization,
https://doi.org/10.1007/978-981-15-3311-2_32

1   Introduction

Let us begin our study with the following customer stories.

> When I was at school, I took a part-time job in Harajuku, the most fashionable neighborhood in Japan, which inspired me to become interested in fashion. I purchased a lot of clothes, eventually acquiring an outrageous purple suit that I could only wear in Harajuku.

> My clothing allowance increased after I got married, so I was able to choose more luxurious brands than I had ever owned before.

How can we interpret these episodes from a marketing viewpoint? They demonstrate that, with consumer behavior, changes to the environment can cause psychological and behavioral changes, leading to changes in fashion behavior. Some aspects of this process have been studied in fields such as fashion psychology, life course theory, and experiential consumption theory. For instance, Solomon and Schopler (1982) and O'Cass (2000) investigated consumer behavior with respect to clothing in a fashion psychology context, and Moschis (2007) studied changes in consumer behavior due to changes and events in terms of life course theory. The Diderot effect (McCracken (1990)) is one of the most famous findings about fashion behavior in the field of experiential consumption theory. However, as far as we know, few studies have attempted to describe the entire consumer behavior process. The aim of our study is therefore to describe the whole picture of consumer fashion behavior using empirical data.

One of the most significant advantages for marketers will be to obtain the entire perspective on such behavior, not just some details of it, so that they can use suitable psychological and environmental triggers to encourage their customers to, for example, visit their stores or purchase their products. Previous research has enabled marketers to understand some of the relationships between environmental and psychological factors and consumer fashion behavior, but not the causal relationships among those factors. Hence, we perform simulations to see how customer behavior would change as a result of particular environmental and psychological triggers. This study aims to increase understanding of the causal relationships between factors involved in fashion behavior by analyzing the results of simulations based on a Bayesian network approach.

After describing the whole picture of fashion behavior, we validate its theoretical consistency by comparing our results with the findings of previous studies. Since most of these studies focused on only one aspect of the bigger fashion behavior picture, we compare them with the corresponding aspects of our picture. On the basis of these validations of particular aspects of the fashion picture, we mainly consider impulse buying behavior in a fashion context and consumer rationality here owing to space limitations because impulse buying is one of the most significant and interesting fashion behaviors (Shima and Ohashi (1994)) and has been widely investigated in previous research.

Finally, we aim to derive some management insights from the results of our network.

2 Literature Review

As discussed above, in this paper, we mainly consider the impulse buying aspect of the broader picture presented in the next section. In order not only to validate this aspect theoretically but also to interpret some relationships between it and other factors involved in fashion behavior, we now review the literature on impulse buying behavior and highlight some of the important findings.

"Impulse buying" is the name given to sudden, unpremeditated purchases where the customer had no previous intention of buying either that specific product or a product in that category. Such behavior occurs after the customer experiences a sudden, irresistible urge to buy and tends to be spontaneous, without much reflection (Beatty and Ferrell 1998). Impulse buying behavior occurs most frequently in situations where the consumer deems such behavior to be appropriate, such as when buying a gift for someone else (Solomon 2002).

Further categorizing impulse buying enables us to understand its characteristics in more detail, and it is typically divided into four types: pure impulse buying, reminder impulse buying, suggestion impulse buying, and planned impulse buying (Stern 1962). Of these, only pure impulse buying represents truly impulsive buying, a novelty or escape purchase that breaks with the customer's normal buying pattern. On the other hand, reminder impulse buying occurs when seeing an item reminds a consumer of information related to that item.

Much research has been done into various aspects of impulse buying. With respect to fashion behavior, an interest in fashion and positive feelings about fashion directly affect fashion-oriented impulse buying (Park et al. 2006). In addition, hedonic consumption has an indirect effect (Park et al. 2006), and impulse buying also helps to satisfy hedonic needs and provide emotional gratification (Hausman 2000; Piron 1991).

Gender affects not only fashion behavior but also impulse buying. Females tend to be more impulsive than males (Dittmar et al. 1995; Wood 1998; Coley and Burgess 2003) and tend to buy symbolic and self-expressive goods that are associated with their appearance and the emotional aspects of their selves. On the other hand, males tend to impulse-buy practical and leisure items that project their independence and activities (Dittmar et al. 1995).

3 Explanatory Analysis of Fashion Behavior

In this study, we train several Bayesian networks based on the fashion behavior of either men or women. Gender is one of the most crucial factors to consider in a fashion context because the gender-related differences in impulse buying behavior have been widely discussed in the fashion behavior literature (Dittmar et al. 1996), (Khan et al. 2016; Pentecos and Andrews 2010; Tifferet and Herstein 2012). Moreover, practically speaking, fashion goods are typically sold in gender-segregated shops, and the types of goods consumers buy are totally different for each gender. We want to include consumer purchase behavior in our approach, so we conduct separate analyses for each gender. Our preliminary network analysis also supported the importance of gender in the network.

3.1 Data

For this study, we used data provided by an anonymous fashion e-commerce site in Japan via the Joint Association Study Group of Management Science. This included two types of data: histories of purchases on the site between April 2015 and March 2016 and data from a survey the site's customers completed during the week of March 17th, 2016. Both datasets included 985 men and 2159 women. Owing to the timing of the survey, we excluded the purchase history data after the survey from consideration.

From these two datasets, we extracted factors related to fashion behavior and divided them into five groups: Demographics, Environment, Consumer sentiment, Fashion sentiment, and Behavior. Table 1 shows the extracted factors and their factor groups.

For the Environment group, we extracted the environmental changes that occurred during the data period from the responses to open questions in the survey by morphological and latent semantic analysis. First, we excluded unnecessary or inappropriate words (e.g., "nothing") from answers to open-ended questions, corrected misspellings and missing words in the answers, standardized the expressions using a dictionary, and extracted morphemes from them using RMeCab (a Japanese morphological analysis package for R). Then, we conducted a latent semantic analysis by deducing the contextual meanings of the words used, extracting the meanings of nouns that were used by more than 2.5% of the respondents because many of the answers consisted only of nouns. Finally, we

Table 1 The five groups of factors related to fashion behavior used in this study and their related data source

Group	Factors	Data source
Demographics	Age, gender, membership	Survey
Environment	Marital status, income, occupation, fashion budget, pregnancy, childcare	Survey
Consumer sentiment	Philosophy of life, happiness, general feelings about consumption	Survey
Fashion sentiment	Thoughts about fashion, problems with fashion, changes in fashion interest	Survey
Behavior	Changes in fashion behavior, purchase timing, purchased product categories, purchase frequency, purchase types (on sale, reserved)	Survey, purchase history

integrated dimensions such that they satisfied the condition that the cumulative sum of the singular values was 0.5.

The most significant difference between the Consumer and Fashion sentiment groups is whether or not the questions focused on feelings specifically related to fashion purchases. Hence, although factors in the Consumer sentiment group represent consumers' tendencies to, for example, cherry-pick and impulse-buy, factors in the Fashion sentiment group deal with consumers' problems and thoughts about fashion.

The types of product purchased were included in the Behavior group in Table 1. We selected the top 70% of the product categories purchased by volume for each gender and created a binary variable for each category indicating whether or not a given consumer purchased a product from that category during the data period.

3.2 Training the Bayesian Network

As discussed in the Introduction, we took a Bayesian network approach in describing the entire range of consumers' fashion behavior. A Bayesian network is a probabilistic graph-based model that represents a set of random variables and their conditional dependencies in terms of a directed acyclic graph. We used the bnlearn package for R, developed by Scutari (2010), to train the Bayesian network and estimate its conditional probability tables using the following procedure:

1. Train 500 networks with resampled data using a bootstrap process.
2. Average the networks and select the most significant paths.
3. Estimate conditional probability table for each node.

In Step 1, we applied a score-based learning algorithm and used the Bayesian Dirichlet equivalent score (BDe) as a goodness-of-fit index for the algorithm. The networks trained in Step 1 were then averaged, and the most significant paths in the averaged network were selected using the "significant threshold" index (Scutari 2013). This resulted in a more robust network with better predictive performance than choosing the single highest scoring network (Nagarajan et al. 2013). Finally, we estimated the conditional probability table for each node in the averaged network using the Bayesian method with a non-informative prior distribution.

The edges in the graph and their directions were learned automatically from the data by following the above steps. However, not all edges and directions are theoretically equally likely from a consumer behavior theory viewpoint. In order to extract a more reasonable and realistic network from the data, we added some path constraints to the network learning process, as shown in Fig. 1. The first constraint was that paths to the Demographics group from the other factor groups were not allowed. Factors in the Demographics group are generally fixed from the start of the data period, so we hypothesized that these would not be affected by the other factors. The second constraint was that no paths to the Environment group were allowed from the Consumer sentiment, Fashion sentiment, or Behavior groups. Over just one

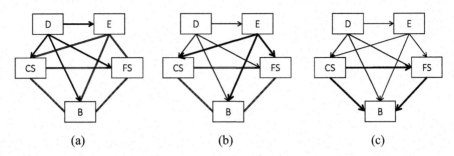

Fig. 1 Possible paths under the constraints in (**a**) Model 1, (**b**) Model 2, (**c**) Model 3. Double lines represent paths in both directions. Bold arrows are possible one-way paths newly added as a restriction in each model. Acronyms in nodes indicate feature groups as follows: D: Demographics, E: Environment, CS: Consumer sentiment, FS: Fashion sentiment, B: Behavior

year, environmental changes are likely to cause changes in sentiment or behavior, but not vice versa, so we believed that this would be a reasonable constraint in this case. The third constraint was that paths were not allowed from the Fashion sentiment group to the Consumer sentiment group or from the Behavior group to either of the other two sentiment groups. This was because consumer behavior theory suggests that general consumer sentiments are not caused by more specific sentiments, and how consumers behave is a consequence of what they think or feel. We trained four models by adding these constraints one by one: the Base model, Model 1, Model 2, and Model 3. The Base model was trained without any constraints, whereas Model 1 included the first constraint, Model 2 included the first two constraints, and Model 3 included all three constraints. Model 3 thus represented the most rational consumer behavior model of the four options. All of the models were trained by following the three steps outlined above.

3.2.1 Validation

We used two indices to select the best model for each gender from the four models listed above. First, we did k-fold cross-validation with $k = 1 + \ln(number\ of\ samples)/\ln(2)$ to calculate the expected log-likelihood loss. Second, we recalculated the BDe score for each averaged network to check its final goodness-of-fit. The best model for each gender was then selected based on these two metrics.

3.3 Results

3.3.1 Model Selection

The box charts in Fig. 2 show the log-likelihood losses for each model, and the goodness-of-fit (BDe) scores are listed in Table 2. For men, both the box charts and

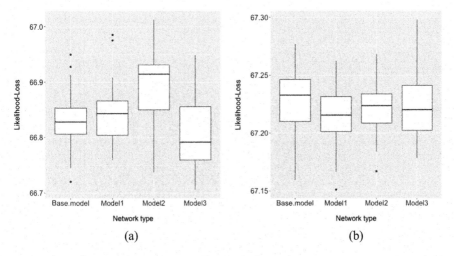

Fig. 2 Log-likelihood loss box charts for men (**a**) and women (**b**)

Table 2 Goodness-of-fit (BDe) values for the four network models

	Base model	Model 1	Model 2	Model 3
Men	−64,373.1	−64,338.5	−64,405.4	−64,273.2
Women	−143,717.3	−143,727.9	−143,677.1	−143,831.9

BDe scores show that Model 3 was the best, so we chose Model 3 to describe men's fashion behavior. For women, on the other hand, no one model was the best with respect to both indices. Using the Base model values as a baseline, only Model 2 improved on the baseline in any way, having the lowest BDe value.

Figure 3 shows an overview of the network selected for men, with the dotted rectangles indicating the factor groups to which the network nodes belonged. (Owing to space limitations, only this network is shown as an example.) In order to explain the differences between the men's and women's networks in detail, Fig. 4 shows network overviews for both genders where the nodes have been divided into interpretable node clusters. In the men's network, the nodes are roughly divided into five clusters, namely, Environment changes, Fashion problems, Life priorities, Consumer and fashion sentiment, and Product category and purchase behavior. The women's network shows six clusters, some with slightly different characteristics from those in the men's network: Environment and fashion sentiment changes, Fashion problems, Life priorities, Consumer and fashion sentiment, Product category, and Purchase behavior. Owing to the constraints on the paths allowed, the flow of the men's network looks more rational, and the nodes in the men's network related to sentiment and purchase behavior were clearer than those in the women's network.

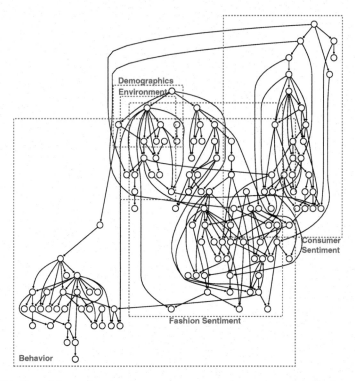

Fig. 3 Overview of the men's network

3.3.2 Implication of the Networks

Next, we investigated the differences between men's and women's networks to obtain some managerial insights.

One of the most interesting differences between men and women is how they think when faced with a fashion-related problem such as, "I cannot buy clothes because I do not have enough money for fashion." Figure 5 shows the relevant parts of the networks for both genders, and the signs indicate the polarities of the node relationships based on the network's conditional probability table: a positive sign means the node is likely to be true if its parent is true, whereas a negative sign means the node is likely to be false if its parent is true. Figure 5 indicates that, for men, this type of problem can be cured by the realistic solution of having a generous budget he can spend freely when he sees nice clothes. On the other hand, women are likely to feel sad when faced with this type of problem and tend to (negatively) restructure their emotional life instead of searching for realistic solutions. These differences imply that fashion purchasing decisions tend to be made functionally by men and emotionally by women. Previous studies support this gender difference in practice (Barrett et al. 1998). For example, NeuroFocus (2009) showed that there were fundamental differences between the male and female brains and that women

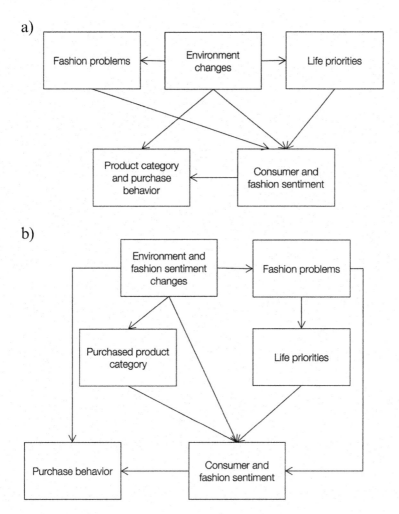

Fig. 4 Node clusters in the men's (**a**) and women's (**b**) networks and their interpretations

have a markedly higher tendency to attach emotional significance to stimuli than men.

Another interesting difference can be seen in the triggers for impulse buying and its consequences. The sub-networks for impulse buying shown in Fig. 6 indicate that men who go shopping without a specific purpose and consider brands when shopping are likely to make impulse purchases. This implies that always carrying mobile devices could make people more brand-aware and that this brand knowledge could become a trigger for impulse buying. Emotional aspects, such as releasing stress, are just a consequence of purchasing behavior. On the other hand, the causes of impulse buying in women are more emotional: women are likely to buy fashion products impulsively because buying clothes is a way to release stress. In addition,

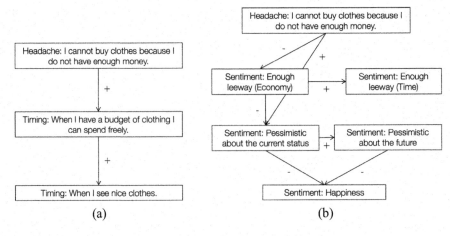

Fig. 5 Men's (**a**) and women's (**b**) sub-networks for fashion-related problems

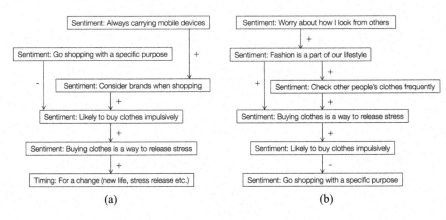

Fig. 6 Men's (**a**) and women's (**b**) sub-networks for impulse buying

women who frequently look at what other people are wearing or who show a strong commitment to fashion as, for example, part of their lifestyle are more likely to buy clothes to release stress and ultimately make impulse purchases.

According to Park et al. (2006), commitment to fashion has a direct, positive impact on impulse buying and also has a positive effect on tendency to hedonic consumption and hence indirectly on impulse buying. We derived the same findings from the women's sub-network in Fig. 6 and could categorize women's impulse buying as "pure impulse buying," as defined by Stern (1962), because the triggers for women's impulse buying were emotional state and the need to escape from stress. On the other hand, should men's impulse buying be categorized as "pure impulse buying"? The cause here was brand awareness when shopping without a specific purpose, and there was no emotional driver in the men's sub-network. This implies that men were more likely to impulse-buy when they were reminded of their need for fashion by seeing, for example, a brand name on a sign, and hence, their behavior should be categorized as "reminder impulse buying." In summary, men's impulse buying was functional and cognitive, whereas women's was emotional. Dittmar et al. (1995) and Coley and Burgess (2003) suggested that men mainly buy practical products on impulse, whereas women focus on products related to appearance, and brands are seen as assuring the quality of goods. Our finding that men's impulse buying is driven by brand awareness agrees well with the findings of these previous studies.

In this study, we have taken a Bayesian network approach to describe the reasons for impulse buying by men and women and have highlighted some of the differences shown by previous studies. Our study's novel contribution to the field is that we also conducted sensitivity analyses of the men's and women's networks to derive managerial insights and plan marketing activities, such as how to promote impulse buying. Sensitivity analysis is also known as belief updating or probabilistic reasoning. In particular, we simulated how the posterior probability of an objective node, the node representing impulse buying tendency, would change given hard evidence. As hard evidence, we chose to set the "Check other people's clothes frequently" and "Consider brands when shopping" nodes to be true or false. Table 3 shows posterior probabilities that the objective (impulse buying tendency) node is true given these pieces of evidence. It also shows the marginal probability of the objective node without any evidence as a baseline for comparison.

When the "Check other people's clothes frequently" node was set to be true, the posterior probability given evidence is higher than the marginal probability, whereas no change for men. This result implies that checking other people's clothes would reinforce women's impulse buying tendencies. Hence, marketers who want their female customers to buy products on impulse might consider designing a system that encourages women to regularly check out other people's clothes. However, such marketing plan would not affect men's sentiment or behavior.

On the other hand, when we set the "Consider brands when shopping" node to be true, the posterior probability is higher than the marginal probability for men and lower for women. When marketers are planning to promote impulse buying, reinforcing the impact of brand names would be effective for men so that male customers can easily recognize them and be reminded of their needs for fashion. Unfortunately, this marketing plan would not work as well for women. These

Table 3 Sensitivity analyses for two pieces of evidence: posterior probability of "impuluse buying tendency"

Evidence	True	False	Evidence	True	False
Men's network			*Women's network*		
Consider brands when shopping	0.616	0.422	Consider brands when shopping	0.612	0.616
Check other people's clothes frequently	0.557	0.557	Check other people's clothes frequently	0.637	0.593
No evidence	0.557		No evidence	0.614	

differences in fashion behavior imply that the optimal marketing approaches to promote impulse buying are different for each gender. Since sensitivity analysis allows us to investigate the extent to which the evidence changes the posterior probability, it allows us to take multiple evidence options and choose the most effective marketing strategy.

4   Conclusion

In this study, we have described consumer fashion behavior holistically by applying a Bayesian network approach to actual business data. Describing this whole picture is one of our study's novel contributions because few studies have attempted to consider the complete picture of fashion behavior. We have also trained four types of Bayesian network with different constraints, finding that the two genders were best described by different types of network. We then validated the theoretical consistency of the selected networks by comparing their results with the findings of previous studies. Owing to space limitations, we mainly focused on fashion-related problems and impulse buying behavior. In addition to these academic contributions, we have shown that, by applying a Bayesian network approach to describing fashion behavior and conducting sensitivity analysis, businesses could allow marketers to simulate different marketing triggers and find ways to change customer behavior.

Our study does, however, have some limitations, which will be addressed in future work. First, our networks only consider gender differences, but fashion behavior is different for a variety of different consumer groups. Future studies could consider other ways of dividing consumers into groups to discover a more sophisticated consumer behavior in the fashion industry.

Second, the factors in the Behavior group could be reconsidered. Our networks mainly considered the product category and consumer purchase behavior (such as frequency and tendency to cherry-pick). One potential approach to derive more actionable business insights could be to break down the product categories to include more specific characteristics, such as newly launched products or brands.

Third, we need not only to validate the theoretical consistency of our model's sub-networks against the findings of previous studies but also to verify the links between these sub-networks in more detail so that we can gain a comprehensive understanding of fashion behavior and discover more effective triggers for changing consumer behavior.

Finally, our networks are static and do not consider dynamic relationships among factors. In reality, the products that customers purchase now could affect factors such as fashion sentiment at later times, as in the episode quoted in the Introduction. Having a total picture of fashion behavior that incorporates its dynamics will give us a tool for planning marketing activities to trigger changes in customer behavior in the longer term.

References

Barrett, L.F., Robin L., Pietromonaco, P.R., Eyssell, K.M.: Are women the "more emotional" sex? evidence from emotional experiences in social context. Cogn. Emot. **12**, 555–578 (1998)

Beatty, S.E., Ferrell, M.E.: Impulse buying: modeling its precursors. J. Retail. **74**, 169–191 (1998)

Coley, A., Burgess, B.: Gender differences in cognitive and affective impulse buying. J. Fash. Mark. Manag. **7**, 282–295 (2003)

Dittmar, H., Beattie, J., Friese, S.: Gender identity and material symbols: objects and decision considerations in impulse purchases. J. Econ. Psychol. **16**, 491–511 (1995)

Hausman, A.: A multi-method investigation of consumer motivations in impulse buying behavior. J. Consum. Mark. **17**, 403–19 (2000)

Khan, N., Hui, L.H., Chen, T.B., Hoe, H.Y.: Impulse buying behaviour of generation Y in fashion retail. Int. J. Bus. Manag. **11**, 144–150 (2016)

McCracken, G.: Culture and Consumption: New Approaches to the Symbolic Character of Consumer Goods and Activities. Indiana University Press, Bloomington (1990)

Moschis, G.P.: Life course perspectives on consumer behavior. J. Acad. Mark. Sci. **35**, 295–307 (2007)

Nagarajan, R., Scutari, M., Lèbre, S.: Bayesian Networks in R with Applications in Systems Biology. Springer, Berlin (2013)

NeuroFocus: Gender and marketing the female brain. In: Nielsen (2009). http://www.nielsen.com/us/en/insights/news/2009/ of subordinate document. Accessed 12 Oct 2017

O'Cass, A: An assessment of consumers product, purchase decision, advertising and consumption involvement in fashion clothing. J. Econ. Psychol. **21**, 545–576 (2000)

Park, E.J., Kim, E.Y., Forney, J.C.: A structural model of fashion-oriented impulse buying behavior. J. Fash. Mark. Manag. **10**, 433–446 (2006)

Pentecos, R., Andrews L.: Fashion retailing and the bottom line: the effects of generational cohorts, gender, fashion fanship, attitudes and impulse buying on fashion expenditure. J. Retail. Consum. Serv. **17**, 4352 (2010)

Piron, F: Defining impulse purchasing. Adv. Consum. Res. **18**, 509–13 (1991)

Scutari, M.: Learning Bayesian networks with the bnlearn R package. J. Stat. Softw. **35**, 1–22 (2010)

Scutari, M., Nagarajan, R.: On identifying significant edges in graphical models of molecular networks. Artif. Intell. Med. **57**, 207–217 (2013)

Solomon, M.R.: Consumer Behavior. Prentice-Hall, Upper Saddle (2002)

Solomon, M.R., Schopler J.: Self-consciousness and clothing. Personal. Soc. Psychol. Bull. **8**, 508–514 (1982)

Stern, H.: The significance of impulse buying today. J. Mark. **26**, 59–62 (1962)

Terashima, K., Ohashi, M.: A quantitative analysis for the classification of consumer goods. J. Jpn. Ind. Manag. Assoc. **45**, 22–29 (1994)

Tifferet, S., Herstein, R.: Gender differences in brand commitment, impulse buying, and hedonic consumption. J. Prod. Brand. Manag. **21**, 176182 (2012)

Wood, M.: Socio-economic status, delay of gratification, and impulse buying. J. Econ. Psychol. **19**, 295–320 (1998)

Part V
Statistical Data Analysis for Social Science

Determining the Similarity Index in Electoral Behavior Analysis: An Issue Voting Behavioral Mapping

Theodore Chadjipadelis and Georgia Panagiotidou

Abstract The aim of the research is to classify all respondents and parties regarding their positions on issues. For the purposes of this paper a dataset from Greek HelpMeVote application (VAA) from 2015 was used. The dataset contains the answers of voters and candidate parties on questions regarding issues on a Likert scale 1–5. Additionally, the objective is to introduce an alternative methodology for similarity metrics in electoral analysis, regarding voter's opinions about issues. By using a 3-step methodology of Hierarchical Cluster analysis with Ward's criteria, chi-square distances between objects and factor analysis of Correspondence we create groups of respondents with similar attitude on issues and all clusters will be placed on the axis system The procedure is described as an issue voting behavioral mapping. All groups of respondents and issues will be positioned on an axis system, creating multiple behavioral contexts and depicting positions and distances among groups and issues. In the end of the analysis we interpret the Greek electoral behavior map and the parties' competition of 2015.

1 Methodology

The objective of this research is to examine the electoral profile of voters during the Greek national elections regarding issues and parties (Carmines and Stimson 1980). A secondary objective is to achieve this while introducing an alternative methodology for similarity metrics in electoral analysis, regarding voter's opinions about issues (Benjamin 2010). The aim is to produce cluster of voters with similar behavior upon issues (similar issue voting profiles) and to link each group of voters to the group of issues which contribute the most to the formation of their group. The procedure will produce an issue voting behavioral map where all relationships between groups of variables will be depicted. For the purposes of the research,

T. Chadjipadeli (✉) · G. Panagiotidou
Aristotle University of Thessaloniki, Thessaloniki, Greece
e-mail: chadji@polsci.auth.g; gvpanag@polsci.auth.gr

© Springer Nature Singapore Pte Ltd. 2020   415
T. Imaizumi et al. (eds.), *Advanced Studies in Classification and Data Science*,
Studies in Classification, Data Analysis, and Knowledge Organization,
https://doi.org/10.1007/978-981-15-3311-2_33

analysis was conducted with data gathered by the Greek Voting Advice Application's results for the latest national elections of 2015 (Voting Advise Application Helpmevote for the Greek national elections of 2015, www.helpmevote.gr). This online digital application was first implemented in Greece and is a independent, joint research initiative of the Laboratory of Applied Political Research of the Department of Political Sciences of the Aristotle University of Thessaloniki and the e-Democracy Center of the University of Aarau (ZDA) of Zurich. Development and technical support was provided by Pit Solutions. The electronic platform was available online to all users eager to participate. The users would answer to 30 thematic questions on a 1–5 Likert scale (totally disagree–totally agree), which had already been answered by the candidate parties. The questions were selected carefully in order to be linked and represent main political, economic, and social issues which reflect the basic axis of political competition in Greece (Left-Right, Libertarian-Authoritarian, ethnocentric-European profile (Chadjipadelis et al. 2011)). After having answered all 30 questions the political profile of the respondent could be estimated and was depicted in a two-dimensional diagram (Chadjipadelis and Andreadis 2012). In the same diagram all candidate parties were also positioned according to their answers to the same set of questions, thus their position on issues (Kessel 2011), so the respondent was informed about his proximity to all available options to vote (Benoit and Laver 1999). In electoral behavior analysis distance metrics and mostly the Euclidian or squared Euclidian distance have been used for the identification of similarities between the answers of two objects, which usually apply on a 1–5 or a 1–7 Likert scale (Bechrakis 1999). In behavioral analysis the key question is what the real distance between two respondents is, so determining the right metric is important (Cho and Endersby 2003). These two sets of respondents do not share

Respondent 1 answers: 1 (strongly disagree)	Respondent 1 answers: 2 (disagree)
Respondent 2 answers: 3 (neither/neither)	Respondent 2 answers : 4 (agree)

the same distance. In electoral behavior analysis all contributing factors should be considered for estimating the proximity between two objects. The similarity index should include a set of behavioral and social factors that contribute to the formation of voter's opinion upon issues. A more efficient similarity index which will be able to translate the similarities or dissimilarities between voters and candidates based on issues is essential and should be researched.

Applying a different methodology to classify objects upon their similarity, we use Hierarchical clustering using Ward's criteria. The classification of all respondents (r) is made regarding their positions on issues (q) (Table 1). The table to be analyzed with HAC is a table of distances between objects. However, the metric we choose for these distances is chi-square, suitable for qualitative variables (Bechrakis 1999). The analysis produces a dendrogram with all clusters (Fig. 1). HAC gives five (5) groups of respondents regarding their position on issues: 1992, 1998, 2001, 2000, and 1999.

Table 1 Contingency table
of all respondents (r) and
their positions on issues (q)

	q1	q2	q3	⋯	qn
i1	2	4	3	⋯	1
i2	3	3	4	⋯	1
i3	2	3	5	⋯	4
⋮	⋮	⋮	⋮	⋮	⋮
in	2	2	4	⋯	3

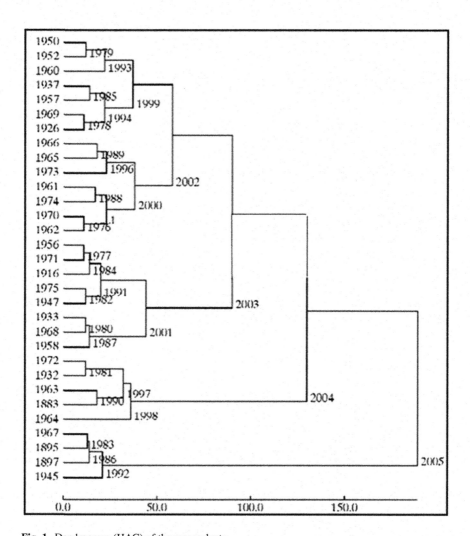

Fig. 1 Dendrogram (HAC) of the respondents

Table 2 Clusters of subjects (respondents)

Cluster	1992	1998	1999	2000	2001
A(I)	1986	1997	1993	1996	1991
B(I)	1945	1964	1994	1995	1987
Count	83	165	153	314	288
	i6	i2	i1	i13	i5
	i63	i529	i30	i706	i25
	i188	i819	i157	i358	i967
	⋮	⋮	⋮	⋮	⋮

Table 3 Updated data table with group. Following the same procedure, adding this time the group in which each subject belongs to (**gr_9** is a new variable indicating the group)

ind	q86	q88	q82	q...	**gr_9**
i1	4	2	2	⋯	1999
i2	1	5	5	⋯	1998
i3	1	2	4	⋯	1999
i4	5	1	5	⋯	1999
i5	3	4	3	⋯	2001
i6	4	5	3	⋯	1992
i7	4	5	4	⋯	1999

In the second step of the analysis we apply the method VACOR (Karapistolis 2002), which helps to identify the contribution of each statistic unit and each variable in each cluster and to the following A and B (Table 2) and which variables contribute to the division of each cluster (Portal for Multidimensional Statistical Data Analysis).[1]

Furthermore, each respondent is linked to a specific group which has been formed according to their answers on the issue questions (Table 3). The same procedure is applied in the new table to create clusters for groups of respondents and their answers. In this application of HAC four (4) groups appear: 314, 312, 315, and 316 (Fig. 2). Each cluster is formed by answers to the questions and groups of respondents.

The analysis shows (Table 4) that cluster 312 concentrates mostly those:

– who have answered 5 (strongly agree) in questions 88, 15, 20, 29, 33, 28, e.g. in favor of the Memorandum and limitation of protests
– Who have answered 1 (strongly disagree) in questions 81, 77, 83, 30, 40, e.g. against public control of banks, in favor of conservatism
– Who belong in group 1998

In the same way all clusters combine the previous clusters of respondents with their answers on specific questions. In Tables 5 and 6 contribution of groups in each option (1–5) to the questions is displayed (Karapistolis 2011). The procedure concludes to catalogue of issue profiles which correspond to the given clusters produced by HAC and their linkage to the answers which they have the high contribution (significant contribution > 2.54)(Table 7). Subsequently, Factor analysis

[1] Karapistolis, D. www.pylimad.gr.

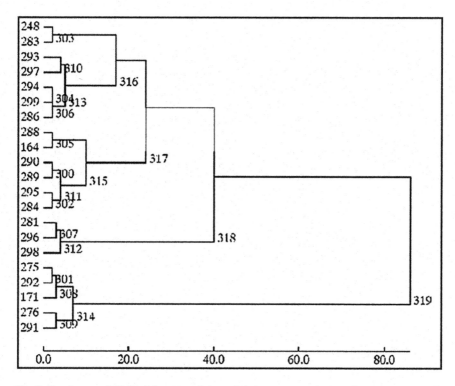

Fig. 2 Dendrogram (HAC) of the respondents and their answers to the questions

of Correspondences is applied (Table 8) in order to investigate the interactions between the variables as divided in clusters (see footnote 1). FACOR is used in order to position clusters and units that clusters are consisted of in a two-dimensional field, as created by the two factors who have the highest contribution (Benzécri et al. 1973/1976). FACOR: This method includes all variables in one system of two factors taking into account the total number of factors (see footnote 1). The whole procedure can produce a semantic map positioning on the same system all questions, answers, and groups of respondents, where all internal relationships are visible (Fig. 3). The behavioral issue voting map can be enriched with other variables such as demographics, past vote, vote intention, education, etc. In this case the same methodology is applied again step by step, adding in the dataset the answers of HelpMeVote 2015 for past vote (national elections of 2012, European elections 2014), intention to vote for 2015, and the confidence for their intention to vote what they declare. The objective in this point is to link all clusters of issues and respondents with the parties and this new information will be also positioned on the semantic map of electoral behavior. HAC, VACOR (Table 9), and FACOR (Table 10) are applied in the same order but for the dataset which contains also the new variables, with the issue groups which were produced in the previous analysis. The final diagram (Fig. 4) shows all clusters of respondents with reference to issues,

Table 4 Clusters consisting of the variable values (position on issues) and the initial cluster of respondents

Cluster	312	314	315	316	303	313
A(I)	306	307	305	303	248	311
B(I)	299	309	310	313	285	308
Weight	0,134	0,1495	0,27385	0,44212	0,04277	0,39935
	q885	q865	q864	q861	q861	q862
	q155	q21	q822	q25	q25	q24
	q205	q821	q22	q135	q135	q54
	q295	q51	q132	gr_91999	gr_91999	q164
	q813	q161	q854	q825	q825	q334
	...	...	...	...	...	...
	q335	q291	q314	q833		q193
	q811	gr_91992	q82	q772		q873
	q771	q815	q292	q284		q863
	q831	q305	q152	q402		q23
	q301	q775	q203	q852		q133
	q285	q835	q313	q142		q73
	gr_91998	q855	q394	q74		q223
	q401	q131	q213	q784		q773
		q145	gr_92001	q794		q823

vote 2012, EU-vote of 2014, intention to vote for 2015, and confidence of vote intention with their exact positions on the two-dimensional scatterplot. This map shows six main clusters and the distances between all objects defining in this way all competitive relationships between groups. As shown in the scatterplot one cluster includes strong voters of party no 4 and group 1998, another cluster consisted of group 2001 which is close to voters of party no 2 in 2012 and intends to vote for party no 1 in 2015. At the same moment we have already linked the groups of voters to their issue preferences and political views with the initial analysis. The result is a map which can indicate all behavioral profiles and their interrelations with issues, vote intention, or past vote. This electoral behavior mapping can give important insight to political competition, voters, and parties mobility. The context of our analysis can take multiple forms:

– mobility over time can be analyzed by using variable of past votes
– mobility on issues and vote
– mobility on other characteristics such as demographics and socioeconomic data

Table 5 Contribution of groups in each variable value (position on issues)

	q861	q862	q863	q864	q865	q881	q882	q883	q884	q885
1992	-9.0192	-13.2294	-8.5867	-1.4044	39.0273	34.005	3.2428	-3.0348	-12.906	-5.9425
1998	-2.4663	4.5245	-0.1349	0.6335	4.8573	-9.3244	-10.5787	-8.9735	-9.7743	28.6589
1999	29.2903	0.3354	-9.9953	-8.1294	-10.2752	4.341	-4.0911	-1.6851	-1.597	2.7078
2000	-2.2142	6.1203	3.8563	-2.2196	-8.379	-6.3247	-1.5524	-0.0654	8.8421	-5.5664
2001	-9.5233	-5.8486	3.1519	6.3395	5.7075	-0.6585	8.3201	6.4131	0.3169	-10.4557

Table 6 Significant contribution of groups in each variable value (position on issues)

	1992	1998	1999	2000	2001
q861			29,1982		
q862		4,4565		6,196	
q863				3,9887	3,3063
q864					6,4359
q865	39,0131				5,6825
q881	34,2566		4,5496		
q882	3,3312				8,2501
q883					6,363
q884				8,8274	
q885		28,6867	2,6318		
q821	40,135				
q822		4,0433			5,2952
q823	2,6681				
q824				4,4712	
q825			16,9626		

Table 7 Clusters linked to the response on the questions

	SD	D	NAnD	A	SA
	1	2	3	4	5
q86	1999	2000	2000	2001	1992
q88	1992	2001	2001	2000	1998
q82	1992	2001	2000	2000	1999
q81	1998	1998	1998	2000	1992
q85	1999	2000	2000	2001	1992
q2	1992	2001	2000	2000	1999
q5	1992	2001	2000	2000	1998
q83	1998	1998	2000	2000	1992
q7	1992	2001	2000	2000	1999
q8	1992	2001	2001	2000	1998
q30	1998	1998	2000	2001	1992
q20	1992	2001	2001	2001	1998
q13	1992	2001	2000	2000	1999
q14	1999	2000	2000	2001	1992

2 Issue Voting Behavioral Map: The Greek National Elections of 2015

Interpretation of the issue voting behavioral mapping procedure for the Greek case of 2015 produces a clear image of interrelations between issues and parties during the national elections. Political competition is reflected on three main axes (Fig. 5): European-anti-European, Left-Right, and Libertarian-Authoritarian (Heath et al. 1994; Evans et al. 1996). Parties which are close to the center of the left-right axis

Table 8 Application of FACOR producing the factors and the scores of the subjects

IND	#F1	COR	CTR	#F2	COR	CTR	#F3	COR	CTR	#F4	COR	CTR
q861	−239	150	3	33	3	1	457	554	35	−32	2	1
q862	−287	472	6	−30	5	1	21	2	1	104	62	9
q863	−123	76	1	−139	97	3	−151	115	5	9	0	1
q864	111	92	1	−30	6	1	−144	151	6	−122	109	11
q865	771	757	21	253	81	6	−114	16	2	46	2	1
q881	742	606	13	159	28	2	328	118	10	113	14	4
q...	...	...	...	...	...	...	...	...	...	...	...	...
gr_91992	1385	809	40	463	90	10	129	7	1	375	59	34
gr-91998	−507	304	10	682	553	44	−239	67	9	−106	13	5
gr_91999	−235	91	2	33	1	0	664	737	69	−129	27	7
gr_92000	−257	312	5	−233	256	9	−32	4	0	220	229	45
gr_92001	295	295	6	−289	281	13	−220	163	14	−220	163	41

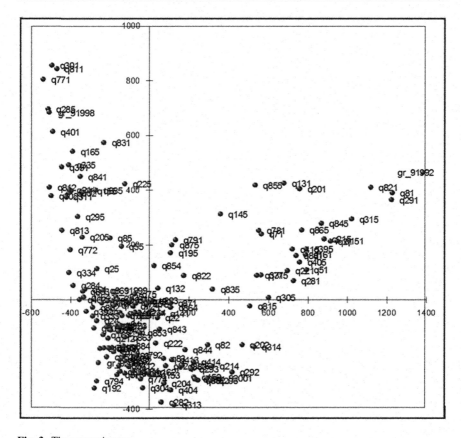

Fig. 3 The semantic map

Table 9 Significant contribution on groups of the second step of analysis in each variable value (voting behavior)

	1437	1458	1463	1478	1479	1484	1485	1486	1487	1489
	19	77	44	98	168	42	111	26	145	18
	2,5%	10,3%	5,9%	13,1%	22,5%	5,6%	14,8%	3,5%	19,4%	2,4%
pvote1		53,4589								
pvote2			1,8298						30,0575	
pvote3			10,9452	20,4266				56,4224		
pvote4						98,2094				
pvote5				5,0266						116,848
pvote6					20,727			4,2917		
pvote7	122,7475			13,831						
pvote8			22,2696		12,2801					
pvote9							35,0617			
euvote1			3,947						34,436	
euvote2		62,7087								
euvote3				8,7495						103,173
euvote4				9,3502				103,085		
euvote5				3,6776	22,9652					
euvote6	136,3368			7,7012						
euvote7						67,2708				
euvote8			28,493		12,9872					
euvote9							34,4757			
vote1									35,3094	
vote2		51,6321								
vote3										118,548
vote4								139,054		
vote5					21,2727					
vote6	141,8553			5,3413						
vote7						60,4951				
vote8			74,1524							
vote9				10,4845			22,8446			
conf0							25,4918			
conf1				33,3698		15,0132				
conf2			3,8654		6,7426					
conf3		4,5244			3,2649					
conf4			6,6844						2,8016	
conf5	27,2986							9,989	8,0758	4,3861
gr_91992	60,474		12,7452						10,2379	
gr_91998					10,5204			33,6913		
gr_91999		10,6738					10,4862			23,7412
gr_92000		8,1789		3,6431			5,0039	4,142		
gr_92001			2,9638						17,9508	

ND (center-right) and PASOK (center-left) create a common cluster focusing on their pro-European position. An antithesis is found between this cluster and the cluster of Golden dawn which represents the Anti-European/authoritarian position and the cluster of anti-European left parties of SYRIZA and the communist party (KKE). An inner antithesis in the second cluster is found showing the contrast between left party of SYRIZA and the communist party, as KKE is strongly positioned to the more left of the axis compared to SYRIZA.

Five groups are shown in the map regarding voter's behavior profiles: First group consists of voters of KKE and SYRIZA or other party (8.3% of total voters). The second group is positioned close to the first one and includes voters of SYRIZA-PASOK or other (28.7%). Both of these groups have the same characteristics (Fig. 6): ethnocentric, libertarian, left and are found in group 1992 and 2001 from the first HAC of the analysis.

Table 10 Application of FACOR producing the factors and the scores of the voting variables and the groups of the second step of the analysis

IND	#F1	CO1	CT1	#F2	CO2	CT2	#F3	CO3	CT3	#F4	CO4	CT4
pvote1	-729	253	48	-673	216	55	-176			455		
pvote2	815	395	70	285			683	277	68	-139		
pvote3	-465			492			-501			-1118	397	105
pvote4	-287			-253			247			667		
pvote5	-438			-2902	476	150	1808			-1722		
pvote6	-226			419			-105			-195		
pvote7	1977	439	107	-1109			-1810	368	123	48		
pvote8	-250			220			-191			-90		
pvote9	28			285			111			471		
euvote1	763	429	79	259			632	295	75	-136		
euvote2	-820	254	49	-670			-252			596		
euvote3	-546			-2336	494	143	1327			-1220		
euvote4	-650			679			-853			-1793	490	167
euvote5	-450			326			-277			-135		
euvote6	2099	408	99	-1273			-2026	380	127	76		
euvote7	-144			-343			384			892		
euvote8	-151			135			-86			-21		
euvote9	-36			255			43			465		
vote1	747	369	67	206			727	350	88	-157		
vote2	-778	250	47	-680			-244			445		
vote3	-620			-3152	501	145	1743			-1739		
vote4	-733			792			-1033			-2210	481	159
vote5	-521			300			-331			-183		
vote6	2259	437	106	-1317			-2050	360	121	35		
vote7	-361			-582			326			775		
vote8	75			245			-63			-364		
vote9	-102			217			-14			340		
conf0	-253			458			-123			590		
conf1	-221			134			-144			371		
conf2	-65			113			-38			148		
conf3	-161			51			89			39		
conf4	-157			-66			106			-114		
conf5	352			-135			-79			-187		
gr1992	1212	424	67	-417			-585			-42		
gr1998	-585	247	36	289			-505			-452		
gr1999	-471			-671	320	55	140			54		
gr2000	-235			78			-10			238		
gr2001	554	398	52	218			405	212	38	7		

Another group which is formed concentrates the voters of (16.5%) PASOK, DHMAR, POTAMI, and others (Fig. 7). All parties mentioned belong in the wider center-left area of the left-right axis and share common characteristics: pro-European, Libertarian, Right, while their voters belong to group 1998.

The next group corresponds to the anti-European, authoritarian, and right voters who vote for the extreme right wing of the axis Golden Dawn and for ND (15.3%). A last but large group of voters (31.3%) follow ND, PASOK, KKE or give no answer, thus are mentioned as unspecified voters. Even though we do not know their voting behavior we can presume their electoral profile by their classification upon issues. In this case these voters are characterized as European, authoritarian, both come from the left or right side of the axis and are found in group 2000 (Fig. 8).

The importance of the behavioral map is found in the classification of objects with ward's criteria and chi-square distances. In this way the map can reflect relationships between variables and objects, in terms of antithesis and identification.

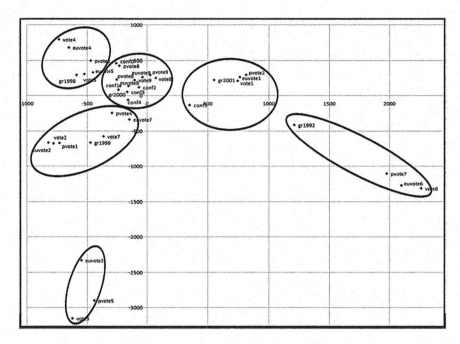

Fig. 4 2-dimensional space with all clusters of respondents (with reference to issues), vote 2012, EU-vote of 2014, intention to vote for 2015 and confidence of vote intention

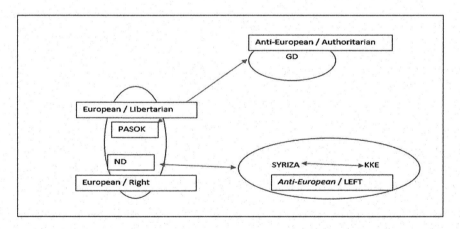

Fig. 5 Political competition on three main axes: European-anti-European, Left-Right and Libertarian-Authoritarian (and inner antithesis KKE-syriza)

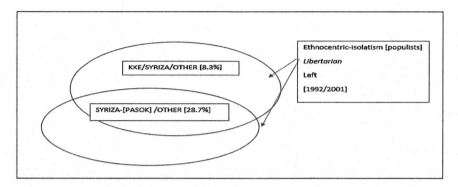

Fig. 6 First and second group of voters with their political characteristics

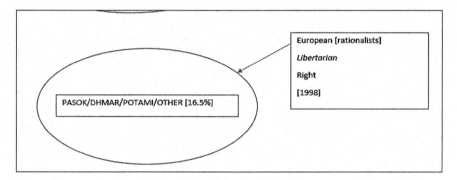

Fig. 7 Third group of voters with their political characteristics

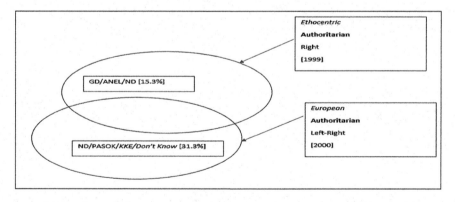

Fig. 8 Fourth and fifth group of voters with their political characteristics

In the Greek case, we can clearly detect the main group of voters and how they relate to each other, regarding their position on issues and their vote. The main components of political competition during Greek elections of 2015, regarding parties and their voters, are three: issues corresponding to the left or right, authoritarian or libertarian, and the highest contribution is found on the axis anti-European/ethnocentric or populists and pro-European, with the issue/dilemma about what should the Greek stance against Memorandum be.

Appendix

Questionnaire (www.helpmevote.gr)

q86) The right for temporary leaves from prison is more important than the risk of escape.

q88) The Memorandum has not caused the economic crisis; the economic crisis has resulted in the Memorandum.

q82) People who break the law should be given stiffer sentences.

q81) Popular demands are today ignored in favor of what benefits the establishment.

q85) Immigrants are good for [country's] economy.

q2) The police should use stricter enforcement measures to protect the property of citizens.

q5) We should have more flexible forms of work in order to combat unemployment.

q83) The government should take measures to reduce income inequalities.

q7) Defense spending should not be reduced to avoid becoming a vulnerable country.

q8) The reduction of corporate taxes would have a positive impact on the development of the economy.

q30) With the Memoranda we accumulate debts without any visible benefits.

q20) It is better for Greece to be in the European Union rather than outside.

q13) The requirements for asylum and citizenship must be tightened.

q14) The existence of multiculturalism in Greece is a positive phenomenon.

q77) The people, and not politicians, should make our most important policy decisions.

q79) Same-sex marriages should be prohibited by law.

q84) People can be better represented by a citizen than by an experienced politician.

q87) Women should be free to decide on matters of abortion.

q41) We should not apply any law that we feel is unfair.

q29) We ought to have done many of the changes provisioned in the Memoranda on our own long ago.

q15) It must be possible to operate non-governmental, non-profit institutions of higher education.

q16) The national health system can become more efficient through partial privatization.

q19) The church and the state should be completely separated.

q21) The economy of Greece would have been better if we had our own currency instead of Euro.

q22) The decision power of the European Parliament should be increased on all matters of internal and foreign policy.

q28) Memoranda of Understanding with the Troika were necessary to avoid the bankruptcy of Greece.

q40) Banks and utilities must be under public control.

q31) We have every right to cancel the debt without consulting anyone else.

q33) There should be legislation to limit protests.

q39) The probability of Grexit should not be considered as a disaster.

q78) Immigrants should be required to adapt to the customs of [country].

References

Bechrakis, T.: Multidimensional Data Analysis. Methods and Applications. Nea Sinora-A.A. Livanis, Athens (1999)

Benjamin, H.: The contextual causes of issue and party voting in American Presidential elections. Polit. Behav. **32**(4), 453–471 (2010)

Benoit, K., Laver, M.: Party Policy in Modern Democracies. Routledge, New York (1999)

Benzécri, J.P., et al.: L' analyse des données. Tome 1: La taxinomie. Tome 2 : Analyse des Correspondances, End ed. Dunod, Paris (1973/1976)

Carmines, E.G., Stimson, J.A.: The two faces of issue voting. Am. Polit. Sci. Rev. Camb. J. **74**(1), 78–91 (1980)

Chadjipadelis, Th., Andreadis, I.: On Proximity Coefficients for a VAA Experiment, VAA Research – State of the Art and Perspectives Schloss Mickeln, Düsseldorf, 2–3 March 2012

Chadjipadelis, Th., Andreadis, I., Panagiotidou, G.: Greek Regional Elections of 2010. Analysis of HelpMeVote Data, 6th Panhellenic Conference for Data Analysis, Thessaloniki, Sept 2011

Cho, S., Endersby, J.W.: Issues, the spatial theory of voting, and British general elections: a comparison of proximity and directional models. Public Choice **114**(3), 275–293 (2003)

Evans, G., Heath, A., Lalljee, M.: Measuring left-right and libertarian authoritarian values in the British electorate. Br. J. Sociol. **47**(1), 93–112 (1996)

Heath, A., Evans, G., Martin, J.: The measurement of core beliefs and values: the development of balanced Socialist/Laissez faire and Libertarian/Authoritarian scales. Br. J. Polit. Sci. **24**(1), 115–32 (1994)

Karapistolis, D.: Software MAD. Data Analysis Notebooks. Thessaloniki **2**(1), 133 (2002)

Karapistolis, D.: Multidimensional Statistical Analysis. Altintzis, Thessaloniki (2011)

Kessel, J.: Comment: the issues in issue voting. Am. Polit. Sci. Rev. **66**(2), 459–465 (2011)

Portal for Multidimensional Statistical Data Analysis. http://www.pylimad.gr

Voting Advise Application Helpmevote for the Greek national elections of 2015. http://www.helpmevote.gr

Well-Being Measures

Mónika Galambosné Tiszberger

Abstract Well-being itself has multiple, very complex definitions, in which the extent of the necessary variables (both objective and subjective) indicates a wide range. It is "officially" recognized since 2009—the year of the Stiglitz Report—that Gross Domestic Product is not an all-powerful indicator. It has a specific purpose, however, it cannot be used to measure such a multidimensional phenomenon as well-being. The construction of an indicator to measure well-being is usually motivated by the aim of making comparisons or that of supporting policy decisions. The role of the indicator drives the amount and type of the collected variables. There are several popular and less frequently used measures for well-being with different contents, concepts, and objectives. This paper tackles the most recent trends in well-being measures through the collection of ten indicators and concepts. The author will compare the content of these measures to indicate supreme (the most relevant) dimensions. The final aim is to give a comprehensive list of those characteristics and criteria that must be taken into account when a complex well-being indicator is proposed.

1 Introduction

The quantification of well-being indicators that cover most countries is a complex challenge, especially when these indicators are expected to be used for comparative purposes. The purpose of this paper is to contribute to the international literature by aiding the development process of such indicators. Well-being itself has a number of complex definitions, in which the extent of the necessary variables (both objective and subjective) indicates a wide range. It is "officially" recognized since 2009, the year of the Stiglitz Report, that Gross Domestic Product is not an all-powerful indicator. It has a specific purpose, however, it cannot be used to measure such a

M. G. Tiszberger (✉)
Faculty of Business and Economics, University of Pécs, Pécs, Hungary
e-mail: tiszbergerm@ktk.pte.hu

© Springer Nature Singapore Pte Ltd. 2020
T. Imaizumi et al. (eds.), *Advanced Studies in Classification and Data Science*,
Studies in Classification, Data Analysis, and Knowledge Organization,
https://doi.org/10.1007/978-981-15-3311-2_34

multidimensional phenomenon as well-being. Section 2 of the paper gives a brief summary of the most important conclusions of this report regarding how well-being ought to be defined. The construction of well-being indicators is usually motivated by the interest in exploring differences between countries, or by the need to support policy related decisions. Naturally, the purpose of any indicator dictates the amount and type of the variables involved. There are several, popular, and less frequently used measures of well-being with different components, concepts, and objectives. Section 3 tackles the most recent trends in well-being measures through the collection of ten indicators and concepts. This study compares the content of these measures and identifies the most relevant dimensions. The final aim is to give a comprehensive list of the characteristics and criteria that must be taken into account when a complex well-being indicator is proposed. It appears that the current international literature on well-being measurement lacks a framework of evaluation criteria, however, it can be very helpful or even crucial during the variable selection to fulfill the purpose of the indicator. The list and detailed description of these criteria is covered in Sect. 4. The paper concludes by highlighting the contributions. (Sect. 5).

2 Beyond GDP

Economic production is clearly not the same and sometimes not even in line with the well-being of nations. This is the main reason why these two phenomena are assessed individually, and perhaps why more and more indicators started putting more emphasis on the latter. Many authors Hajdu and Hajdu (2014), Easterlin (2013), Deaton (2008), Leigh and Wolfers (2006) come to the conclusion that there is no linear relationship between life satisfaction and wealth, regardless how the latter is measured: using disposable income or GDP (Gross Domestic Product) per capita. It also underlines the need for a more accurate measurement of the aspects of well-being other than wealth. The well-known Stiglitz report, based on academic research and initiatives developed around the world, suggests a multidimensional definition that describes well-being in Recommendation 5. It proposes a set of dimensions that should be considered simultaneously (Stiglitz et al. 2009, pp. 14–15):

1. Material living standards (income, consumption, and wealth);
2. Health;
3. Education;
4. Personal activities, including work;
5. Political voice and governance;
6. Social connections and relationships;
7. Environment (present and future conditions);
8. Economic and physical insecurity.

Sustainability and inequality are both crucial points that are missing from this list, even though the report distinguishes between the two concepts. Sustainability (which depends on long-term effects) should be measured separately by well-defined figures, whereas inequality is not a single-dimension variable. It is, in a way, derived from a composite of all dimensions mentioned earlier, however, its measurement requires individual level data, which is not always available.

Objective and subjective indicators are both very important in measuring these dimensions. According to the report, it is important to highlight that subjective indicators are not only those that refer to abstract ideas such as life satisfaction, happiness or quality of life, but almost every aspect of well-being can be understood from a subjective point of view, as one's self-reported health status, their satisfaction with the quality of education or other similar factors also play an important role in understanding the level of individuals' well-being.

Objective information, on the other hand, is often easier to collect and measure, as they are typically standard components of official statistics. Moreover, they are more reliable and also tend to be available in most countries with more or less well-harmonized definitions and methodologies. The case of subjective indicators, however, is less straightforward, as there are many ways to capture such information, each of which may result in different numerical outcomes. This area of measurement is a rather new challenge to official statistical institutions. There are several ways to gather information concerning subjective issues, including variants of the Likert-scale, and ratio scales, but if we try to capture "observable" details like how often someone smiled or laughed the day prior to responding to the questionnaire, it would constitute lots of guesses from the side of individuals, and the responses would depend on the respondents' disposition at the time. Discrepancies rooted in culture and tradition may also prevent the quantification of unbiased estimates, therefore, a seemingly simply comparison between countries can pose its own, substantial challenges (Kroll 2013).

3   Well-Being Measures

Competition is a natural part of being human. At the same time, ranking can only rely on precise and harmonized information, measurement or data, and measuring or ranking social progress, human development or well-being as a complex phenomenon has a relatively short history. The first attempts to adjust or amend GDP took place in the 1970s Gáspár (2013). Later, new approaches have appeared with the purpose of creating more suitable indicators to measure well-being. Some professionals created composite indices to replace GDP; others applied multivariate statistical methods such as factor analysis or cluster analysis to provide better "weightings" of the components. By the year 2000, subjective measures have been introduced in certain methodologies. The observation and inclusion of inequality and sustainability have become a prevalent issue in the following years. Defining the purpose of such indicators is the first and perhaps the most essential part of the design process of these indicators and their corresponding monitoring systems Galambosné (2016). The purpose of an indicator can be to:

- Compare results in space (countries, regions) and in time to monitor differences and changes.
- Monitor the performance of a given country over time.
- Identify "best practises," learn from those who rank the highest.
- Define underdeveloped areas and allocate financial or professional support to the proper locations.
- Identity fields (dimensions), where development is necessary.
- Measure or monitor the effect of certain changes in policies, institutions, and actions (at country or regional level).
- Disseminate information to a broad audience including residents.

Each one of these objectives may require a particular focus and framework. The objectives define the level, scope, frequency, and accuracy of the necessary data input and of the indicators. Hence the designation of the aims needs to be the first step when measuring the well-being of a nation. The list of objectives clearly indicates that there is not a single-best indicator that would be capable of answering all questions. In other words, in most cases, we have one particular objective and we would like to create a measure or measurement system that only supports that specific objective.

The following section gives an overview of the currently used indicators. My aim is to analyze what type of factors were taken into consideration during the formulation of these indicators. The list includes indicators from across the world: international—country level, old—new, weighted—unweighted, as well as popular ones and ones that are less frequently used. The goal is not to give a detailed overview of them, but to simply describe their purpose and their main features, and to compare the dimensions they measure. The order of the different indicators is roughly based on that of the first edition. The list begins with the "oldest" indicator.

- *Human Development Index* (HDI): This composite index is calculated by the United Nations since 1990. Since its first announcement results in the form of ranking and a complete report with special focus are available each year. Several methodological revisions have been introduced during its 27-year history, however, the main idea is still the same. It is built on three main dimensions: health (life expectancy at birth in years), education (expected years of schooling and mean years of schooling), and standard of living (gross national income per capita in 2011 PPP in USD). These dimensions are transformed and aggregated into one single indicator, ranging from 0 to 1. Countries are ranked by their corresponding HDI figures. There are several other indicators to support a more overall picture of human development, but this main index is the most well-known outcome. According to the most recent report, 188 countries are included in the analyses (UN Development Program 1990; Human Development Report 2016).

- *Measures of Australia's Progress* (MAP): This set of indicators, collected and published by the Australian Bureau of Statistics, is designed to help Australian citizens address the question: "Is life in Australia getting better?" It provides a set of indicators in four main categories: society, economy, environment, and governance. In the online dissemination of the data and the results the changes are highlighted. It focuses on making Australians aware of their own situation and development. It presents data at both national and regional granularity (Trewin 2002).
- *Happy Planet Index* (HPI):The HPI (also known as a global index of sustainable well-being) was created by the New Economics Foundation (NEF) in the United Kingdom (UK) and was first published in 2006. They declare that "it measures what matters: sustainable well-being for all." It is also said that it provides a compass to guide nations and shows that it is possible to live good lives without costing the Earth. Currently, four years' data is available (2006, 2009, 2012, and 2016). In the most recent report, HPI covered 140 countries (http://happyplanetindex.org. Accessed 20 Sept 2017). The innovation of NEF lies in the direct utilization of inequality and sustainability, and at the same time, in the indicator itself. The index considers three main areas: life expectancy (health), experienced well-being (on a 10-point scale, measuring self-reported well-being), and ecological footprint (sustainability). The first two variables are corrected with inequality (based on the distribution of life expectancy and well-being data) on a national level. This indicator results in a very different rank order compared to the HDI. Wealthy western countries, often seen as the standards of success, rank relatively low on HPI, compared to other well-being indicators. The Latin American and the Asia Pacific region lead the rankings because of their substantially smaller ecological footprint (Happy Planet Index 2016).
- *Set of Quality of Life Indicators* (SQLI): The European Union (EU) had already had a "Beyond GDP" conference in 2007, two years before the Stiglitz Report was published, to discuss how to better measure the progress of societies and their well-being and how to sustain quality of life in the future. It took some years to develop a framework of indicators, which is under continuous improvement. 8+1 dimensions of quality of life had been defined based on academic research and several initiatives. These dimensions are suggested to be considered simultaneously, because of potential trade-offs between them. The "+1" dimension is the overall experience of life. It is highlighted separately as it covers subjective well-being. The necessary data for this dimension is gathered within the EU-SILC (European Union Statistics on Income and Living Conditions) data collection, as part of the official statistics across the EU member states (http://ec.europa.eu/eurostat/statistics-explained/index.php/Quality_of_life_indicators_-_measuring_quality_of_life#8.2B1_dimensions_of_quality_of_life. Accessed 2 Nov 2017).
- *Measuring National Well-being* (MNW-UK): The Measuring National Well-being programme was launched and led by the UK Office for National Statistics in 2010, supported by the UK Government, with a clearly national focus. This particular set of domains and indicators aims to measure the quality of life of

UK citizens. There are ten domains supported by 41 measures, including both objective and subjective measures. Since 2010, detailed datasets and reports have been published to inform the society, to support policy makers and to monitor recent changes (Self 2017).

* *Canadian Index of Well-being* (CIW): After 12 years of research, the CIW Network found its permanent home at the University of Waterloo in 2011 and released the first complete version of the CIW national composite index. The vision of the creators was to enable Canadian citizens to take aim at the highest well-being status by developing and publicizing statistical measures that offer clear and valid reporting on the progress toward well-being goals. The target group of the indicator and the reports are the citizens, policy makers, and the government leaders of Canada. The eight domains are rooted in core Canadian values and are broken down into 75 detailed measures (http://uwaterloo.ca/ canadian-index-wellbeing. Accessed 10 Aug 2017).

* *Measuring National Well-being* (MNW-J): This initiative was published by The Commission on Measuring Well-being in 2011 in Japan as a result of intensive research and discussions by a study group of leading researchers of Happiness Studies in Japan. Their framework involves three domains of life: socio-economic situation, health, and relatedness. They make a clear distinction between happiness and well-being. Happiness is understood as a state of content and pleasantness on a daily basis. They apply well-being to capture how well people are doing in all aspects of life. Altogether nine dimensions and more than 130 potential indicators have been incorporated. Detailed description is given of the main domains, the purpose of each indicator such as target group of the questions, sources, what the considerations are behind them, and whether there are examples of use overseas (Measuring National Well-being - Proposed Well-being Indicators 2011).

* *Better Life Index* (BLI): The Organisation for Economic Co-operation and Development (OECD) proposed the Better Life Initiative in 2011. Since then, four reports have been published. Each of them covers 35 OECD countries and 6 partner countries. They work with 11 topics that the OECD has identified as essential in the areas of material living conditions and quality of life. Altogether, 50 indicators are collected and published, including both objective and subjective ones. On their interactive homepage, anyone can create their own preferences by assigning arbitrary weights to each of the 11 topics. The status of the person's home country and the best country based on the personal preferences update real-time (www.oecdbetterlifeindex.org. Accessed 12 Sept 2017).

* *Equitable and Sustainable Well-being* (BES after the Italian acronym): The Italian government says that they are the first country in the European Union to include equitable and sustainable well-being indicators besides GDP in their economic plans in 2017 (www.mef.gov.it. Accessed 29 Oct 2017). The BES initiative was introduced in 2013 by the Italian National Institute of Statistics. They defined 12 main topics, among which we find original ideas such as the quality of services, landscape, and heritage. They collect more than 120 indicators, including subjective measures for the in-depth analysis of each topic.

Their aim is to inform the public and to include the results and trends in the process of decision making on a national level (BES 2015).

- *Indicator System of Well-being in Hungary* (ISW): The Hungarian Central Statistical Office introduced a publication of the status of well-being of Hungarian citizens with eight dimensions and 40 variables. Statisticians and academics from various fields cooperated to establish the framework and content of the indicator system. They have built upon the recommendations of the Stiglitz Report and created their own understanding of well-being: Well-being includes, in its widest sense, all the information about the quality, conditions, and comfort of the citizens' everyday life that can be captured through statistical data. The objective of the system and the report is to publicize the information about the state of Hungary and its citizens (A jóllét magyarországi indikátorrendszere 2013).

Table 1 serves as a summary of the domains (dimensions, factors) that are included in each individual well-being measure. The heading contains the acronyms of the indicators along with the corresponding country or organization, in the same order as they were addressed before. The top section (above the double line) is based on the list of dimensions in Recommendation 5 from the Stiglitz Report. The remaining rows, marked with a + sign are the additional ones. The value 1 means that the given measure includes in a certain way the dimension of the row. Empty cells mean the lack of that dimension in the given indicator. The last column of Table 1 (Total) indicates the frequency (sum of the ones), so how many times the given domain appears (out of 10). Here, a dimensioned is considered as supreme if its corresponding frequency is eight or more. Those with a frequency of six or seven are the important dimensions, and the rest I refers to as complementary dimensions.

- *Supreme dimensions*: Unquestionably, the ultimate "winner" is health. This is the only dimension that is part of each of the ten indicators. According to Deaton (2008): "Without health, there is very little that people can do and, without income, health alone does little to enable people to lead a good life" (p62). According to the results, the dimension of health is the most universally defined concept: in most of the cases, it is measured by life expectancy at birth. Material living standards (income, consumption and wealth, basic needs, economic well-being) and education (learning, knowledge, skills) are included in nine indicators. Work (work-life balance, job opportunities, free time, and employment) and political voice and governance (civic/democratic engagement, politics and institutions, societal involvement) play role in eight indicators. All of the supreme dimensions are in line with the recommendations of the Stiglitz Report.
- *Important dimensions*: Subjective well-being is included in seven indicators. Considering that the collection of data on subjective measures of life satisfaction has a rather short history compared to the well-operationalized objective indicators, its scope is indeed considerably broad. Social connections and relationships (community connections, social interactions) and environment (natural environment, landscape) are also part of the important dimensions.

Table 1 Summary of well-being indicators and their components

Domains	HDI (UN)	MAP (AUS)	HPI (NEF)	SQLI (EU)	MNW (UK)	CIW (CDN)	MNW (J)	BLI (OECD)	BES (I)	ISW (HU)	Total
Material living standards (income, consumption, wealth)	1	1		1	1	1	1	1	1	1	9
Health	1	1	1	1	1	1	1	1	1	1	10
Education	1	1		1	1	1	1	1	1	1	9
Personal activities including work		1		1	1	1	1	1	1	1	8
Political voice and governance		1		1	1	1	1	1	1	1	8
Social connections and relationships		1		1	1	1	1	1	1		7
Environment		1		1	1	1		1	1		6
Insecurity (economic and physical)		1		1				1	1		4
Inequality	1	1	1								3
Sustainability			1				1			1	3
+ Subjective well-being			1	1	1		1	1	1	1	7
+ Housing				1	1		1	1		1	5
+ Economy		1			1						2
+ Research and innovation									1		1
+ Quality of services									1		1

Source: self-edition

- *Complementary dimensions*: Housing (home or primary place of residence) is covered by five indicators. Insecurity (economic and physical, safety, personal security) has a frequency of four. Inequality and sustainability are both important according to the international researches, but in most cases these are not or just partially included in the measurement system. The three remaining, apparently the most marginal, dimensions are the economy, research, and innovation, as well as the quality of services.

4 Evaluation Criteria

Many studies deal with the definition and content of well-being, but much fewer intend to give detailed overview of what makes a good indicator, i.e. there is little discussion regarding the evaluation of the set of necessary variables.

Similarly to other statistical institutions, Eurostat (the statistical office of the European Union) has a thorough system ensure the high-quality official statistics. These principles are to be respected by every statistical institution in the member states. The most recent version of Quality Assurance Framework of the European Statistical System (2015) includes the quality principles of statistical output. It also describes institutional environment (importance of commitment to quality, the requirement of statistical confidentiality) and considerations about the required level of quality in the statistical process (appropriate methodology and procedures, non-excessive burden on respondents, and cost effectiveness) and output. These principles are:

- Relevance: This aspect is driven by the data users. It can be guaranteed by consulting the end users, by monitoring the usability of existing statistics in terms of how effectively they meet their needs and their emerging priorities.
- Accuracy and reliability: Accuracy means that the data source, the intermediate results, and the statistical outputs are systematically assessed and validated. The principle of reliability makes it important that all sampling and non-sampling errors are measured systematically and that there is continuous effort to improve statistical processes.
- Timeliness and punctuality: These aspects refer to the frequency of the data collection together with the preferably short time period between data collection and publication. A published release calendar covering all statistics ensures that timing strategy and timeliness can be followed by the data users.
- Coherence and comparability: Coherence means that there must be consistency between preliminary and final data, between microdata and aggregated data, between annual, quarterly, and monthly data, between statistics and National Accounts. Comparability is meant over a reasonable period of time, and calls for the proper handling of methodological changes. Cross-national comparability is ensured by the European Statistical System.

- Accessibility and clarity: Statistics and the corresponding metadata are published and archived in a form that facilitates proper interpretation and meaningful comparisons from the users' end. Microdata is accessible for research purposes that meet the corresponding regulations on data protection.

We can also read some details about evaluation criteria in Costanza et al. (2009). The authors refer to these as data barriers and they also talk about two methodological barriers:

- Data reliability: "whether a change in an indicator is an accurate signal of change in the system it is supposed to measure."
- Timeliness: the frequency of the surveys and publications of the underlying data.
- Data scope: how many things, items are covered by the indicator.
- Data scale: the granularity of data, its collection and publication (e.g. settlement, county, country).
- Methodology standardization: existence of harmonized measurement criteria and methodology behind the indicator.
- Values embedded in methodology: Through the way of measurement and choice of indicators we implicitly define the goals. As society changes over time, so may the relevance of the components associated with well-being, which demands adjustments in the indicators, as well.

Most of these criteria (namely: relevance, accuracy and reliability, scope, scale, accessibility, and clarity) are important and are to be taken into consideration in the process of well-being indicator development. If we focus on national data, comparability is not a necessary aspect, however, the understanding and observation of national specifics are substantial. Upon a thorough review of all the indicators (several hundreds) that are involved in the ten different well-being measures described in Sect. 3, it appears that there are many more specific features and characteristics to be analyzed before a variable is selected. I have created an additional system of evaluation criteria to support decisions in the selection process of potential well-being measures.

1. *Pace of change*: Several policy decisions or even environmental change have "only" long-term effects on a given population. It means that certain, "slowly" moving variables should be observed only once every two or three years. There might be others that indicate rapid changes so the frequency of data collection and dissemination might be higher even within a year. For example, the reforms in the system of education have slow, long-term outcomes. The PISA (Programme for International Student Assessment) test results will not indicate substantial changes within a year. However, change in the social transfer system or the taxation policies might have strong short-term effects. I suggest considering the pace of change of the potential indicators in order to determine how often new information should be collected.
2. *Impact on everyday life*: Similarly to the "pace of change," the problem this criterion highlights is the nature of the effect of changes in certain policies. How closely and how evenly does it affect the life of an individual? On the one hand,

new regulations on certain gas emissions affect the overall status of a country or a region (where the air is going to be less polluted), but individuals may not be able to report about the change that they may or may not notice. On the other hand, the costs of food for children in daycare that is paid for by the parents has a very strong and direct effect on the everyday life of individuals and this effect is distributed quite unequally.

3. *Objective/subjective nature*: Subjective measures are usually connected with overall life satisfaction, happiness or the quality of one's life. However, many aspects that have been observed traditionally in an objective manner can also be measured in the form of subjective indicators. For instance, life expectancy at birth is a popular indicator of the overall health status of a population. At the same time, we can also inquire about how a person feels about their health status. This second approach would be a subjective indicator, and would provide us with additional information about the health status of a country, and even inequality could be measured by doing so. Therefore, in the case of most variables, researchers should consider the type of the question. Should the focus be put on objective or subjective aspects, or would both be necessary to collect all relevant intelligence in a given dimension.

4. *Suggestibility*: This criterion approaches the question of how impressionable a certain variable is from the perspective of the human life. There are factors with high importance that are only present at lower levels. For instance, the dimension of social connections (friends, family) is a very determining part of our lives, however, it is too "personal," too close to the individual level to be the subject of a national or even a regional policy. Social connections can be influenced by the settlement, or more likely, by decisions and institutions at lower levels of the community. Therefore, one may consider addressing policies and monitoring data on low territorial levels.

5. *Measurement level*: Based on the previously discussed aspects and considerations, the measurement level of a given indicator is determined by its nature. In each case, the statistician has to choose between individual, household, settlement, regional, and national granularity. This decision affects the characteristics of the surveys and makes the information more valuable and effective.

6. *Relevant population*: Similarly to the measurement level criterion, the relevant population of certain indicators has to be defined, as well. Certain topics only influence certain groups of people. Pension system affects older generations. Child care is crucial for families with small children. Unemployment supports influence those without a job. These are all different clusters of the population, which can be separated with well-established demographic or other characteristics. Obviously, the surveys should focus on the relevant group of people.

7. *Availability*: As new and ill-defined aspects arise, we may not be able to rely on the official statistics in every case. If data has to be gathered from multiple sources, institutions or agencies, the availability is not always clear. Obviously, the final aim should be to incorporate all necessary information in the system of official statistics, but it may take too long or may not be feasible in general. This possibility should not be ignored, either.

8. *Desirable order of magnitude*: This criterion raises the question whether we have any knowledge about the ideal value of the indicator. Do we simply wish to increase it or decrease it, is there optimal value, or a certain amount of change to attain? For example, the number of prisons per settlement is a suggested variable in the Japan initiative, but it is not clear if a higher or a lower per settlement density is more preferable. In case of average years spent in education there may not be a single optimal value. Can there be a point where additional years become counterproductive? Or, ideally, how long should we aspire to prolong free time? If closeness to family is measured by the number of non-immediate family members who live in the same settlement, what should we consider optimal? All these types of questions have to be considered if we plan to draw conclusions about the values or the dynamics of the data. Strategic planning of actions should have clear target values so that proper arrangements can be made and the effect of policies can be accurately evaluated.

5 Conclusions

The first section of this paper describes the multidimensional approach of how well-being can be measured and highlights some dilemmas and challenges regarding its measurement. The subsequent sections aim to contribute to the development of such indicators in two ways. Firstly, the paper indicates supreme dimensions to be included in any well-being measure through the detailed analysis and comparison of ten well-being indicators from across the world. Secondly, a comprehensive list of evaluation criteria is established in order to aid the selection process of variables when a well-being measurement framework is motivated by predefined goals.

In conclusion, we propose that the development process of any well-being indicator should prioritize its objectives over the evaluation criteria that are to be considered throughout each step of the process.

Acknowledgments Supported by the ÚNKP-17-4-III New National Excellence Program of the Ministry of Human Capacities.

References

A jóllét magyarországi indikátorrendszere, 2013. Központi Statisztikai Hivatal, Budapest (2014)

BES 2015: The Equitable and Sustainable Well-being, Summary. Italian National Institute of Statistics (2015). www.istat.it/en/archive/180526. Accessed 10 Oct 2017

Costanza, R., Hart, M., Posner, S., Talberth, J.: Beyond GDP: The Need for New Measures of Progress. The Pardee Papers No. 4. Boston University The Frederick S. Pardee Center for the Study of the Longer-Range Future (2009)

Deaton, A.: Income, health and wellbeing around the world: evidence from the gallup world poll. J. Econ. Perspect. **22**, 53–72 (2008). https://doi.org/10.1257/jep.22.2.53

Easterlin, R.: Happiness, growth and public policy. Discussion Paper Series, Forschungsinstitut zur Zukunf der Arbeit, No. 7324 (2013)

Galambosné Tiszberger, M.: Jól(l)ét és fejlettség mérés. Kísérlet regionális szintu mutató készítésére. In: Erdos, K., Komlósi É.: Tanítványaimban élek tovább. Emlékkötet Buday-Sántha Attila tiszteletére. Pécs: PTE KTK Regionális Politika és Doktori Iskola (2016)

Gáspár, T.: A társadalmi-gazdasági fejlettség mérési rendszerei. Statisztikai Szemle **91**(1), 77–92 (2013)

Hajdu, T., Hajdu, G.: Income and subjective well-being: how important is the methodology? Hungarian Stat. Rev. **18**, 110–128 (2014)

Happy Planet Index 2016, Methods Paper. New Economics Foundation (2016). https://static1.squarespace.com/static/5735c421e321402778ee0ce9/t/578dec7837c58157b929b3d6/1468918904805/Methods+paper_2016.pdf. Accessed 4 Nov 2017

Human Development Report 2016 - Human Development for Everyone. United Nations Development Programme (2016)

Kroll, C.: A happy nation? Opportunities and challenges of using subjective indicators in policymaking. Soc. Indic. Res. **114**, 13–28 (2013). https://doi.org/10.1007/s11205-013-0380-1

Leigh, A., Wolfers, J.: Happiness and the human development index: Australia is not a paradox. Aust. Econ. Rev. **39**(2), 176–184 (2006)

Measuring National Well-being - Proposed Well-being Indicators. The Commission of Measuring Well-being, Japan (2011). http://www5.cao.go.jp/keizai2/koufukudo/pdf/koufukudosian_english.pdf. Accessed 29 July 2017

Quality Assurance Framework of the European Statistical System. Version 1.2 European Statistical System. (2015)

Self, A.: Quality of life measurement and application to policy: experiences from the UK office for national statistics. Soc. Indic. Res. **130**, 147–160 (2017). https://doi.org/10.1007/s11205-015-1140-1

Stiglitz, J. E., Sen, A., Fitoussi, J.-P.: Report by the commission on the measurement of economic performance and social progress (2009). https://www.uio.no/studier/emner/sv/oekonomi/ECON4270/h09/Report. Accessed 15 Oct 2017

Trewin, D.: Measuring Australia's Progress. Australian Bureau of Statistics, Canberra (2002)

UN Development Program: Human Development Report: Overview. Oxford University Press, New York (1990)

The Relationship Between Household Assets and Choice to Work: Evidence From Japanese Official Microdata

Shinsuke Ito and Takahisa Dejima

Abstract This research uses individual data from Japan's National Survey of Family Income and Expenditure to examine the impact of residential area and real estate prices on employment, i.e. individuals' choice to work. The results show a significant negative impact of real assets on employment. This result indicates a theoretical possibility that an accumulation of household assets induces non-working. This research also finds that the influence of real assets on employment is different depending on the area. In the Kanto, Kitakyushu, and Fukuoka metropolitan areas there is a negative influence of real assets on employment. This result reflects the differences in real estate and land prices in these areas, and suggests that if there is the negative effect of real assets on employment in metropolitan areas, tax benefits on land and housing could have an adverse effect on the labor market in these areas.

1 Introduction

The availability of non-labor income including asset income and labor income from other household members can reduce household members' willingness to work. In general, when non-labor income increases, the demand for leisure increases due to the income effect, and as a result labor supply decreases. Existing empirical research about housing and labor supply includes (Fortin 1995; Johnson 2014); (Yoshikawa and Ohtake 1989). Ito and Dejima (2016) examined the influence of asset and rental income on the employment behavior of youths using anonymized microdata from

S. Ito (✉)
Faculty of Economics, Chuo University, Tokyo, Japan
e-mail: ssitoh@tamacc.chuo-u.ac.jp

T. Dejima
Faculty of Economics, Tokyo, Japan
e-mail: t-dejima@sophia.ac.jp

© Springer Nature Singapore Pte Ltd. 2020   445
T. Imaizumi et al. (eds.), *Advanced Studies in Classification and Data Science*,
Studies in Classification, Data Analysis, and Knowledge Organization,
https://doi.org/10.1007/978-981-15-3311-2_35

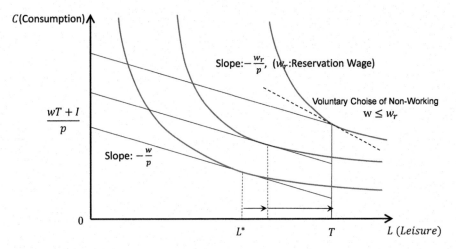

Fig. 1 Non-labor income (I) and labor supply (h)

Japan's National Survey of Family Income and Expenditure from the years of 1989, 1994, 1999, and 2004. However, this anonymized microdata did not contain detailed information on household assets, which limited this research's ability to perform a detailed analysis of employment and household assets. The present research utilizes original data from Japan's National Survey of Family Income and Expenditure (which contains information on household attributes including income, expenditure, and household assets), and based on this more detailed data examines the impact of household assets (including financial assets and real estate) on individuals' choice to work.

Figure 1 illustrates the relationship between non-labor income and labor supply using the constraint that labor supply (h) is the subtraction of leisure (L) from available hours (T). From a microeconomic perspective, leisure is a normal good so as non-labor income increases, demand for leisure increases, and work time decreases. This suggests that the income effect influences the supply of labor.

2 Data and Descriptive Statistics

In this research, original microdata from Japan's National Survey of Family Income and Expenditure from the year of 2009 was used. Sample size of the survey is about 50,000 households excluding one-person households, and the survey covers items such as monthly household accounts, household assets, and individual and household attributes including employment information. The survey is conducted

Table 1 Descriptive statistics

Variables	Average	Standard deviation	Minimum	Maximum	N
Working status	0.967	0.179	0.000	1.000	23970
Home ownership	0.736	0.441	0.000	1.000	23970
Male	0.894	0.308	0.000	1.000	23970
Age	44.673	9.050	20.000	59.000	23970
Labor income of spouse of head of household(ln)	92.683	157.780	0.000	1918.000	23970
Income from assets (ln)	874.565	11280.888	0.000	572068.000	23970
Savings amount (ln)	1002.560	1359.921	0.000	28780.000	23970
Ordinary deposit(ln)	123.573	326.043	0.000	8810.000	23970
Fixed deposit(ln)	266.280	698.465	0.000	12628.000	23970
Life insurance(ln)	295.300	498.376	0.000	16316.000	23970
Money trust(ln)	5.820	74.237	0.000	3200.000	23970
Stocks(ln)	50.367	289.454	0.000	17800.000	23970
Bond(ln)	24.432	174.436	0.000	6000.000	23970
Risk asset ratio	0.031	0.104	0.000	1.000	23970
Liabilities amount(ln)	661.804	1095.981	0.000	50000.000	23970
Amount of residential and land assets(ln)	1694.079	1776.266	0.000	15204.420	23970
Three persons	0.273	0.445	0.000	1.000	23970
Four persons	0.351	0.477	0.000	1.000	23970
Five persons and over	0.177	0.382	0.000	1.000	23970
Kanto metropolitan area	0.210	0.407	0.000	1.000	23970
Chukyo metropolitan area	0.065	0.246	0.000	1.000	23970
Kinki metropolitan area	0.137	0.343	0.000	1.000	23970
Kitakyushu-Fukuoka metropolitan area	0.036	0.187	0.000	1.000	23970

every five years, and is based on a cross-sectional design, which limits the ability to create panel data and conduct panel data analysis. The focus of this analysis is on individuals between 20 and 59 years of age excluding corporate executives, self-employed, workers in family businesses, and persons with a second job.

Table 1 presents the descriptive statistics for the variables used in this analysis. Figures 2, 3, 4, and 5 show the histograms of the amount of residential and land assets for the four metropolitan areas in Japan.[1] The graphs show that the

---

[1]The four metropolitan areas used in this paper are Kanto metropolitan area (Tokyo and surrounding prefectures), Chukyo metropolitan area (Nagoya and surrounding prefectures), Kinki metropolitan area (Osaka and surrounding prefectures), and Kitakyushu–Fukuoka metropolitan area. These four metropolitan areas were selected for this research due to their concentration of economic activity and significantly higher land prices.

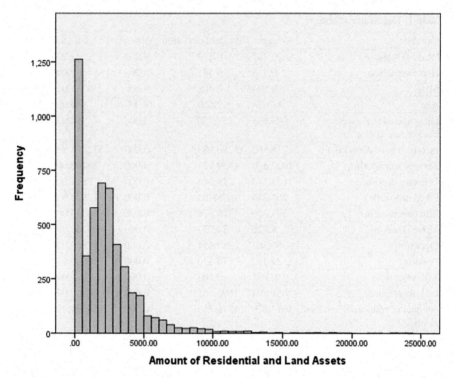

Fig. 2 Histogram of household assets for four metropolitan areas in Japan: Kanto Metropolitan area (Tokyo and surrounding prefectures)

distribution of residential and land assets is skewed, with the distribution in the Kanto metropolitan area (Tokyo and surrounding prefectures) having a longer tail compared to the other areas. This suggests that there are more individuals with extensive real estate ownership in the Kanto area.

3 Methodology

Binary logit regression analysis was used to establish the impact of household income, household assets, and individual and household attributes on household members' employment status. Two logit regression models were created. The first model is a model on individuals' choice to work and household assets (including the ownership of risky assets). The key variables used as independent variables in

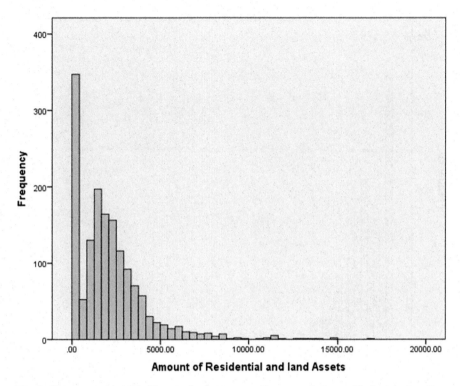

Fig. 3 Histogram of household assets for four metropolitan areas in Japan: Chukyo metropolitan area (Nagoya and surrounding prefectures)

Model 1 were home ownership, gender,[2] age, labor income of spouse of head of household, income from assets, total savings, risk asset ratio,[3] amount of residential and land assets, total liabilities, number of household members, and category of metropolitan area.[4]

The second model is a model on individuals' choice to work and detailed household assets. Key variables used as independent variables in Model 2 were home ownership, gender, age, labor income of spouse of head of household, income from assets, financial assets (ordinary deposits, fixed deposits, life insurance, money trusts, stocks, bonds), amount of residential and land assets, total liabilities, number of household members, category of metropolitan area, and others.

---

[2]In Japan, labor supply differs significantly according to gender. Therefore, it was necessary to introduce gender as one of the controlled variables to the two logit regression models.

[3]While there are several definitions of risky assets (Japanese government bonds are usually not included in risky assets), in this research, risk asset ratio were defined as the ratio of stock and money trust holdings to total savings, and corporate and other bonds were included.

[4]In this research, the top four metropolitan areas were selected based on population size. For Model 1, the dummy variable for each of the four metropolitan areas was used.

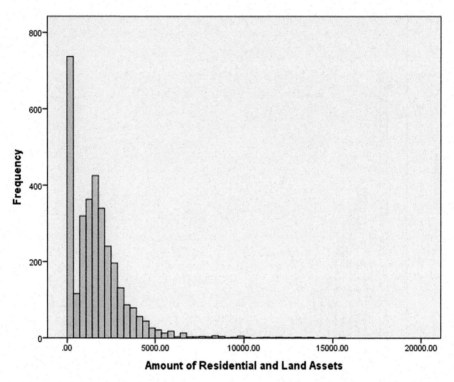

Fig. 4 Histogram of household assets for four metropolitan areas in Japan: Kinki metropolitan area (Osaka and surrounding prefectures)

There are outliers specific to household assets which can cause bias. To control this issue, average plus 3 sigma and average minus 3 sigma were set as thresholds for outliers and records outside these thresholds were deleted in both Model 1 and Model 2.

4 Key Results

Tables 2, 3, 4, 5, 6, and 7 contain the results of the binary logit regression analysis for Model 1 and Model 2 using 2009 data. Results are separated by gender. The results show that in Model 1, the coefficient of risk asset ratio is significantly negative for individuals' employment. This indicates that the result is consistent with the household theory of economics. Results also show that in Model 2, the coefficient of stocks and bonds is not significant for individual employment. This result is likely a measurement error due to subdivision of variables for household assets. Results also show that the coefficient of total household liabilities is significantly positive for individual employment in both Model 1 and Model 2. This indicates that the

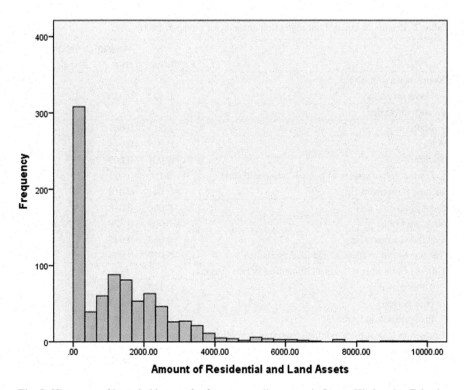

Fig. 5 Histogram of household assets for four metropolitan areas in Japan: Kitakyushu–Fukuoka metropolitan area

lifetime budget constraint is close to zero, which results in a negative income effect as stipulated by microeconomic theory. The coefficient of income from assets is significantly negative for individual employment in both Model 1 and Model 2. This suggests that individuals who hold significant amounts of risky assets such as stocks are more likely to choose not to work.

The coefficient of residential and land assets is significantly positive for employment. This indicates that a positive income effect under budget constraint results in

Table 2 Results of binary logit regression analysis for model 1, 2009

	Coefficient	Standard error	Significance level
Home ownership <No>			
Home ownership	0.331	0.246	
Gender <Female>			
Male	1.713	0.087	***
Age	0.059	0.044	
Square of age	−0.001	0.000	**
Labor income of spouse of head of household (ln)	0.181	0.021	***
Income from assets (ln)	−0.041	0.018	**
Savings amount (ln)	0.080	0.018	***
Risk asset ratio	−0.565	0.328	*
Liabilities amount (ln)	0.087	0.015	***
The amount of residential and land assets (ln)	−0.079	0.034	**
Dummy of number of household members<Two persons>			
Three persons	0.143	0.090	
Four persons	0.706	0.117	***
Five persons and over	0.633	0.149	***
Located in largest metropolitan area<cities other than four largest metropolitan areas>			
Kanto metropolitan area	−0.248	0.099	**
Chukyo metropolitan area	0.462	0.218	**
Kinki metropolitan area	−0.418	0.106	***
Kitakyushu-Fukuoka metropolitan area	−0.165	0.191	
Constant	1.202	0.956	
Pseudo R^2 (Cox and Snell)	0.054		
LR χ^2	5663.255		
−2lnL	1322.709		
N	23970		

Note: Reference group in <brackets>
*10% significance level; **5% significance level; ***1% significance level

more leisure time and less work. For women, the coefficient of real estate assets as non-labor income is significantly negative for employment, and a household size of 5+ has a negative effect on employment. This suggests that female heads of household who live with parents and siblings and own real estate are more likely to choose not to work.

Among both men and women, the coefficient of life insurance (amount) is significantly positive for individual's choice to work. A possible reason is reverse causality, i.e. the possibility that working persons tend to take out higher life insurance. The coefficients of the dummy variables of the Kanto metropolitan area in Model 1 and Model 2 are significantly negative for employment. A possible explanation is that land values tend to be more stable in the Kanto metropolitan area.

Table 3 Results of binary logit regression analysis (men) for Model 1, 2009

	Coefficient	Standard error	Significance level
Home ownership <No>			
Home ownership	−0.257	0.372	
Age	0.112	0.066	*
Square of age	−0.002	0.001	***
Labor income of spouse of head of household (ln)	0.159	0.023	***
Income from assets (ln)	−0.090	0.020	***
Savings amount (ln)	0.119	0.024	***
Risk asset ratio	−0.505	0.462	
Liabilities amount (ln)	0.155	0.021	***
The amount of residential and land assets (ln)	0.043	0.051	
Dummy of number of household members<Two persons>			
Three persons	0.534	0.128	***
Four persons	0.978	0.150	***
Five persons and over	1.131	0.198	***
Located in largest metropolitan area<cities other than four largest metropolitan area>			
Kanto metropolitan area	0.040	0.140	
Chukyo metropolitan area	0.868	0.328	***
Kinki metropolitan area	−0.206	0.150	
Kitakyushu-Fukuoka metropolitan area	0.397	0.331	
Constant	1.748	1.441	
Pseudo R^2 (Cox and Snell)	0.025		
$LR\chi^2$	3283.049		
−2lnL	550.645		
N	21425		

Note: Reference group in <brackets>
*10% significance level; ***1% significance level

Tables 8 and 9 contain the results of the binary logit regression analysis for income from assets and amount of residential and land assets for Model 1 for the four metropolitan areas in 2009. In the Kanto and Kitakyushu–Fukuoka metropolitan areas, the coefficients of residential and land assets are significantly negative for individuals' choice to work. This indicates the possibility that the more stable real estate values in these areas have a negative effect on individual's choice to work. Furthermore, for male heads of households living in the Kanto metropolitan area, the coefficient of income from assets is significantly negative for individuals' choice to work. This is likely due to the larger number of high net-worth individuals in the Kanto metropolitan area (which includes Tokyo) compared to other areas.

Table 4 Results of binary logit regression analysis (women) for Model 1, 2009

	Coefficient	Standard error	Significance level
Home ownership <No>			
Home ownership	−0.257	0.372	
Age	0.112	0.066	*
Square of age	−0.002	0.001	***
Labor income of spouse of head of household (ln)	0.159	0.023	***
Income from assets (ln)	−0.090	0.020	***
Savings amount (ln)	0.119	0.024	***
Risk asset ratio	−0.505	0.462	
Liabilities amount (ln)	0.155	0.021	***
The amount of residential and land assets (ln)	0.043	0.051	
Dummy of number of household members<Two persons>			
Three persons	0.534	0.128	***
Four persons	0.978	0.150	***
Five persons and over	1.131	0.198	***
Located in largest metropolitan area<cities other than four largest metropolitan areas>			
Kanto metropolitan area	0.040	0.140	
Chukyo metropolitan area	0.868	0.328	***
Kinki metropolitan area	−0.206	0.150	
Kitakyushu-Fukuoka metropolitan area	0.397	0.331	
Constant	1.748	1.441	
Pseudo R^2 (Cox and Snell)	0.025		
LRχ^2	3283.049		
−2lnL	550.645		
N	21425		

Note: Reference group in <brackets>
*10% significance level; ***1% significance level

We would like to mention two qualifications for the results of this analysis. First, land and housing prices are estimates calculated using official microdata that reflect the official land prices in three closest locations. Therefore, there is a possibility that the price of real estate assets used in this research fails to reflect the specific characteristics of owned land and housing. Second, there are gender-based differences when it comes to housing. Male heads of households tend to purchase real estate via housing loans, while female heads of household tend to own real estate without debt. This could explain the gender-based differences in the results.

Table 5 Results of binary logit regression analysis for Model 2 and 2009

	Coefficient	Standard error	Significance level
Home ownership <No >			
Home ownership	0.307	0.248	
Gender <Female>			
Male	1.747	0.087	***
Age	0.050	0.044	
Square of age	−0.001	0.000	**
Labor income of spouse of head of household (ln)	0.179	0.021	***
Income from assets (ln)	−0.036	0.018	**
Ordinary deposit (ln)	−0.020	0.016	
Fixed deposit (ln)	0.013	0.014	
Life insurance (ln)	0.077	0.015	***
Money trust (ln)	−0.065	0.044	
Stocks (ln)	−0.022	0.020	
Bond (ln)	−0.024	0.027	
Liabilities amount (ln)	0.082	0.015	***
The amount of residential and land assets (ln)	−0.069	0.034	**
Dummy of number of household members<Two persons>			
Three persons	0.131	0.090	
Four persons	0.680	0.117	***
Five persons and over	0.585	0.149	***
Located in largest metropolitan area<cities other than four largest metropolitan areas>			
Kanto metropolitan area	−0.248	0.099	**
Chukyo metropolitan area	0.462	0.218	**
Kinki metropolitan area	−0.418	0.106	***
Kitakyushu-Fukuoka metropolitan area	−0.165	0.191	
Constant	1.202	0.956	
Pseudo R (Cox and Snell)	0.054		
LRχ^2	5663.255		
−2lnL	1322.709		
N	23970		

Note: Reference group in <brackets>
**5% significance level; ***1% significance level

Table 6 Results of binary logit regression analysis (men) for Model 2 and 2009

	Coefficient	Standard error	Significance level
Home ownership\<No\>			
Home ownership	−0.281	0.376	
Age	0.104	0.066	
Square of age	−0.002	0.001	
Labor income of spouse of head of household (ln)	0.155	0.023	***
Income from assets (ln)	−0.086	0.021	
Ordinary deposit (ln)	−0.021	0.021	
Fixed deposit (ln)	0.028	0.019	
Life insurance (ln)	0.092	0.020	***
Money trust (ln)	−0.077	0.056	
Stocks (ln)	0.007	0.028	
Bond (ln)	−0.018	0.035	
Liabilities amount (ln)	0.152	0.022	***
The amount of residential and land assets (ln)	0.051	0.051	
Dummy of number of household members\<Two persons\>			
Three persons	0.509	0.128	***
Four persons	0.944	0.150	***
Five persons and over	1.055	0.198	***
Located in largest metropolitan area\<cities other than four largest metropolitan areas\>			
K.anto metropolitan area	0.079	0.140	
Chukyo metropolitan area	0.887	0.328	***
Kinki metropolitan area	−0.186	0.150	
Kitakyushu-Fukuoka metropolitan area	0.376	0.331	
Constant	2.372	1.463	
Pseudo R^2 (Cox and Snell)	0.026		
LRχ^2	3276.802		
−2lnL	556.893		
N	21425		

Note: Reference group in \<brackets\>
***1% significance level

Table 7 Results of binary logit regression analysis (women) for Model 2 and 2009

	Coefficient	Standard error	Significance level
Home ownership <No>			
Home ownership	1.031	0.366	
Age	−0.022	0.063	
Square of age	0.000	0.001	
Labor income of spouse of head of household (ln)	0.458	0.092	***
Income from assets (ln)	0.053	0.034	
Ordinary deposit (ln)	−0.014	0.024	
Fixed deposit (ln)	−0.011	0.021	
Life insurance (ln)	0.068	0.022	***
Money trust (ln)	−0.075	0.071	
Stocks (ln)	−0.074	0.032	**
Bond (ln)	−0.010	0.042	
Liabilities amount (ln)	−0.018	0.021	
The amount of residential and land assets (ln)	−0.185	0.050	***
Dummy of number of household members<Two persons>			
Three persons	−0.107	0.127	
Four persons	0.139	0.182	
Five persons and over	−0.475	0.232	**
Located in largest metropolitan area<cities other than four largest metropolitan areas>			
K.anto metropolitan area	−0.596	0.145	***
Chukyo metropolitan area	−0.015	0.303	
Kinki metropolitan area	−0.575	0.153	***
Kitakyushu-Fukuoka metropolitan area	−0.602	0.243	**
Constant	2.777	1.358	**
P seudo R^2 (Cox and Snell)	0.050		
LRχ^2	2127.699		
−2lnL	132.638		
N	2545		

Note: Reference group in <brackets>
**5% significance level; ***1% significance level

Table 8 Result of binary logit regression analysis for income from assets for Model 1 (four metropolitan areas, 2009)

	Coefficient	Significance level
Kanto metropolitan area	−0.134	**
Kanto metropolitan area for men	0.055	
Kanto metropolitan area for women	−0.268	***
Chukyo metropolitan area	0.272	
Kinki metropolitan area	−0.041	
Kitakyusyu-Fukuoka metropolitan area	−0.321	**

**5% significance level; ***1% significance level

Table 9 Result of binary logit regression analysis for amount of residential and land assets and for Model 1 (four Metropolitan areas, 2009)

	Coefficient	Significance level
Kanto metropolitan area	−0.060	***
Kanto metropolitan area for men	−0.121	
Kanto metropolitan area for women	0.102	
Chukyo metropolitan area	0.061	
Kinki metropolitan area	−0.031	
Kitakyusyu-Fukuoka metropolitan area	−0.077	

***1% significance level

5 Conclusion

There is a significant negative impact of ownership of risky assets on individuals' employment. This indicates the possibility that an accumulation of risky assets that generate non-labor income can contribute to individuals' decisions not to work. Furthermore, the influence of residential and land assets on employment differs depending on geographic area. In the Kanto and Kitakyushu/Fukuoka metropolitan areas there is a negative influence of residential and land asset ownership on employment. This result possibly reflects the more stable real estate and land prices in these areas, and suggests that if there really is a negative effect of residential and land asset ownership on employment in metropolitan areas, tax benefits on real estate ownership could have an adverse effect on the labor supply in these areas.

The qualifications indicated in the above section present an important topic in themselves, and further work is required to address these tasks.

References

Fortin, N.M.:. Allocation inflexibilities, female labor supply, and housing assets accumulation: are women working to pay the mortgage? J. Labor Econ. **13**(3), 524–557 (1995)

Ito, S., Dejima, T.: Influence of non-labor income on youth unemployment in Japan: are youths in households with larger budgets less likely to work. J. Econ. **57**(1–2), 1–22 (2016)

Johnson, W.R.: House prices and female labor force participation. J. Urban Econ. **82**, 1–11 (2014)

Yoshikawa, H., Ohtake, F.: An analysis of female labor supply: housing demand and the saving rate in Japan. Eur. Econ. Rev. **33**(5), 997–1023 (1989)

Visualization and Spatial Statistical Analysis for Vietnam Household Living Standard Survey

Takafumi Kubota

Abstract In this study, the data of Vietnam Household Living Standard Survey in 2006 (VHLSS2006) were used as microdata level data to find out some relations among characteristics, medical cares, and work styles in Vietnam. VHLSS2006 were aggregated by province level to a spatial data set to find out spatial characteristics and to apply spatial statistical models. The R package shinydashboard was applied to present the data interactively as dashboards for representative values, tables, and maps. In spatial statistical analysis, the focus was on working variables such as working rate, employment rate, self-employed in agriculture or non-agriculture rate in each province. To detect spatial dependence the Moran's I Statistics were applies these working related objective variables and maps of Vietnam. The shape files of Vietnam in GADM database of Global Administrative Areas were used to calculate neighborhood information of provinces in Vietnam and to draw Choropleth map. Conditional autoregressive model was applied to explain spatial dependences by provinces and characteristics of the relation between working status and characteristics such as gender, age, education, and medical cares.

1 Introduction

In this study, the data of Vietnam Household Living Standard Survey in 2006 (hereinafter referred to as VHLSS2006 by the first letter of words plus year of the survey) were used as microlevel data to find out some relations among characteristics, medical cares, and work styles in Vietnam. Vietnam has achieved economic growth with the Doi Moi policy. However, regional education and health

T. Kubota (✉)
Tama University, Tokyo, Japan
e-mail: kubota@tama.ac.jp

© Springer Nature Singapore Pte Ltd. 2020
T. Imaizumi et al. (eds.), *Advanced Studies in Classification and Data Science*,
Studies in Classification, Data Analysis, and Knowledge Organization,
https://doi.org/10.1007/978-981-15-3311-2_36

problems remain unsolved. Therefore, modeling and visualization are carried out to obtain analytical results that provide evidence, in particular, for enforcing regional education and health policies. The method examined the candidate of the hot spot. VHLSS2006 were aggregated by province level to a spatial data set to find out spatial characteristics and to apply spatial statistical models.

The R package shinydashboard was applied to present the data interactively as dashboards for representative values, tables, and maps. In spatial statistical analysis, the focus was on household living survey such as income, education, and health expenditure in each province. To detect spatial dependence the Moran's I Statistics were applied these working related objective variables and maps of Vietnam. The shape files of Vietnam in GADM database of Global Administrative Areas were used to calculate neighborhood information of provinces in Vietnam and to draw Choropleth map.

Ordinary least squares regression (OLS) model and conditional autoregressive (CAR) model were applied to explain not only explanatory variables but also spatial dependences by provinces and characteristics of the relation between income and expenditures such as education and health. This paper showed a comparative study of the results of ordinary least squares regression model and conditional autoregressive model of VHLSS2006.

2 Data

In this study, household summary file TTCHUNG of VHLSS2006 was used for analysis. The data file TTCHUNG is the household-level summary data, which derived from data files. In this study, following variables are selected to apply OLS and CAR models.

- Response variable
  - income: income per capita per month (thousand dong)
- Explanatory variables
  - ttnt: urban/rural (1: urban, 2: rural)
  - hsize: house hold size
  - edu: education expenditure per month (thousand dong)
  - hea: total of health expenditure per month (thousand dong)

TTCHUNG was grouped by province and transformed to ttc2 (hereinafter referred to as ttc2 by the first three letters of TTCHUNG for version 2). In this data, province Ha Tay located in south west of Ha Noi was merged with Ha Noi

Table 1 Summary of ttc2

Variable	Mean	sd	Min	Max
Income	681	231	323	1651
hsize	4.31	0.46	3.30	5.65
edu	1321	590	426	3318
hea	1342	503	397	2271

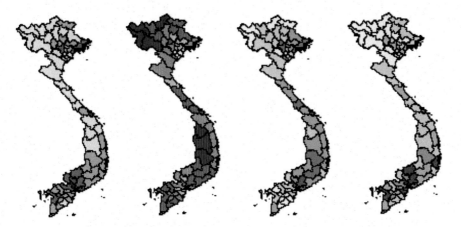

Fig. 1 Choropleth maps of income, hsize, edu, and hea (from left to right in this figure) of ttc2

municipality. Table 1 shows summary of ttc2 including mean, standard deviation, minimum and maximum values of variables.

Figure 1 shows histograms of income, hsize, edu, and hea of ttc2. From the histograms, income, edu, and hea have long tails to the right.

For drawing map and calculate contiguity matrices, shape file of GADM database of Global Administrative Areas (Hijmans 2017) in Vietnam was used. Figure 2 shows choropleth maps of income, hsize, edu, and hea of ttc2. Income (income) is high for large cities such as Hanoi in the north and Ho Chi Minh in the south, but low in other areas such as the northwest, north-central, and south-central regions. However, the near to large cities, the higher the expenditure values are. Education expenditure (edu) and health expenditure (hea) have the same tendency as income, but house hold size (hsize) shows that the tendency of ups and downs is reversed. From the choropleth maps, the spatial characteristics between hsize and other variables are different. Therefore, spatial dependences were checked by Moran's I statistics.

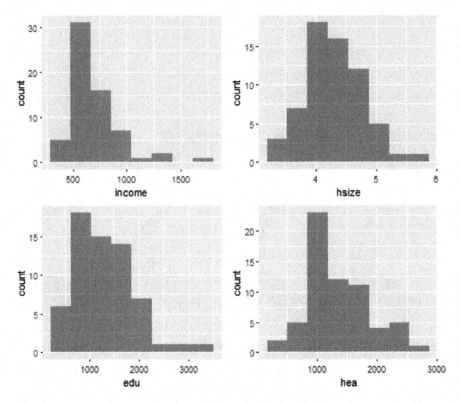

Fig. 2 Histograms of income (top left), hsize (top right), edu (bottom left), and hea (bottom right) of ttc2

3 Data Analysis

In this section, ordinary least squares regression (OLS) model and conditional autoregressive (CAR) model were applied to ttc2. Moran's I statistics were also calculated to compare spatial dependencies. Then, the results of OLS model and CAR models were compared.

3.1 OLS

First of all, OLS model was applied to ttc2, as following regression equation:

$$income = ttnt + hsize + edu + hea. \tag{1}$$

Table 2 Result of OLS

	Coefficient	Std.error	t.value	p.value
(Intercept)	2444	351	7.0	<.001
ttnt	−808	130	−6.2	<.001
hsize	−140	34	−4.2	<.001
edu	0.05	0.03	1.4	0.153
hea	0.15	0.03	4.4	<.001

Table 2 shows the results of OLS model. From the result, coefficient of education expenditure was not effective to explain income while other variables were effective. Adjusted R squared value of the result was 0.77.

Root mean squared error (RMSE) which measures the average magnitude of the error. It is the square root of the average of squared differences between prediction ($\hat{y}$) and actual observation (y_i) as follows:

$$\text{RMSE} = \sqrt{\frac{1}{n}\sum_{i=1}^{n}(\hat{y}_i - y_i)^2}. \tag{2}$$

In this result, the RMSE value is 13.5.

Figure 3 shows dash boards of the results of OLS model. From the results, there was high (positive) residual province (Ha Noi) and were other low (negative) residual provinces. However, almost all fitted values have small residuals.

3.2 Moran's I Statistics

To check spatial dependences, Moran's I statistics were calculated. Table 3 shows the result of Moran's I statistics and p-values of income, edu, and hea. From the results, every variables have global dependence. Then, local Moran's I statistics are also calculated. Figures 4, 5, and 6 show local Moran's I statistics vs actual values of income, edu, and hea, respectively.

From these results, there are spatial dependence in income, edu, and hea. The first quadrant which has high actual values and high local Moran's I corresponds to candidate of hotspots. From these figures Ho Chi Minh and Binh Doung are the candidate of hotspots of income, while Dong Nai, Ba Ria—Vung Tau, Ho Chi Minh, and Da Nang are the candidate of hotspots of edu, and Ho Chi Minh, Dong Nai, and Binh Duong are the candidate of hotspots of hea. There candidates were different provinces while only Ho Chi Minh was appeared in all variables.

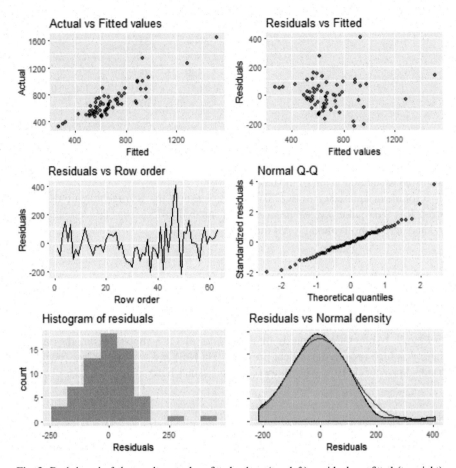

Fig. 3 Dash board of the result; actual vs fitted values (top left), residuals vs fitted (top right), residuals vs row order (middle left), normal Q-Q plot (middle right), histogram of residuals (bottom left), and residuals vs normal density (bottom right)

Table 3 Global Moran's I statistics of income, edu, and hea

Variable	Global Moran's I	p-value
Income	0.575	<0.01
edu	0.538	<0.01
hea	0.541	<0.01

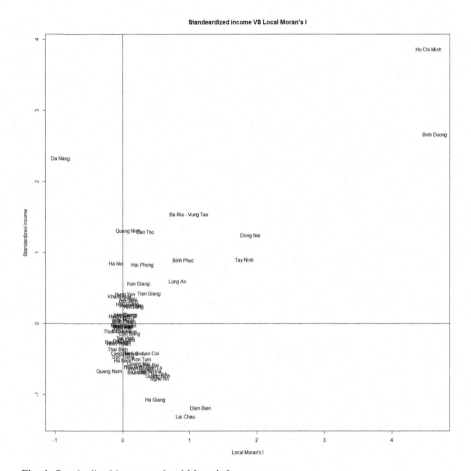

Fig. 4 Standardized income vs local Moran's I

Therefore, it was found that Ho Chi Minh City has high actual values for both income and health expenditures and education expenditures, but it is not high for education achievements compared with the surroundings. Also, Dong Nai did not have high actual value or Moran I for Imcome, but it also turned out that health expenditure and education expenditure are high. On the other hand, Binh Duong has high income but low Education expectations.

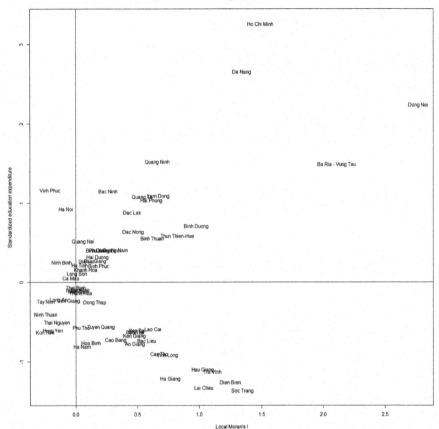

Fig. 5 Standardized edu vs local Moran's I

3.3 CAR

Then, CAR model which the spatial dependence parameter rho be fixed rather than being estimated in the model (CAR model 1) was applied to ttc2. S.CARleroux (Leroux et al. 2000) was applied to the data. Table 3 shows the results. In the results, Geweke.diag shows the diagonostics of Geweke (1992). Figure 7 shows the result of CAR model 1, and the result of the RMSE value is 14.2.

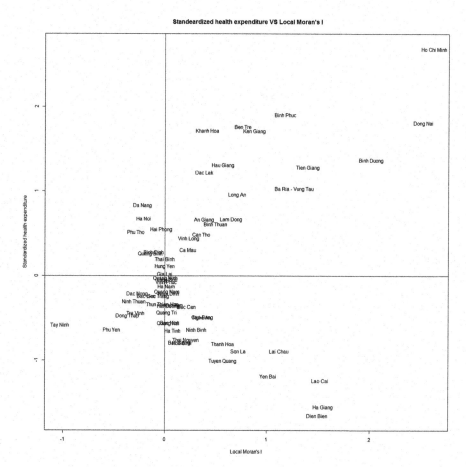

Fig. 6 Standardized hea vs local Moran's I

Then, CAR model which the spatial dependence parameter rho be also changed being estimated in the model (CAR model 2) was applied to ttc2. Table 4 shows the results. Figure 8 shows the result of CAR model 2, and the result of the RMSE value is 14.3.

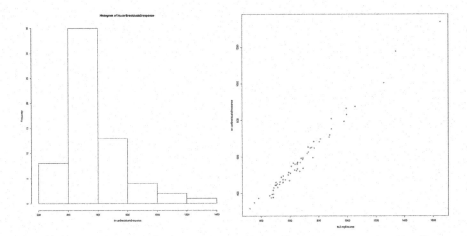

Fig. 7 The results of CAR model 1; histogram of residuals (left) and actual vs predicted values of income (right)

Table 4 Result of CAR model1

	Median	2.5%	95.5%	n.sample	Percent accept	n.effective	Geweke.diag
(Intercept)	342	−508	1229	3000	100	3160	−1.0
ttnt	−135	−514	230	3000	100	3000	1.0
hsize	−22	−145	108	3000	100	3000	0.2
edu	0.03	−0.07	0.13	3000	100	3000	−0.4
hea	0.04	−0.08	0.15	3000	100	3000	0.4
nu2	366,000	244,000	571,000	3000	100	3000	−1.0
tau2	342	−508	1229	3000	100	3160	−1.0
rho	1	1	1	NA	NA	NA	NA

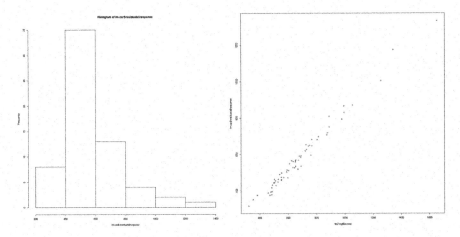

Fig. 8 The results of CAR model 2; histogram of residuals (left) and actual vs predicted values of income (right)

Table 5 Result of CAR model1

	Median	2.5 percent	95.5 percent	n.sample	Percent accept	n.effective	Geweke.diag
(Intercept)	351	−574	1275	3000	100	3160	0.3
ttnt	−139	−517	230	3000	100	3000	0.1
hsize	−22	−148	104	3000	100	3000	0.2
edu	0.03	−0.07	0.13	3000	100	3000	0.0
hea	0.04	−0.08	0.16	3000	100	2729.4	−2.3
nu2	368,000	245,000	561,000	3000	100	3000	−1.7
tau2	342	−508	1229	3000	100	3160	−1.0
rho	0.009	0.002	0.100	3000	100	555	−1.2

4 Summary and Future Studies

From the results, OSL was better result which mean small RMSE than CAR model 1 or CAR model 2. Between CAR model 1 and 2, the model CAR model 1 was better. For Future studies, it will be checked the convergence of edu in CAR model 2. Also the relation between high R squared and global Moran's I, to find out the reason why not using CAR is better results, as shown in Table 5.

Acknowledgments The data for this analysis, name of VHLSS 2006, was provided by Statistical Information Institute for Consulting and Analysis (Sinfonica) as International Official Statistical Micro Database (IOSMDB).

References

Geweke, J.: Evaluating the accuracy of sampling-based approaches to calculating posterior moments. In: Bernardo, J.M., Berger, J.O., Dawid, A.P., Smith, A.F.M. (eds.) Bayesian Statistics, vol. 4. Clarendon Press, Oxford (1992)

Hijmans, R.: GADM database of global administrative areas (2017). http://www.gadm.org

Leroux, B., Lei, X., Breslow, N.: Estimation of disease rates in small areas: a new mixed model for spatial dependence. In: Halloran, M., Berry, D. (eds.) Chapter Statistical Models in Epidemiology, the Environment and Clinical Trials, pp. 135–178. Springer, New York (2000)

Changes in the Gendered Division of Labor and Women's Economic Contributions Within Japanese Couples

Miki Nakai

Abstract Despite the continuing rise in Japanese women's rates of participation in the economy, gender division of labor has been accepted as "normal" and still strong. The aim of this paper is to examine whether and how the determinants of married women's labor force participation have changed. Based upon national sample in 1985, 1995, 2005, and 2015, we analyze change/stability of the factors that differentiate dual-income couples from husband sole provider couples and how these associations have changed over time. Results show that dual-income couples have increased but it is not at a constant rate: it increased at a slow pace until around 2005, and then increased dramatically recently. The results also show that women's own human capital has not been a determinant of labor participation for married women until recently. Husband's low income have a significant positive effect on labor force participation of married women, suggesting that high occupational resources of husband drive wife out of the labor market, which has been found in conservative and Mediterranean welfare regimes.

1 Introduction

1.1 Background

A clear division of paid and unpaid work along gender lines in households is found in every country of the world, but a continuing trend towards dual-earner families, where both husband and wife are the family breadwinners, can be detected in recent years in many advanced industrial societies. The percentage of male breadwinner families steadily decreases and it may reflect a rise in the numbers of more gender-equal couples.

M. Nakai (✉)
Department of Social Sciences, College of Social Sciences, Ritsumeikan University, Kyoto, Japan
e-mail: mnakai@ss.ritsumei.ac.jp

© Springer Nature Singapore Pte Ltd. 2020
T. Imaizumi et al. (eds.), *Advanced Studies in Classification and Data Science*,
Studies in Classification, Data Analysis, and Knowledge Organization,
https://doi.org/10.1007/978-981-15-3311-2_37

However, despite the continuing rise in women's participation in the economy over the period of industrialization and beyond in Japan as well as many Western societies, gender division of labor has been still accepted as "normal" and strong in Japanese society. While the number of households with wives entirely dependent on their spouses' income has dramatically declined, most women in dual-earner households still earn much less than their spouses, and the number of households in which wives earn more than their husbands are very few.

As gender inequalities in the division of labor within households are closely related to gender inequalities in society at large, particularly in the labor market, understanding what determines the division of labor within couples is a key to understanding other aspects of gender stratification. Many studies have argued that women's economic dependency on men is an important attribute of stratification systems and essential force in the maintenance of gender inequality (e.g., Sorensen and McLanahan 1987). In this paper, we examine what differentiates dual-income couples from husband sole provider couples, and how these associations have changed over the past three decades in Japan.

1.2 Hypotheses

It is widely acknowledged that the incentives and restrictions that affect women's employment are: (a) woman's human resources, (b) economic need for household, (c) availability of resources for balancing work and family, and (d) values (e.g., Oppenheimer 1982; Treas 1987). Do these relationships also apply to Japan?

Based on some previous studies in Table 1, hypothesis are as follows.

1.2.1 Human Resource Hypothesis

First, we hypothesize that the women's education may have positive effects on married women's labor force participation. Substantial studies have shown that women with higher educational resources have a higher participation rate in the labor market in all the Western industrialized countries (e.g., Sweet 1973; Blossfeld and Drobnič 2001). Highly educated women's risk of employment exit tends to be considerably lower and the re-entry rate is higher. In general, highly educated women appear to combine work and family by reducing their working time rather than by exiting from employment. Educational expansion and the resulting improvement in women's educational opportunities have led to increasing female labor force participation, undermining social norms favoring the male breadwinner households in many post-industrial societies. However, the effect of a woman's educational attainment on her employment has not been significant in Japan. It has been pointed out that Japan is an anomaly, where women are highly educated but typically barred from making full use of their education in economic and political fields (e.g., Brinton 1993; Shirahase 2003; Nakai 2009).

Table 1 Some major findings regarding the determinants of women's participation in paid work

		United States	West Germany	East Germany	Liberal regime	Continental regime	Social democratic regime	Southern regime
Age	25–34	(reference)	+	ns	(reference)			
	35–44			–	+	+	+	+
	45–54	–			–	+	+	–
	55–64				–	–	–	–
Wife's education	1 (lowest)	–	–	ns	(reference)			
	2	(reference)	(reference)	(reference)				
	3	+	(reference)	(reference)				
	4	+	+	ns	+	+	+	+
	5 (highest)	+	+	+	+	+	+	+
Difference of education	Husband > wife	–	(reference)	(reference)				
	Equal	(reference)	+	ns				
	Husband < wife	+	+	ns				
Number of children	0	(reference)			(reference)			
	1	–		+	+	+	+	
	2	–		–	–	–	–	
	3 or more	–		–	–	–	–	
Preschool children	Yes	–	–	–	–	–	–	–

Sources: Raley et al. (2006), Hofacker et al. (2013), Cipollone et al. (2013)

1.2.2   Supplement Household Income Hypothesis

Second, husband's socio-economic status may have negative effects on married women's labor force participation. Married women may be more likely to enter the labor market when their husbands' income is low, so that their earnings can supplement household income. According to some empirical research, a married woman's labor participation is negatively associated with her husband's socio-economic status (Nakai 2011).

However, it has been found that the impact of husbands' resources on their spouses' employment differs according to the institutional context since around the end of the twentieth century and these differences correspond to the welfare state regimes (Esping-Andersen 1990, 1999; Sainsbury 1999; Blossfeld and Drobnič 2001; Stier and Mandel 2009; Hofacker et al. 2013). For example, in continental conservative welfare states, the effect of husband's socio-economic status on wife's labor force participation is negative: husbands' high occupational resources suppress spouses' participation in paid work, showing the traditional division of labor within couples and increasing dependency of married women on their spouses over the life course. On the other hand, in social democratic welfare states, the effect of husband's socio-economic status is positive, which means that men's occupational resources increase their spouses' labor market activity. Positive effect implies that economic resource at the household level facilitates a woman's employment also because it helps balancing work and family. More and more advanced post-industrial economies see the positive effects of husband's occupational resources on their partner's participation rates in recent year. Women married to well-educated husbands as well as women with high-income partners are less likely to leave the labor market than women with low-resource partners.

There are a wide variety of discussions on the characteristics of Japanese welfare state as well as on East Asian welfare model (e.g., Goodman and Peng 1996; Esping-Andersen 1997). In some classification, Japan's welfare system has often been classified as a sub-category of the conservative welfare states regime inclined towards the liberal regime or referred to as a combination of key elements of both the conservative and liberal welfare models (Esping-Andersen 1997, 1999). Japan model is regarded to fall closer to conservative welfare states, especially Southern Europe, in the sense that the welfare state is committed to traditional familialism. It implies that the family and the local community are the natural and ideal loci of welfare provision and the state's role should be limited. If this framework can be applied, there may be a negative effect of husband's resources on wife's participation in paid work as some previous research showed (Blossfeld and Drobnič 2001). On the other hand, turning to the role of market-provided welfare and very limited social expenditures for social services, Esping-Andersen identifies Japan closer also to the liberal welfare regime, which is associated with the dual-earner/market career family model. Under welfare states similar to the liberal welfare regime, the effect of husband's resources on their wives' employment transitions might be different from that in countries belonging to the conservative and Mediterranean welfare states.

1.2.3 Modernization Hypothesis

Thirdly, we also hypothesize that values and attitudes toward the family and gender roles may affect women's participation in labor market. Inglehart and Norris (2003) argue that the twentieth century gave rise to profound changes in traditional sex roles. But the force of this "rising tide" has varied among rich and poor societies. They demonstrate that richer, post-industrial societies support the idea of gender equality more than agrarian and industrial societies and intergenerational differences in values are largest in post-industrial societies and relatively minor in agrarian societies, suggesting that the former are undergoing intergenerational changes in values. They also argue that cohort change in gender-role attitudes in post-industrial societies is unidimensional, with newer cohorts consistently more egalitarian than older cohorts. This "increasing egalitarianism in gender-role attitudes" is attributed to modernization and generational replacement. Given that younger cohorts are more egalitarian than older cohorts, it may lead to the rise in married women's labor force participation.

1.2.4 Cohort Hypothesis

Suppose a particular cohort experiences a change in labor market institutions that significantly affects the employment opportunities for women, such as the Equal Employment Opportunity Law (EEOL) or the Maternity Leave Act, and thus, the employment patterns of that cohort. Then, certain cohort may possess different gender norms, or more gender-egalitarian values, and tend to participate in employment more than other birth cohorts (e.g., Elder 1975, 1994; Shorrocks 2016).

The life course perspective focuses on the interplay of human lives and historical times. Especially in rapidly changing societies, differences in birth year expose individuals to different historical world, with their constraints and options. Individual life courses may well reflect these different times.

2 Data and Methods

2.1 Data

Data for this study were obtained from the past three decades of four waves of cross-sectional data: the 1985, 1995, and 2005 Social Stratification and Social Mobility (SSM) surveys of Japanese society, and the 2015 Stratification and Social Psychology (SSP) survey in Japan. All the surveys were conducted with similar approach: face-to-face interviews with a special focus on social stratification and inequality in contemporary Japan. All the surveys selected national representative respondents through multiple-stage sampling. The subjects of these surveys were men and women, aged between 20 and 69 for the surveys in 1985, 1995, and 2005, and between 20 and 64 for the 2015 SSP survey. Data were collected from 1248 men and 1405 women in 1985, 2490 men and 2867 women in 1995, 2660 men and 3082 women in 2005, and 1644 men and 1931 women in 2015. The response rates were 67.9%, 66.0%, 44.1%, and 43.0% in 1985, 1995, 2005, and 2015, respectively.

To make data comparable across the four datasets, we limit our analysis to the working-age couples: Specifically, those who are married and wives' age is between 25 and 54, $N = 9067$ (994 in 1985, 3180 in 1995, 2862 in 2005, and 2031 in 2015).

2.2 Measurement of Variables

2.2.1 Dependent Variable

We focus on within-couple inequality in the family. We use a concept of wives' contribution to household income as an aspect which reflects within-couple inequality, which is defined as: (a) income provision-role type and (b) wives' contribution to total household income. In this study, we analyze (a) income provision-role type as a dependent variable.

Income provision-role type is measured based on whether a dominant provider exists and identifies who she/he may be. We use a five-group classification: (1) husband sole provider, (2) husband provides majority, (3) equal providers, (4) wife provides majority, (5) wife sole provider (Raley et al. 2006). Husband sole provider category consists of couples where only husband is employed. Husband provides majority category consists of couples where husbands' earnings represent 60% or more of the combined total income of the husband and wife. Equal providers category identifies couples where wife's earnings represent somewhere from 40% to 60%, meaning that each partner contributes between 40 and 60% of total household income. Wife provides majority category consists of couples where wives' earnings represent 60% or more of the combined total income of the husband and wife. Wife sole provider category consists of couples where only wife is employed.[1]

We analyze which factors differentiate dual provider couples (three dual-income groups) from husband sole provider couples.

---

[1]Wives' contribution to total household income (b) is measured as the proportion of the sum of wives' income and husbands' income that comes from the wife. This relates to measures used to proxy of wives' economic dependency on their spouses in some prior studies (e.g., Bianchi et al. 1999; Sorensen and McLanahan 1987).

2.2.2 Independent Variables

To capture the effects of human resources of women, we include wife's education. Wife's education is collapsed into four categories: (1) less than high school, (2) high school graduate, (3) two-year college, and (4) four-year tertiary education or more. Wife's age is coded into six categories: (1) 25–29, (2) 30–34, (3) 35–39, (4) 40–44, (5) 45–49, and (6) 50–54, where 30–34-year-old group is the reference category. Wife's birth cohort is coded into five categories: (1) 1931–1943, (2) 1944–1953, (3) 1954–1963, (4) 1964–1973, and (5) 1974–1995. The birth cohort group of 1931–1943 were born before WWII. The birth cohort group of 1944–53 were born in early postwar period and this group includes the first baby boomer generation. The group who were born in 1954–1963 is a cohort who entered the labor market prior to the Equal Employment Opportunity Law (EEOL) enforcement, whereas the group who were born in 1964–1973 is a post-EEOL cohorts, those who were 22 or younger in 1986, when EEOL went into effect in Japan. The group who were born in the period 1974–1995 are the most recent cohort and often referred to as the second baby boom generation and post-bubble who finished schooling in the recession period, when the labor demand was weak.

Married couples division of labor within household may vary systematically also with regards to household level characteristics. The household level explanatory variables include age and the number of children within a household, husband's income, and the couples' relative education. We create variables to indicate the number of children and the presence of a preschooler. The four categories measuring number of children are: (1) no, (2) one, (3) two, and (4) three or more children, where no children is the reference category. Husband income level is measured by income decile (ten groups) in each survey year. The couples' relative education-level variable measures whether wife has higher or lower education than her spouse and has three categories: (1) husband and wife have equal educational attainment, (2) wife has higher education than her spouse (hypogamy), and (3) husband has higher education than his spouse (hypergamy), where equal educational attainment is the reference category.

2.3 Method

Although the level of the original dependent variable is four ordinal categories except wife sole provider category, we treat the response variable as dichotomous, whether the married couple is dual provider or husband sole provider couple. The focus here is on as to which factors are determinants for wives in playing an important role by contributing to household income because not all research has consistent findings with regard to the impact of various factors on women's participation until recently. We examine what differentiates dual-income couples, which consist of three dual-income groups (where the husband provides the majority of income, equal providers, and the wife provides the majority), from husband sole provider couples. This is very important since the traditional division of roles within households where a man is the sole provider and a woman is the main care provider for the family remains relatively unchanged despite the increasing rate of female labor force participation in Japan. The effects of individual-level and household level characteristics on the pattern of couple's income provision role, or breadwinner type, are analyzed using logistic regression analysis. Since we want to take account of differences across birth cohorts with respect to the effect of wife's education on couples income provision type, assuming that associations of women's human capital with work increase over time, we add interaction terms to our models.

3   Results

3.1   Descriptive Statistics

We first examined how families are trying to allocate time to market work and household production within couples. Table 2 shows how bread-winning patterns among married couples have changed over the past three decades. Until 2015, an overwhelming majority of couples were dual providers. The proportion of the households with husbands as sole provider have declined from 42.8% in 1985 to 29% of the observations in 2015. However, compared to Western post-industrial societies, excluding a couple of continental European countries such as Spain and Italy, male breadwinner is still much higher (Harkness 2010). Moreover, equally shared income provisioning portrayed only 14% in Japan even in 2015.

The table also shows how married women's earnings contribution of household income changed between 1985 and 2015. As expected, the percentage of earnings that comes from wives has increased quite dramatically recently, but still roughly only a quarter (25.6% and 23.1%) of family income in 2015.

Table 2 Trends in percent distribution of household types of couples and wife's economic contribution to household income: 1985–2015

	1985	1995	2005	2015
Household type				
Husband sole provider	42.8%	42.0%	41.3%	29.0%
Husband provides majority	46.8%	47.7%	44.7%	51.4%
Equal providers	8.9%	8.7%	11.2%	14.1%
Wife provides majority	1.6%	1.2%	1.9%	5.2%
Wife sole provider	0.0%	0.4%	0.9%	0.3%
Wife's economic contribution				
All age	14.0	15.1	18.6	25.6
Wife aged between 25–54	15.1	14.9	17.8	23.1

3.2 Determinants of Wife's Participation in Paid Work

Table 3 displays the results of a logistic regression estimating the likelihood of being in paid employment for wives of married couples. It investigates whether employment of married women is influenced by their own human capital, whether household level factors have a significant influence, and whether the effects have changed over time. Women's human capital measured by their own education is not a determinant of labor participation for married women in Japan. Even having a college degree does not lead to women's higher participation in the labor force.

Table 3 Logistic regression estimates for the likelihood that a couple is dual provider: 1985–2015

		β	S.E.	Exp(β)
Age (ref: 30–34)	25–29	−0.156	0.126	0.865
	35–39	0.117	0.100	1.124
	40–44	0.222*	0.126	1.249
	45–49	0.360**	0.149	1.434
	50–54	0.040	0.179	1.041
Wife's education (ref: high school)	Less than high school	0.007	0.103	1.007
	Two-year college	0.081	0.082	1.085
	Four-year college	0.080	0.164	1.084
Couple's relative education (ref: equal)	Husband > wife	−0.336***	0.067	0.714
	Husband < wife	0.144	0.091	1.155
Husband's income decile		−0.117***	0.012	0.889
Number of children (ref: 0)	1	0.015	0.120	1.015
	2	0.417***	0.111	1.517
	3 or more	0.508***	0.122	1.662
Preschool children (ref: no)	Yes	−1.089***	0.088	0.337
Birth cohort (ref: 1954–1963)	1931–1943	−0.135	0.181	0.874
	1944–1953	0.123	0.108	1.131
	1964–1973	−0.106	0.108	0.900
	1974–1995	0.118	0.184	1.125
Four-year college * Birth cohort	1931–1943	−0.390	0.519	0.677
	1944–1953	−0.548**	0.272	0.578
	1964–1973	0.529**	0.242	1.697
	1974–1995	0.146	0.271	1.158
Survey year (ref: 1985)	1995	−0.414***	0.121	0.661
	2005	0.065	0.174	1.067
	2015	0.611***	0.234	1.842
Intercept		1.143***	0.172	3.136
Observations	9067			
Nagelkerke (Pseudo) R^2	0.140			

*$p < 0.10$, **$p < 0.05$, ***$p < 0.01$

Having said that, a four-year college degree may have somewhat different meaning by cohorts. Cohort itself does not make differences in the likelihood of being dual provider household, but, according to the interaction term, highly educated women born in the postwar period may have different gender-role attitudes in terms of participation in paid work and division of labor within household across birth cohorts. Women who were born between 1944 and 1953 may have relatively stronger gendered division of household labor than other birth cohorts and their economic participation tend to be inactive. Many of them got married and settled down in the 1960s and 1970s, when Japan's economy grew. Being housewife symbolized middle-class status back then and many female baby boomers stayed home as housewives. In contrast, highly educated women who were born between 1964 and 1973 tend to commit to their career than other cohorts. They experienced a social context that was significantly different from that of the earlier period in their youth, when economic growth stagnated after the oil shock in 1973. These women came of age in the Bubble Era and entered the labor market as the Equal Employment Opportunity Law was being implemented, which might lead the college graduates to active participation in the labor force.

Age effects show a different pattern from Western countries. It is not reverse U-shaped or hump-shaped pattern, which implies the highest employment level in mid-career, or in their 30s, and lower probability for older age group. In Japan, wives of older age group, or in their 40s, have higher probability of being in paid work than those in their 30s.

Compared to the women's own variables, household level and spouses' variables have significant effects. We find significant negative effects of husbands' income on wives' labor force participation. This means that high occupational resources of husband drive wife out of the labor market, which is the relationship known as "Douglas-Arisawa effect," which hypothesizes that the decision as to whether or not a woman becomes a paid worker is influenced by her husband's income; the lower the husband's income, the more the wife tends to choose paid work (Douglas 1934; Arisawa 1956). It is sometimes claimed nowadays that the argument of negative association between husband's income and wife's participation in paid work does not hold true anymore. However, in reality, the results of the analysis show that the relationship mentioned in the "Douglas-Arisawa effect" is still valid in Japan. This relationship has been found in conservative and Mediterranean welfare states in some previous research (Blossfeld and Drobnič 2001). Unlike in the case of Social democratic welfare states, whose tax system promotes wives' employment as taxes are individual-based rather than household-based, the tax and pension policies in Japan are geared toward discouraging wives to work.

Quite a few studies have investigated the associations between wives' labor force participation and income inequality of society level (e.g., Treas 1987; Breen and Salazar 2010; Breen and Andersen 2012; Shin and Kong 2015). Past studies have referred that wives' earnings once decreased income inequality among households in a society, but the increasing positive association between spouses' earnings due to educational assortative mating, which helps to maintain greater economic equality within marriage on the one hand, contributes to growing earnings inequality among married couples in recent years (Schwartz 2010; Shin and Kong 2015). In Japan, however, the traditional division of labor in the family has not changed a lot and therefore wives' earnings still help to reduce income inequality across married couple families.

As for relative education level of a couple, wives with higher educational attainment than their husbands are not necessarily more likely to participate in work, but wives' labor participation is restricted among couples where wives have lower educational level than their husbands. This suggests that values related to household context influence gendered arrangement for work and care in the household. Educational hypergamous couples, in which women marry men of higher status than themselves, may prefer more traditional marriage practices, with women being mainly responsible for caring for children and housework, with men having bread-winning role.

The presence of preschool children strongly negatively affects wife's labor participation. However, mothers' participation increases as the number of child increases. In Western European countries and the United States, there has been a steadily increasing trend toward paid employment by married women, especially among those with young children. On the other hand, the employment rate for women with young children in Japan is currently one of the lowest among advanced industrial countries (OECD 2016).

4 Conclusion and Discussion

To summarize, household level and partner variables have stronger influences, whereas women's own human capital has little influence on women being in paid work, which would suggest a more traditional division of labor.

After all, unlike in the case of Western advanced societies, Japanese women seem to adjust their working style primarily in response to household needs. These findings are consistent with previous studies (Nakai 2009, 2011). In the Western advanced societies, wife's college degree pushes a couple toward dual providing, but in Japan, women's own education has not been a determinant of labor participation for married women until recently.

Analyzing cohort differences more thoroughly in future research could enrich theory and evidence about how introduction of policy package might affect employment of married women, especially mothers of preschool children, as well as societal level of gender equality. Further research must explore the factors that differentiate couples where the husband provides the majority of the couple's income from equal providers.

Acknowledgments This work is supported by JSPS Grant-in-Aid for Scientific Research (No. 26380658, No. 17K04103), and (No. 16H02045) as part of the SSP Project. The author thanks both the 2005 Social Stratification and Social Mobility (SSM) Committee and the 2015 Stratification and Social Psychology (SSP) Project for the permission to use the SSM and SSP data.

References

Arisawa, H.: Structure of wages and structure of economy. In: Nakayama, I. (ed.) Basic Survey of Wages, pp. 40–57. Toyokeizaishinposha, Tokyo (1956)

Bianchi, S.M., Casper, L.M., Peltola, P.K.: A Cross-national look at married women's earnings dependency. Gender Issues, Summer, 3–33 (1999)

Blossfeld, H.P., Drobnič, S.: Careers of Couples in Contemporary Societies. From Male Breadwinner to Dual Earner Families. Oxford University Press, Oxford (2001)

Breen, R., Andersen, R.B.: Educational assortative mating and income inequality in Denmark. Demography **49**, 867–887 (2012)

Breen, R., Salazar, L.: Has increased women's educational attainment led to greater earnings inequality in the United Kingdom? A multivariate decomposition analysis. Eur. Sociol. Rev. **26**(2), 143–157 (2010)

Brinton, M.C.: Women and the Economic Miracle: Gender and Work in Postwar Japan. University of California Press, Berkeley (1993)

Cipollone, A., Patacchini, E., Vallanti, G.: Women labor market participation in Europe: novel evidence on trends and shaping factors. IZA Discussion Paper No. 7710 (2013). Available at http://ftp.iza.org/dp7710.pdf. Cited 1 June 2020

Douglas, P.H.: The Theory of Wages. Macmillan, New York (1934)

Elder, G.H. Jr.: Age differentiation and the life course. Annu. Rev. Sociol. **1**, 165–190 (1975)

Elder, G.H. Jr.: Time, human agency, and social change: perspectives on the life course. Soc. Psychol. Q. **57**(1), 4–15 (1994)

Esping-Andersen, G.: The Three World of Welfare Capitalism. Princeton University Press, Princeton (1990)

Esping-Andersen, G.: Hybrid or unique?: The Japanese welfare state between Europe and America. J. Eur. Soc. Policy **7**(3), 179–189 (1997)

Esping-Andersen, G.: Social Foundations of Postindustrial Economies. Oxford University Press, Oxford (1999)

Goodman, R., Peng, I.: The East Asian welfare states. In: Esping-Andersen, G. (ed.) Welfare States in Transition, pp. 192–224. Sage, London (1996)

Harkness, S.: The contribution of women's employment and earnings to household income inequality: a cross-country analysis. LIS Working Paper Series, No. 531 (2010). http://hdl.handle.net/10419/95534. Cited 1 Nov 2017

Hofacker, D., Stoilova, R., Riebling, J.R.: The gendered division of paid and unpaid work in different institutional regimes: comparing West Germany, East Germany and Bulgaria. Eur. Sociol. Rev. **29**(2), 192–209 (2013)

Inglehart, R., Norris, P.: Rising Tide: Gender Equality and Cultural Change Around the World. Cambridge University Press, Cambridge (2003)

Nakai, M.: Occupational segregation and opportunities for career advancement over the life course. Jpn. Sociol. Rev. **159**(4), 699–715 (2009)

Nakai, M.: Trends in women's career patterns and occupational mobility in Japan: analysis of the social stratification and mobility survey 1985–2005. Jpn. J. Res. Househ. Econ. **89**, 11–21 (2011)

OECD: The labour market position of families (LMF) (2016). http://www.oecd.org/els/family/. Cited 1 Nov 2017

Oppenheimer, V.K.: Work and the Family : A Study in Social Demography. Academic Press, New York (1982)

Raley, S.B., Mattingly, M.J., Bianchi, S.M.: How dual are dual-income couples? Documenting change from 1970 to 2001. J. Marriage Fam. **68**(1), 11–28 (2006)

Sainsbury, D.: Gender and Welfare State Regimes. Oxford University Press, New York (1999)

Schwartz, C.R.: Earnings inequality and the changing association between spouses' earnings. Am. J. Sociol. **115**(5), 1524–1557 (2010)

Shin, K.-Y., Kong, J.: Women's work and family income inequality in South Korea. Dev. Soc. **1**, 55–76 (2015)

Shirahase, S.: Wives' economic contribution to the household income in Japan with cross-national perspective. LIS Working Paper Series, No. 349 (2003). http://hdl.handle.net/10419/95413. Cited 1 Nov 2017

Shorrocks, R.: A feminist generation? Cohort change in gender-role attitudes and the second-wave feminist movement. Int. J. Public Opin. Res. (2016). https://doi.org/10.1093/ijpor/edw028. Cited 1 Nov 2017

Sorensen, A., McLanahan, S.: Married women's economic dependency, 1940–1980. Am. J. Sociol. **93**, 659–687 (1987)

Stier, H., Mandel, H.: Inequality in the family: the institutional aspects of women's earning contribution. Soc. Sci. Res. **38**, 594–608 (2009)

Sweet, J.A.: Women in the Labor Force. Seminar Press, New York (1973)

Treas, J.: The effect of women's labor force participation on the distribution of income in the United States. Annu. Rev. Sociol. **13**, 259–288 (1987)

Employment Structures vs. Educational Capital in the European Union Regions

Elżbieta Sobczak and Beata Bal-Domańska

Abstract The study aims at answering the research question whether high level of educational capital quality constitutes an important determinant of employment structure in European regions NUTS 2. The research used the methods of multivariate statistical analysis. Due to the growing significance of knowledge and innovation the analysis covered employment structures in economy sectors identified based on the intensity of knowledge (defined regarding manufacturing sector by expenditure on research and development, and in case of services by the level of tertiary educated persons) including high and medium high-technology manufacturing, mid-low and low-technology manufacturing, knowledge-intensive services, less knowledge-intensive services. The research covered the period 2008–2016.

1 Introduction

Now the significance of economy sectors, based on the implementation of knowledge and innovation, keeps growing (Bishop 2008; Aslesen and Isaksen 2007). Research studies focus on the role of workforce structure in economy sectors defined as the relation of expenditure on R&D against added value of manufacturing sector (Hatzichronoglou 1996).

A continuous increase in the importance of education for the development of national and regional economies is observed (Oancea et al. 2017, The Role of Universities .., 2014). Knowledge becomes the key resource stimulating economic transformations. It is related to the growing importance of human capital. The long-term regional development should be based on the increasing resources of tertiary education population, developing innovative and creative potential of regions.

In order to assess the impact of educational capital quality on the spatial concentration level of employment structure, it is necessary to define the set of

E. Sobczak · B. Bal-Domańska (✉)
Department of Regional Economics, Wroclaw University of Economics and Business, Jelenia Góra, Poland
e-mail: elzbieta.sobczak@ue.wroc.pl; beata.bal-domanska@ue.wroc.pl

© Springer Nature Singapore Pte Ltd. 2020   485
T. Imaizumi et al. (eds.), *Advanced Studies in Classification and Data Science*,
Studies in Classification, Data Analysis, and Knowledge Organization,
https://doi.org/10.1007/978-981-15-3311-2_38

geographical areas and the economic phenomenon (variable) in the cross-section of employment structure in technological intensity sectors and variables related to educational capital quality, the spatial distribution of which shall become the subject of the conducted analysis.

The purpose of the study is to assess the relation between the level of educational capital quality and the employment structure in technological intensity sectors. The study covered 265 European Union regions (NUTS-2) in 2016. In addition, changes in selected indicators of educational capital in the period of 2008–2016 were taken into account.

2　The Background Information and Methodology of the Research

Now the significance of economy sectors, based on the implementation of knowledge and innovation, keeps growing (Bishop 2008; Aslesen and Isaksen 2007). Research studies focus on the role of workforce structure in economy sectors defined as the relation of expenditure on R&D against added value of manufacturing sector (Hatzichronoglou 1996).

A continuous increase in the importance of education for the development of national and regional economies is observed (Oancea et al. 2017, The role of Universities .., 2014). Knowledge becomes the key resource stimulating economic transformations. It is related to the growing importance of human capital. The long-term regional development should be based on the increasing resources of tertiary education population, developing innovative and creative potential of regions.

In order to assess the impact of educational capital quality on the spatial concentration level of employment structure, it is necessary to define the set of geographical areas and the economic phenomenon (variable) in the cross-section of employment structure sectors and variables related to educational capital quality, the spatial distribution of which shall become the subject of the conducted analysis.

Workforce structure constitutes the reference basis of conducted analyses, in the cross-section of the following technological intensity sectors, prepared by Eurostat and OECD: *HMTM*—high and medium high-technology manufacturing, *LTM*—low and medium low-technology manufacturing, *KIS*—knowledge-intensive services, *LKIS*—less knowledge-intensive services, *OTH*—other sectors (farming, hunting, forestry, fishing, mining, production and supply of electricity, gas, water, construction).

The identification of educational capital quality in the European Union regions was conducted using the below presented indicators: *EL*—early leavers from education and training aged 18–24 (%), *LL*—participation rate in education and training (last 4 weeks) of people aged 25–64 (%), *ETER*—population with tertiary education (levels 5–8) (%), *SE*—% scientists and engineers of active population.

The study covered 265 European Union regions selected based on NUTS 2 (*The Nomenclature of Territorial Units for Statistics*) classification. Due to the unavailability of statistical data the analysis did not cover 11 NUTS2 regions.

The statistical information, required for the empirical research, was obtained based on Eurostat database. Time range of research covered the period 2008–2016. Since 1 January 2008 the updated NACE classification (NACE Rev. 2) and the definition of high-tech manufacturing and knowledge-intensive services have changed.

The following research procedure was applied:

1. The identification of educational capital quality indicators.
2. The classification of the European NUTS 2 level regions with regard to employment sector structures in 2016.
3. The assessment of relationships occurring between employment sector structures in 2016 and the level of educational capital quality in the years 2008–2016 in NUTS 2 regions of the European Union Member States.

In order to classify regions according to employment sector structures in 2016 the following procedure was performed (Anderberg 1973; Hartigan 1975; Sneath and Sokal 1973):

• the specification of diversification between studied regions using squared Euclidean distance,
• hierarchical classification of regions into homogenous classes using Ward's method,
• the selection of optimal classification using classification quality indicator suggested by Mojena,
• the presentation of classification results on a dendrogram and the diagram of distance integration in relation to integration stages,
• the presentation of the obtained classes of regions' composition and their characteristics by applying basic descriptive parameters.

3 Empirical Analysis Results

Figure 1 illustrates the hierarchical classification results of the analysed regions' employment structure for 2016 using spanning trees and integration distance diagrams with regard to classification stages. Mojena indicator points to optimal classification for k in the range 2.75–3.50. On their basis a variant division of 265 regions into four classes, representing relatively homogenous sector structure of employment was suggested.

Figure 2 presents spatial classification results of the EU regions employment structure for 2016.

Composition, number and profile of regional classes for optimal division in 2016 was presented in Table 1.

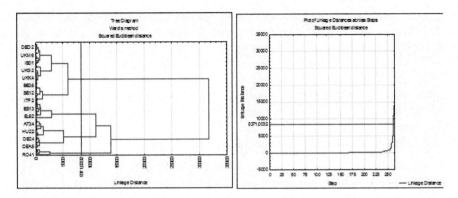

Fig. 1 Dendrogram of connections, integration distances and classification stages using Ward method for NUTS2 level regions in 2016. Source: authors' compilation based on Eurostat data by applying STATISTICA 13.1 PL statistical package

Fig. 2 Spatial classification results of the EU regions employment structure in 2016. Source: authors' compilation based on Eurostat data applying ArcGIS and EuroGeographics for administrative boundaries (map)

Table 1 Classification results of the European regions using Ward's method with regard to sector structure of employment in 2016

Class no.	Specific nature of classes	Regions no.	Descriptive parameters	Sectors				
				HMTM	LTM	KIS	LKIS	OTH
1	Regions presenting low level of smart smart specialization	16	$\bar{x}$	2.6	11.3	26.0	26.4	33.8
			Me	2.5	11.1	27.8	25.7	31.3
			V	79.4	32.0	22.0	17.3	24.9
2	Service regions (*LKIS* and *KIS*)	43	$\bar{x}$	2.8	8.7	34.4	39.0	15.1
			Me	2.6	9.4	34.3	37.1	15.1
			V	80.9	35.9	10.7	15.6	22.2
3	Regions specializing in *KIS*	135	$\bar{x}$	4.3	7.4	46.0	30.4	12.0
			Me	4.1	7.2	45.5	30.5	11.4
			V	44.3	37.5	11.7	10.9	29.1

where: $\bar{x}$—arithmetic mean; Me—median; V—variation coefficient (in %)
Source: authors' compilation based on Eurostat data using STATISTICA 13.1 PL statistical package

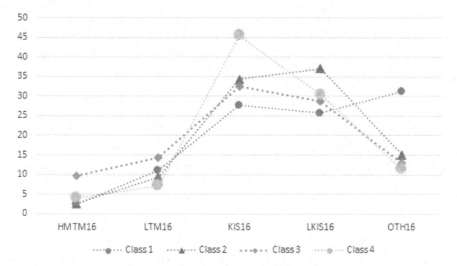

Fig. 3 Average share of employment in particular economy sectors for classes of regions distinguished in 2016. Source: authors' compilation based on Eurostat data base applying STATISTICA 13.1 PL statistical package

Figure 3 illustrates mean values of employment share in particular economy sectors for the distinguished classes of regions.

The largest class 4 covers 135 European regions characterized by highly developed specialization in knowledge-intensive services. It includes regions featuring definitely the highest share of employment in knowledge-intensive services sector (approx. 46% on average). This class covers all Swedish, Finnish, British, Irish, French, Dutch, Danish and Belgian regions, a dozen or so German regions and also

Luxemburg, Malta, Estonia and individual regions from the other European Union countries (the total of 23 regions from the European Union Member States). This class does not list any Croatian or Slovak regions, nor Lithuania, Latvia or Cyprus.

Class 3 is the next most numerous one as it covers 71 industrial regions. The regions included in this class are characterized by the highest employment share in both high and medium high-technology sector (approx. 9.6% on average) and in low and medium low-technology sector (approx. 14.4% on average). This class lists regions from 14 European Union countries, including 8 countries from the latest accession (excluding Lithuania, Latvia, Estonia, Malta and Cyprus) and also many German regions.

The second class covers service oriented regions characterized by the domination of employment in less knowledge-intensive services (on average 39% of employment) and in knowledge-intensive services (on average 34.4% of total employment). This class predominantly includes Spanish, Italian, Greek, Portuguese regions as well as Lithuania and Latvia (the total of 43 regions).

The first class, covering 16 regions and characterized by low level of smart specialization, turned out to be the least numerous. It covered the regions featuring definitely the highest share of employment in the so-called other sectors (approx. 33.8% on average) and definitely the lowest share of employment in knowledge-intensive services (26%), less knowledge-intensive services (26.4%) and also in high and medium high-technology manufacturing (2.6%). This class listed 7 Greek, 5 Romanian and 4 Polish regions.

In each of the identified classes the highest diversification was characteristic for the share of employment in manufacturing sectors, both high and medium high-technology and low and medium low-technology sectors.

The next step of the analysis is the assessment of relationships occurring between employment sector structures in 2016 and the level of educational capital quality in 2008–2016. Figure 4 presents median values of educational capital quality indicators for the distinguished classes of regions.

The analysis of information presented on Fig. 4 results in the following conclusions. Class 4 made up of regions specializing in *KIS* is characterized by definitely the highest median values of *LL* variables among all identified classes in the entire analysed period—participation rate in education and training (4 weeks) of people aged 25–64 (%), *ETER*—population with tertiary education (levels 5–8) (%) and *SE*—scientists and engineers in % of active population.

The regions characterized by the low level of smart specialization, included in the first class, present, in turn, definitely the lowest level of the listed indicators of educational capital quality. In the second class, covering service oriented regions, with the majority of less knowledge-intensive services, both specialists and engineers have low employment share, just like in the first class. The variable— employment rate of young people aged 15–34 not in education and training takes similarly low values. In both discussed classes, however, the following indicators take the highest values of *EL*—early leavers from education and training aged 18–24. This indicator takes the lowest values in the third and the fourth class covering

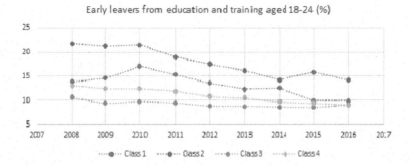

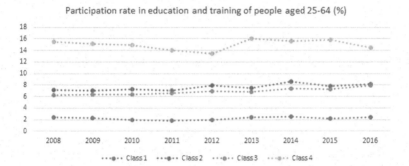

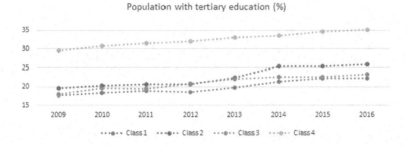

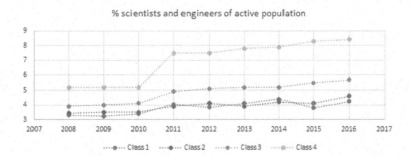

Fig. 4 Median values of educational capital quality indicators for the distinguished classes of EU NUTS 2 regions. Source: authors' estimations and compilation based on Eurostat database

Table 2 Significant correlations between educational capital quality indicators and the share of employment in technological intensity sectors in 2016

Class no.	Specific nature of classes	Regions no./r*	Indicators	Sectors				
				HMTM	LTM	KIS	LKIS	OTH
1	Regions representing low level of smart specialization	16/0.49	EL			−0.76		0.65
			LL			0.73		
			ETER			0.85		−0.69
			SE	0.61				
2	Service regions (LKIS and KIS)	43/0.30	EL		−0.31		0.31	
			LL	0.40				−0.43
			ETER					
			SE	0.37		0.36	−0.31	−0.32
3	Industrial regions (including HMTM)	71/0.23	EL				0.27	
			LL	0.24	−0.27	0.54		−0.66
			ETER		−0.39	0.48		−0.29
			SE	0.28	−0.47	0.45		−0.33
3	Regions specializing in KIS	135/0.169	EL			−0.23	0.32	
			LL			0.29	−0.40	
			ETER	−0.31	−0.61	0.72		−0.39
			SE	−0.20	−0.64	0.66		−0.43

where: r*—the critical value of correlation coefficient for significance level $\alpha = 0.05$
Source: authors' compilation based on Eurostat data

regions specializing in the development of knowledge-intensive services and in high and medium high-technology manufacturing.

Table 2 presents the statistically significant values of correlation coefficients between the educational capital indicators and the share of employment in technological intensity sectors in 2016 in the identified classes of regions.

Predominantly, the educational capital presented a statistically significant correlation against the knowledge-intensive services sectors. However, the strength of the relationship varied between the classes of regions distinguished by technological intensity. Employment in the knowledge-intensive services sector was characterized by the strongest correlation in the first class presenting low level of smart specialization with the level of education (ETER) and the tendency of employees towards lifelong learning (LL), and in the fourth class specializing in KIS with educational level (ETER) and knowledge capital resources in the form of specialists and engineers of active population (SE). It shows that for most of the identified classes of regions, the development of KIS sector, regardless of its importance in the region, is associated with a relatively higher share of tertiary education population. Only in case of the second class, covering regions specializing in both high and low education level services, the statistically low or insignificant correlation of educational capital with employment in the knowledge-intensive sector was recorded. This observation is confirmed by the negative value of correlation coefficient between the share of employment in KIS sector and the share of early school leavers (*EL*), which was statistically significant, however, negative in two classes (class 1 and 4).

The development of *LKIS* sector was not clearly linked to any indicator of educational capital. The observed correlations show only a low, positive relationship with the early school leavers (classes 2, 3 and 4), and a negative, low correlation with specialists and engineers of active population (class 2) and the participation rate in education and training (last 4 weeks) of people aged 25–64 (class 4).

Similar conclusions, indicating both low and negative (or the absence of) correlation with educational capital can be attributed to *LTM* sector. An unfavourable correlation between the level of low-technology sector development in industry and education capital can be observed among industrial regions (including *HMTM*) covered by class 3 and the regions specializing in KIS included in class 4. Higher share of employment in industry resulted in average (correlation coefficient in class 3: 0.39–0.47 and class 4: above 0.6), negative correlation of tertiary education population share and specialists and engineers share in the region (*SE*).

In case of high and medium technology sector (*HMTM*) a low or average positive correlation was identified with the tendency towards lifelong learning (*LL*) (classes 2 and 3) and the share of specialists and engineers in active population in a region (*SE*) (classes 1–3). Class 4 was characterized by a low, negative correlation with two measures of educational capital, tertiary education (*ETER*) and the share of specialists and engineers in active population in a region (*SE*).

In 2016 the lest favourable sector of economy, identified in terms of technological intensity was the sector employing relatively large numbers of population in enterprises outside high, medium and low-technology and knowledge-intensive sectors (*OTH* sector). The vast majority of correlations identified in *OTH* sector were negative, indicating that the share of employment in this sector was associated with the lower level of educational capital. The highest (over 0.6) negative correlation of employment and education in this sector was identified in class 1 (*ETER*) and in class 3 with the tendency to participate in lifelong learning (*LL*). At the same time, higher share of employment in *OTH* sector, in case of class 1 regions covering the regions presenting low level of smart specialization, can be correlated with the higher early school leaving tendency (*EL*).

4   Conclusions

The obtained research results facilitate defining the specific nature of regional groups identified with regard to sector structure of employment in the EU regions, as well as the assessment of relationships occurring between the selected indicators of educational capital quality and the typology of employment structure. The conducted analysis shows that the share of specialists and engineers among the professionally active population, the share of population with tertiary education and lifelong learning stimulate the development of employment structures characterized by smart specialization, thus the development of high and medium high-technology manufacturing and also knowledge-intensive services, acting as an incentive for socio-economic development of regions and national economies in the modern world.

The structure of employment characterized by low level of smart specialization, with the majority of employment in such sectors as farming, hunting, forestry, fishing, mining, production and supply of electricity, gas, water, construction and also in the sector of less knowledge-intensive services result in high shares of early leavers from education and training aged 18–24.

The carried out correlation analysis shows the existing relationship between educational capital and the development of knowledge-intensive services sector. However, the majority of considered correlations between educational capital and the level of economic sectors' development, identified in terms of technological intensity, adopted values below 0.5.

In most classes of regions, the higher share of employment in *KIS* was accompanied by the higher share of tertiary education population, higher percentage of specialists and engineers and higher tendency towards improving qualifications (*LL*), along with the lower percentage of early school leavers. *LTM*, *LKIS* and *OTH* sectors proved to be particularly unfavourable for educational capital.

Acknowledgments The study was funded within the framework of the National Science Centre project 2015/17/B/HS4/01021.

References

Anderberg, M.R.: Cluster Analysis for Application. Academic Press, New York (1973)

Aslesen, H.W., Isaksen, A.: New perspectives on knowledge-intensive services and innovation. Geogr. Ann. Ser. B Hum. Geogr. **89**(suppl 1), 45–58 (2007)

Bishop, P.: Spatial spillovers and the growth of knowledge intensive services. Tijdschr. Econ. Soc. Geogr. **99**, 281–292 (2008)

Hartigan, J.A.: Clustering Algorithms. Wiley, New York (1975)

Hatzichronoglou, T.: Revision of the High-Technology Sector and Product Classification. OECD, Paris (1996)

Oancea, B., Pospíšil, R., Drăgoescu, R.M.: Higher Education and Economic Growth. A Comparison between Czech Republic and Romania. Prague Econ. Pap. **26**(4), 467–486 (2017). https://doi.org/10.18267/j.pep.622

Sneath, P.H., Sokal, R.R.: Numerical Taxonomy. Freeman, San Francisco (1973)

The role of Universities and Research Organisations as drivers for Smart Specialisation at regional level, European Commission (2014). https://www.researchgate.net/publication/289124398. Cited 28 Nov 2017

The IPUMS Approach to Harmonizing the World's Population Census Data

Matthew Sobek

Abstract IPUMS integrates census microdata from around the world into one consistent database, harmonizing variable codes across countries over a 60 year period. The variables in the files provided by National Statistical Offices utilize a wide variety of classification schemes. To enable comparative research requires coping with this empirical reality of pre-classified microdata. The aim is to impose order while minimizing loss of detail. This paper describes the general principles IPUMS follows in the harmonization process. It discusses the challenge of applying international standards to demographic data produced over a long period of time by dozens of statistical offices. The preferred strategy involves the use of composite codes in which leading digits for a variable apply generally across countries and trailing digits retain details not universally available. That approach is not always viable, however, and the paper describes a number of other strategies to cope with the variety of challenges posed by international population data harmonization.

1 Introduction

IPUMS is the world's largest collection of population microdata available for research and education. The project integrates census data from 85 countries into one consistent database. The signature feature of IPUMS is to harmonize variables across countries and over a 60 year period, so the same code has the same meaning in all times and places. The aim is to facilitate comparative research by reducing the cognitive and logistical burden on researchers, enabling them to focus on analysis.

The anonymized microdata files provided to IPUMS by National Statistical Offices arrive coded into a wide variety of classification schemes dating from when they were originally processed in the different countries. Most variables simply report the categories that were listed as response options on the particular census

---

M. Sobek (✉)
IPUMS Center for Data Integration, University of Minnesota, Minneapolis, MN, USA
e-mail: sobek@umn.edu

© Springer Nature Singapore Pte Ltd. 2020
T. Imaizumi et al. (eds.), *Advanced Studies in Classification and Data Science*,
Studies in Classification, Data Analysis, and Knowledge Organization,
https://doi.org/10.1007/978-981-15-3311-2_39

questionnaire from which they were derived. There is no standardization across countries and little consistency within countries over time. Some countries adhere to international classifications when such standards exist, while others modify or ignore them. To enable comparative research requires coping with this empirical reality of pre-classified microdata.

This paper begins with a brief description of IPUMS and the concept of harmonization. This is followed by a discussion of data standardization and the default IPUMS approach to variable harmonization: using a multi-digit coding structure to impose a common classification scheme across countries while retaining all the original category detail. The balance of the paper describes situations where alternative harmonization approaches are needed because of unresolvable incompatibilities in the original classification schemes or deep conceptual differences among census questions.

2 About IPUMS

IPUMS is composed of census microdata: each record is a person, and all of their individual characteristics are known. Microdata allow researchers to create tabulations never envisioned by the collectors of the data, and they enable sophisticated multivariate modeling. IPUMS currently includes data for 672 million individuals recorded in over 300 censuses taken since 1960 (Ruggles et al. 2015). Most countries provide multiple censuses, enabling study of change over time both nationally and internationally. IPUMS data are samples, typically comprising 1–10% of the national population. Prospective users must apply for access, and over 14,000 researchers have been registered.

A web dissemination system allows users to browse the contents of the database and construct custom data extracts that pool data from multiple countries and time periods into a single file. The user downloads the file—typically containing some millions of records and twenty to thirty variables—to their desktop for analysis. Through the web system, researchers have access to detailed documentation for each variable, including comparability discussions, codes, frequencies, and other information (Sobek et al. 2011). The website is accessible at https://international.ipums.org.

A critical characteristic of microdata lies in the categorical detail it retains at the individual level. It is this detail that makes it feasible to harmonize the data across countries and over time. The tabulated data that are the traditional product of each census often cannot be meaningfully harmonized cross-nationally because of decisions built into their construction. With IPUMS, researchers can devise custom tabulations using the full detail of the microdata while imposing consistent population universes across samples. Microdata will also support the kinds of multivariate analyses conducted by most academic and policy researchers. The data are cross-sections in time; it is not possible to link people across censuses.

The IPUMS samples incorporate most of the detail from the original census questionnaires. All censuses have basic demographic information such as age, sex, and marital status. Nearly as universal are socioeconomic variables, such as education, employment status, and occupation. There is considerable topical variation beyond these, but questions on migration, ethnicity, disability, and fertility are also broadly asked (Sobek 2016). Most censuses, particularly in the developing world, have information about the dwelling as well, such as construction materials, plumbing, utilities, and household assets.

3   Harmonization Overview

IPUMS harmonizes variables across the entire database. There are three elements to variable harmonization: applying consistent codes across samples, devising labels for those codes, and collating integrated variable descriptions that speak to issues not sufficiently conveyed by codes and labels.

The central harmonization challenge is to equate codes that have the same meaning for a variable that is common across samples. This is fundamentally a metadata issue. One must understand the meaning of the codes, which is conveyed by their labels, by the coding structures, and by the deeper context of the census questionnaire text and enumerator instructions. Each of those elements poses challenges. The labels provided with census files are often shorthand for more complex concepts or combinations of items. They may have been created ad hoc during processing, and in most cases they have been translated out of their original language into English at some cost to their precise meaning. Coding schemes often have structure, where the meaning of a particular category can only be understood in the broader context of the classification. This is especially true for residual categories, such as "Other relatives," whose meaning is defined by the other categories that are enumerated in the classification. Finally, much meaning is embedded in how the census question was worded and in the instructions given to the census enumerators regarding the question. For example, some countries restrict the status of being "married" to only legal marriages, while others make allowance for "common law," custom, or other variations. Those distinctions are often not reflected in the value labels and may only be discoverable from the questionnaire or instructions.

Population data harmonization ultimately depends on informed human judgment. Computers can help greatly with the logistics, but they can provide only limited leverage equating the meanings of international census data, which depend so much on context. IPUMS has nevertheless written a great deal of software to assist with the harmonization process. In most cases, researchers manipulate metadata to standardize and harmonize the data, with the software being driven by the metadata. A description of that process follows below.

The census data provided to IPUMS come in many formats with varying documentation in many languages. The categorical variables in recent censuses often reflect the influence of international standards and recommendations, but countries may choose to modify or ignore them. The older data are less regularized in every respect. Harmonizing data from such disparate source material is a complicated process, and we break it down into a series of discrete steps to make it manageable and efficient. To the extent possible, we strive for an industrial as opposed to a craft model of production.

4 Data Standardization

Before data processing can commence, one must understand the data structure. We require basic metadata to interpret the files: the relationship between data records, the linking keys, names and locations of variables, and labels for categorical variables. These metadata must be translated into English, as necessary, before we begin.

IPUMS metadata development begins with the creation of a data dictionary for each dataset. An IPUMS data dictionary is much like a codebook, but it contains more information and in a more structured format suited to machine processing. IPUMS software is designed to read this metadata structure. Table 1 shows a small part of a data dictionary. It records each source variable's name, location in the data file, labels for variables and values, frequencies for each value, universe of

Table 1 Data dictionary

Var	Column	Width	Variable label	Universe	Value	Value label	Freq
SEX	129	1	Sex	All persons			
					1	Male	1,516,951
					2	Female	1,595,079
MAR	130	1	Marital status	All persons			
					1	Married	1,324,684
					2	Widowed	171,370
					3	Divorced	277,068
					4	Separated	45,470
					5	Never married or under age 15	1,294,438
SCH	132	1	School enrollment	Persons age 3+			
					1	No, not in last 3 months	2,240,086
					2	Yes, public school or public college	637,353
					3	Yes, private school	138,062
					Blank	N/A(less than 3 years old)	97,529

respondents, and any other fields needed to fully document the data or control data processing, such as indicating string fields or implied decimal places. Some of these fields may not be immediately known, but are added later during processing and diagnostic analysis.

The first stage of processing is to convert the source datasets into a common format. We turn all datasets into fixed-format ASCII files with a hierarchical structure: each household record is followed by multiple person records representing its members. We receive data in many formats, which might require merging separate household and person files, converting out of native SPSS or Stata format, reorganizing files with complex geographic hierarchies, or other manipulations. Custom programming is often required at this stage, because unique situations commonly arise and errors may be uncovered. In the process of regularizing the data structures we also create some common technical variables useful for our system. As we modify the data, any changes to variables or record layout are recorded in the data dictionary, which evolves to stay in sync with the data file. Once formatting is completed, the data is in a form understood by the rest of our data transformation, diagnostic, and web software.

The final part of variable standardization involves connecting the source variables via metadata with their associated text in the census questionnaire and instructions. This information is necessary to fully understand the variable and is crucial during harmonization. The task is to convert pdfs and other static documentation into usable, machine-actionable metadata. To this end, all census questionnaires and instructions are translated into English and converted into a custom XML format. Having systematized this material, it can be compiled on demand using software, for both internal use and in the web dissemination system.

5 Variable Harmonization

At the highest level, harmonization requires determining which variables are conceptually the same across datasets (Esteve and Sobek 2003). Beyond variable names and labels, such determinations may require referring to codes, value labels, text of census questions, category frequencies, or other metadata. This is sometimes a judgment call for the harmonizer, who must ask whether combining variables with differing shades of meaning is likely to mislead researchers trying to interpret the data. Even if the concepts appear equivalent, an additional issue concerns the fundamental compatibility of the classifications. For example, continuous variables may be coded into incompatible value ranges, or different censuses may group response items in overlapping ways that defy harmonization.

The signature activity of data integration is to harmonize variable codes and labels across data samples. Our primary device for achieving this is a "translation table" like the one for Marital Status depicted in Table 2. The leftmost columns contain the harmonized output values and their labels. Each column on the right side documents every value that exists in one of the input datasets being harmonized: in

Table 2 Translation table (marital status)

Harmonized data		Input data		
Code	Label	Bangladesh 2011	Mexico 1970	Kenya 1999
100	Single	1=Unmarried	8=Single	1=Never married
200	Married/in union	2 = Married		
210	Married, formally			
211	Civil		2 = Married, civil	
212	Religious		3 = Married, religious	
213	Civil and religious		1 = Marr., civil, & relig	
214	Monogamous			2 = Monogamous
215	Polygamous			3 = Polygamous
220	Consensual union		4 = Consensual union	
300	Divorced or separated	4 = Divorced/separ		
310	Separated		7 = Separated	6 = Separated
320	Divorced		6 = Divorced	5 = Divorced
400	Widowed	3 = Widowed	5 = Widowed	4 = Widowed

this case census samples from three developing countries: Bangladesh, Mexico, and Kenya. Note that the full translation table for this IPUMS variable contains over 300 samples. Each row in the translation table contains items that are conceptually the same and that thus receive the same codes in the output. The work is performed by a researcher using the tools we have developed specifically for this process. In broad strokes, the process is as follows: a researcher identifies the source variables, a program directly inserts the input values into the translation table from the appropriate data dictionaries, and a researcher then aligns the codes and assigns output codes and labels (the "harmonized data" columns on the left). Thus, the original codes "1: Unmarried," "8: Single," and "1: Never married" are all aligned and will be recoded to the internationally harmonized IPUMS output code "100: Single." This sort of semantic integration is intellectual labor that no computer program can perform. It requires a holistic view of the universe of codes for each sample and consideration of the underlying questionnaire text, especially for some of the more challenging. Our harmonization of variables is designed to meet two goals: (1) retain all the detail provided in the original samples and (2) provide a truly integrated database, in which identical categories in different samples always receive identical codes. We employ several strategies to achieve these competing goals. In cases where original variables are compatible and recoding is straightforward, we write documentation noting any subtle distinctions between samples. For some variables, it is impossible to construct a single uniform classification without losing information from samples that are detail-rich. In these cases, we construct composite coding schemes. The first one or two digits of the code provide information available across all samples. The next one or two digits provide additional information available in a broad subset of samples. Finally, trailing digits provide detail only rarely available.

The classification scheme for marital status in Table 2 illustrates the composite-coding approach. In this example, the first digit of marital status has four categories consistently available in all samples: (1) single, (2) married/in union, (3) divorced or separated, and (4) widowed. The distinction between divorced and separated is not maintained in all samples, so these categories are combined at the fully comparable first digit. At the second digit, we distinguish divorced and separated persons in the samples with that information, as well as formal marriages and consensual unions. The third and final digit differentiates among types of marriages (civil, religious, polygamous) available for select countries only. The one-digit and multi-digit versions of the composite variables can be accessed as their own distinct variables in the IPUMS database. For many researchers, the single-digit version is sufficiently detailed and offers the assurance that most comparability issues are resolved.

We refer extensively to the questionnaire text and instructions to inform the integration process. If the actual question wording for a variable indicates a significant conceptual difference between samples, we create a separate variable to minimize the likelihood of user error.

Our approach to variable harmonization demonstrates an underlying principle in our integration methods. Our entire system represents what might be termed a metadata-centric approach, in which the research staff manipulates relatively simple but highly structured documents that drive the data processing and web software. From these documents we generate a unique XML markup that identifies all elements necessary to guide the recoding and documentation of variables and to associate each variable with its relevant enumeration materials. The data, documentation, and dissemination systems are all driven by the same metadata, which ensures that they always remain synchronized.

The translation tables exemplify this metadata approach to data management and dissemination. We do not write recode statements in code, except in exceptional circumstances. We write software to read our metadata. Simply moving an item from one cell to another in the translation table accomplishes the recode. The benefits are significant: a researcher can readily interpret the coding decisions while seeing all the associated labels with their codes and frequencies. If a new code is needed to handle some variation introduced by a sample, the researcher simply adds a row in the table and aligns the appropriate input codes to it. The translation tables also help with sustainability: reorganizing the codes to accommodate a new sample is quite easy compared to editing a mass of impenetrable logical assignment statements. Thus our system is far less error-prone and is much more adaptable than what could be achieved in a statistical package or simplistic approach to data processing. IPUMS is a living project, and we can never know the full universe of labels and coding structures that will need to be incorporated into the existing harmonized variables in future. The metadata-driven translation tables provide a practical solution to this challenge.

The custom IPUMS data conversion program reads the translation tables to produce the integrated output data. There are, of course, some instances where translation tables cannot accommodate the logic required to recode a variable, and variable-level programming is required; for example, for recoding continuous numeric variables like income into categories or combining multiple input variables. The data conversion program has the capacity to manipulate the data in any way required.

6 Harmonized Documentation

Variable harmonization involves more than harmonizing codes. New documentation must be written for each integrated variable and made accessible to users. Because integrated variables have time and space dimensions, a key aspect of the documentation is to highlight any comparability issues that arise across samples. One area of focus is to indicate for users wherever changes in question wording may potentially cause subtle differences in meaning, even where the codes and labels look otherwise compatible. Changes in the universe of people who were asked the question are another common source of comparability issues. In these cases the primary aim of the description text is to direct the user's attention to the collated questionnaire text or universe statements for the variable. Our goal is to empower the researcher, who must ultimately decide if the issues that remain after harmonization are relevant to their analysis.

Because variable documentation is so critical for the proper use of harmonized data, the IPUMS web dissemination system is an integral component of our approach. There is no avoiding the reality that harmonized data are simply more complex than discrete datasets. Users need better tools than pdf files and labels to understand and properly use the data. The IPUMS system lets them filter only the samples of interest and browse variables in an information-rich environment. One cannot force researchers to avail themselves of the potential of the web system to inform their work, but we strive to make it as easy as possible. Figure 1 shows the Marital Status variable page in the web dissemination system. The series of tabs allow the user to explore all the metadata associated with the harmonized variable from one viewing pane.

7 Harmonization Challenges

The Marital Status example above exemplifies our approach to harmonization, but situations arise in the global census data that require alternative strategies. The remainder of this paper describes some of those scenarios and how we address them.

When we harmonize a variable we refer to any international standards that might exist for that topic. We particularly find useful the United Nations Principles and Recommendations for Population and Housing Censuses, which influences how many countries choose to ask certain questions (United Nations 2007). Unfortunately, many countries ignore this advice, and others appear to adopt it only loosely. But any standards are welcome. For our purposes, the UN principles also provide guidance about the salient features around which a harmonized variable might be organized. Some more complex census items, like occupation and industry, have well established international classifications used by a subset of countries every decade. Over the years, however, even those classifications have evolved, so there is never a time-invariant system for our purposes.

IPUMS greatly appreciates the application of standards in census questions and classification, but in the final analysis we must deal with the empirical reality of the data we are given. It is a truism that the least detailed classification among the input variables dictates the overall coding scheme of the harmonized variable. You can often recode more detailed variables to match simpler classifications, but one cannot add detail to variables that do not have it. In practice, this means the first digit of most harmonized variables is governed by the simplest classifications. But applying pure logic to harmonization can sometimes lead to variables that are hard to understand and use. Perhaps a sample(s) must be left out of the variable, or a category must be coded in a way that requires some caveat in the documentation.

Literacy Some variables require virtually no recoding to harmonize categories. Literacy and School Attendance are simple binary variables in nearly all countries, and there is little to do other than align the "no" and the "yes" responses. Despite their simplicity, however, there are definitional differences that cannot be conveyed via category labels. For example, some censuses define literacy as the ability to read and write a small paragraph, some use a threshold of years of schooling, and other censuses impose no objective standard. Whether such differences are important for a particular analysis is a question for the researcher. The only practical way to indicate such nuance is via the variable's comparability discussion and the feature to compile the questionnaire text.

Employment Employment Status offers a similar challenge, but with a more concrete definitional difference. The variable is amenable to composite coding such as we employ for Marital Status: the first digit indicates employed, unemployed,

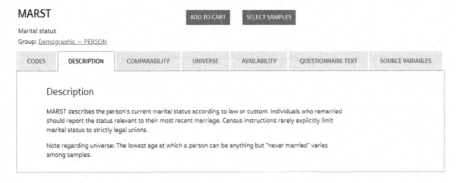

Fig. 1 Harmonized variable page in web system: marital status

and inactive persons, and trailing digits retain detailed categories whose availability varies across samples. But underlying the coding structure are differences in the reference period between countries. Different censuses assess employment status at the moment of the census, over a period of a week, or as an average over a longer time span.

For some issues, like seasonal variations, the reference period can be important. The variable documentation must carry this information. The alternative is to create a set of parallel variables for the differing reference periods, but that would impose different costs on users.

Disability Disability Status poses a more difficult challenge and example where an emerging world classification standard suggests an evolving harmonization approach. Like literacy, disability is essentially a set of one or more binary variables (blindness, mobility impairment, etc.). But there are clearly cultural and census instruction differences at work within the data. Many censuses provide instructions regarding what constitutes a disability, such as how to interpret loss of one eye or the need for a hearing aid; but other censuses provide little or no guidance. At some point, responses presumably depend on cultural norms. Even within countries where there are no discernable changes in question wording, the incidence of disability in a population can vary notably from one census to the next. In short, full comparability of disability statistics is difficult to attain under any circumstances.

To further complicate matters, there was a shift in the early twenty-first century toward adoption of the U.N. Washington Group set of questions on disability (Madans et al. 2011). The new questions are intended to provide tightly comparable data across countries, aiming to identify functional limitations that produce social exclusion. Many countries in the 2010 census round have adopted the new question wording, which employs the terminology of "some" or "a lot" of "difficulty" doing the particular activity. The creation of a world standard is laudable and should produce better and more comparable statistics among adopting countries. But equating degree of difficulty with older censuses is difficult, even within a single country. Statistical analysis suggests a disjuncture occurs when the new questions are imposed, which can yield much higher disability rates. In IPUMS, our approach has been to combine disability variables, interpreting "a lot" of difficulty as most comparable to the traditional questions. We include strong language in the variable comparability discussion warning researchers to be careful. However, as the Washington Group adherents have become more numerous, and as we have learned more about the issue, we now think we should create a distinct set of disability variables that adhere to the new approach. This would emphasize their difference from older samples and countries still using traditional questions, and it would highlight the high degree of comparability among the set of countries using the new standard.

Dwellings Housing variables pose some of the more difficult harmonization challenges. Dwelling materials for floors, walls, and roofs can be highly localized, with terminology varying by language. Material types can be grouped together in ways that straddle groupings in other countries, defying prospects for strict logically nesting. For floors, a few major materials contain much of the variation: wood, concrete, stone, brick. Some terms, like "tile," are ambiguous. It is also not always clear where an unlisted material might be combined with others in the source data, given the limitations of the labels and number of categories available on the questionnaire.

Despite these issues, researchers would surely benefit from being able to manage a single variable with a lot of variation as opposed to many individual variables. The IPUMS approach in these cases might be termed partial harmonization. We concluded that the most useful distinction for most users—and which is easily achievable in terms of consistent classification—is to make the first digit distinguish only between unfinished (dirt) floors and finished floors. That binary distinction at the first digit captures key variation in terms of sanitation and socioeconomic status. The types of finished floors are grouped together as best as possible, but we do not use the second digit to suggest there is any structure to the 35 categories of finished floors. Thus, users have the data for all countries in one variable and access to all the original labels, with minimal modification. Table 3 shows a snippet of the Floor variable codes page in the IPUMS dissemination system—with each of the columns on the right representing a census, and the "X"s indicating the availability of the category for each sample. For analyses that require distinctions beyond finished–unfinished, the burden is on the user to group the codes as necessary. We take a similar approach with walls, roofs, and cooking fuel, roughly grouping categories and leaving the full original category labels in place. One might call this "nominal" harmonization, in that categories with the same label are assigned the same code, but their full unspoken contents may differ somewhat even within those categories.

Dwelling water supply poses the challenge of dueling concepts among the source variables. The various censuses are oriented to a number of differing considerations: exclusive access to the water supply, piped water into the dwelling versus outside it, public piped water, and the ultimate source of the water (e.g., lake, river, well). The key distinction IPUMS harmonizes around is access to piped water, and secondarily whether distinctions can be made regarding exclusive use and the location of the spout on the property. The codes page for Water Supply is shown in Table 4.

It was not possible within a single variable to accommodate all the concepts in water supply. And it is not indicated in most datasets when "piped" indicates clean water. In future, we intend to create another variable on the ultimate source of the water, for those samples that offer that detail, and perhaps we will be able to identify "clean" water in samples that will support that distinction.

Complex Variables IPUMS takes a different approach with key education and work variables: we coerce them into a classification intended to roughly follow international standards. The shoe-horning of categories into major groups can be uneven, and much detail in the original samples is sacrificed. The product is a simple, fairly consistent variable, but with a degree of noise. For the variables discussed above, we largely concede to the empirical reality of the categories we are presented with, and we fashion our harmonized classification in reaction to that, with some consideration of existing census standards and recommendations.

Table 3 Floor variable codes page (partial)

Code	Label	AR 1980	AR 1991	AR 2001	AR 2010	BO 1976	BO 1992	BO 2001	BW 1981	BW 1991	BW 2001	BW 2011	BF 1980	BF 1996	BF 2006
000	NIU (not in universe)	X	X	X	.	X	X	X	X	X	X	X	X	.	.
100	None/unfinished(earth)	X	X	X	X	X	X	X	X	X	X	X	X	X	X
110	Sand	.	.	.	.	.	.	.	.	.	.	.	.	X	X
120	Dung	.	.	.	.	.	.	.	.	.	.	.	.	.	.
200	Finished	.	.	.	.	.	.	.	.	.	.	.	.	.	.
201	Cement, title, or brick	.	.	.	.	.	.	.	X	X	.	.	.	.	.
202	Cement	.	.	.	.	X	X	X	.	.	X	.	X	X	X
203	Concrete	.	.	.	.	.	.	.	.	.	.	.	.	.	.
204	Cement screed	.	.	.	.	.	.	.	.	.	.	.	.	.	.
205	Cement tile	.	.	.	.	.	.	.	.	.	.	X	.	.	.
206	Paving stone, cement tile	.	.	.	.	.	.	.	.	.	.	.	.	.	.
207	Stone	.	.	.	.	.	X	.	.	.	.	.	.	.	.
208	Brick	.	.	.	.	X	.	X	.	.	.	.	X	.	.
209	Brick or stone	.	X	X	.	.	.	.	.	.	X	X	.	.	.
210	Brick or cement	X	X	X	X	.	.	.	.	.	.	.	.	.	.
211	Block	.	.	.	.	.	.	.	.	.	.	.	.	.	.
212	Terrazzo	.	.	.	.	.	.	.	.	.	.	.	.	.	.
213	Wood	X	.	.	.	X	X	X	X	X	X	X	X	.	.
214	Palm, bamboo	.	.	.	.	.	.	.	.	.	.	.	.	.	.
215	Parquet	.	.	.	.	.	.	.	.	.	.	.	.	.	.
216	Parquet, tile, vinyl	.	.	.	.	.	.	.	.	.	.	.	.	.	.
217	Parquet, tile, marble	.	.	.	.	.	.	.	.	.	.	.	.	.	.
218	Ceramic, marble, granite	.	.	.	.	.	.	.	.	.	.	.	.	.	.
219	Ceramic, marble, tile, or vinyl	.	.	.	.	.	.	.	.	.	.	.	.	.	.
220	Marble	.	.	.	.	.	.	.	.	.	.	.	.	.	.

An 'X' indicates the category is available for that sample

Table 4 Water supply codes page (partial)

Code	Label	AR 1980	AR 1991	AR 2001	AR 2010	AM 2001	AM 2011	AT 1981	AT 1991	AT 2001	BY 1999	BY 2009	BO 1976	B 1992
00	NIU (not in universe)	X	X	X	.	.	.	X	X	X	.	X	X	X
10	PIPED WATER	.	X	X	.	.	.	X	X	X	.	X	.	.
11	Piped inside dwelling	X	.	.	X	X	X	.	.	.	.	.	X	X
12	Piped, exclusively to this household	.	.	.	.	.	.	.	.	.	.	.	.	.
13	Piped, shared with other household	.	.	.	.	.	.	.	.	.	.	.	.	.
14	Piped outside the dwelling	.	.	.	.	.	.	.	.	.	.	.	.	.
15	Piped outside dwelling, in building	.	.	.	.	X	X	.	.	.	.	.	.	.
16	Piped within the building or plot of land	.	.	.	X	.	.	.	.	.	.	.	X	X
17	Piped outside the building or lot	X	.	.	.	.	.	.	.	.	.	.	X	X
18	Have access to public piped water	.	.	.	.	X	X	.	.	.	.	.	.	.
20	NO PIPED WATER	X	X	X	X	X	X	X	X	X	X	X	X	X
99	UNKNOWN	.	X	.	.	.	X	.	.	.	X	X	.	.

An 'X' indicates the category is available for that sample

With work and education we are much more aggressive. We are motivated to do so because few censuses provide income information; thus, education and occupation are the key socioeconomic status indicators typically available. They are critical control variables for many kinds of analyses. For education, we identify primary, secondary, and tertiary level completion. We roughly aim to identify people with 6 to 8, 11 to 12, or 15 to 16 years of education. The organization is broadly reflective of the ISCED 1997 classification (United Nations 1997). Table 5 presents the 1-digit and 3-digit versions of Educational Attainment while displaying case-counts for each sample.

The internationally harmonized work and education variables lose much detail and are sometimes an imperfect fit for the national systems. This can be problematic where education is a key explanatory variable in a researcher's analysis. In recognition of education's importance, we create a separate harmonized variable for each country that is true to its specific education system. No attempt is made to apply a standard, only to harmonize around the classifications the country provides. This is harder than it sounds, as most countries have undergone changes or even complete reorganizations of their systems over the decades covered by IPUMS. Each country is therefore its own harmonization puzzle writ small, often requiring a good deal of research. Not surprisingly, educational attainment is one of the subjects upon which users most often provide feedback or identify errors.

For occupation, we collapse the typical 100–300 categories in the original samples into a 9-category variable intended to mimic the major groupings in the 1988 ISCO standard as closely as possible (International Labour Office 2012). We do something similar in mapping industry using ISIC as a general guide (United Nations 2002). Due to its importance, we take an additional step with occupation. The ISCO occupation classification is used by many countries, and it can provide fully comparable detailed occupation data for all the countries that subscribe to it. Historically, these were more often developing countries, but in recent censuses developed countries are using it as well. ISCO has undergone several iterations. We make harmonized variables for the critical mass of samples providing 3-digit detail in both the ISCO-1968 and ISCO-1988 classifications, which are available for 27 and 57 samples, respectively.

For occupation and industry, we also make the full original classifications available through single cross-national variables that do not actually harmonize the codes. Thus, for the OCC (Occupation) variable, the codes for one sample mean entirely different things from another. The data are all organized into one place for user convenience, but there are no value labels with the data (they are available online).

Geography Geography poses a unique set of issues for harmonization. Most countries have undergone changes in their administrative units over the past several decades, through merging, splitting, or moving a boundary. The goal of IPUMS is to harmonize subnational units spatially, so a province or district has the same spatial footprint in all time periods. This requires GIS boundary files, and IPUMS has created them from paper maps in all cases where digital versions were not available.

Table 5 Educational attainment codes pages: general and detailed versions

Code	Label	Argent 1970	Argent 1980	Argent 1991	Argent 2001	Argent 2010	Armenia 2001
0	NIU (not in universe)	46,954	346,793	266,113	200,072	29,655	27,986
1	Less than primary completed	205,409	1,177,161	1,487,528	1,091,158	1,243,101	66,162
2	Primary completed	168,683	910,975	1,778,148	1,451,686	1,625,587	36,061
3	Secondary completed	28,666	38,403	112,606	146,923	203,999	43,141
4	University completed	6,562	38,403	112,606	146,923	203,999	43,141
9	Unknown	10,618	.	55,615	.	.	.

Code	Label	Argent 1970	Argent 1980	Argent 1991	Argent 2001	Argent 2010	Armenia 2001
000	NIU (not in universe)	46,954	346,793	266,113	200,072	29,655	27986
100	LESS THAN PRIMARY COMPLETED	.					.
110	No schooling	34,231	232.157	337,955	327,822	462,905	22,460
120	Some primary	171,178	945,004	1,149,573	763,336	780,196	.
130	Primary (4 years)						43,702
	PRIMARY COMPLETED, LESS THAN SECONDARY						
	Primary completed						
211	Primary (5 years)						
212	Primary (6 years)	135,689	706,279	1,303,024	1,026,463	1,052,586	.
	Lower secondary completed						
221	General and unspecified track	9189	58,370	475,124	425,223	573,001	36,061
222	Technical track	23,805	146,326				
	SECONDARY COMPLETED						
	General or unspecified track						
311	General track completed	5891	43,849	356,182	440,572	410,078	93,699
312	Some college/university	6681	30,119	95,163	120,118	202,948	5608
320	Technical track			.	.	.	
321	Secondary technical degree	16,094	97,097	.	.	.	9488
322	Post-secondary technical education		23,317	135,092	175,574	250,877	44,414
400	UNIVERSITY COMPLETED	6562	38,403	112,606	146,923	203,999	43,141
999	UNKNOWN/MISSING	10,618	.	55,615	.	.	.

The process of harmonization requires overlaying each census's boundaries on each other and combining units as necessary to create entities that contain all the changes for an area within them. A researcher using these harmonized geographies knows they are holding space constant as they examine the attributes of the people and dwellings within those spaces. Spatial harmonization is essentially a least-common-denominator approach: if two units are combined in one census, they are combined in all of them. Detail is sacrificed to the goal of comparability during harmonization, but IPUMS also provides the unaltered original geography for each census.

8 Summary

In the final analysis, harmonization involves cost–benefit analysis. The goal is to make comparative research easier to conduct without obscuring the complications and thereby encouraging errors. Part of the job involves predicting how researchers are likely to use the data. Harmonizers must therefore have some subject matter expertise to strategize solutions effectively. But researchers are endlessly inventive, and a multi-purpose database will inevitably be used in ways that we cannot anticipate. Thus, a degree of conservatism is warranted while providing enough documentation to allow users to exercise informed judgment. Because of the inherent limitations of harmonization, IPUMS also makes the unharmonized source variables available to users, ensuring all original detail is retained.

An unfortunate reality of internationally harmonized data is the burden it places on the user. Both variable availability and the categories within those variables differ across samples. Using the most generalized versions of compositely coded variables resolves many comparative issues, but certain definitional or population universe issues can still persist. And the composite-coding approach is not applicable to all variables. In sum, researchers are obligated to pay more attention to the metadata than they may be accustomed to, and it tends to be more complex. An ongoing challenge of our web dissemination system is to find better ways to convey the most important information without overwhelming users with details until they need them.

IPUMS is committed to harmonizing without losing information, but we see a role for least-common-denominator variables and intend to develop them in the future. These will only offer categories that are fully comparable across all samples, and they will apply the most restrictive universe of people who answered the question among the available samples. In essence, the least detailed sample and the sample with the most restrictive universe will dictate the nature of these simplified harmonized variables. The main impetus from our perspective is the utility of such variables in our online tabulator, but we also expect many users who download data will employ these highly comparable simplified variables as controls in their models.

From our perspective, international population data harmonization is a puzzle whose subtleties are mostly amenable to human problem-solving rather than automation. But automation helps, and there is always more scope for it. At some point the costs outweigh the benefits, but the balance will continue to shift in future as machine learning and other data science tools improve. We have already developed many utility programs that take advantage of semantic and coding similarities among data collections that are more coherent than the international censuses. For the foreseeable future, however, population microdata harmonization is bound to retain a significant component of human judgment.

References

Esteve A., Sobek M.: Challenges and methods of international census harmonization. Hist. Methods **36**, 66–79 (2003)

International Labour Office. International Standard Classification of Occupations: Structure, Group Definitions and Correspondence Tables, Geneva (2012)

Madans, J., Loeb, M., Altman, B.: Measuring disability and monitoring the UN convention on the rights of persons with disabilities: the work of the Washington group on disability statistics. BMC Public Health **11**(suppl 4) (2011). https://doi.org/10.1186/1471-2458-11-S4-S4

Ruggles, S., et al.: The IPUMS collaboration: integrating and disseminating the world's population microdata. J. Demogr. Econ. **81**, 203–216 (2015)

Sobek, M.: Data prospects: IPUMS-International. In: White, M. (ed.) International Handbook of Migration and Population Distribution, pp. 157–174. Springer, New York (2016)

Sobek, M., et al.: Big data: large-scale historical infrastructure from the Minnesota population center. Hist. Methods **44**, 61–68 (2011)

United Nations. International Standard Classification of Education: ISCED 1997. United Nations Educational, Scientific and Cultural Organization, New York (1997). http://www.unesco.org/education/information/nfsunesco/doc/isced_1997.htm

United Nations. International Standard Industrial Classification of All Economic Activities (ISIC), Revision 3.1. Department of Economic and Social Affairs, Statistics Division, New York (2002)

United Nations. Principles and Recommendations for Population and Housing Censuses (Revision 2). Department of Economic and Social Affairs, Statistics Division, New York (2007)

A Supervised Multiclass Classifier as an Autocoding System for the Family Income and Expenditure Survey

Yukako Toko, Kazumi Wada, Seigo Yui, and Mika Sato-Ilic

Abstract Coding is a task that classifies an object to a corresponding code (or class). This is often required for survey data processing in the field of official statistics. Since the governmental survey has large number of objects and codes (or classes), and the release time of the survey result has to be strictly observed, the autocoding system is a key solution for improving data processing. For this autocoding system, mainly two types of methodologies have been developed. One is the use of the supervised classification methods including machine learning techniques and the other is rule-based methods. For the supervised classification method, we have developed a supervised multiclass classifier using machine learning which has the advantages of simplicity and practical calculation time. In this paper, we present an application of the proposed method for the Family Income and Expenditure Survey in Japan with a comparison of the accuracy and the efficiency of the rule-based method.

1 Introduction

In the field of official statistics, there are some types of answer columns for statistical survey questionnaires such as a column selected as a choice between two alternatives, a column selected as a choice among multiple alternatives, and a free filling column. In survey data processing, survey items answered as a choice between two or multiple alternatives can be processed easily. For example, as the

Y. Toko (✉) · K. Wada · S. Yui
National Statistics Center, Shinjuku-ku, Tokyo, Japan
e-mail: ytoko@nstac.go.jp; kwada@nstac.go.jp; syui@nstac.go.jp

M. Sato-Ilic
National Statistics Center, Shinjuku-ku, Tokyo, Japan

University of Tsukuba, Tsukuba, Ibaraki, Japan
e-mail: msato@nstac.go.jp; mika@risk.tsukuba.ac.jp

© Springer Nature Singapore Pte Ltd. 2020
T. Imaizumi et al. (eds.), *Advanced Studies in Classification and Data Science*,
Studies in Classification, Data Analysis, and Knowledge Organization,
https://doi.org/10.1007/978-981-15-3311-2_40

survey item "sex" is generally answered as the choice between two alternatives in a statistical survey; thus obtaining the total number of male and female respondents is simple.

However, survey items answered as free descriptions, such as occupation, industry, and various items related to family income and expenditure are difficult for summarizing data, since the data will be represented as textual descriptions and those descriptions have a wide variety. Therefore, in order to summarize this type of data for producing statistical tables, these descriptions are needed to translate into a given classification code. This task of translating a textual description into a corresponding classification code is referred to as "coding" in the field of official statistics.

Traditionally, the coding was done manually in which humans engaged in the coding task with their specialized knowledge of the classification. However, manual coding needs a certain amount of expert human resources and the national statistics offices are strongly required to release quickly the results of statistical surveys.

Therefore, recently, with the improvement of computer technology, the studies of the autocoding system have been developed in the field of official statistics. For example, Hacking and Willenborg (2012) illustrated coding tasks for governmental surveys in Netherland including their study of an autocoding system. Methods for automated occupation coding which are mostly based on statistical learning have been proposed by Gweon et al. (2017). The National Statistics Center of Japan (NSTAC) was also launched an initial study that focused autocoding system in 1992 (Yui 2017) as a rule-based method, and after it, a multiclass classifier using simple machine learning algorithm has been developed (Shimono et al. 2018; Toko et al. 2017). The algorithm is simple and this causes strong advantages for the governmental survey data, such as adaptability for big data classification with a large number of codes (or classes).

This paper is organized as follows: A general explanation of autocoding systems is given in Sect. 2, the proposed autocoding system using machine learning is presented in Sect. 3. The experiments and the results of the autocoding system using machine learning with comparisons of a rule-based autocoding system are described in Sect. 4, and conclusions and suggestion for future work are presented in Sect. 5.

2 Autocoding Systems

Autocoding systems provide an object (or a textual description) the most promising classification code automatically, and such systems can be divided into two types of systems: one is a system employing rule-based algorithm and the other is a system employing machine learning algorithm. Both of these systems have advantages and disadvantages, respectively.

A rule-based autocoding system assigns a classification code using manually prepared if-then classification rules. Here, experts create rule dictionary by utilizing their classification knowledge. A rule-based autocoding system matches each rule to a text description and assigns the corresponding classification code. The greatest advantage of a rule-based autocoding system is that it works without a training dataset. In addition, it is easy to interpret the output from the system because classification rules are maintained manually. Meanwhile, the disadvantage of a rule-based autocoding system is the significant maintenance cost of the classification rules by experts. Furthermore, in order to obtain the high accuracy of the classification, generally the number of the rules have to be increased, then the maintenance cost is accumulated and the system itself becomes more complicated. This causes instability of the solution of the system.

On the other hand, a machine learning autocoding system finds latent patterns (i.e., classification rules) in data automatically and assigns classification codes according to those patterns. The greatest advantage of machine learning autocoding systems is their low maintenance cost. As the system creates coding rules automatically according to the latent structure of given data, maintaining huge numbers of rules is easy. In addition, the system can find hidden patterns that experts may not be able to identify. However, this system requires a sufficiently sized training dataset for learning and this is a core issue for obtaining satisfactory classification accuracy as the result. Basically, as the number of classification codes increase, the required size of the training dataset will also increase.

3   Autocoding System Using Simple Classifier Based on Machine Learning

The proposed autocoding system has been developed for the Family Income and Expenditure Survey in Japan. This monthly survey is conducted by the Statistics Bureau of Japan. Approximately 9000 selected households are required to keep daily accounts of all transactions related to their income and expenditure in Japanese. The experts of the survey classification have sorted a variety of entries that contain orthographical variance and local dialects into approximately 600 labels for survey data processing. The proposed autocoding system (Shimono et al. 2018; Toko et al. 2017) comprises of training and evaluation processes. In the training process, the system performs feature extraction and tabulates the extracted features along with the given classification codes into a feature frequency table (see Table 1). In the present study, the object is a short textual description. We employed the word-level N-gram model for feature extraction. We performed the following processes: Firstly, tokenizing each description by a morphological analyzer. We used MeCab (Kudo et al. 2004), which is a dictionary-attached morphological analyzer, to divide text descriptions into constituent words. Secondly, we took word-level N-grams ($N = 1, 2, 3 \ldots$) from the word sequences of text descriptions. In the present paper, we take 1-grams (any word), 2-grams (any sequence of two consecutive words), and the whole-word sequence as features.

Table 1 Feature frequency table

Features		Classes						
		1	2	3	...	k	...	K
Features	f_1	n_{11}	n_{12}	n_{13}	...	n_{1k}	...	n_{1K}
	f_2	n_{21}	n_{22}	n_{23}	...	n_{2k}	...	n_{2K}
	f_3	n_{31}	n_{32}	n_{33}	...	n_{3k}	...	n_{3K}
	⋮	⋮	⋮	⋮	⋮	⋮	⋮	⋮
	f_j	n_{j1}	n_{j2}	n_{j3}	...	n_{jk}	...	n_{jK}
	⋮	⋮	⋮	⋮	⋮	⋮	⋮	⋮
	f_J	n_{J1}	n_{J2}	n_{J3}	...	n_{Jk}	...	n_{JK}

For example, suppose an object of the textual description consists of two words as "A" and "B." In this case, the features are set as "A," "B," and "A B." For all of the objects in the training dataset, we create the features $f_1, \cdots, f_J$ shown in Table 1. According to the supervised classification, the frequency of j-th feature to k-th code (or class), n_{jk} ($j = 1, \cdots, J$, $k = 1, \cdots, K$), is calculated by using the training dataset, where the number of codes (or classes) is K.

Let multinomial classes C take values in $\{ 1, \ldots, K \}$, and let $\mathbf{F} = (F_1, \ldots, F_J)$ be a J-dimensional random variable whose elements take values 0 or 1, which, respectively indicate the absence or presence of a particular feature. Then, as each feature is assumed to be conditionally independent of any other features given C, the conditional probability of the features of $\mathbf{F}$ given class C can be written as follows:

$$P(F_j = f_j, j = 1, \ldots, J | C = k) = \prod_{j=1}^{J} P(F_j = f_j | C = k)$$

$$= \prod_{j=1}^{J} p_{jk}^{f_j} (1 - p_{jk})^{1-f_j}, \qquad (1)$$

where $p_{jk} = p(F_j = 1 | C = k)$ for $k = 1, \ldots, K$.

Then, let n_k be the number of objects in the training dataset in a class k, and let n_{jk} be the number of objects in the training dataset in a class k with $f_j = 1$. The maximum likelihood estimate of p_{jk} can be written as:

$$\hat{p}_{jk} = \frac{n_{jk}}{n_k}. \qquad (2)$$

We add α to the denominator and β to the numerator in order to prevent $\hat{p}_{jk}$ from being equal to 0 or 1:

$$\hat{p}_{jk} = \frac{n_{jk} + \beta}{n_k + \alpha}. \qquad (3)$$

Then, under the assumption that $P(C = k) = p_k$, the posterior probability $P(C = k|F_j = f_j, j = 1, \ldots, J)$ is proportional to

$$p_k \prod_{j=1}^{J} p_{jk}^{f_j} (1 - p_{jk})^{1-f_j}. \quad (4)$$

Under $\beta_{jk} = \log \frac{p_{jk}}{1-p_{jk}}$, the posterior probabilities have been described as follows:

$$P(C = k|F_j = f_j, j = 1, \ldots, J) = \frac{\exp(\sum_{j=1}^{J} f_j \beta_{jk})}{\sum_{l=1}^{K} \exp(\sum_{j=1}^{J} f_j \beta_{jl})}, \quad (5)$$

when prior probabilities $p_k \propto \{\prod_{j=1}^{J}(1 - p_{jk})\}^{-1}$ for $k = 1, \ldots, K$ (Taguchi 1997).

In previous studies (Shimono et al. 2018; Toko et al. 2017), we have simplified the process of assigning classes. First, we considered the following:

$$\arg\max_{k} P(C = k|F_j = f_j, j = 1, \ldots, J) \propto \arg\max_{k} \prod_{\{j|f_j=1\}} \frac{p_{jk}}{1 - p_{jk}}. \quad (6)$$

From (2), (3), and (6), it can be seen that the maximum probability $P(C = k|F_j = f_j, j = 1, \ldots, J)$ when $f_j = 1$ over K classes is influenced by only the amount of n_{jk}. Therefore, we have defined new $\tilde{p}_{jk}$ as follows:

$$\tilde{p}_{jk} = \frac{n_{jk} + \beta}{n_j + \alpha}, \quad n_j = \sum_{k=1}^{K} n_{jk}, \quad (7)$$

where we set heuristically $\alpha = -0.111111$ and $\beta = -0.444444$ in the present study.

4 Experiments and Results

4.1 Setting

We prepared the dataset of the Family Income and Expenditure Survey in Japan for performance evaluation. We used approximately 5.4 million objects (approximately 163 MB of data) as a training dataset because a certain volume of data is required owning to the number of requisite classes. Furthermore, since each class contains a variety of text descriptions, including orthographical variance and local dialects, we needed to use as much data as we could prepare to cover those descriptions.

Meanwhile, we used approximately 43,300 objects for an evaluation dataset because only that number of objects was adjusted for the comparison with the rule-based autocoding system. Each object of this dataset has one text (or one word) in Japanese, the corresponding classification code from approximately 600 kinds of codes, and subsidiary information such as an amount of income, an amount of expenditure, a payment option (cash, credit card, or loan), and a quantity of the purchase including its unit.

Although the target objects are short textual descriptions, we use income and expenditure data separately to improve classification accuracy, because there are several textual descriptions which are the same name but different meanings in income and expenditure. For example, a description "public pension" can be considered to have two meanings: a pension receipt and a payment to pension fund, also a description "savings" can be considered two meanings: deposits and withdrawal. Similarly, we also need to distinguish between payments by credit card and payments by other methods. Because some objects are assigned different classification code according to payment option. In this case, we put a particular symbol to each feature in the feature set whose information of payment option indicated "payment by credit card."

4.2   Results

The proposed autocoding system assigned a classification code to each object of the evaluation dataset. We also applied the same data to the rule-based autocoding system to compare the classification accuracy.

The rule-based autocoding system (Yui 2017) was applied to 2014 survey of the National Survey of Family Income and Expenditure whose survey items are similar to the Family Income and Expenditure Survey. The experts in our office have put the knowledge of this experience for developing a rule-based autocoding system for the Family Income and Expenditure Survey. This rule-based autocoding system has approximately 15,000 rules that are prepared and maintained manually. The system includes the following steps for coding process:

(Step 1)   If the inputted text description is matched a classification rule which has clear corresponding codes, then the code (or class) is assigned based on the rule-based dictionary definitions.

(Step 2)   If the remained inputted text description is matched a classification rule which corresponds to multiple codes, then the code (or class) is assigned based on the additional information which is the amount of income or expenditure.

Table 2 shows the comparison of the classification efficiency and accuracy between the autocoding system using simple classifier based on machine learning and the rule-based system. In Table 2, "Coverage," "Accuracy," and "Efficiency" are

Table 2 Comparison of classification efficiency and accuracy

	The number of total objects	The number of assigned objects	The number of matched objects	Coverage	Accuracy	Efficiency
Autocoding system used machine learning	42,820	42,595	39,204	99.5%	92.0%	91.6%
Rule-based autocoding system	42,820	30,274	30,225	70.7%	99.8%	70.6%

defined as follows:

$$\text{Coverage } (\%) = \frac{\text{The number of assigned objects}}{\text{The number of total inputted objects}} \times 100,$$

$$\text{Accuracy } (\%) = \frac{\text{The number of matched objects}}{\text{The number of assigned objects}} \times 100,$$

$$\text{Efficiency } (\%) = \frac{\text{The number of matched objects}}{\text{The number of total inputted objects}} \times 100,$$

where, the number of total inputted objects is the number of total objects in the evaluation dataset. The number of assigned objects means the number of objects in which the system can assign classification codes for these objects, but it may include the number of uncorrected classification. The number of matched objects shows the number of objects that the system gives correct classification codes.

From Table 2, it can be seen that the value of "efficiency" which shows how much correctly assigned objects exist out of the inputted total number of objects for the autocoding system based on machine learning is much better than the value of "efficiency" for the rule-based system, although the value of "accuracy" of the autocoding system based on machine learning is lower than the value of "accuracy" of the rule-based system. This is caused by the high coverage for the autocoding system based on machine learning when compared with the rule-based system.

In fact, Fig. 1 shows values of the following M_j for the income data which consists of 57,477 features and 37 codes (or classes).

$$M_j = \sum_{k=1}^{37} \tilde{p}_{jk}^2, \quad j = 1, \ldots, 57,477, \tag{8}$$

where $\tilde{p}_{jk}$ is shown in (7) when $\alpha = \beta = 0$. Note that Fig. 1 was plotted after sorting the values of M_j into ascending order. From the definition of $\tilde{p}_{jk}$ shown in (7), $\tilde{p}_{jk}$ satisfies the following conditions when $\alpha = \beta = 0$:

$$\tilde{p}_{jk} \in [0, 1], \quad \sum_{k=1}^{37} \tilde{p}_{jk} = 1, \quad j = 1, \ldots, 57,477.$$

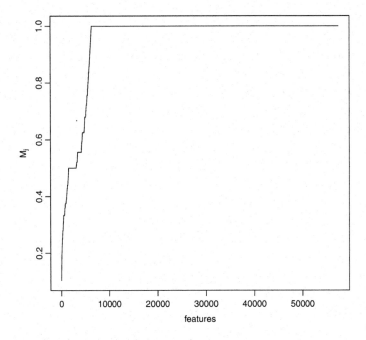

Fig. 1 Values of M_j for the income data

Therefore, M_j satisfies the following:

$$\frac{1}{37} \le M_j \le 1, \quad j = 1, \ldots, 57{,}477.$$

The value of M_j shows status of classification of j-th feature for 37 codes (or classes). The clearer classification of each feature has a larger value of M_j. From Fig. 1, we can see that most of the features are clearly classified to the codes.

We also compare classification efficiency and accuracy by sector. Table 3 shows an overview of the evaluation dataset by sector. As seen in the table, there are distribution biases between each sector because the frequency of transactions differs by sector.

Table 4 shows the result of classification efficiency and accuracy by sector for the proposed autocoding system based on machine learning and Table 5 is the result of the classification efficiency and accuracy for the rule-based autocoding system. From the comparison of the results between Tables 4 and 5, values of "efficiency" used proposed autocoding system based on machine learning were better than the values of the rule-based coding system. Moreover, the coverage of each sector was stable with high ratio in the proposed system used machine learning, whereas the accuracy differed by sector. For example, the system was particularly good at

Table 3 Evaluation dataset by sector

Sector names	Sector codes	The number of evaluated objects
Income	A	290
Receipts other than income	B	592
Foods	C	28,889
Housing	D	59
Fuel, light, and water charges	E	368
Furniture and household utensils	F	1359
Clothing and footwear	G	437
Medical care	H	924
Transportation and communication	I	937
Education	J	41
Culture and recreation	K	1259
Other consumption expenditures	L	1210
Non-consumption expenditures	M	287
Disbursements other than expenditures	N	678
Others	O	5490

Table 4 Classification efficiency and accuracy by sector for autocoding system used machine learning

Sector codes	The number of total objects	The number of assigned objects	The number of matched objects	Coverage	Accuracy	Efficiency
A	290	287	212	99.0%	73.9%	73.1%
B	592	591	584	99.8%	98.8%	98.6%
C	28,889	28,737	26,354	99.5%	91.7%	91.2%
D	59	59	40	100.0%	67.8%	67.8%
E	368	368	363	100.0%	98.6%	98.6%
F	1359	1341	1155	98.7%	86.1%	85.0%
G	437	435	293	99.5%	67.4%	67.0%
H	924	919	857	99.5%	93.3%	92.7%
I	937	934	897	99.7%	96.0%	95.7%
J	41	41	19	100.0%	46.3%	46.3%
K	1259	1244	1066	98.8%	85.7%	84.7%
L	1210	1190	1029	98.3%	86.5%	85.0%
M	287	287	272	100.0%	94.8%	94.8%
N	678	675	639	99.6%	94.7%	94.2%
O	5490	5487	5424	99.9%	98.9%	98.8%

assigning correct codes for sector codes B, E, and O; however, it was not good at assigning correct codes for sector codes A, G, and J. The primary reason for which the system was not good at assigning correct codes in this case because it learned an insufficient amount of information for autocoding. In order to assign correct codes to those data, the system will be required to learn family information,

Table 5 Classification efficiency and accuracy by sector for rule-based autocoding system

Sector codes	The number of total objects	The number of assigned objects	The number of matched objects	Coverage	Accuracy	Efficiency
A	290	64	60	22.1%	93.8%	20.7%
B	592	569	569	96.1%	100.0%	96.1%
C	28,889	20,403	20,368	70.6%	99.8%	70.5%
D	59	3	3	5.1%	100.0%	5.1%
E	368	348	346	94.6%	99.4%	94.0%
F	1359	648	646	47.7%	99.7%	47.5%
G	437	84	84	19.2%	100.0%	19.2%
H	924	567	566	61.4%	99.8%	61.3%
I	937	547	546	58.4%	99.8%	58.3%
J	41	1	1	2.4%	100.0%	2.4%
K	1259	600	600	47.7%	100.0%	47.7%
L	1210	533	532	44.0%	99.8%	44.0%
M	287	166	165	57.8%	99.4%	57.5%
N	678	458	455	67.6%	99.3%	67.1%
O	5490	5284	5284	96.2%	100.0%	96.2%

such as household structure, type of dwelling (i.e., rents for dwelling public or rents for dwelling private), and type of school (public or private).

On the other hand, for the rule-based autocoding system, the coverage of nearly half of the sectors was less than 50%, and the accuracy of most sectors was more than 99%. In addition, the accuracy of each sector was stable with high accuracy, whereas the coverage differed by sector. For example, the system assigned codes for a large portion of data for sector codes B, E, and O; however, it assigned codes for a small portion on data in sector code J. It is curious that both the proposed and rule-based systems were particularly good at assigning codes with high accuracy for a large portion of data for sector codes B, E, and O. Although sectors where both accuracy and coverage were extremely high were the same in both systems, the tendency of the other sectors appears to differ.

We also applied the autocoding system using machine learning to data that the rule-based autocoding system could not classify. Table 6 shows the result of the classification efficiency and accuracy by sector. In this table, each efficiency and accuracy value was less than that of the proposed system used machine learning shown in Table 4. However, we found that the proposed system assigned correct codes at a certain accuracy for a large portion of data that the rule-based system could not classify.

Table 6 Classification efficiency and accuracy of the proposed classifier using machine learning according to each sector for data the rule-based system left unclassified

Sector codes	The number of total objects	The number of assigned objects	The number of matched objects	Coverage	Accuracy	Efficiency
A	226	223	152	98.7%	68.2%	67.3%
B	23	22	15	95.7%	68.2%	65.2%
C	8486	8354	6247	98.4%	74.8%	73.6%
D	56	56	37	100.0%	66.1%	66.1%
E	20	20	17	100.0%	85.0%	85.0%
F	711	695	520	97.7%	74.8%	73.1%
G	353	351	212	99.4%	60.4%	60.1%
H	357	352	297	98.6%	84.4%	83.2%
I	390	387	351	99.2%	90.7%	90.0%
J	40	40	19	100.0%	47.5%	47.5%
K	659	644	474	97.7%	73.6%	71.9%
L	677	657	502	97.0%	76.4%	74.2%
M	121	121	107	100.0%	88.4%	88.4%
N	220	219	188	99.5%	85.8%	85.5%
O	206	203	141	98.5%	69.5%	68.4%

5 Conclusions

For the Family Income and Expenditure Survey in Japan, this paper presents an application of autocoding method using simple machine learning technique with a comparison of the efficiency and the accuracy of a rule-based method. From this comparison, it can be seen that the autocoding system using machine learning and the rule-based autocoding system have unique advantages, respectively. The rule-based system can assign classification codes at extremely high accuracy; however, it yields a certain volume of unclassified data. Also, the proposed system based on machine learning can assign classification codes for a large portion of the dataset; however, the accuracy is not equivalent to the accuracy of manual coding, although the efficiency which shows the ratio of the number of correctly classified objects out of the number of total inputted objects is satisfactory higher than the value of the ratio of the rule-based autocoding system.

From this viewpoint, it may be worthwhile to develop the hybrid system (Yui 2017) to improve the effectiveness of coding tasks. And this may work for not only the Family Income and Expenditure Survey but other statistical surveys in the field of official statistics.

For future studies, the proposed classifier yields different accuracy results according to each sector, and the primary reason for this is that there is a certain volume of data that requires more information than it learned to assign a correct code. In order to improve classification accuracy of such data without compromising the simplicity of the system structure, we are considering that assigning unified codes for those data in the proposed classifier and processing them in another process using the respective household information to break down those unified codes. Applying our autocoding system to other coding tasks of governmental surveys is also one future study. In addition, the detailed investigation for the lower value's features in Fig. 1 and the use of values of M_j shown in (8) for the reliability of classification of each feature is a topic for future study.

Acknowledgments We are grateful to Dr. Tusbaki, H., Director-General of the Institute of Statistical Mathematics for helpful comments for this research.

References

Gweon, H., Schonlau, M., Kaczmirek, L., Blohm, M., Steiner, S.: Three methods for occupation coding based on statistical learning. J. Off. Stat. **33**(1), 101–122 (2017). https://doi.org/10.1515/jos-2017-0006

Hacking, W., Willenborg, L.: Method series theme: coding; interpreting short descriptions using a classification. In: Statistics Methods. Statistics Netherlands (2012). Available at: https://www.cbs.nl/en-gb/our-services/methods/statistical-methods/throughput/throughput/coding. Cited 16 Nov 2017

Kudo, T., Yamamoto, K., Matsumoto, Y.: Applying conditional random fields to Japanese morphological analysis. In: Proceedings of the 2004 Conference on Empirical Methods in Natural Language Processing (EMNLP-2004), pp. 230–237 (2004)

Shimono, T., Wada, K., Toko, Y.: A supervised multiclass classifier using simple machine learning algorithm for autocoding. Res. Mem. Off. Stat. **75**, 41–60 (2018) (in Japanese)

Taguchi, G.: Mathematical for quality engineering – 7. Signal-to-noise ratio for chemical and biological systems. In: Quality Engineering Forum **5**(2), 3–9 (1997) (in Japanese)

Toko, Y., Wada, K., Kawano, M.: A supervised multiclass classifier for an autocoding system. J. Rom. Stat. Rev. **4**, 29–39 (2017)

Yui, S.: Application of a statistical learning algorithm to autocoding system. Statistics, Jan. 2017. Japan Statistical Association (2017) (in Japanese)